£24.95

D1141915

LEABHARLANN CHONTAE ROSCOMAIN

in

This book should be returned not later than the last date shown below. It may be renewed if not requested by another borrower.

Books are on loan for 14 days from the date of issue.

392590(4)

333

7516

DATE DUE	DATE DUE	DATE DUE	DATE DUE
24.			

EARTHSCAN

London • Sterling, VA

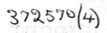

ISBN: 1-84407-154-5   paperback
         1-84407-153-7   hardback

Typesetting by Composition and Design Services
Printed and bound in the UK by Bath Press, Bath
Cover design by Andrew Corbett

For a full list of publications please contact:

Earthscan
8–12 Camden High Street
London, NW1 0JH, UK
Tel: +44 (0)20 7387 8558
Fax: +44 (0)20 7387 8998
Email: earthinfo@earthscan.co.uk
Web: **www.earthscan.co.uk**

22883 Quicksilver Drive, Sterling, VA 20166-2012, USA

Earthscan is an imprint of James & James (Science Publishers) Ltd and publishes in association with
the International Institute for Environment and Development

A catalogue record for this book is available from the British Library

Library of Congress Cataloging-in-Publication Data has been applied for

Printed on elemental chlorine-free paper

Contents

Part 1 The Forest Resource

Part 2 Forests and Livelihoods

Part 3 Threats and Opportunities

Part 4 The Challenge of Sustainable Management

Part 5 The Way Forward: Forestry for the Future

Foreword

It never fails. Whenever someone introduces me to their friends or relatives the topic eventually gets around to what I do. When I tell them I study tropical forests, they usually say something like, 'Oh, the rainforests? It is terrible the way loggers cut down those trees and don't even plant new ones. My son [or daughter] is always telling me we should use recycled paper. A couple of years ago we gave a donation to conserve the animals, but I haven't followed things that much lately. It is good to know there are still people like you working to save the forest'.

After that, I never know how to respond. After all, these fine people are only being polite and the last thing they want is a lecture. Still, some of their ideas are pretty off base. Anyway, where to start? Should I tell them that farmers, ranchers and land speculators often cause more problems than loggers do? Perhaps I should explain that it is good to plant trees and recycle paper, but that it won't solve the main problems. I might even try to clarify that it is not just about saving forests, but of making sure that people have food, medicine, housing and fuel. That one is really difficult. Most middle class urbanites simply cannot imagine people getting their food, medicine and incomes from the forest.

Who can blame them? After all, how much do I know about stocks, insurance or retail sales? Practically, the only times these people hear about tropical forests are when they go to the zoo, watch the Discovery channel, see some article about a species going extinct or get an appeal from some conservation group. They are not experts and we cannot expect them to be.

Still, what these people think about rainforests actually matters. Politicians pay attention to them and try to look good by doing things that will make the public believe they are working to save the rainforests. By giving speeches, creating parks and planting trees the politicians can respond to public concerns without aggravating anybody or running up a big bill. This simple logic shapes much of the global discussion about tropical forests.

Surely, though, one would imagine that even if the public and politicians don't really understand tropical forest problems and how to solve them, the experts do. In effect, foresters, biologists and social scientists often tell more sophisticated stories. They say things like, 'We need to make it more profitable to use forests productively; otherwise they will be converted to other uses.' Or, 'We need to give policy makers evidence about how much forests are worth, so they will protect them.' Many think that getting governments to simply get serious about enforcing the forestry laws will do the trick. Others put their hopes in the free market and the ability of forest certification, bioprospecting and carbon trading to give companies incentives to be more environmentally friendly.

Most point to a vicious circle whereby poverty forces people to cut trees in order to grow crops or cook their food, which makes them even poorer because by destroying the forests they also destroy all the benefits that come with them.

Even though these ideas are more complex and sophisticated than those of the general public, in both cases they tend to be taken more as dogmas and articles of faith than hypotheses to be tested. And when, as often happens, the facts on the ground fail to bear out their predictions, that rarely dampens their enthusiasm or convictions. While some of the things the general public and the experts believe are true, others are not, but as no one ever really puts them to the test, people keep on believing them.

While one can easily blame the poor understanding of politicians and the public on lack of information and experience, with the experts the explanation is more complex. In many cases they have built their careers and institutional budgets by promoting these ideas. Most find the idea of trade-offs between growth, equity and the environment, or of conflicts between the interests of different groups that lay claims to forests, uncomfortable. These ideas imply there might be losers as well as winners, and the losers may not be too happy.

One thing experts, the public and politicians all have in common is that they all depend heavily on 'policy narratives' to make sense of things. Policy narratives are simple (or some might say simplistic) explanations of what the problems are, who and what causes them and what can be done to solve them. Without such narratives they would be overwhelmed by the diversity and complexity of situations and have no way to process or respond to them.

So it is easy to see why people tend to have simplistic ideas about tropical forests and hold on to them even without much evidence or proof. One can also understand why conservationists, forestry companies and development planners might have different visions about forest issues. Still, if we are ever to actually make progress towards conserving biodiversity, improving the lives of forest dweller and solving other forest problems we are going to need a much more realistic sense of the magnitudes and causes of the problems and of the likely effects of various practices and policy options. That implies an approach that is based more on empirical measurement, hypothesis testing, comparative analysis and constant questioning of the assumptions. It also needs an approach that does not try to paint over the trade-offs between objectives and conflicts between interest groups. That is what this book's editor, Jeffrey Sayer, encouraged when he directed the Center for International Forestry Research (CIFOR) and what I have tried to maintain as his successor. It is the approach that underlies this book.

The main reason that we need to take such a critical, realistic and science-based approach is because other approaches simply have not worked very well. Despite considerable efforts, tropical forests continue to be cleared and degraded at a rapid rate. There has been surprisingly little progress in reducing poverty in forested regions, particularly in Africa and Asia. Huge sums of forest revenues that could have propelled economic development have ended up in luxury consumption and foreign bank accounts. Previous forestry and conservation projects have failed to achieve the expected levels of impact, and new funding is being cut back partly as a result. If we are ever to move beyond this situation it won't be with 'business as usual' – it will only happen by learning from experience, questioning conventional wisdom, constructing a new set of policy narratives and developing institutional mechanisms that allow you to adapt those narratives to local realities.

This book goes a surprisingly long way in that direction. It certainly reflects some of the best information and analysis about tropical forests, and it does not shy away from providing clear guidance about what needs to be done to achieve certain objectives. Still, it is anything but dogmatic. It accepts that groups' objectives may legitimately differ and that experts can also interpret the same sets of data quite differently. Perhaps most importantly, rather than treating this as a concern, this Reader treats this difference of objectives and interpretations as a source of strength; a pluralism of ideas and positions that can enhance our understandings and maintain some balance.

Most readers will find a lot of surprises. While many tropical forests are being lost, others are coming back and being restored. Macroeconomic, agricultural and transportation policies influence deforestation much more than forestry policies. Small farmers and communities now own many more forests than a few decades ago. Forests are very important for poor people, but may not help them escape poverty. Illegal logging causes all sorts of problems, but enforcing the existing forestry laws could also have rather negative effects. Commercial hunting of wild animals is an important source of income and nutrition, as well as a threat to biodiversity. Large national parks are not the only way, and often not the best way, to protect biodiversity. Some tree crop systems are excellent options for conserving many types of biodiversity. Fuelwood usage is still increasing rapidly in sub-Saharan Africa, but has begun to decline in other tropical regions. New, more efficient wood processing technologies may increase pressure on forests, rather than reduce it. While carbon trading may provide some marginal benefits, it is unlikely to have a major impact on the world's forests or provide a major source of funds for forestry activities for the foreseeable future.

These are precisely the types of surprises that only serious empirical research and critical thinking can give you. And it is the type of information we need if we want to solve problems.

So, whether you are a seasoned conservationist or forestry professional, an aspiring student or simply a concerned citizen, this book is for you. You don't have to be an expert to read and understand it. If you do happen to be an expert, you will probably find many of your most cherished beliefs seriously challenged. That is as it should be.

David Kaimowitz
Bogor
April 2005

Acknowledgements

This Reader was compiled whilst I was working for the Forests for Life programme of WWF-International in Switzerland. I have benefited in selecting articles from the discussions that I have had with my colleagues here in Gland and with others throughout the WWF network around the world. I would particularly like to thank Chris Elliot, Mark Aldrich, Nils Hager and Leonardo Lacerda with whom I have frequent discussions about emerging conservation issues and whose advice helped in shaping this book. Colleagues at IUCN have also contributed ideas and Stewart Maginnis and Sandeep Sengupta adapted their paper on the global forest situation to meet the needs of the Reader. Simon Rietbergen, editor of the last Reader, a decade ago (Rietbergen, 1993), has also offered advice in the preparation of this edition.

Much of my own thinking on strategic forest issues emerged during the period that I headed the Center for International Forestry Research in Bogor, Indonesia. Neil Byron, Dennis Dykstra, Manuel Ruiz-Perez, Reidar Persson, Doug Sheil, Carol Colfer, Lini Wollenberg, Bruce Campbell, Sven Wunder and Inna Bangun were all valued colleagues who influenced my thinking. Many of them are authors of papers reproduced in this volume. My good friend David Kaimowitz, who succeeded me as Director General of CIFOR, has generously contributed a foreword to this volume and continues to be a source of inspiration on all forest issues. My thanks to all of the above plus all of the other CIFOR staff and board members with whom I worked for nine enjoyable years.

I would also like to thank the large number of people who responded to my emailed request for nominations for papers that they considered to have been groundbreaking. They are too numerous for me to be able to thank them all but I would like to single out Peter Holmgren and his colleagues of the FAO Forestry Division. Alan Pottinger, editor of the *International Forestry Review*, has not only made suggestions but has also allowed us to use papers from his journal. I would also like to thank the authors, editors and business managers of all the other journals whose papers are reproduced here.

I would like to especially thank Hosni el Lakhani and the staff of the FAO Forestry Division for allowing us to reprint material from their journal *Unasylva*, from the Proceedings of the 12th World Forestry Congress in Quebec in 2003.

Rob West and the staff of Earthscan in London have been patient and extremely helpful in dealing with all of my requests and in guiding this whole process through to completion. My sincere thanks go to them.

In extending my sincere thanks to all those who have contributed, I do need to stress that the final choice of chapters was entirely my own and probably reflects a rather personal bias towards a certain view of world forestry. I hope that the result will be useful and enjoyable for readers.

Jeffrey Sayer
Gland
April 2005

References

1   Rietbergen, S. (1993) *Earthscan Reader in Tropical Forestry*. Earthscan, London.

List of Sources

Chapter 2 Sengupta, S. and Maginnis, S. 2004. Forests and Development: Where Do We Stand? Adapted from an unpublished IUCN Global Situation Analysis, IUCN, Gland, Switzerland.

Chapter 3 Geist, H. J. and Lambin, E. F. 2002. Proximate causes and underlying driving forces of tropical deforestation. *Bioscience*, vol 52(2), pp143–149.

Chapter 4 White, A. and Martin, A. 2003. *Who Owns the World's Forests? Forest Tenure and Public Forests in Transition*, Forest Trends and the Center for Environmental Law, Washington, DC.

Chapter 5 Wunder, S. 2001. Poverty alleviation and tropical forests – What scope for synergies. *World Development*, vol 29, pp1617–1833.

Chapter 6 Arnold, J. E. M. and Ruiz-Perez, M. 2001. Can non-timber forest products match tropical forest conservation and development objectives? *Ecological Economics*, vol 39, pp437–447.

Chapter 7 Byron, N. and Arnold, M. 1999. What future for the peoples of the tropical forests? *World Development*, vol 27(5), pp789–805.

Chapter 8 Kaimowitz, D. 2003. Forest law enforcement and rural livelihoods. *International Forestry Review*, vol 5(3), pp199–210.

Chapter 9 Brown, D. and Williams, A. 2003. The case for bushmeat as a component of development policy: Issues and challenges. *International Forestry Review*, vol 5(2), pp148–155.

Chapter 10 Arnold, M. and Persson, R. 2003. Reassessing the fuelwood situation in developing countries. *International Forestry Review*, vol 5(4), pp379–383.

Chapter 11 Goldammer, J. G. 2004. Forest fires: A global overview. Adapted from An Overview of Vegetation Fires Globally. Produced with inputs from the Working Group on Wildland Fire, United Nations International Strategy for Disaster Reduction. Further information from fire@fire.uni-freiburg.de.

Chapter 12 Kaimowitz, D., Byron, N. and Sunderlin, W. nd. Public policies to reduce inappropriate tropical deforestation. Adapted from an article that originally appeared in the World Bank Development Review.

Chapter 13 Sayer, J., Vanclay, J. K. and Byron, N. 1997. Technologies for Sustainable Forest Management: Challenges for the 21st Century. CIFOR Occasional Paper, Bongor Barat, Indonesia

Chapter 14 Pearce, D., Putz, F. E. and Vanclay, J. K. 2003. Sustainable forestry in the tropics: Panacea or folly? *Forest Ecology and Management*, vol 172, pp229–247.

Chapter 15 Zuidema, P., Sayer, J. and Dijkman, W. 1997. Forest fragmentation and biodiversity: The case for intermediate sized conservation areas. *Environmental Conservation*, vol 23(4), pp290–297.

Chapter 16 Smith, J., Mulungoy, K., Persson, R. and Sayer, J. 2000. Harnessing carbon markets for tropical forest conservation: Towards a more realistic assessment. *Environmental Conservation*, vol 27(3), pp300–311.

Chapter 17 Wilshusen, P. R., Brechin, S. R., Fortwangler, C. L. and West, P. C. 2002. Reinventing a square wheel: Critique of a resurgent 'protection paradigm' in international biodiversity conservation. *Society and Natural Resources*, vol 15, pp17–40.

Chapter 18 Sheil, D. and van Heist, M. 2000. Ecology for tropical forest management. *International Forestry Review*, vol 2(4), pp261–270.

Chapter 19 Alden Wily, L. 2003. From meeting needs to honouring rights: The evolution of community forestry. Proceedings of the 12th World Forestry Congress, Quebec, Canada.

Chapter 20 Edmunds, D. and Wollenberg, E. 2001. A strategic approach to multi-stakeholder negotiations. *Development and Change*, vol 32, pp231–253.

Chapter 21 Sayer, J., Elliot, C. and Maginnis, S. 2003. Protect, manage and restore: Conserving forests in multi-functional landscapes. Proceedings of the 12th World Forestry Congress, Quebec, Canada.

List of Acronyms and Abbreviations

BIBEX	Biomass Burning Experiment
BOLFOR	Bolivia Forest Conservation and Sustainable Management Programme
CBD	Convention on Biological Diversity
CBPR	community-based property right
CCN	Cloud Condensation Nuclei
CDM	Clean Development Mechanism (of the Kyoto Protocol)
CIFOR	Center for International Forestry Research
CL	conventional logging
COFORD	The National Council for Forest Research and Development
CONAP	National Council of Protected Areas
CPF	Collaborative Partnership on Forests
CPRD	Common Property Resource Digest
CSERGE	Centre for Social and Economic Research on the Global Environment
DEWA	Division of Early Warning and Assessment
DFID	Department for International Development (UK)
ECOSOC	Economic and Social Council of the United Nations
ECTF	Edinburgh Centre for Tropical Forest
EKC	Environmental Kuznets Curve
ENSO	El-Niño-Southern Oscillation
ESMAP	Energy Sector Management Assistance Programme
ESRC	Economic and Social Research Centre (UK)
FAO	Food and Agriculture Organization
FDI	foreign direct investment
FLEG	forest law enforcement and governance
FLR	forest landscape restoration
FORIS	Forest Information System
FRA	Forest Resources Assessment
FSC	Forest Stewardship Council
FSU	Former Soviet Union
GDP	gross domestic product
GEO	Global Environment Outlook
GFMC	Global Fire Monitoring Center
GFP	Global Fire Partnership
GFTN	Global Forest and Trade Network

GOFC/GOLD	Global Observation of Forest Cover/Global Observation of Land Dynamics
GTOS	Global Terrestrial Observing System
IAE	International Academy of the Environment
IAF	International Arrangement on Forests
ICDP	integrated conservation and development project
IEA	International Energy Agency
IFF	Intergovernmental Forum on Forests
IGAC	International Global Atmospheric Chemistry (Project)
IIED	International Institute for Environment and Development
IISD	International Institute for Sustainable Development
InBio	Biodiversity Institute of Costa Rica
INPE	Institute for Remote Sensing Research
IPCC	Intergovernmental Panel on Climate Change
IPF	Intergovernmental Panel on Forests
ISDR	International Strategy for Disaster Reduction
ITTA	International Tropical Timber Agreement
ITTO	International Tropical Timber Organization
IUCN	World Conservation Union
IUFRO	International Union of Forest Research Organizations
JFM	Joint Forest Management
LFCC	low forest cover country
LULUCF	land use, land-use change and forestry
MDF	medium-density fibreboard
NAFTA	North American Free Trade Association
NASA	National Aeronautic and Space Administration
NGO	non-governmental organization
NRM	natural resource management
NTFP	non-timber forest product
NTV	non-timber values
NWFP	non-wood forest product
OECD	Organization for Economic Cooperation and Development
OSB	oriented strand board
PEFC	Programme for the Endorsement of Forest Certification
PRSP	Poverty Reduction Strategy Papers
RAIPON	Khabarovsk Regional Association of Indigenous Peoples of the North
RIL	reduced impact logging
SAFARI	Southern Africa Fire-Atmosphere Research Initiative
SAP	structural adjustment programme
SFM	sustainable forest management
STARE	Southern Tropical Atlantic Regional Experiment
STM	sustainable timber management
TEV	total economic value
TMCF	tropical montane cloud forest
TNC	The Nature Conservancy

TOF	trees outside forests
TTP	territory of traditional natural resource use
UN	United Nations
UNCCD	United Nations Convention to Combat Desertification
UNCED	United Nations Conference on Environment and Development
UNCTAD	United Nations Conference on Trade and Development
UNDP	United Nations Development Program
UNEP	United Nations Environment Programme
UNFCCC	United Nations Framework Convention on Climate Change
UNFF	United Nations Forum on Forests
USAID	United States Agency for International Development
US EPA	United States Environmental Protection Agency
WCD	World Commission on Dams
WCFSD	World Commission on Forests and Sustainable Development
WHO	World Health Organization
WMO	World Meteorological Organization
WRCF	world's remaining natural closed forest
WRI	World Resources Institute
WSSD	World Summit for Sustainable Development
WTA	willingness to accept
WTO	World Trade Organization
WTP	willingness to pay
WWF	World Wide Fund for Nature

Challenging the Myths: What is Really Happening in the World's Forests

Jeffrey Sayer

In July 2003 Earthscan approached me to see if I would be interested in compiling and editing a Reader on forestry. This was to follow up on, and complement, the successful *Reader on Tropical Forestry* edited by Simon Rietbergen (Rietbergen, 1993). This seemed like a fairly straightforward task – I simply had to locate about 15 of the more interesting articles that had appeared since Rietbergen's Reader was published, get permission to reprint them and write an introduction and conclusion. However, identifying the most significant publications on forestry to have appeared in recent years proved an interesting and somewhat controversial task. The answer depends very much on where one's interests lie and on what sort of a personal vision for forests one holds. My own vision has been formed from a professional career working to conserve forests in developing countries. But during this career I have worked for organizations spanning those promoting economic development to those whose role, primarily, is to conserve biodiversity. My own activities have focused very much on how one can reconcile these two, often conflicting, objectives. The Reader is biased towards writings that help to unravel the complexity of these conservation and development trade-offs.

I decided to introduce a degree of objectivity into the process of selection of the chapters. I therefore canvassed the opinion of people that I knew who were active in the field of forestry and development. I sent an email to about 250 of these people with the request that they nominated papers that they thought had been significant in either influencing thinking on forests or that provided a good synthesis of an important issue. The replies came as something of a surprise. Many respondents said that although there were lots of interesting technical papers, they could not immediately think of any that really qualified in terms of innovation or impact. However, about 27 people did reply with specific nominations – between them they proposed 35 papers. Again there were surprises. I had assumed that there would be a general consensus that some really outstanding papers would have been published in the past few years. But there was little consensus; in fact only three papers were nominated more than once – and each of them only twice. The other surprise was that less than 50 per cent of the papers came from forestry journals; the majority came from social science publications or were published by international organizations.

I presented the results of this survey at a side event during the World Forestry Congress in Quebec, Canada, in September 2003 and my findings provoked an interesting discussion. The debate centred on the issue of how science and scientific publications influence forestry policies and practice. The answer is that there is no simple relationship.

The classic pathway of scientists making a discovery, publishing it in a peer-reviewed journal – where it is read and its findings are adopted by forestry practitioners – no longer seems to hold up. Perhaps this simple vision of the relation of science to practice never did really exist. This conclusion was not new; Spilsbury and Kaimowitz (2000) had earlier conducted a rather more rigorous study of the influence of publications on forest policies and had come to similar conclusions. Their study suggested that policy papers produced by prominent organizations or individuals had the greatest impact. Spilsbury and Kaimowitz focused on the issue of how scientific innovations could better be packaged and marketed to reach their ultimate audiences.

My reflections on this issue over the past year lead me to see the problem in a slightly different way. It appears to me that conventional wisdom on forest issues is constructed out of intense personal and electronic interactions amongst networks of specialists interested in forests. These networks are catalysed by international processes such as the United Nations Forum on Forests, the Convention on Biological Diversity, the FAO commissions and committees dealing with forestry, the various networks of the International Union of Forest Research Organizations (IUFRO) and a number of networks formed amongst non-governmental organizations (NGOs). Scientists form part of these networks and the results of their work are already communicated amongst their peers before they get to the stage of being published. Insightful work gets discussed at workshops and conferences and already becomes part of the 'forest narrative' long before it grinds its way through the lengthy processes of peer review and journal publication. This raises questions about the role of journals. My conclusion is that journals have the role of verifying, through peer review, and archiving the results of research so that the methods, analyses and assumptions become part of the accumulated body of formal knowledge. Journal papers are not the primary mechanism for communicating results to users.

This raises the question of the value of bringing together a set of journal papers that will inevitably be several years old. If these papers were relevant to the forest narrative, presumably their contents would already have been incorporated into the conventional wisdom of the professional networks concerned.

However, it is also clear that there is not a single, uniformly accepted 'forest narrative'. There are several influential networks of scientists, activists and practitioners who all construct their own narratives. Forsyth (2003) has pointed out how there is often surprisingly little communication between these networks, with the result that they each construct and defend their own vision of the world of forests. Forest science is not value-neutral – a whole range of myths, special interests, cultures and assumptions influence it, and these vary amongst the networks. The divergent nature of these networks is at once apparent when one considers the different cultures of those composed of industrial foresters, those advocating local people's interests and those concerned with biodiversity. The divergences and lack of communication between these networks is one reason that forestry issues become so embroiled in controversy.

My analysis of why forest science is so controversial led me to conclude that the quest for an objectively selected set of the 'best' papers on forest issues was itself problematic. My own views on what was significant or important would clearly be influenced by the networks with which I was associated and my email sample of professionals was clearly heavily biased towards these networks. However, my career and my present position do bring me into contact with a broad range of people interested in forests.

Perhaps the best that I could do was to exploit my access to several different forest networks and select those papers that provided the most pertinent insights on the broadest possible vision of forests. This approach to the task interested me personally because much of my work in recent years has been on trying to strengthen the scientific basis for holistic, integrated management of forest systems (Sayer and Campbell, 2004). I have become a strong advocate of what has come to be called the 'Ecosystem Approach' to forests. This is an approach that attempts to optimize management to achieve the best possible balance between the interests of different stakeholders and to understand the impacts of management at different spatial and temporal scales. My work has been mostly in developing countries and the overall dilemma has been to reconcile the often-conflicting objectives of improving local livelihoods and conserving forest values of global significance.

My problems in compiling this selection of papers did not stop there. The original assumption was that the target audiences were undergraduates and graduate students studying forestry, together with some established forestry professionals who might like to get themselves up to date on the literature. This posed two problems. First, those who are involved in formal forestry are constantly flooded with publications, CD compilations, conference papers and so on. One might assume that if the papers that I selected were truly noteworthy then many of these people would already have seen them. The second, and more interesting, problem was that decisions about forests are no longer the unique preserve of forestry professionals. Forest science now embraces a whole range of disciplines; decisions are made with inputs from a wide range of organizations and with a heavy dose of pressure from civil society. It seemed, therefore, to make more sense to target the Reader at all those who take decisions that impact on forests.

The fact that such a diverse range of people are concerned about forests and that there is so much disagreement amongst them suggested another solution to the dilemma that I faced in selecting papers. Given the number of forest issues about which there is disagreement, perhaps the most useful contribution would be to collect a series of papers that give a balanced overview of these points of contention. This approach has the added advantage that it addresses one of the fundamental sources of dysfunction in the forest debate: the fact that many of the protagonists are heavily focused on a single issue. The reality of forests is that their management is complex and it is rarely safe to take decisions on the basis of a narrow, single-sector vision of problems and solutions.

Having chosen this path of focusing on papers that provide thoughtful analyses of controversial issues enables me to address a problem that has long been of concern to many who are professionally involved in forests. This is the fact that many important decisions are not based upon scientific evidence at all. The politicians and bureaucrats who allocate funds and approve ambitious development and conservation programmes are as likely to be influenced by the popular media as by the latest scientific findings. The fact that the public is interested in forests is, of course, an excellent thing; the fact that its main source of information is often an over-simplified account in a newspaper or on the television is a cause for concern. This brings us back to the complexity and interconnectedness of forest issues and the fact that perceptions of them are so much influenced by our cultural and economic background. The chapters by Kaimowitz on the issues of forest law enforcement (Chapter 8) and of Brown and Williams on bush-meat in Africa (Chapter 9) are excellent examples of work that shows how simple knee-jerk political responses to problems can easily yield solutions that are not sustainable. So I selected my chapters in the hope that some of my readers might be environmental journalists or the communications staff of some activist conservation organizations. I also hope that some of those members of 'civil society' who give money or volunteer their time to support the work of conservation organizations might also find it useful to read these works.

This collection of chapters nonetheless remains a rather arbitrary selection from the huge literature on forest issues that is currently available. They all provide excellent lists of references that will help readers to access further literature on the subjects that they address. In addition there are a number of excellent sources of information on forests that are readily accessible through the Internet or from good libraries. The websites of the Food and Agriculture Organization (FAO), the World Bank, the International Union for Conservation of Nature, WWF, the United Nations Environment Program (UNEP) World Conservation Monitoring Centre, the Center for International Forestry Research and the European Forest Institute all contain abundant material on current forest issues. The recent World Bank Forest Strategy contains an excellent overview of forest issues in developing countries and countries with economies in transition. It comes with a CD that includes a full range of background papers that are a rich source of information on contemporary forestry (World Bank, 2004). FAO supports a global Forest Information System that brings together information from a consortium of organizations on a wide range of forestry topics. Many national forest services have websites with a range of forest information. FAO's quarterly review, *Unasylva*, provides excellent concise, synthetic papers on important forest themes.

The papers in the Reader are divided into five sections. The first section deals with forest resources, how much forest there is, what its values are, how these resources are changing and who owns forests. Even these issues are surprisingly contentious. How you measure a forest, or even how you define a forest, depends very much on who you are. Much of the confusion results from different definitions and interpretations driven by different visions of what constitutes a forest. The opening chapter by Sengupta and Maginnis (Chapter 2) was commissioned specially for this Reader. It draws heavily on the FAO Forest Resources Assessment (FAO, 2003) but the authors have also included information from other sources where these shed light on areas of uncertainty. By the time this Reader is published the Millennium Assessment (www.millenniumassessment. org) will have produced its final reports and Chapter 21 of the Condition and Trends Assessment will provide an excellent balanced overview of the current state of forest ecosystems.

The second section deals with forests and people's livelihoods. I have selected four texts that provide a scientific perspective on some of the myths that have been widely held on the role of forests in alleviating global poverty. A lot of poor people live in, and around, forests and a lot of these people are highly dependent on forest resources. But this does not mean that the quickest route out of poverty for these people will be through better forestry. Similarly the peoples of the forests often have cultures and ways of life that are fascinating to outsiders and that many of us would like to see conserved. The people themselves may not always share our views. They also want to enjoy the improved health care and material wellbeing that they can often only attain by moving out of the forest.

Section three, 'Threats and Opportunities', addresses a number of issues that are prominent on the present international forest agenda where there are a number of narratives and counter-narratives. Illegal activities, poaching of bushmeat, collecting fuelwood, and fire have all been widely portrayed as the major threats to forests. Hundreds of millions of people, however, have an interest in how these so-called problems are addressed. Sometimes the solutions that emerge from international negotiations may be quite different to those that would be chosen by the people directly involved in these activities.

The fourth section presents some of the more thoughtful recent publications on the challenges of achieving sustainable management. Which approaches to management will yield the best results for the maximum number of concerned people? Which approaches are technically feasible? Where are the intervention points for conservation investments? The chapters span the entire spectrum from government policies to local actions – as any comprehensive strategy for sustainable forests must.

The concluding section deals with some promising ways forward. It returns the focus to the people who depend upon the forests. It places them more firmly in the driving seat; it is consistent with the global tendency towards decentralization of control over forests. It concludes with a chapter that sets out the policy of two of the world's leading forest conservation bodies: the WWF and the World Conservation Union, who argue that the future lies in finding the right balance between protection, management and restoration of forest resources.

The original intention to provide a Reader for graduate and undergraduate students of forestry also had to be revisited. A little reflection shows that forest departments in universities are endangered institutions; not many students now study pure forestry. In the past such departments or free-standing forestry training institutions – for instance in France and Germany – produced the forest engineers who staffed forest departments. Nowadays forestry requires specialized skills from a wide range of disciplines and there is less demand for the generalist 'forester'. Those people who do complete courses in forestry are as likely to end up working for an environmental NGO or a major corporation with forestry interests as for a conventional forest department. Fewer people now study 'forestry' in the traditional form of silviculture; more people study 'environment' or 'natural resources management'. All of this seems to me to be a very favourable development. We are moving towards forestry institutions that take a broad view of environmental issues and that can assemble teams of specialized staff able to deal with the complex social and biophysical problems that forests present us with. I like to think that this makes this Reader even more useful. People from a wide range of disciplines who are involved in some way with forests will find a lot of useful material here. The collection is eclectic and few people will have come across the full range of these papers – at least not in any formal university course.

Forestry is entering a new era (Kohm and Franklin, 1997). It is becoming more science-based and certainly more holistic in its view of forests. Forest departments, like many natural resource agencies, are no longer the 'command and control' structures of earlier times, imposing their own expert view on forest managers and owners. Forest departments are now facilitating, experimenting, negotiating and learning (Sayer and Campbell, 2003). They are, or should be, more decentralized and more responsive to local needs. Their staff need to be more knowledgeable and better able to exercise judgement.

In spite of the interest in criteria and indicators and 'model forests' it is not possible to manage forests according to predefined formulae. Every forest is different and what constitutes good forestry in one place today will not be appropriate at another location, or at the same location, at some time in the future. All of this means that those who are professionally involved in forestry face challenging times. Managing forests is becoming even more demanding and offers careers that will be rich and fulfilling – but they will not be the same predictable careers as those of the foresters of the past.

The term 'ecosystem approaches' is now widely used to describe approaches to the management of the social-ecological systems to which our forests belong. Ecosystem approaches require that we have an understanding of all of the factors and interest groups that are influencing our forests. The concept has been developed and endorsed by the Convention on Biological Diversity and the 180 or so countries that are signatories to that convention. This Reader is designed to be read by anyone who wishes to be part of the practice of ecosystem approaches to forestry. The chapters cover the widest possible range of forest issues and the authors are people who have broad credibility in their respective specialized fields. I have deliberately sought out material that challenges conventional wisdom. The popularization of forest issues and the attention given to them by the media have created a fertile ground for the establishment and propagation of myths. There are many widely held ideas about forests and the problems of their conservation and sustainable use that are simply wrong. Forests are so much a part of our culture and our subconscious that we readily accept any explanation of a forestry situation that conforms to our preconceived ideas. But perceptions that emerge from the culture of Western Europe may not transfer well to the depths of the Amazon. A conservationist in a rich industrialized country sees a problem where a poor farmer on the fringes of a tropical forest may see an opportunity.

So the chapters in the Reader are a selection of those that provide an honest and rigorous overview of issues where there is controversy or where I consider that conventional wisdom needs to be challenged. I have given heavy emphasis to material dealing with people, and especially poor people in developing countries or countries in transition. Formal forest departments have not been good at addressing the interests of these people – they have been more concerned with the interests of forest industries and of governments. One of the most welcome developments in recent years has been this shift of emphasis in forestry towards the people who live in and around the forests. These are the people whose welfare is most closely linked to the condition of the forest. They may derive benefits from the extraction of commodities such as logs, but often they are the ones who genuinely see forests as 'systems' and who have often exercised stewardship over the full range of forest goods and services.

References

1   FAO (2004) Forestry web site. www.fao.org/forestry/index.jsp
2   Forsyth, T. (2003) 'Critical political ecology'. *The Politics of Environmental Science*. Routledge, London.
3   Kohm, K. A. and Franklin, J. F. (1997) 'Creating a forestry for the 21st century'. *The Science of Ecosystem Management*. Island Press, Washington, DC.

4   Rietbergen, S. (1993). *The Earthscan Reader in Tropical Forestry*. Earthscan, London.
5   Sayer, J. A. and Campbell, B. (2003) *The Science of Sustainable Development: Local Livelihoods and the Global Environment*. Cambridge University Press, Cambridge.
6   Spilsbury, M. J. and Kaimowitz, D. (2000) 'The influence of research and publications on conventional wisdom and policies affecting forests'. *Unasylva*, vol 203, pp3–10. www.fao.org/DOCREP/X8080e/x8080e02.htm#P0_0
7   World Bank (2004) 'Sustaining forests: A development strategy'. *World Bank Forest Strategy*, The International Bank for Reconstruction and Development, Washington, DC.

Part 1

The Forest Resource

Forests were placed firmly on the political agenda by President Carter's Global 2000 initiative in 1980. WWF, IUCN and UNEP gave prominence to forests when they launched the World Conservation Strategy in 1981. A number of international conservation organizations embarked upon forest conservation programmes at that time and the World Resources Institute released its Tropical Forestry Action Plan at the World Forestry Congress in Mexico in 1985. Throughout this period the media and political decision makers have been hungry for information on what exactly was happening to the forests. There was a widely held perception that forests were being destroyed, but a lack of agreement on where and how fast this was happening and, especially, on what was responsible for the decline. Numerous publications have made conflicting claims about exactly what was occurring in the forests; to the lay observer the contradictions amongst these claims have been quite confusing.

Since 1980 FAO has maintained a Forest Resources Assessment programme that has produced major global reports at the end of each decade. These reports have often been criticized but they are the only sources of information that are based upon clear definitions and for which the origins of the information are presented in a transparent manner. Criticisms of FAO forest data are often based upon comparisons drawn with sources of information that use different definitions or incompatible methodologies. FAO figures remain the best ones available but patience and some technical knowledge are needed to fully understand and interpret them.

Recently, remote sensing technologies have improved considerably and this has enabled a number of other independent forest assessment schemes to produce interesting results. The European Union Joint Research Institute at Ispra in Italy has done some first-class work on deforestation. On the other side of the Atlantic cooperation between the University of Maryland and the NASA has also yielded new insights. Chapter 2 of this Reader by Sengupta and Maginnis provides an overview of what is known. It is based largely on FAO material but also draws upon a number of the newer sources that have appeared up until 2004. The chapter also weaves together the stories of how the resource is changing and the factors that are causing these changes. It summarizes the international initiatives that are underway to counter those changes in forests that are thought to be harmful.

The causes of changes in forests, and especially of deforestation in the tropics, have also been beset by controversy. In the early 1980s industrial logging was seen to be the major threat. Subsequently, 'slash and burn' agriculture was identified as the prime culprit.

More recently, planned expansion of agriculture has been recognized as the main cause of deforestation. In Chapter 3 Geist and Lambin examine the different proximate and underlying causes of forest change and demonstrate the full complexity of this issue. The theme is taken up again in Chapter 12, which examines the policy measures that are needed to counter undesirable deforestation.

Quite a lot of confusion has been generated by the question of who owns forests and who should own them. Forests have traditionally been owned by nation states and many forest values are public goods to which all people feel that they have a right. But forests are increasingly managed and owned by private individuals, corporations and communities. What is more, the ownership of forests is changing rapidly. Governments are devolving ownership to the private sector, individuals or communities. Understanding who owns forests and who has rights to them is fundamental in understanding how forests could and should be managed. Chapter 4 by White and Martin presents a new analysis of this issue and shows that governments are less important than one might have thought and that their role in many countries is declining in significance. This issue is taken up again in Part 2, 'Forests and Livelihoods' and in Part 5, 'The Way Forward'.

2

Forests and Development: Where Do We Stand?[1]

Sandeep Sengupta[2] and Stewart Maginnis[3]

Introduction

Forest ecosystems play multiple roles at global as well as local levels and provide a range of important economic, social and environmental goods and services that impact on the wellbeing of poor rural communities, local and national economies and global environmental health. It is estimated that at the global level, forestry formally contributes some 2 per cent to world GDP or more than US$600 billion per annum (FAO, 1997; Lomborg, 2001). However, the actual contribution of forests to the world economy may be much higher, though extremely difficult to quantify. A 1997 study in the journal *Nature* estimated the global value of the goods and services that forest ecosystems provide – from timber to climate regulation to water supply to recreation – at some US$4.7 trillion a year, or more than a quarter of that year's world GNP of US$18 trillion (Constanza et al, 1997; World Bank, 2002). As the *State of the World's Forests 2003* report emphasizes, forests can help in important ways to reduce food insecurity, alleviate poverty, improve the sustainability of agricultural production and enhance the environment in which many impoverished rural people live all over the developing world (FAO, 2003).

A number of global assessments of forests have been carried out over the past three decades.[4] While differing in their definitions of forest cover, methodology and specific results, making detailed comparisons difficult, these assessments nonetheless reinforce each other in their overall depiction of declining forest area and continued degradation of forest ecosystems (UNEP, 2002). This situation analysis puts together the main findings from some of the major global forest assessments and studies that have been carried out in recent years in order to provide an overview of the current status of forests in the world today. First we look at the current physical status of forests and the broad trends in use, management and ownership of forest resources in different parts of the world. Then we discuss some of the main proximate and underlying drivers of forest related land use change. The final section analyses the key current issues and emerging themes that are, and will continue to remain, of relevance to the forest sector in the coming years. Among other resources, this document draws substantially on the *Global Forest Resources Assessment 2000* (FAO, 2001),

Note: Reprinted from *Forest and Development: Where Do We Stand?* Sengupta, S. and Maginnis, S. Adapted from an unpublished IUCN Global Situation Analysis, Copyright © (2004), with permission from IUCN

Table 2.1 *Forest area distribution by region, 2000*

Region	Land area million ha	Total forest (natural forests and forest plantations)				Natural forest million ha	Forest plantation million ha
		million ha	% of land area	% of all forests	Net change 1990–2000 million ha/year		
Africa	2978	650	22	17	−5.3	642	8
Asia	3085	548	18	14	−0.4	432	116
Europe	2260	1039	46	27	0.9	1007	32
North and Central America	2137	549	26	14	−0.6	532	18
Oceania	849	198	23	5	−0.4	194	3
South America	1755	885	51	23	−3.7	875	10
World total	13,064	3869	30	100	−9.5	3682	187

Source: FAO, 2001

GEO: Global Environment Outlook-3 (UNEP, 2002), *World Resources 2000–01* (WRI, 2000) and *The State of the World's Forests* 2001 and 2003 reports (FAO, 2001a; FAO, 2003).

Current Status of the World's Forests

Using a single definition of forests for the first time,[5] FAO's *Global Forest Resources Assessment 2000* concluded that:

The total area covered by forests worldwide is approximately 3869 million hectares, almost a third of the world's land area, of which 95 per cent is natural forest and 5 per cent is planted forest; 17 per cent is in Africa, 14 per cent in Asia, 27 per cent in Europe, 14 per cent in North and Central America, 23 per cent in South America and 5 per cent in Oceania (Table 2.1; Figure 2.1).

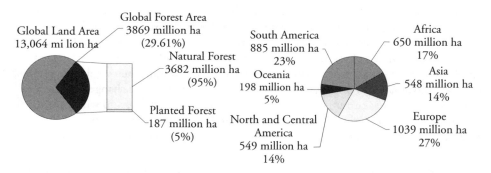

Figure 2.1 *Global distribution of forests*

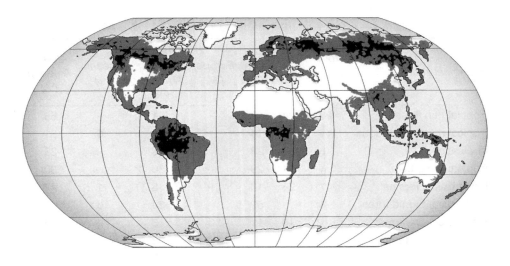

Ecological Distribution of Forests

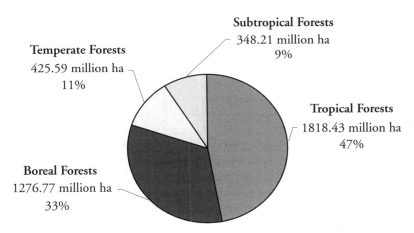

Note: On the map, black represents closed forest, more than 40 per cent covered with trees more than 5m high; dark grey represents open (10–40 per cent coverage) and fragmented forest; light grey represents other woodland, shrubland and bushland.

Source: FAO, 2001a

Figure 2.2 *Map of forest cover, 2000*

Using a combination of new global maps and statistical data, FRA 2000 also estimated the distribution of forest area by ecological zones: 47 per cent is in the tropics, 33 per cent in the boreal zone, 11 per cent in temperate areas and 9 per cent in the subtropics (Figure 2.2). The detailed distribution of forests by ecological zones in shown in

Table 2.2 *Forest area distribution by ecological zone, 2000*

Ecological zone	Total forest (%)	Africa (%)	Asia (%)	Europe (%)	North and Central America (%)	Oceania (%)	South America (%)
Tropical rain forest	28	24	17	–	1	–	58
Tropical moist deciduous	11	40	14	–	9	6	31
Tropical dry	5	39	23	–	6	–	33
Tropical mountain	4	11	29	–	30	–	30
Total tropical forests	47	28	18	–	5	1	47
Subtropical humid forest	4		52	–	34	8	6
Subtropical dry forest	1	16	11	30	6	22	14
Subtropical mountain	3	1	47	13	38	–	1
Total subtropical forests	9	2	42	7	37	7	5
Temperate oceanic forest	1	–	–	33	9	33	25
Temperate continental forest	7	–	13	40	46	–	–
Temperate mountain	3	–	26	40	29	5	–
Total temperate forests	11	–	17	39	39	4	2
Boreal coniferous forest	19	–	2	74	24	–	–
Boreal tundra woodland	3	–	–	19	81	–	–
Boreal mountain	11	–	1	63	36	–	–
Total boreal forests	33	–	2	65	34	–	–
Total forests	100	17	14	27	14	5	23

Notes: Distribution of percentages does not exactly tally with other area statistics because of distortions in the remote sensing classification of forests in the global forest cover map. Only zones with forest are included.

Source: FAO, 2001a

Table 2.2. Tropical and subtropical dry forests are concentrated in Africa (containing 36 per cent of the world total), South America (30 per cent) and Asia (21 per cent). The majority of tropical rainforests are located in South America (58 per cent), but a large proportion (24 per cent) is also found in Africa; most of the rest is in Asia (17 per cent). Nearly all temperate and boreal forests are located in Europe and North and Central America. Mountain forests are found mainly in Europe (40 per cent) and North and Central America (34 per cent), respectively (FAO, 2001a).

Two-thirds of the world's forests are located in ten countries alone: the Russian Federation, Brazil, Canada, the US, China, Australia, the Democratic Republic of Congo, Indonesia, Angola and Peru (Figure 2.3). Only 22 countries have more than 3 hectares of forest per capita, and only about 5 per cent of the world's population lives in these countries – mostly in Brazil and the Russian Federation. Three-quarters of the world's

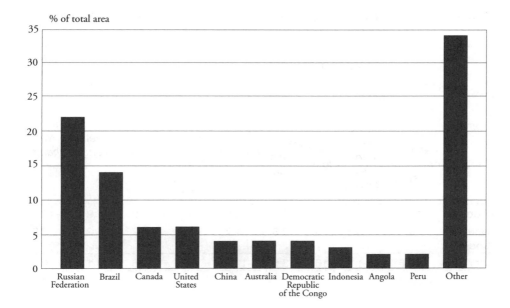

Source: FAO, 2001a

Figure 2.3 *Countries with the largest percentage of the world's forests*

population, on the other hand, lives in countries with less than 0.5 hectares of forest per capita, including most of the densely populated countries in Asia and Europe (FAO, 2001a). The proportion of total land area under forest varies significantly by region and country. About half the land area of South America and Europe is covered by forest, but only a sixth of Asia's land is forested. Africa, North and Central America and Oceania fall in between, each with about a quarter of its land covered by forest. According to FRA 2000 data, 56 countries are 'low forest cover countries (LFCCs)' with less than 10 per cent of their land covered by forest, and 20 countries have more than 60 per cent of their land under forest (FAO, 2001a; FAO, 2003).

The FRA 2000 data indicate that the world's natural forests continued to be converted to other land uses at a very high rate during the 1990s. Although there has been a slightly lower rate of net forest loss in the 1990s than in the 1980s, the worldwide loss of natural forests has continued at roughly comparable high levels over the last two decades (FAO, 2001a). An estimated 16.1 million hectares of natural forest worldwide was lost annually during the 1990s (14.6 million hectares through deforestation and 1.5 million hectares through conversion to forest plantations). Of the 15.2 million hectares lost annually in the tropics, 14.2 million hectares was converted to other land uses and 1.0 million hectares was converted to forest plantations. In non-tropical areas, 0.9 million hectares of natural forest was lost per year, of which 0.5 million hectares was converted to forest plantations and 0.4 million hectares to other land-use classes (Table 2.3). All geographical regions (Table 2.1), with the exception of Europe, witnessed a net loss

Table 2.3 *Forest area change in tropical and non-tropical areas, 1990–2000*
(million hectares per year)

| Domain | Natural forest | | | | | Forest plantations | | | Total forest |
| | Losses | | | Gains | | Gains | | | |
	Defore-station (to other land use)	Conver-sion to forest plan-tations	Total loss	Natural expan-sion	Net change	Conver-sion from natural forest (reforesta-tion)	Affore-station	Net change	Net change
Tropical	−14.2	−1	−15.2	1	−14.2	1	0.9	1.9	−12.3
Non-tropical	−0.4	−0.5	−0.9	2.6	1.7	0.5	0.7	1.2	2.9
World	−14.6	−1.5	−16.1	3.6	−12.5	1.5	1.6	3.1	−9.4

Source: FAO, 2001

of forests in this decade, with Africa and South America accounting for the greatest forest loss.

Against the gross annual loss of 16.1 million hectares of natural forests worldwide, there was a gain of 3.6 million hectares as a result of the natural expansion of forest, giving a balance of −12.5 million hectares as the annual net change of natural forest area globally. Of this 3.6 million hectares, 2.6 million hectares was in non-tropical areas – primarily Europe and North America, while 1.0 million hectares was in the tropics (FAO, 2001a). Much of the gain in natural forest area was the result of natural forest succession on abandoned agricultural land.[6]

Gains in forest area also occurred through the expansion of forest plantations. The average rate of successful plantation establishment over the decade was 3.1 million hectares per year, of which 1.9 million hectares was in tropical areas and 1.2 million hectares in non-tropical areas (FAO, 2001a). As shown in Table 2.3, 1.6 million hectares was the result of afforestation on land previously under non-forest land use, whereas 1.5 million hectares resulted from the conversion of natural forests. Of the estimated 187 million hectares of plantations worldwide, Asia has by far the largest area, accounting for 62 per cent of the world total (Figure 2.4), with plantations making up over a fifth of all forests in Asia (ibid). FRA 2000 identified the ten countries with the largest plantation development programmes (as reported by percentage of the global plantation area) as China, 24 per cent; India, 18 per cent; the Russian Federation, 9 per cent; the United States, 9 per cent; Japan, 6 per cent; Indonesia, 5 per cent; Brazil, 3 per cent; Thailand, 3 per cent; Ukraine, 2 per cent and the Islamic Republic of Iran, 1 per cent. These countries account for 80 per cent of the global forest plantation area. The extent of plantations in industrialized countries was less clear than in developing countries since many industrialized countries make no distinction between planted and natural forests in their inventories. In terms of composition, *Pinus* (20 per cent) and *Eucalyptus* (10 per cent) remain

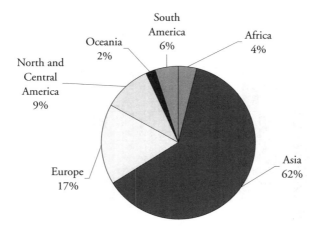

Source: FRA, 2001

Figure 2.4 *Distribution of forest plantations by region*

Table 2.4 *Annual plantation rates and plantation areas by region and species group*

Region	Total area 000 ha	Annual rate 000 ha/yr	Plantation areas by species groups (000 ha)							
			Acacia	Eucalyptus	Hevea	Tectona	Other broadleaf	Pinus	Other conifer	Unspecified
Africa	8036	194	346	1799	573	207	902	1648	578	1985
Asia	115,847	3500	7964	10,994	9058	5409	31,556	15,532	19,968	15,365
Europe	32,015	5	–	–	–	–	15	–	–	32,000
North and Central America	17,533	234	–	198	52	76	383	15,440	88	1297
Oceania	3201	50	8	33	20	7	101	73	10	2948
South America	10,455	509	–	4836	183	18	599	4699	98	23
WORLD TOTAL	187,086	4493	8317	17,860	9885	5716	33,556	37,391	20,743	53,618

Source: FAO, 2001a

the dominant genera worldwide (Table 2.4), although the diversity of species planted is increasing (ibid).

Combining the global deforestation rate of 14.6 million hectares per year and the rate of forest area increase of 5.2 million hectares per year (natural expansion plus

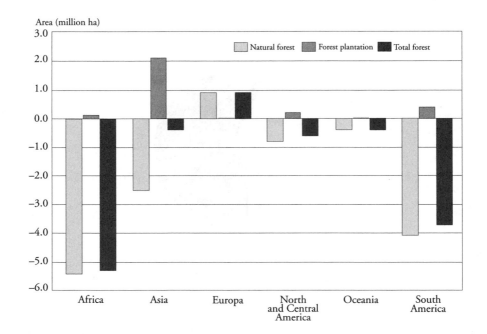

Source: FAO, 2001a

Figure 2.5 *Annual net change in forest area by region, 1990–2000*

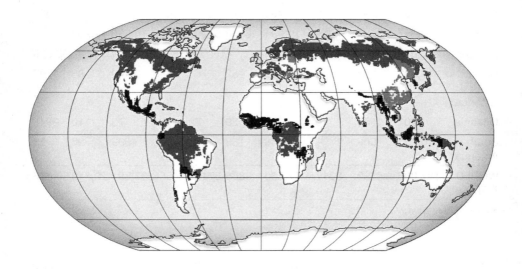

Source: FAO, 2001a

Figure 2.6 *Areas showing significant net change in forest area*

afforestation), the net loss in forest area at the global level during the 1990s was thus an estimated 94 million hectares – an area larger than Venezuela and equivalent to 2.4 per cent of world's total forests (FAO, 2001a). However, as shown in Figure 2.5, the global figures obscure significant differences in forest cover change among regions and countries. Net deforestation rates were highest in West Africa and South America. The loss of natural forests in Asia was also high, particularly in South-East Asia, but was significantly offset (in terms of area) by forest plantation establishment. This resulted in a more moderate rate of change of total forest area in the region. In contrast, the forest cover in the other regions, which are largely made up of industrialized countries, increased slightly (FAO, 2001a). The areas which registered the highest and lowest rates of net change in forest areas are shown in Figure 2.6. Table 2.5 shows the top ten net losers and gainers of forest cover in the 1990s using data compiled from the FRA 2000 datasets. The countries with the highest net loss of forest area between 1990 and 2000 include Brazil, Indonesia, Sudan, Zambia, Mexico and the Democratic Republic of Congo. Those with the highest net gain of forest area during this period were China, the United States, Belarus, Kazakhstan and the Russian Federation. Overall, the world today contains around 6000m^2 of forest for each person, but this is being reduced by approximately 12m^2 every year (ibid).

Along with the physical coverage or *quantity* of forests, it is equally important to consider the *quality* of forests. The main limitation of the emphasis on forest area coverage is that this is not necessarily a good qualitative indicator of the health of a forest ecosystem since much of the world's forests are very fragmented and face high human pressure (FAO, 2001; UNEP, 2001). Although deforestation is widely recognized as a major conservation issue, the related issue of habitat fragmentation often receives less attention. As human pressure increases in both temperate and tropical forests, areas that were once continuously forested have become more fragmented. In the Brazilian Amazon alone, the area of forest that is now fragmented (with forests less than 10,000 hectares in area) or prone to edge effects (less than 1km from clearings) is more than 150 per cent greater than the area that has actually been deforested (FAO, 2003). Recent research indicates that small fragments have very different ecosystem characteristics from larger areas of forest, containing more light-loving species, more trees with wind- or water-dispersed seeds or fruits and relatively few understorey species. The smaller fragments also have a greater density of tree falls, a more irregular canopy, more weedy species and unusually abundant vines, lianas and bamboos. Thus, they preserve only a highly biased subset of the original flora and fauna, which is adapted to these conditions (Laurance, 1999, 2000). In a 1997 study, the World Resources Institute (WRI, 1997) coined the term 'frontier forests' to describe the world's remaining large intact natural forest ecosystems that are relatively undisturbed by human activity and are large enough to maintain all of their biodiversity, including viable populations of the wide-ranging species associated with each forest type. According to this study, frontier forests constitute about 40 per cent of the total remaining forest area,[7] but are heavily concentrated in three large blocks – two areas of boreal forest (in Canada, Alaska and Russia) and one relatively contiguous area of tropical forest spanning the north-western Amazon Basin and Guyana Shield (in Brazil, Peru, Venezuela and Colombia). Additional important outliers can still be found in Central Africa (Congo) and Papua New Guinea. However, 39 per cent of these remaining frontier forests are estimated to be under moderate or high threat of ongoing degradation and fragmentation (ibid).

Table 2.5 *Top ten losers and gainers of forest cover in 1990–2000*

Country	Total forest	Total forest	Forest cover change (1990–2000)	
LOSERS	1990 (000 ha)	2000 (000 ha)	Annual change (000 ha)	Annual rate of change (%)
Brazil	566,998	543,905	–2309	–0.4
Indonesia	118,110	104,986	–1312	–1.2
Sudan	71,216	61,627	–959	–1.4
Zambia	39,755	31,246	–851	–2.4
Mexico	61,511	55,205	–631	–1.1
Dem. Rep. of Congo	140,531	135,207	–532	–0.4
Myanmar	39,588	34,419	–517	–1.4
Nigeria	17,501	13,517	–398	–2.6
Zimbabwe	22,239	19,040	–320	–1.5
Argentina	37,499	34,648	–285	–0.8
GAINERS	1990 (000 ha)	2000 (000 ha)	Annual change (000 ha)	Annual rate of change (%)
China	145,417	163,480	1806	1.2
United States	222,113	225,993	388	0.2
Belarus	6840	9402	256	3.2
Kazakhstan	9758	12,148	239	2.2
Russian Federation	850,039	851,392	135	n.s.
Spain	13,510	14,370	86	0.6
France	14,725	15,341	62	0.4
Portugal	3096	3666	57	1.7
Vietnam	9303	9819	52	0.5
Uruguay	791	1292	50	5
TOTAL WORLD	3,963,429	3,869,455	–9391	–0.2

Source: FAO, 2001

Another recent study using globally comprehensive and consistent satellite data estimated that the extent of the world's remaining natural closed forests (WRCF) – defined as those forests having a crown cover or canopy density of more than 40 per cent – was around 2870 million hectares in 1995, ie about 21.4 per cent of the land area of the world (UNEP, 2001). Such a level of canopy closure is considered vital if the forest is to be considered healthy and able to perform all its known environmental and ecological functions effectively. Such forests are also home to some of the world's rarest and most unique species including the elusive clouded leopard of Russia and the lion-tailed macaque of the Western Ghats in India. About 80.6 per cent of the WRCF are concentrated in 15 countries. Ranked in the highest to lowest order in Table 2.6, they are Russia, Canada, Brazil, the United States, Democratic Republic of Congo, China, Indonesia, Mexico, Peru,

Table 2.6 *Distribution of the world's remaining closed forests in the top 15 countries*

Countries	Population	Total land area (000 hectares)	Area under CF[a] (000 hectares)	% of CF Total Area	Population density (people/1000 ha)
Russia	148,292,000	1,681,414.4	669,651.8	39.83	88
Canada	27,791,000	983,400.2	368,650.9	37.49	28
Brazil	148,002,000	850,063.3	361,597.2	42.54	174
United States	254,106,000	940,626.9	236,683.3	25.16	270
Democratic Republic of the Congo (Zaire)	37,405,000	233,814.5	116,204.2	49.70	160
China	1,176,397,000	940,234.9	111,578.9	11.80	1,251
Indonesia	182,812,000	188,748.2	92,753.4	49.14	969
Mexico	83,226,000	195,378.4	60,107.7	30.76	426
Peru	21,569,000	129,554.8	59,312.2	45.78	166
Colombia	32,596,000	114,115.9	51,931.9	45.51	286
Bolivia	6,573,000	108,868,2	41,942.9	38.53	60
Venezuela	19,502,000	91,408.4	40,709.0	44.54	213
India	850,793,000	315,440.8	37,952.2	12.03	2,697
Australia	16,888,000	768,639.9	35,548.5	4.62	22
Papua New Guinea	3,839,000	45,929.1	32,422.3	70.59	84
Total	3,009,790,000	7,587,637.9	2,317,046.4	30.54	397
World	5,368,000,000	13,405,362.7	2,872,363.8	21.43	400

a CF: Closed Forest
Source: UNEP, 2001

Colombia, Bolivia, Venezuela, India, Australia and Papua New Guinea. Three countries – Russia, Canada and Brazil – contain about 49 per cent of the WRCF. Fifty-four countries have over 30 per cent of their land area under closed forests. The continental distribution of total area under the WRCF is estimated at 9.65 per cent in Africa, 6.23 per cent in Australia and Pacific, 37.93 per cent in Europe and Asia, 24.32 per cent in North and Central America and 21.87 per cent in South America. However, only about 9.4 per cent of the WRCF have been accorded some sort of a formal protection status (ibid).

By virtue of their importance as habitats, forests figure prominently in efforts to conserve biological diversity. Estimated to contain at least half of the world's total biological diversity, natural forests have the highest species diversity and endemism of any ecosystem type. According to the UNEP Global Biodiversity Outlook, about 60 per cent, and possibly up to 90 per cent, of all species are found in moist tropical forests alone, despite the fact that they cover little more than 7 per cent of the world's land surface and around 2 per cent of the surface of the globe (UNEP, 2001a). In 1997, Conservation International identified 17 megadiversity countries that contain within their borders more

Table 2.7 *Forests in protected areas*

Region	Forest area in 2000 million ha	Forest in protected areas million ha	Proportion of forest in protected areas %
Africa	650	76	11.7
Asia	548	50	9.1
Oceania	198	23	11.7
Europe	1039	51	5.0
North and Central America	549	111	20.2
South America	886	168	19.0
Total	3869	479	12.4

Source: FAO, 2001a

than two-thirds of the planet's biological wealth – Brazil, Colombia, Indonesia, China, Mexico, South Africa, Venezuela, Ecuador, Peru, the United States, Papua New Guinea, India, Australia, Malaysia, Madagascar, the Democratic Republic of Congo and the Philippines. However, the threat to global biological diversity remains serious. According to recent figures released by IUCN a total of 12,259 species of plants and animals today face a high risk of extinction in the near future, mainly as a result of human activities (IUCN, 2003). This includes 24 per cent of mammal species and 12 per cent of bird species. Habitat loss and degradation, especially in lowland and mountain tropical rainforests has been identified as the primary reason for this loss of biodiversity, affecting 89 per cent of all threatened birds, 83 per cent of threatened mammals and 91 per cent of the threatened plants assessed (ibid).

Unsurprisingly, interest in forests for conserving biological diversity has increased considerably during the past decade. FRA 2000 estimates that about 12.4 per cent of the world's forests or 479 million hectares currently enjoy protected area status as per IUCN classifications. Results by ecological domain indicate that tropical and temperate regions have the highest proportion of those forests in protected areas, whereas only 5 per cent of boreal forests are located in protected areas. The North and Central America region has the largest proportion of its forests under protected area status, followed by South America (Table 2.7). Figure 2.7 shows the location of the major areas of forest under protection status. The number of transboundary conservation areas (TBCAs) has also grown significantly in recent years, from 59 in 1998, mainly in Europe and North America, to 169 in 2001, spread all around the world, with the International Tropical Timber Organization (ITTO) alone supporting transboundary conservation projects spanning 10 million hectares of tropical forest across eight countries (ITTO/IUCN, 2003). With TBCAs currently covering at least 10 per cent of the total global protected area, existing and proposed transboundary complexes together offer significant potential opportunities for forest biodiversity conservation (Zbicz, 1999, cited in FAO, 2001a).

Forest ownership patterns vary considerably worldwide. A recent study conducted by Forest Trends (White and Martin, 2002) presents the official perspective of forest ownership in 24 countries representing 93 per cent of the world's remaining natural

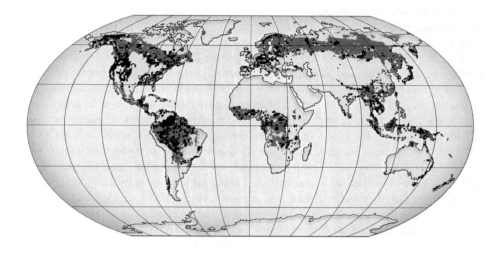

Source: FAO, 2001

Figure 2.7 *Major areas of forest under protection status shown in black*

forests, including 14 of the 17 megadiversity countries in the world and the top seven of the ten leading industrial roundwood-producing countries. The study indicates that about 77 per cent of the world's forest is (as per national laws) owned and administered by governments, at least 4 per cent is reserved for communities, at least 7 per cent is owned by local communities, and approximately 12 per cent is owned by individuals. It shows the importance of community reserves and ownership to be even higher when considering developing countries alone, with community-owned and administered forests totalling at least 377 million hectares, or at least 22 per cent of all developing country forests – thrice the amount owned by private individuals or industry. Interestingly, the analysis by White and Martin shows that while 71 per cent and 7 per cent of forests in developing countries are owned by the government, and private individuals and firms respectively, the corresponding figures for developed countries are 81 per cent and 16 per cent. Thus, both government and private-ownership of forests is higher in developed countries than in developing countries, while the opposite is the case for community ownership and access. However, as the study points out, there are important exceptions to these generalizations. For instance, in the United States, private individuals and firms own 55 per cent of the country's forests. Sweden, Finland and Argentina are other examples where over two-thirds of forests are privately owned by individuals and firms. At the opposite end of the spectrum are countries like Mexico and Papua New Guinea, where indigenous and other local communities own some 80 per cent and 90 per cent of forests respectively. Also, as the FAO points out, about 2.5 per cent of all forest and other wooded land, or 62 million hectares, in industrialized countries belongs to indigenous and tribal peoples, as defined in the International Labour Organization's Indigenous and Tribal Peoples Convention. Most of this land is in Australia. There are, however, serious political discussions in a number of countries, including Canada and New Zealand, about giving or returning ownership of very large areas of land, much of which is forest, to indigenous peoples (FAO, 2001a).

Forest ownership patterns are undergoing considerable change today around the world. The area owned and administered by communities doubled between 1985 and 2000, and this trend looks likely to continue over the next several decades as major forest countries, including those with highly centralized systems such as Indonesia and Russia, begin to actively engage in forest sector reforms and decentralization processes (White et al, 2004). Transition in forest ownership is also occurring in several Central and Eastern European countries, as forest land is restituted to its former owners or is privatized. This is a long and complex process, involving major legal and practical issues. In Africa, most countries with available data show no land areas officially reserved for communities and no privately held forest, community or individual. However, as White and Martin conclude, more research is needed to systematically document and analyse forest tenure and ownership trends in Africa, as some countries there have started reforming their land laws and recognizing customary use of forest resources.

Wood supply and production remains the focus of most forest inventories (UNEP, 2002). This reflects the economic importance of wood to many forest owners, public and private. FRA 2000 estimated the biomass and volume of wood (growing stock) in forests worldwide. Total standing wood volume and above-ground woody biomass in forests were estimated for 166 countries, representing 99 per cent of the world's forest area. The world total standing volume in the year 2000 was 386 billion m^3 of wood (FAO, 2001). The world total of above-ground woody biomass in forests is about 420 billion tonnes, of which more than a third is located in South America and about 27 per cent in Brazil alone (FAO, 2001a). Figure 2.8 shows the countries with the highest total forest woody biomass. The worldwide average above-ground woody biomass in forests is 109 tonnes per hectare. South America has the highest average biomass per hectare, at 128 tonnes per hectare. The countries with the highest standing volume per hectare include many Central American and Central European countries, the former having high-volume tropical rainforests and the latter having well-stocked temperate forests (ibid).

According to an FAO study by Bourke and Leitch, there has been substantial growth in demand for wood products in the last 25 years – both of industrial wood and fuelwood (FAO, 2000). Their estimates show that global consumption of total roundwood reached 3391 million m^3 in 1997 from around 2400 million m^3 in 1970 – a 41 per cent rise. Fuelwood consumption, which accounted for over half of the total roundwood production, and as much as 80 per cent of developing country production, expanded more rapidly than industrial roundwood consumption, growing by 55 per cent to 1853 million m^3 in 1997. Industrial roundwood consumption, on the other hand, grew by 21 per cent to 1537 million m^3 in 1997, although actually declining from a high of 1730 million m^3 in 1990 (ibid). While about 90 per cent of the fuelwood was produced and consumed in developing countries, industrial roundwood production was dominated by developed countries, which together accounted for 79 per cent of total global production (UNEP, 2002). The overall trend for industrial roundwood production was relatively flat during the 1990s. This was a significant change from the rapid growth that occurred prior to 1990. A global trend towards greater reliance on

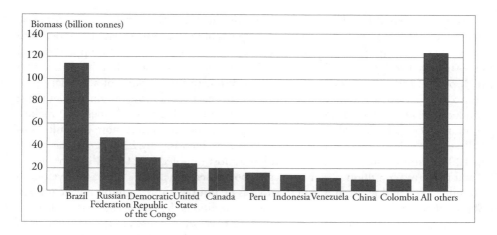

Source: FRA, 2000.

Figure 2.8 *Countries with highest above-ground woody biomass*

plantations as a source of industrial wood has also been noted, with industrial plantations (producing wood or fibre for supply to wood processing industries) accounting for
48 per cent of the global forest plantation estate. Non-industrial (eg for provision of
fuelwood or soil and water protection) and unspecified plantations accounted for a further 26 per cent each of the global forest plantation estate (FAO, 2001). Sawnwood
production in the last 15 years has remained static, with non-coniferous production rising and coniferous production falling since 1990. In the case of wood-based panels,
growth in production has been substantial, both in developed and developing countries
(FAO, 2000).

Forestry has become increasingly global over recent years. As Bourke and Leitch point
out, global trade in primary forest products has been expanding over the past decades in
both volume and value, and also as a proportion of global production, though the vast

Table 2.8 *Product share in global exports*

Category	Percentage by value	
	1970	*1997*
Industrial roundwood	15	8
Sawnwood	21	19
Wood-based panels	9	13
Wood pulp	20	13
Paper and paperboard	35	50
Total	100	100

Source: FAO, 2000

majority of all wood-based production is still destined for consumption in the domestic markets of producing countries. However, exports of most products have expanded, and the value of primary forest products (ie logs, sawnwood, panels and pulp and paper) exported reached US$136 billion in 1997, although that was a decline from the 1995 high of US$148 billion (FAO, 2000). In addition to this, exports of secondary processed products such as mouldings, doors, furniture, etc, and those of non-timber forest products (NTFPs), such as rattan, rubber, Brazil nuts, oils and medicines, also make a substantial contribution, although estimating their contribution is more difficult. Globally, export volumes of industrial roundwood have increased by 22 per cent since 1970 to 120 million m³ in 1997; sawnwood has almost doubled to 113 million m³, as has wood pulp (to 35 million metric tonnes); wood-based panels have increased fivefold to 50 million m³; and paper and paperboard has quadrupled to 87 million metric tonnes. Within this, as markets have changed, resource conditions altered, product forms changed and domestic consumption levels altered, there have been changes in the importance of different countries as exporters (ibid). It must be emphasized, however, that in many cases the global picture hides considerable regional and country variation. For example, Asian exports of some products have declined recently while at the same time those of South America have risen. Further, in some cases the regional variation actually reflects changes in only one or a few countries (eg Malaysia, Indonesia and Brazil) rather than all countries in a region. Overall, trade in forest products is currently estimated to make up 2–3 per cent of the total global merchandise trade (WTO, 2002; WRI, 2003; FAO, 2004).[8]

While trade in all forest product categories has been growing, the relative export importance of some of the products has been changing (Table 2.8). Although there is still substantial trade in unprocessed logs and woodchips by some countries, there has been a move from logs and, to a lesser extent, sawnwood towards plywood and secondary wood products. While expanding in absolute terms, industrial roundwood's share of the value of global forest product exports has dropped from 15 per cent to 8 per cent, largely since the early 1980s. The share of wood pulp has declined steadily to 13 per cent, and sawnwood has also declined slightly from 21 per cent to 19 per cent. By contrast, over this period the share of wood-based panels increased from 9 per cent to 13 per cent, while the greatest change has been in paper and paperboard which moved from 35 per cent to 50 per cent. Additionally, exports of furniture components, mouldings and so on have also increased substantially (FAO, 2000).

In 1997 developed countries accounted for the bulk of the forest products trade – both imports and exports (Table 2.9). Only plywood exports were dominated by developing countries of which Indonesia accounted for 41 per cent and Malaysia a further 20 per cent (FAO, 2000). Five countries accounted for 55 per cent of world exports of forest products and ten accounted for 70 per cent, with Canada and the United States alone accounting for a third. However, it must be noted that the export figures of some countries may not necessarily come from domestic production alone. On the import side, five countries accounted for 47 per cent and ten countries for 64 per cent of total world imports of forest products, with the United States and Japan being responsible for 28 per cent (ibid). In value terms, developed countries accounted for about 75 per cent of total exports and imports (down from the 1970 level of 87 per cent), their dominance being greatest for pulp, paper and paperboard – all with over 75 per cent (from 84 per

Table 2.9 *Top 14 global importers and exporters of forest products in 1997*

Importers	(US$million)	Exporters	(US$million)
USA	24,134.45	Canada	25,647.81
Japan	16,684.40	USA	19,835.07
Germany	10,916.46	Finland	10,414.17
United Kingdom	9,992.62	Sweden	10,295.37
Italy	6,823.03	Germany	9,828.22
France	5,866.29	Indonesia	5,142.29
China (Main)	5,661.41	France	4,663.72
Netherlands	3,739.72	Malaysia	3,951.83
Canada	4,657.66	Austria	3,834.62
China (Hong Kong)	3,976.01	Russian Federation	3,007.70
Korea, Republic of	3,835.96	Netherlands	2,667.47
Spain	3,739.72	Italy	2,650.94
Taiwan Province of China	3,719.66	Brazil	2,647.47
Switzerland	3,144,211	Bel-Lux	2,446.37
	2,211.77	China (Hong Kong)	2,267.27
World	144,980.52	World	138,280.84

Source: FAO, 2000

cent of the wood pulp exports and 76 per cent of the imports, to 90 per cent of the paper and paperboard exports and 75 per cent of the imports).

While international trade flows in forest products continue to be geared towards markets in the United States, Canada, Japan and Europe, major new players have emerged in the last few years. Foremost of these have been China for both imports and exports, and Russia for exports (FAO, 2004). Driven by its strong economic growth, low per capita endowment of wood, and policy constraints to domestic production from natural and plantation forests, China has grown from being the seventh largest importer of forest products to the second largest in the last ten years, and is now also the top importer of industrial roundwood worldwide (Sun et al, 2004). Between 1997 and 2002, Chinese forest product imports rose 75 per cent from US$6.4 billion to US$11.2 billion, and preliminary figures suggest that they reached almost US $13 billion in 2003 (ibid). Its forest product exports have also grown rapidly as global demand for finished Chinese wood products, such as furniture, have increased. However, developed or non-tropical countries still dominate exports in all forest product categories. Currently, the three main exporting regions are the former USSR, followed by Europe and the United States (FAO, 2004). The case of the former USSR is notable. Following the collapse of the centrally planned Soviet system, low production costs, a weak currency and abundant natural resources have enabled Russia to increase its roundwood exports by as much as 14 per cent (in 2002). However, even with greatly expanded exports, Russian production is still well below its annual allowable cut and with rapid increases in its productive capacity, its exports are expected to have a major impact on European and Asian markets in the future (ibid).

Developing countries are the main exporters of raw tropical timber, although there is some trade by developed countries that are either further processing tropical wood or are acting as trans-shipment points. Tropical products only represent a small share of the overall trade in forest products. In 1997, tropical industrial roundwood production was estimated to represent about 18 per cent of world industrial roundwood production; accounting for varying, but generally small, shares of the total exports of different products – 17 per cent of the industrial roundwood exported; 8 per cent of sawnwood; less than 10 per cent of pulp and paper and paperboard products; and 36 per cent of wood-based panels. It, however, accounted for about 70 per cent of global plywood exports (FAO, 2000). Overall, the proportion of timber that enters international trade from tropical countries is quite small, about 5 per cent of the total roundwood harvested (FAO, 2004). However, this hides considerable variation between countries as in a number of countries such as Indonesia, Malaysia and Cameroon, the proportion traded is substantially higher. While in the case of industrial roundwood, 28 per cent of the total industrial roundwood felled in tropical countries currently enters the international market, often constituting a significant portion of developing country exports. For non-industrial roundwood, which accounts for as much as 80 per cent of developing country production, only about 6–8 per cent enters international trade as little fuelwood and charcoal production is traded internationally (FAO, 2000; 2004). On the whole international trade is increasing for both tropical and non-tropical countries, although at a faster rate in the latter, and this trend is projected to continue. However, it must be remembered that the bulk of the trade in forest products still occurs in the form of intra-regional trade (FAO, 2004).

The concept of sustainable forest management (SFM) is becoming increasingly important in the forestry sector. A total of 149 countries, whose combined forest area equals 97.5 per cent of the total global forest area, are currently involved in a total of nine ecoregional criteria and indicator processes for SFM (FAO, 2003). A study based on information compiled for FRA 2000, and updated in 2002, indicates that 89 per cent of forests in all industrialized countries and countries with economies in transition are being managed 'according to a formal or informal management plan'. National statistics on forest management plans were not, however, available for many developing countries, including several of the larger countries in Africa and some key countries in Asia. Nevertheless, preliminary results from 49 developing countries for which information was available indicate that at least 255 million hectares, or about 12 per cent of the total forest area of all developing countries, were covered by a 'formal, nationally approved forest management plan covering a period of at least five years' as of the end of 2002. Considerable progress has been made in this regard over the last one and half decades. According to a study by the International Tropical Timber Organization (Poore et al, 1989), a maximum of 1 million hectares of forest in 17 tropical timber producing countries was being managed sustainably for wood production purposes in 1988. In 2002, more than 141 million hectares of forests in these 17 countries were reportedly 'managed in accordance with a formal forest management plan'. However, it must be emphasized that some areas covered by a management plan may not be sustainably managed, whereas other areas that are not under any formal management plan may still be.

Information on forest certification was also collected for FRA 2000 as an instrument to confirm SFM. A number of international, regional and national forest certification schemes now exist, focusing primarily on forests managed for timber production

Table 2.10 *Distribution of certified forests by region and by scheme (million hectares)*

Region	FSC	PEFC	Other	Total	% Certified Area
EU	22.01	41.01	–	63.02	35
Russia	2.12	–	–	2.12	1
Non EU	3.23	11.33	–	14.56	8
North America	9.50	–	79.33	88.83	48
Latin America	6.58	–	0.95	7.53	4
Africa	1.86	–	–	1.86	1
Asia & Oceania	1.71	–	4.33	6.04	3
Totals	47.01	52.34	84.61	183.96	100

Certification Scheme	Total Area	Europe	North America	Rest of World	% Share
ATFS	10.93	–	10.93	–	6
CSA	28.40	–	28.40	–	15
FSC	47.01	27.36	9.50	10.15	26
PEFC	52.34	52.34	–	–	28
SFI	40.00	–	40.00	–	22
Other	5.28	–	–	5.28	3
Totals	183.96	79.70	88.83	15.43	100

Source: FSC, 2004; PEFC, 2004; Phillips, 2004

purposes. Depending on how the term 'area certified' is defined, the area of certified forests worldwide as of the end of 2000 was estimated by FRA 2000 to be about 80 million hectares, or about 2 per cent of total global forest area. While some important wood-producing countries in the tropics have forests certified under existing certification schemes or are in the process of developing new schemes, 92 per cent of certified forests are located in temperate, industrialized countries. At the end of 2000, only four countries with tropical moist forests (Bolivia, Brazil, Guatemala and Mexico) were listed as having more than 100,000 hectares of certified forests, giving a combined total of 1.8 million hectares (FAO, 2001a). However, forest certification has been expanding rapidly over the last few years (Table 2.10). Latest figures show that the area of certified forests has increased to 184 million hectares (about 5 per cent of the world's forests), close to the World Bank/WWF Alliance target of having 200 million hectares of certified forests by 2005 (ITTO, 2002; FSC, 2004; Phillips, 2004).[9] The Forest Stewardship Council (FSC), which until recently registered all the world's certified forests, today has about 26 per cent of the forest certification market share while the Programme for the Endorsement of Forest Certification (PEFC) has about 28 per cent. The growth in forest certification over the last few years is shown in Figure 2.9. The breakdown of FSC certified area by forest type is also shown in Figure 2.10. There are currently more than 10,000 certified wood product lines in the market, and more than 800 companies in 19 countries that are participating in the Global Forest and Trade Network (GFTN), a buyers group promoted by WWF (Molnar,

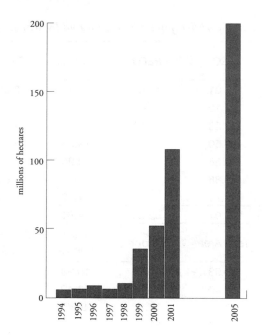

Source: ITTO, 2002

Figure 2.9 *The world's certified forests in 1994–2002*

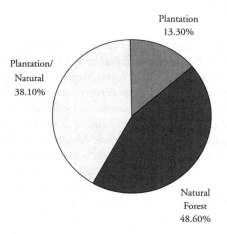

Source: FSC, 2004

Figure 2.10 *Percentage of total FSC-certified area by forest type*

2003; FAO 2004). Certification of forest products, both from natural forests and plantations, will continue to grow in importance, though care needs to be taken to ensure that rival certification schemes do not undermine consumer confidence.

Main drivers of forest-related land-use change

As seen in the previous section, the world's natural forests continue to be converted to other land uses at a significant rate although net loss has slowed down somewhat compared with the last decade. Nevertheless, the major problem forests face today is that of deforestation and degradation of the forest ecosystem, including fragmentation and biodiversity loss. There are a number of drivers – both proximate and underlying – that are responsible for this.[10] However, forest-related land use change is a complex socioeconomic, historical, cultural and political phenomena and it is erroneous to attribute it to a simple cause–effect relationship or assume that such a relationship will remain valid across all spatial or temporal dimensions (Contreras-Hermosilla, 2000).

A review carried out by Geist and Lambin (2002) on the proximate and underlying causes of tropical deforestation in 152 subnational case studies concludes that tropical deforestation is caused by identifiable regional patterns of several different, predominantly governance-related factors acting together to impact forest-land use change, of which the most prominent are economic factors, institutional arrangements, national land-use policies and remote influences (at the underlying level), which drive agricultural expansion, wood extraction and infrastructure development (at the proximate level). Their findings also reveal that prior studies have given too much emphasis to population growth and shifting cultivation as primary causes of deforestation.

Forest-related land-use change can also be seen as the result of actions by a number of agents (Contreras-Hermosilla, 2000). Agents are individuals, groups of individuals or institutions that directly convert forested lands into other uses or that intervene in forests without necessarily causing deforestation but substantially reduce their productive capacity. Agents include cultivators, private logging companies, mining and farming corporations, forest concessionaires and ranchers among others, and it is their action, influenced by underlying drivers, that shape the immediate proximate causes of forest loss and degradation. However, these agents are seldom independent of one another, and it is necessary to understand the motivations and incentives that shape their interactions if forest decline is to be effectively understood and addressed. Forest-related land-use change also occurs due to natural causes such as insect pest attacks, disease, fire, alien invasive species and extreme climatic events. However, even these are very often influenced and driven by underlying anthropogenic factors (FAO, 2001; UNEP, 2002).

Proximate causes of forest loss and degradation

Agricultural expansion

Over the years, researchers have identified agricultural expansion as a common direct factor in almost all studies on deforestation. Agricultural expansion (including permanent cropping, cattle ranching, shifting cultivation and colonization agriculture) was identified as the leading land-use change associated with 96 per cent of the deforestation cases that were reviewed by Geist and Lambin (2002). Almost 70 per cent of the total area that was deforested in the 1990s was converted to agriculture, however, predominantly under permanent rather than shifting systems (UNEP, 2002). There are also important regional differences to note. In Latin America most such conversion has been large-scale and permanent. In Africa small-scale agricultural enterprises have predominated.

In Asia, the changes have been more equally distributed between permanent agriculture (both large- and small-scale) and areas under shifting cultivation. However, as Barraclough and Ghimere (2000) caution, while agricultural expansion appears to be a significant factor in explaining deforestation in some countries, it need not always be the case. For example, in Brazil, Malaysia and China, the main causes of deforestation have been a combination of state-driven infrastructure development, industrial policies, growing manufacturing sectors and rapidly expanding urban populations, rather than agricultural expansion per se. To shed light on whether there is a clear relationship in the dynamics between forested and agricultural areas, FAO analysed qualitative temporal change trends on the basis of global statistics (FAO, 2003).

Preliminary findings (Figure 2.11) indicate that agricultural land is expanding in about 70 per cent of countries, declining in 25 per cent and roughly static in 5 per cent. In two-thirds of the countries where agricultural land is expanding, forest area is decreasing, but in the other third forests are expanding. In 60 per cent of the countries where agricultural land is decreasing, forests are expanding. In most of the rest (36 per cent), forests are decreasing. Historically, much of the increase in food production has been at the expense of hundreds of millions of hectares of forest. Although there are no solid estimates of how much farm and grazing land was originally under forest, the point remains that a large portion was cleared for agriculture. Additional forest land is expected

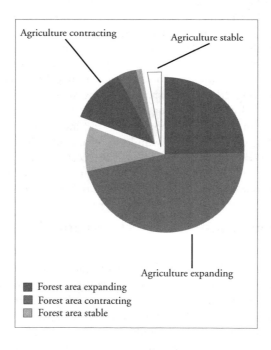

Source: FAO, 2003

Figure 2.11 *Expansion and contraction of agriculture and forests: Proportion of global area*

to be cleared in the future as well and it is important to acknowledge this reality and plan for it in advance (FAO, 2003; Poore 2003). However, it is equally important to recognize that many technological innovations to intensify agricultural production since the green revolution have had a positive impact on forest area. Indeed, as the argument goes, the more agriculture is intensified on a sustainable basis, the less pressure there will be to deforest in order to provide new areas for agriculture. As Victor and Ausubel (2000) observe, a number of developed countries ranging from the United States to France to New Zealand are actually witnessing a transition from deforestation to reforestation today, thanks to technological innovation and efficiency that has driven both high-yield agriculture and forestry in those countries. If appropriate mechanisms are put in place to lock in these gains, this may well set a trend for large-scale forest restoration to take place in the future, as technology reaches developing countries as well.

Infrastructure development

Infrastructure development (road construction, dams, mining, power stations and so on) is another important proximate cause of forest-related land-use change. Among all forms of infrastructure expansion, road construction is by far the most frequently reported cause of deforestation (Geist and Lambin, 2002). Though many of the areas where new road networks pose a threat to forests are among the poorest and have some of the lowest road densities in the world – and it can be argued that road construction can provide communities with better access to markets and overall development – the problem is that road construction is seldom properly planned. Invariably, it has opened up areas of undisturbed, mature forests to pioneer settlements, logging and clearance, sometimes to unsuitable forms of agriculture. The ensuing fragmentation increases the exposure of forests to the dangers of poaching, alien invasive species, fires and pest outbreaks (WRI, 2000). This is evident, for example, from the increased fragmentation and bushmeat trade that is taking place in Central Africa, where forest areas are being opened up to logging (Figure 2.12).

In the Brazilian state of Pará, deforestation due to road construction increased from 0.6 per cent to 17.3 per cent of the state's area between 1972 and 1985 (Contreras-Hermosilla, 2000, cited in UNEP, 2002). In a later study of the Brazilian Amazonia, undertaken by the Institute for Remote Sensing Research (INPE) it was found that of 90,000km^2 of forest lost between 1991 and 1996, 86 per cent was within 25km of an area of previous pioneer deforestation along major roads (Alves, 1999). More recent studies (Carvalho et al, 2001; Laurance et al, 2001a; Cattaneo 2002) suggest that deforestation rates in the Brazilian Amazon could increase sharply in the future (by 15 per cent in the short run and by 40 per cent in the long run) as a result of over US$40 billion in investments in highway paving and major new infrastructure projects being planned in the region under the Avança Brasil (Forward Brazil) programme (Figure 2.13). However, these figures have been contested by the Brazilian government. Infrastructure development activities like dam building and mining also have an adverse impact on forests. The World Commission on Dams (WCD) has noted that large dams have often led to the loss of forests and wildlife habitat, the loss of species populations and the degradation of upstream catchment areas due to inundation of the reservoir area (WCD, 2000). In Ecuador, Peru and Venezuela, mining corporations and individual miners have cleared large areas of forest, and mining has negatively impacted forest cover in a number of Asian countries as well

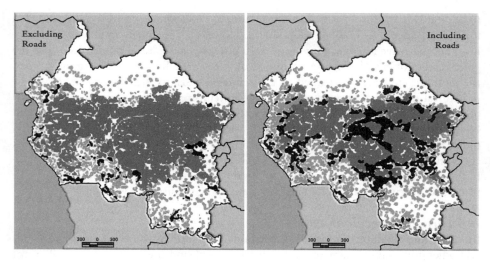

Source: WRL, 2003

Figure 2.12 *Forest fragmentation by roads in Central Africa*

Source: Landsat data 1975, 1986, 1999; USGS Data Center/UNEP, 2002

Figure 2.13 *Deforestation in Rondônia, Brazil, caused by road construction followed by colonization, logging, cattle ranching and crop farming*

(UNEP, 2002). However, a recent study released by CIFOR, *Oil Wealth and the Fate of the Forest: A Comparison of Eight Tropical Countries,* argues that in some cases the macro-level effects of incomes from oil and mining activities may actually reduce the loss of tree cover in tropical countries (Wunder, 2003).

Wood extraction

Despite the growing importance of plantations as a source of wood supply, wood extraction in the form of commercial timber, poles, fuelwood and charcoal – both legal and

Table 2.11 *Industrial roundwood production and illegal logging in select producer countries*

Country	Ind. Roundwood production in 2000 (m cu m)	Estimated illegal harvest as % of total production
Russian Federation	105.8	20
Brazil	103	80
Indonesia	31.4	73
D.R. Congo	3.7	na

Source: Scotland and Ludwig, 2002

illegal – continues to degrade mature natural forests in much of the developing world (WRI, 2000a). In the case of commercial logging, logging methods are often destructive and unsustainable, and especially damaging on steep slopes and in sensitive ecosystems such as mangroves (UNEP, 2002). Though many tropical countries in Africa, Asia and Latin America rely on logging for export earnings, illegal logging costs forest country governments at least US$10–15 billion a year – an amount greater than total World Bank lending to client countries and greater than total annual development assistance in public education and health (White and Martin, 2002). The regions particularly vulnerable to this threat are the Amazon Basin, Central Africa, South-East Asia and the Russian Federation (Table 2.11). For example, illegal logging costs Indonesia approximately US$600 million annually (Baird, 2001). Direct annual fiscal/financial losses to the governments of Honduras and Nicaragua due to clandestine logging have been estimated at US$11–18 million and US$4–8 million, respectively (Richards, 2003). It is also important to remember that illegal logging is not confined to developing countries. The Russian Federation is a major timber producer and exporter, and estimates of the extent of illegal logging range from 20–30 per cent for the country as a whole, to around 40–50 per cent in particular areas of Siberia (Brack et al, 2002). Increased poverty and loss of traditional communist-era livelihoods have also put protected areas and forests in Central and Eastern European countries, and in the former USSR, under severe pressure from illegal tree felling, and in some places have pushed rare species to the brink of extinction (UNEP, 2002).

Fuelwood and charcoal consumption are also important proximate drivers of forest land-use change in developing countries, though the lack of uniform terminology, definitions, units and conversion factors make it difficult to aggregate and compare the data that exists on this (FAO, 2003b). Estimates arising from new analytical models developed by FAO indicate that the global annual consumption of fuelwood peaked in the mid-1990s at about 1600 million m^3 and may now be slowly declining. However, in contrast the global consumption of charcoal is still growing rapidly and wood utilized for charcoal has more than doubled in the last three decades to about 270 million m^3 per annum in 2000 (Arnold et al, 2003). On aggregate, total woodfuel consumption (fuelwood plus wood for charcoal) is still rising but at a declining rate and substantially less rapidly than the equivalent growth in population. Also it is important to note that at a global scale, forests supply only about a third of woodfuels – the

Table 2.12 *FAO projections of woodfuel consumption in main developing regions to 2030*

	1970	1980	1990	2000	2010	2020	2030
Fuelwood (million m³)							
South Asia	234.5	286.6	336.4	359.9	372.5	361.5	338.6
South-East Asia	294.6	263.1	221.7	178.0	139.1	107.5	81.3
East Asia	293.4	311.4	282.5	224.3	186.3	155.4	127.1
Africa	261.1	305.1	364.6	440.0	485.7	526.0	544.8
South America	88.6	92.0	96.4	100.2	107.1	114.9	122.0
World	1444.7	1572.7	1611.6	1616.2	1591.3	1558.3	1501.6
Charcoal (million tonnes)							
South Asia	1.3	1.6	1.9	2.1	22	2.4	2.5
South-East Asia	0.8	1.2	1.4	1.6	1.9	2.1	2.3
East Asia	2.1	2.3	2.3	2.2	2.1	2.0	1.8
Africa	8.1	11.0	16.1	23.0	30.2	38.4	46.1
South America	7.2	9.0	12.1	14.4	16.7	18.6	20.0
World	21.2	27.0	35.8	45.8	55.8	66.3	75.6

Source: Broadhead et al (2001) cited in Arnold et al (2003)

balance being obtained from other sources such as farm-based tree lands, roadsides, community woodlots and wood industry residues (WRI, 2000a). Further, it is important to note that there are significant regional and subregional differences in the consumption of both fuelwood and charcoal (see Table 2.12 for regional trends). Thus woodfuel extraction continues to be a significant underlying cause of forest and wooded area clearance in some regions in Africa and South Asia where woodfuels contribute to up to 80 per cent of the primary energy supply (WRI, 2000a).[11]

Forest fires

Fires are a natural phenomenon and in some ecosystems, such as tropical dry forests, savannahs and temperate and boreal forests, they are a crucial ecological process. However, over 90 per cent of all wildland fires in forests and savannahs worldwide today are due to human action and cause significant forest loss (WRI, 2000a). On average, fires burn between 6 and 14 million hectares of forest per year worldwide – equal to the amount of forests cleared by logging and agricultural land conversion combined (GFP, 2003). For ecosystems not adapted to fire, the result can be catastrophic, leading to enormous economic losses; damage to environmental, recreational and amenity values; and even loss of life (FAO, 2001b). In 1997–1998, loss from forest fires was estimated at US$9 billion worldwide, equivalent to 20 per cent of current total global spending on overseas development aid (GFP, 2003). Expenditure for fire fighting is also rising with the incidence of fires, with fire fighting costs in 1997–1998 estimated at over US$2 billion (ibid). WRI (2000a) also cites new satellite data and studies that demonstrate that fires are increasingly affecting moist tropical forests that have not burned in the past. For instance, major fires occurring in the El Niño years 1982–1983, 1987, 1991, 1994 and 1997–1998 devastated large areas of forest and caused significant economic losses, both in Indonesia where most fires occurred and in neighbouring countries.

The area burned in Indonesia in the 1997–1998 fires is estimated at 9.7 million hectares of both forested and non-forested land, with some 70 million people affected by smoke, haze and the fires themselves. Impacts included damage to health, loss of life, property and reduced livelihood options. The economic and ecological losses to Indonesia have been estimated to exceed several billion dollars (FAO, 2001b; CIFOR, 2003). Serious forest fires have also occurred in recent years in Australia, China, Brazil, Mongolia, West Africa and in Europe and the wider Mediterranean region where it is estimated that on an average 500,000 hectares of forests have burned each year, predominantly as a result of human action (UNEP, 2002).

The boreal forests of North America and the Russian Federation represent the greatest expanse of relatively undisturbed forest remaining outside the tropics. Fires in the boreal region are believed to damage more forest land than logging or other activity therein. Figure 2.14 compares the annual area burned in the Russian Federation and North American boreal forest regions. On average, according to these statistics, 2.4 million hectares per year burned in North America and a little over 900,000 hectares per year burned in the Russian Federation. However, some Russian scientists and foresters claim that the figures for the Russian Federation are an underestimate and that these are actually much higher (WRI, 2000a).

Alien invasive species[12]

As the global movement of people and products expands, so does the movement of plant and animal species from one part of the world to another. When a species is introduced into a new habitat – for example, oil palm from Africa into Indonesia, *Eucalyptus* species from Australia into California, and rubber from Brazil into Malaysia – the alien species typically requires human intervention to survive and reproduce. Indeed, many of the most popular species of trees used for agroforestry are alien or non-native and prosper in

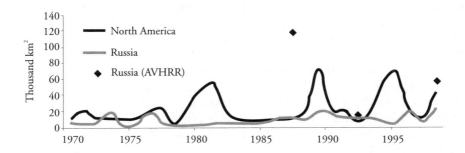

Note: AVHRR = Advanced Very High Resolution Radiometry
Source: Kasischke et al, 1999 in WRI, 2000a

Figure 2.14 *Annual area burned in the boreal forest region, 1970–1998*

their new environments partly because they no longer face the same competitors, predators and pests as in their native environment. Such alien species are economically very important and enhance the production of various forest commodities in many parts of the world. In some cases, however, species introduced intentionally become established in the wild and spread at the expense of native species, affecting entire ecosystems. Notorious examples of such invasion by alien woody species include the introduction of kudzu (*Pueraria lobata*) from Japan and China into the United States, where it now infests over 2 million hectares; the ecological takeover of the Polynesian island of Tahiti by *Miconia calvescens*; the spread of various species of Northern Hemisphere pine and Australian acacia in southern Africa; and the invasion of Florida's Everglades National Park by *Melaleuca* species from South America.

Of the 2000 or so species that are used in agroforestry, perhaps as many as 10 per cent are invasive. Although only about 1 per cent are highly invasive, they include popular species such as *Casuarina glauca*, *Leucaena leucocephala* and *Pinus radiata* (Richardson, 1999). Perhaps even worse are invasive alien species that are introduced unintentionally, such as disease organisms that can devastate an entire tree species (eg Dutch elm disease and chestnut blight in North America) or pests that can have a major effect on native forests or plantations (eg gypsy moths and longhorned beetles). The economic impact of such species amounts to several hundred billion dollars per year (Perrings, Williamson and Dalmazzone, 2000), much of it in forested ecosystems, even within well-protected national parks. However, as global trade grows, so does the threat from devastating invasive species of insect and other pathogens. These could fundamentally alter natural forests and wipe out tree plantations, the latter being especially vulnerable because of their lower species diversity. Efforts related to both conservation of biological diversity and sustainable forest management need to clearly recognize the issue of invasive alien species and address it.

Climate change

Scant attention has been paid so far to the changes that are likely to occur to the world's forest biodiversity over the next few decades due to climate change. Forest managers will need to pay more attention to incorporating climate change into their decisions. According to a recent estimate reported in *Nature*, well over a million species may be threatened with extinction due to climate change by 2050 (Thomas et al, 2004). Climate change has differential impacts on different forest types and tree species. For example, while enhanced photosynthesis and/or tree growth has been observed in some regions of the world, permafrost thawing in central Alaska threatens natural lowland birch forest (FAO, 2003). Also, deciduous forests will probably move northwards and to higher altitudes, replacing coniferous forests in many areas (see Figure 2.15 for a scenario outlined for beech forests in North America). Some tree species will probably be replaced altogether, jeopardizing biological diversity in several places (UNEP/US EPA, 1998).

Higher temperatures and changes in rainfall also threaten tropical montane forests, boreal forests and Mediterranean-type, fire-prone forests. Further, the effects of greenhouse gases can impact the phenology of forest trees, affecting such processes as budding, flowering, fruiting, leaf senescence, frost hardiness, wood quality, branching and insect susceptibility, in a highly species-specific manner (Jach, Ceulemans and Murray, 2001 cited in FAO, 2003).

Source: UNEP/US EPA, 1998

Figure 2.15 *Current and projected ranges of beech forests in North America*

While pest and disease infestations are part of the natural forest cycle, the risk of serious outbreaks increases with more changes in climate change. Several North American conifer pests have now been introduced into Europe, where they have no natural enemies. In the Russian Federation, insects were responsible for 46 per cent of the forest loss that occurred in the mid-1990s, and in Canada over 6.3 million hectares of boreal forests were affected by insect defoliation (UNEP, 2002). Although many other forests have proved relatively resilient to past climate changes, today's fragmented and degraded forests are more vulnerable, with up to 30 per cent of forests likely to be affected by 2050 by climate change (IPCC, 2001). Also extreme weather events, such as droughts and floods, pose other risks to forest ecosystems. The storms that struck Europe and India in 1999, for example, caused significant forest and tree loss. Apart from these, air pollution is also an important contributing factor to forest degradation in Europe and North America (UNEP, 2002).

Underlying drivers of forest loss and degradation

As discussed above, agricultural expansion, infrastructure development, wood extraction, fires, alien invasive species and climate change constitute most of the important immediate causes of forest-related land-use change. However, these causes are usually linked to broader underlying drivers. Also known as root causes, these underlying drivers are the national and international processes that trigger a chain of events that eventually put into motion or accelerate deforestation. While such root causes are similar the world over, the direction and magnitude of the deforestation processes they drive are usually country- and site-specific (Gutman, 2001). Also, as the analysis of Geist and Lambin (2002) points out, in most cases three or four underlying causes act together to drive two to three proximate causes. Further, in over a third of the cases of tropical deforestation, forest decline is being driven by the full interplay of economic, institutional, technological, cultural and demographic variables.

The main underlying drivers of forest related land-use change include the following:

Economic factors and market failure

Deforestation and forest degradation are ultimately the result of decisions made by various agents such as farming communities, shifting cultivators, private entrepreneurs, corporations, and so on. Forests are often perceived to have lesser economic value (or higher opportunity cost) if left standing, and the immediate financial rewards that can be gained from converting them to more economically competitive land uses or forms like agriculture or timber act as powerful incentives for these agents to clear and degrade forests. Further, the inherent 'public good' nature of the many acknowledged services that forests provide, such as watershed protection or carbon sequestration, means that society does not currently have to pay a price for these services and, consequently, the agents responsible for forest land-use change have no incentive to consider them in their land-use equations. This results in a classic situation of market failure.

Other economic drivers (sometimes a result of perverse policy incentives) such as higher agricultural prices, commercialization of timber markets, more rural credit, lower rural wages, increased opportunities for land titling and demand from urban-industrial centres, and so on also contribute to increased deforestation in different parts of the world. For example, the economic crises in Bolivia and Cameroon, and more generally in Latin America and Africa, in the 1980s and the structural adjustment programmes (SAPs) that followed caused widespread deforestation due to expansion of land for both agricultural cash crops and logging for export (Gutman, 2001). In Java, Indonesia, the financial crisis of the late 1990s and the subsequent restructuring encouraged significant forest clearance by small farmers, who reacted to the crisis by increasing their holdings of rubber and other tree crops to ensure future income security (Sunderlin, 2001, cited in Gutman, 2001). A global workshop on addressing the underlying causes of deforestation in 1999 has also recognized international trade and over-consumption in developed countries as one of the key underlying factors that drive forest-related land-use change in developing countries (IISD, 1999).

Institutional factors and government policies

Misguided policy interventions and perverse incentives are also an important underlying cause of forest decline. These policies do not necessarily have to originate in the forest sector alone. Very often extra-sectoral policies that determine, for example, exchange rates, urban employment, trade, infrastructure development and so on, can have an equal or in fact a higher impact on the wellbeing of forests (Wunder and Verbist, 2003). Further, these extra-sectoral policies themselves may be driven by other underlying motives. For example, road construction has been shown to be one of the most important proximate drivers of forest loss in the frontier areas of Latin America and Africa. Yet the road system of a country is largely determined by government transportation policies. Though the natural conclusion from this is that road policies are the underlying cause of deforestation, the real underlying cause, however, is often the pre-existing desire to deforest on the part of some politically powerful group that is able to influence government policy in order to access good commercial opportunities (Contreras-Hermosilla, 2000).

In some cases, government action or policy may also be driven by a genuine interest to develop infrastructure or encourage colonization in frontier areas to generate greater national employment or economic growth. However, the problem then is akin to that of market failure since this judgement is often based on an underestimation of the true value of forests. Government subsidies to the agriculture and forestry sectors are also a major underlying cause of deforestation. OECD estimates that around US$35 billion goes in subsidies to the forest industry each year globally (Pearce, 2002). Especially damaging are those subsidies that are granted to timber concessionaires, who encourage unsustainable wood extraction in countries like Guyana, Indonesia and Ghana, where logging companies aggressively seek concessions from local governments at very low rates. Also, many of these subsidies originate in developed countries and it is the action of those countries then that is the real underlying cause of deforestation. The case of agricultural subsidies provided to farmers in developed countries is similar. By increasing supply and driving down the global prices of primary agricultural products, such subsidies often compel poorer tropical counties dependent on agricultural exports to clear larger areas of forest land in order to raise sufficient foreign exchange to meet their international debt commitments (WRI, 2000a).

Corruption is also an important underlying cause of deforestation. However, while corruption and illegal logging may, and often does, exist because of pure criminality, it can, in some situations, be driven by inappropriate governance structures that turn legitimate concerns or entitlements into illegal activities. For example, in one Central American country in the early 1990s one of the main causes for bribery associated with log transport permits was not that loggers wanted to move illegally harvested trees but rather that they wanted to avoid long bureaucratic delays in attaining permission that would leave legally harvested trees deteriorating in forest loading yards.

Demographic factors

One of the most frequently cited underlying causes of forest decline is population pressure. However, this link remains uncertain and the available evidence shows that there is no fundamental relationship between population growth and density that will always necessarily cause forest decline (Contreras-Hermosilla, 2000). Even though in parts of Africa and South Asia, where the main agents of deforestation are peasants and fuelwood collectors, and population growth and density can cause increased pressure on forests, population growth and density is generally not a cause by itself and has to be seen in the context of government policies driving road construction, colonization, agricultural subsidies, tax incentives and so on that cause the migration of people into forest areas, which may not have occurred otherwise (Culas and Dutta, 2002). Importantly, demographic factors associated with mortality and morbidity, particularly where the HIV/AIDS pandemic is concerned, may be just as, if not more, significant when it comes to forest-related land-use change.

Poverty

Like population, poverty is also often cited as a major immediate, intermediate and root cause of world deforestation. However, as Kaimowitz and Angelsen (1998) point out, poverty is rarely integrated into economic models in general, and deforestation models in particular, and existing models provide weak and conflicting evidence on this relationship. Thus rural poverty should not be seen per se as a factor driving world defor-estation, especially when recent research in Africa and elsewhere shows that market and policy changes are far more important drivers of deforestation.

The empirical evidence for the historical relationship between economic growth, growing middle-class consumption and forest decline is perhaps a little better under-stood but also remains weak and fragmented. Though there is some evidence that an Environmental Kuznets Curve (EKC)[13] does exist in relation to tropical forests (Neu-mayer, 1999; Contreras-Hermosilla, 2000), in reality there are often several different causal relationships, and these need to be better understood. More reassuringly, there is some, yet again fragmented, evidence that no single trajectory is necessarily predeter-mined and that forest resources, under a range of circumstances, can be managed and utilized in such a way as to contribute to poverty reduction while keeping future options open to retain more and lose less forest biodiversity. However, as Contreras-Hermosilla (2000) argues, it must also be considered that deforestation and forest degradation is not always undesirable, and there can be situations where environmental losses may be more than compensated by economic gains and the improved wellbeing of the poor.

Key Emerging Themes in the Forest Sector

In recent years, the forest sector has undergone fundamental changes, largely as a result of restructuring, shifts in ownership patterns and wider recognition of the multiple benefits that forests provide. Emphasis is also being increasingly placed on addressing the underlying causes of forest loss and degradation rather than just tackling its external manifestations. Some of the key emerging issues and themes that are expected to remain of relevance in the forest sector in the coming years include the following.

Forests and poverty reduction

An issue that has attracted renewed attention in recent years is the potential of forests to reduce and prevent poverty, particularly in developing countries, and to contribute towards the Millennium Development Goal of halving extreme poverty in the world by 2015. This has repeatedly emerged at key international workshops and meetings, includ-ing those held recently in Tuscany, Edinburgh, Tuusula and Bonn.[14] The reason for this increased emphasis is due to the fact that although not all forested areas are poor and not all poverty is found in forested areas, there is nonetheless a significant overlap between the forest and poverty maps of the world (FAO, 2003). According to the World Bank, forest resources directly contribute to the livelihoods of 90 per cent of the 1.2 billion people living in extreme poverty and indirectly support the natural environment that nourishes agriculture and the food supplies of nearly half the population of the develop-ing world (World Bank, 2002). As Sunderlin et al (2003) point out, forests have an important role to play in alleviating poverty worldwide in two senses: first, by serving as a vital 'safety net' to help rural people avoid poverty, and those who are poor to mitigate poverty; and second, through its untapped potential, to actually lift some rural people out of poverty.

Forest products are important sources of cash income and employment for the rural poor, and very often the poorest households, women and children depend almost entirely on forests to meet their daily subsistence needs (Oksanen et al, 2002). A study by Cavendish (1997) on drylands in Zimbabwe, for example, showed that forests contribute about 35 per cent of an average household income.[15] Forests also underpin the wellbeing of the agricultural sector, which is critical for attaining rural prosperity, by providing onsite environmental goods and services such as the maintenance of water supply, protection against soil erosion and the provision of soil nutrients and animal fodder. However, there is limited awareness among macroeconomists, treasury officials and higher-level policy makers of the real contribution that forests make towards achieving sustainable livelihoods, poverty reduction and national economic development. This is primarily due to poor forest statistics and valuation, but also to a lack of effective advocacy. There is, therefore, an urgent need to recognize the contribution, and potential, of the forestry sector with regards to poverty reduction efforts and to integrate it into mainstream national poverty reduction processes – such as the Poverty Reduction Strategy Papers (PRSPs) – by developing poverty-focused conservation strategies. However, it should also be appreciated that forests are not the only way to reduce poverty. In fact, as Wunder (2001) argues, a general conclusion is that, in many settings, natural forests tend to have little comparative advantage for the large-scale alleviation of poverty, especially when compared with agriculture. Hence it will be necessary to be flexible and realistic while calculating trade-offs.

Non-timber forest products

Non-timber forest products (NTFPs) constitute a critical component of food security and an important source of income for the poor in many developing countries. They cover a wide range of forest products, which are utilized in very different contexts and play different roles in household livelihood strategies (Angelsen and Wunder, 2003). The sustainable management and correct valuation of NTFPs is thus a topic that is of increasing importance to the forestry sector as more attention begins to be paid to the potential of forests in reducing or preventing poverty. As Neumann and Hirsch (2000) point out, there is overwhelming evidence that the poorest segments of societies around the world are populations principally engaged in NTFP extraction. By fostering greater value-addition and easier access to markets, NTFPs can thus make a significant contribution to both poverty reduction and poverty prevention efforts. However, at a more general level, it has been argued that NTFPs are economically inferior goods and that NTFP-based projects, if not carefully planned or well targeted, can unintentionally create poverty traps since the very characteristics that make them important and attractive to the poor also limit their potential to increase income and bring about socioeconomic advancement (Neumann and Hirsch, 2000; Campbell, 2002; Angelsen and Wunder, 2003). Further, there is also a danger that when certain NTFPs become more valuable, they then attract the more powerful external interest groups that had previously ignored them (Angelsen and Wunder, 2003). All these challenges will therefore have to be recognized and carefully considered while developing a future strategy for the sustainable and equitable use of NTFPs.

Community forest management and decentralized governance

National governments and international organizations are today increasingly favouring decentralization and democratization as a means of fostering both development and sustainable management of natural resources in the developing world. This has emerged from an increasing convergence of both the economic development and environmental protection agendas worldwide, and a growing recognition that without secure tenure rights, indigenous and other local groups lack long-term financial incentives for converting their forest resources into economically productive assets for their own development (White and Martin, 2002). According to a World Bank study, more than 80 per cent of all developing countries and countries with economies in transition are currently experimenting with some form of decentralization or the other (Manor, 1999). Already, in many parts of Asia and Africa, there is widespread acknowledgement that governments and public forest management agencies have not been good stewards of public forests. Consequently many countries have taken steps towards developing a more participatory and collaborative form of forest governance and management that recognizes local communities as primary stakeholders in forest conservation and offers them incentives, in the form of ownership/user rights, benefit-sharing mechanisms and so on, to participate in forest protection and management. Over 50 communities worldwide have received forest management certificates or chain-of-custody certification, and many other forest communities have been brought into the decision making process as stakeholders in the certification of public and private forests (Molnar, 2003).

Community-owned or administered forests are conservatively expected to at least double to 700–800 million hectares by 2015 (White et al, 2004). It is expected that 40 per cent of the world's forests will be managed or owned by communities and individuals by 2050 (FAO, 2003). What is even more compelling is the fact that, at a time when investment in the forest sector is generally declining, communities are emerging as the largest investors in forests. A recent study by Forest Trends entitled '*Who Conserves the World's Forests*' shows that community investment in sustainable forest management in developing countries has now exceeded both Overseas Development Assistance flows to the forest sector and the public expenditure made by governments (Molnar, Scherr and Khare, 2004). However, it is important to note that there are countries where public ownership is still effective in managing forests. Furthermore, decentralization itself should not be considered as a silver bullet and has its own risks and challenges. For example, in Indonesia, new decentralization policies intensified the pressures on forests in some areas (Ribot, 2002; Larson, 2004). Decentralization also has the problem of being only as effective and equitable as the existing underlying social and political structures allow it to be. Thus, to be meaningful, it has to be accompanied by effective democratic structures that ensure that less powerful groups, such as women and the poor, are not excluded or further marginalized and a mechanism is available that enables fair compensation when national and local interests diverge.

Forest law enforcement and governance (FLEG)

Past and current attempts to halt and reverse forest loss and degradation clearly reveal that the challenge is not primarily a technical one. More important is the framework of laws, policies and incentives that shape the behaviour of individual actors and institutions with respect for forest management and land-use change. While this has been known for some time it has only been more recently that governments and other actors have realized the magnitude of the problem that arises from weak forest governance and inadequate (and inequitable) law enforcement. WWF estimates that in Russia alone illegal activities may account for losses in national revenue of at least US$1 billion per year (WWF, 2004). In addition, past experience, particularly from the Asia and Africa FLEG processes, has clearly demonstrated that illegal logging is too a big a problem for governments to address alone and that an active and constructive input from civil society and the private sector is critical if government resolve is to be translated into effective action. If local civil society is to play an active and constructive role in the FLEG process then this must mean taking a long-term view. First it means building trust between different stakeholders – notably, but not exclusively, between governments and local NGOs, indigenous peoples' organizations and community-based organizations. Civil society will need to feel that its voice will not only be listened to but that it will be considered and included as part of the solution. Equally governments need to be assured that their democratic mandate will not be usurped by special interest groups and that civil society is engaging not solely as a critic, but also as a player. It is also important to ensure that local civil society is equipped with the necessary information so that it can deliver its own perspectives (and possible solutions) rather than having large international organizations act as a proxy.

Forest landscape restoration

There is growing acceptance today that forest management decisions need to be made beyond the scale of the primary management unit or stand, and that the sustainability of forest ecosystems has to be considered from the point of view of the larger economic and social landscapes within which they exist. Recognizing that tree cover no longer dominates many tropical forest landscapes, and that local land-use patterns have led to a dramatic and detrimental reduction in the availability of forest goods and services both locally and beyond, forest landscape restoration (FLR) is emerging as a pragmatic and forward-looking approach to deal with this loss in a realistic manner. Instead of solely promoting increased tree cover at a particular location, FLR focuses on how to restore forest functionality, ie the goods, services and ecological processes that forests can provide, at the broader landscape level (Maginnis and Jackson, 2003). FLR recognizes the reality that a typical forest landscape today is more likely to be a mix of primary forest, managed forest, secondary forest, plantations and degraded forest lands interspersed with extensive areas of non-forest land uses. Consistent with the ecosystem approach, it acknowledges that land management is a matter of societal choice and that the livelihood and land-use strategies of the communities living in various landscapes will be determined by trade-offs, rather than by unrealistic aspirations to return forest landscapes to their original pristine state. FLR is, hence, an approach that seeks to put in place forest-based assets that are good for both people and the environment, and it brings together a number of existing rural development, conservation and natural resource management principles and approaches to restore multiple forest functions to degraded landscapes. However, unlike previous landscape planning approaches, there is no top-down blueprint advocated for achieving FLR. Restoring forest functionality to a landscape has to be built on a collaborative process of negotiating land-use trade-offs and learning and adaptive management among all the stakeholders within that landscape.

Trees outside forests

Trees outside forests (TOF) are also gaining in importance. Studies conducted in South and South-East Asia and Africa show that many forest products originate from TOF (Janz and Persson, 2002). Increasingly, trees are being planted to support agricultural production systems, community livelihoods, poverty reduction and to provide the rural poor with access to a secure food supply. For instance, between 1986 and 2000, Mali's agrosilvicultural and silvopastoral activities consisted of the planting of 4000km of shelterbelts, 14,000 hectares of woodlots and 5000 hectares around water points and in pastures. Mali is also noted for its parkland agroforestry based on natural trees, a formation that covers 39 per cent of the country. Namibia has developed similar parkland systems (FAO, 2003).

Communities and smallholder investors (including individual farmers) are today growing trees as shelterbelts, home gardens and woodlots, as well as under a diverse array of agroforestry and farm forestry systems that provide a valuable supply of wood, non-wood forest products, fuelwood, fodder and shelter. Acknowledging the important role that TOF play in sustaining the livelihoods of rural populations, especially of women, FRA 2000 was the first of FAO's global assessments that attempted to take them into consideration.[16] Despite the fact that most of the information on TOF is currently site-specific and scattered among different institutions and sectors, and it is impossible to draw conclusions regarding their exact status (FAO, 2001), it is expected that TOF will become an important forestry focus in the years ahead, integrating with the broader forest landscape restoration approach.

Forest valuation and markets for ecosystem services

There is today an increased recognition of the importance of correctly valuing the goods and services provided by forests. The valuation of forest ecosystem services such as carbon sequestration, biodiversity conservation, watershed protection and ecotourism, and the development of markets and payment systems for these 'non-extractive' forest uses, are becoming particularly important as their potential to provide incentives to conserve forests and generate new sources of income to support rural livelihoods increases (Pagiola et al, 2002). A review carried out by Landell-Mills and Porras (2002) found almost 300 examples of such market-based mechanisms worldwide. According to a more recent study by ITTO, though the total value of direct ecosystem service payments is presently modest, it has grown dramatically over the past decade and has significant potential to benefit low-income producers in the future (Scherr et al, 2004). The emerging role of three main forest ecosystem services – carbon sequestration, watershed protection and biodiversity conservation – is considered below.

Forests and carbon sequestration
Climate change has emerged as one of the most important concerns of the 21st century. Sea level rise, warming temperatures, uncertain effects on forest and agricultural systems,

and increased variability and volatility in weather patterns are expected to have a significant and disproportionate impact in the developing world, where the world's poor are concentrated and are much more vulnerable to the potential damages and uncertainties inherent in a changing climate. Since the negotiation of the Kyoto Protocol, the importance of land use, land-use change and forestry (LULUCF) activities in both preventing and adapting to climate change has been a topic of hot debate. Whether one is a protagonist or an antagonist on the issue, decisions taken within the UN Framework Convention on Climate Change (UNFCCC) have significant implications for the forest sector.

Forests are both a source of carbon dioxide (CO_2) when they are destroyed or degraded, and a sink when conserved, managed or planted sustainably (FAO, 2003). Forest vegetation and soils currently hold almost 40 per cent of all carbon stored in terrestrial ecosystems. Much of this is stored in the great boreal forests of the northern hemisphere and in the tropical forests of South America and Africa (WRI, 2000a). Forest regrowth in the northern hemisphere currently absorbs carbon dioxide from the atmosphere, creating a 'net sink'. However, in the tropics, forest clearance and degradation are together acting as a 'net source' of carbon emissions (WRI, 2000). Though growth in plantations is expected to absorb more carbon, the likely continuation of current deforestation rates means that the world's forests will remain a net source of carbon dioxide emissions and a contributor to global climate change. Figure 2.16 shows the areas that are most responsible for this. However, there are significant opportunities to develop innovative carbon sequestration forestry projects that generate positive synergies between forest restoration, mitigation of climate change and livelihood improvements for the poor through mechanisms like the Clean Development Mechanism

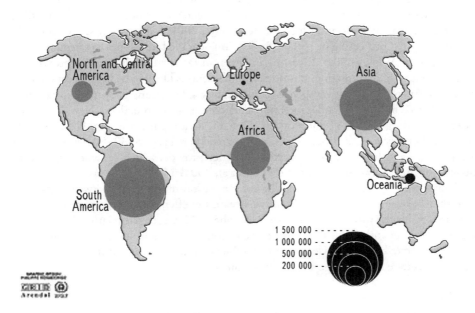

Figure 2.16 *CO_2 emission hotspots from land-use change*

(CDM) of the Kyoto Protocol and the Bio-Carbon Fund of the World Bank. According to a recent estimate, the CDM is expected to raise up to US$300 million annually for afforestation and reforestation activities over the first commitment period (2008–2012) of the Protocol (Scherr et al, 2004). However, the success of such carbon trading initiatives will depend on the resolution of many outstanding issues, including defining who really owns the carbon, the environmental and social implications of sequestration projects, and the outcome of the negotiations for the next commitment period, which are scheduled to begin in 2005. Nonetheless, in many countries, the prospect of trading carbon credits generated by the forestry sector has already started influencing decision making in the forest industries of those countries, and can affect future global timber markets. From the point of view of adapting to climate change, conserving and restoring forests assets at the landscape level can play an important role in reducing the risks from climate-related floods and droughts, especially for the poor, and helping to protect human welfare by minimizing the loss of life and damage to properties and other assets.

Forests and water linkages

Forests play an important role in supplying fresh water. With water shortages increasing in many parts of the world, and with over a third of the world's largest cities currently depending on forest areas for their water supply, the importance of this link is being rapidly realized today (Dudley and Stolton, 2003). Twenty-eight per cent of the world's forests are located in mountains and these forests are the source of some 60–80 per cent of the world's fresh water resources. They are also natural barriers to landslides, torrents and floods (ibid; FAO, 2003).

Tropical montane cloud forests (TMCFs), which have unique hydrological values and high rates of species endemism, are today being lost faster than any other major forest ecosystem. Nearly 30 per cent of the world's major watersheds have lost more than three-quarters of their original forest cover (WRI, 2000a). However, some countries have now started protecting or replanting trees on degraded hill slopes to safeguard their water supplies (WRI, 2000a). Investing in 'green infrastructure' through improved forest management and restoration in upstream catchments is likely to be highly cost-effective relative to structural alternatives such as dams and dykes. For example, by investing approximately US$1 billion in land protection and conservation practices, New York City avoided spending US$4–6 billion on filtration and treatment plans. Similarly, in Costa Rica, landholders in critical watershed areas are paid between US$30–50 per hectare per year for the watershed protection services that they provide, which generates significant income for these rural communities (Scherr et al, 2004).

Developing appropriate payment or compensation mechanisms between upstream watershed service providers and the downstream beneficiaries is expected to be a key area of work for the forestry sector in the years ahead. However, the relationship between upstream forests and downstream watershed services is a complex one, and there is a need to generate better knowledge on this subject, as also on the nature of institutions that can facilitate and manage such transactions.

Biodiversity conservation

As discussed in this chapter, biodiversity loss is a matter of great concern, especially in tropical areas. Tropical deforestation is expected to be responsible for the loss of an estimated 5–15 per cent of the world's species between 1990 and 2020, a rate unparalleled in modern history (Reid and Miller, 1989, cited in Landell-Mills and Porras, 2002). However, given the limited willingness of governments and international donor agencies to pay for biodiversity conservation today, innovative solutions including the development of markets for biodiversity protection services have started to emerge (ibid). These include mechanisms like bioprospecting rights, debt-for-nature swaps, conservation easements and biodiversity friendly products like shade-grown coffee, and so on. According to a recent McKinsey-WRI-The Nature Conservancy (TNC) study, the international finance market for conservation, of which forests constitute a significant part, is currently estimated at US$2 billion per annum. This is presently made up predominantly of development banks and foundations based in the United States and Europe (cited in Scherr et al, 2004). A recent global survey found 72 cases of biodiversity markets in 33 countries, of which 63 were in 28 tropical countries (Landell-Mills and Porras, 2002). However, these markets are still at a relatively nascent stage, and developing them in a practical and equitable manner will be a future challenge for the forestry sector. Implementing other poverty-focused and community-based conservation strategies concurrently to conserve biodiversity and benefit the livelihoods of the poor will also be a key area of focus for the sector.

Forest plantations and wood supply

Wood, and wood products, are by far the most energy efficient and environmentally friendly raw material when compared with competing products such as steel, aluminium and concrete (UNFF, 2003). Although plantation forests constitute less than 5 per cent of global forest cover, they provide more than 35 per cent of the world's wood supply today (ibid). In the future, planted forests will have an increasing role in wood and non-wood forest products supply as natural forest areas available for this purpose decrease owing to deforestation or designation for conservation or other uses. According to a forecast by MacGregor (2003), consumption and production of all timber products will rise by an estimated 10 per cent by 2010 and 15 per cent by 2020; the percentage of plantation wood will rise by 50 per cent by 2020 and 70 per cent by 2040; trade/output ratios will continue to rise; and the capture of non-timber values associated with forests will also increase.

The development of forest plantations in some countries has already had a major impact on wood production. In Chile and New Zealand, for example, the establishment of extensive areas of plantations has enabled these countries to meet almost all their domestic wood requirements and also to support a significant export industry with supplies from plantations (FAO, 2001). This is despite the fact that the global plantation estate is quite recent and half of all plantations in the world are less than 15 years old (UNEP, 2002). New Zealand, particularly, is about to face a 'wall-of-wood' scenario wherein the annual harvest will boom from the current 18 million tonnes annually to 28.5 million tonnes within three years. By 2006 it will reach more than 30 million tonnes – a dramatic 66 per cent rise in six years. This supply, 30 per cent of which is owned privately, is expected to significantly affect regional and global timber markets (Woodmetrics, 2003). Other significant developments include: rising private sector investment in plantations in developing countries; increasing foreign direct investment

(FDI) in plantations especially in Asia; and an expansion of 'outgrower' schemes whereby communities or small landowners produce trees for sale to private companies (FAO, 2001). Although no aggregate global figures are currently available on the total FDI in the forest plantation sector, its increasing importance can be judged from the fact that, in Vietnam alone, it is estimated that the sector has nearly 400 FDI projects with a total capital investment of about US$2.2 billion, accounting for 11.6 per cent of the total projects nationwide (Van Minh, 2003). Plantations are also expected to become increasingly popular in carbon sequestration projects, though some have raised concerns about the environmental and sociocultural impacts that these may have on local communities, especially in developing countries.

New forest technologies

Biotechnology has become a rapidly advancing field of scientific research on plants and offers potential benefits – and risks – to forestry. Modern biotechnologies currently used in forestry fall into three broad categories: the use of molecular markers in tree breeding; technologies that enhance vegetative propagation; and genetic modification of forest trees (FAO, 2001).

Genetic modification of forest tree species using recombinant DNA techniques has been contemplated for addressing traits such as virus resistance, insect resistance, lignin content and herbicide tolerance. However, there has been no reported commercial production of transgenic forest trees, although 116 field trials, in 17 countries and involving 24 tree species, have been reported (Owusu, 1999, cited in FAO, 2001). Further, it is acknowledged that biosafety aspects of genetically modified trees need careful consideration, especially because of the long generation time of trees and the potential for the dispersal of pollen and seed over long distances. Global research has also resulted in the development of new forestry technologies that have far-reaching implications for the way wood is supplied and processed (UNEP, 2002). One such example is the Indurite™ process, which impregnates softwood with an organic cellulose compound giving it the physical properties of hardwood. This technology, thus, has the potential to substantially reduce the pressure on tropical hardwood species (Sylt, 2003). Using sustainably managed plantation species, the process transforms softwood right through to the heart, not only leaving the wood hardened throughout but coloured to the desired finish all the way through. Substitutions in various materials and products have also occurred in many regions. Markets for reconstituted panels – particleboard, medium-density fibreboard (MDF) and oriented strand board (OSB) – have expanded rapidly and captured some of the plywood and sawnwood market. Some of the substitution has affected tropical timber. Further, the emphasis on sustainable forest management has resulted in greater attention to environmentally sound timber harvesting practices, often referred to as reduced impact logging (RIL).

Recently developed codes of practice for forest harvesting call for the use of environmentally sound timber harvesting practices or RIL, and substantial work has been done on testing and using RIL in the field. However, though many countries have initiated research, training and implementation of RIL, it has still not been widely adopted (FAO, 2003).

International arrangements on forests

In the years following the United Nations Conference on Environment and Development (UNCED), ongoing and often intensive international debate on forest policy issues has taken place. This has brought forest issues to the fore and raised awareness of the significant contributions that forests make to the health of the planet and its inhabitants (FAO, 2003).

In October 2000, the Economic and Social Council of the United Nations (ECOSOC) established the United Nations Forum on Forests (UNFF) to pursue international negotiations on the management, conservation and sustainable development of all types of forests, including elements contained in the UNCED Forest Principles and in the outputs of the Intergovernmental Panel on Forests (IPF) and the Intergovernmental Forum on Forests (IFF) (FAO, 2003). The Convention on Biological Diversity (CBD) is also a major international forum for the forest sector, with its sixth Conference of the Parties adopting an expanded work programme on forest biodiversity in 2002.[17] While the CBD has generally acknowledged the work of IPF, IFF and UNFF, some have noted that the CBD negotiations do not reflect the post-UNCED evolution of forest discussions and follow-up action at the international, regional and national levels (FAO, 2003). However, the CBD and UNFF consider their roles as complementary and recognize the need to strengthen collaboration. One recent effort in this direction has been a move by both fora to seek clarification on the relationship between the 'ecosystem approach' as adopted by the CBD and that of sustainable forest management as advocated by the UNFF. Opportunities also exist for building stronger linkages and exploring synergies between the CBD and UNFF and other forest-related agreements, including the UN Convention to Combat Desertification, the UN Framework Convention on Climate Change and its Kyoto Protocol, and the International Tropical Timber Agreement.

Key trends shaping the international forest dialogue today include a shift from negotiation of new commitments to formation of partnerships for implementation, giving rise to many new voluntary initiatives such as the Global Partnership on Forest Landscape Restoration, the Congo Basin and Asia Forest Partnerships, and a new regionalism – reflected in the FLEG and Low Forest Cover Country (LFCC) processes. Also evident is the increasing reliance of governments on international organizations to support implementation of forest-related commitments and proposals for action. There are growing expectations that the Collaborative Partnership on Forests (CPF) – a voluntary partnership of international organizations, including IUCN, charged with supporting the work of the UNFF – will assume a larger role in this regard. The UNFF and CPF together form the International Arrangement on Forests (IAF) established by ECOSOC in 2000. Governments will be reviewing the effectiveness of the IAF in 2005. This review is expected to address the future institutional framework and mandate of the UNFF and its position within the United Nations system.

Translating words into action, however, remains a constant challenge for all international forest policy arenas. While the role of forests in sustaining livelihoods, contributing to food security and reducing poverty has been increasingly recognized in recent years at key international forestry fora, operationalizing these linkages in practice has been less successful. A common criticism is that all international forest discussions have spent considerable time on procedural matters and revisiting existing commitments, even though all post-UNCED fora have repeatedly emphasized the need to move from dialogue to action on the ground. Nonetheless there has been some progress. More than 100 countries today have revised national forest policies and developed national forest programmes; 149 countries are involved in international initiatives concerning criteria and indicators for sustainable forest management; the area under official forest management plans has increased to 89 per cent in developed countries and some 12 per cent in developing countries; 184 million hectares of forests have been certified; 22 per cent of all developing country forests are either owned or administered by communities; and over 10 per cent of the world's forests now fall within protected areas. In addition, restoration initiatives with community involvement are underway in many parts of the world, and ministers from many African, Asian and donor countries have endorsed declarations and illustrative actions on forest law enforcement and governance.

Yet many challenges remain. The growing trend of globalization and trade liberalization, combined with increasing population, greater urbanization and rising incomes is expected to result in continued strong growth in global consumption of most products, including forest products. It is anticipated that consumption will grow most rapidly in developing countries where many countries may move from being net exporters to net importers in some forest product categories (FAO, 2004). These demands will need to be met in a sustainable manner.

The recent Doha WTO Ministerial Conference has brought a number of new issues onto the WTO agenda. Market access and agricultural subsidies are among the most important and contentious trading issues between developed and developing countries and the outcomes of these negotiations too will have significant implications for the global forest sector in the future. However, as problems of weak forest management infrastructure, inadequate capacities, unsustainable and/or unregulated conversion of forests, corruption, illegal logging and illicit trade endure, the more pressing challenge today is that of ensuring good governance. International forest fora and CPF members can act as catalysts to help countries improve their forest governance and implement sustainable forest management by providing them with policy guidance, technical advice, capacity-building and funding support, and platforms for lesson learning. However, it will remain the responsibility, and right, of the countries themselves and the forest owners and managers within them to conserve and sustainably manage the forest and biodiversity resources that exist within their national boundaries in a manner that is consistent with their overall development objectives. Finally, governments alone cannot be expected to provide all the solutions on their own – local communities and the private sector will also have to play their part.

Notes

1   Adapted from a Global Situation Analysis produced by the IUCN Forest Conservation Programme in 2003, IUCN, Gland, Switzerland.
2   Project Officer, IUCN Forest Conservation Programme.
3   Head, IUCN Forest Conservation Programme.
4   These include FAO and UNEP, 1982; FAO, 1995; FAO, 1997; FAO, 2001; UNEP, 2001; and WRI, 1997.

5   Where forests are defined as land areas of at least 0.5 hectares having a tree crown cover of more than 10 per cent and where the trees reach a height of at least 5m. This definition includes both natural forests and forest plantations. It however excludes some stands of trees established primarily for agricultural production, for example fruit tree plantations or some of those planted in agroforestry systems.

6   Expansion of forest has now been occurring for several decades in many industrialized countries, especially where agriculture is no longer an economically viable land use (Victor and Ausubel, 2000; FAO, 2001a).

7   According to this study, the total remaining area of forest is 3336 million hectares, of which 1350 million hectares are frontier forests (WRI, 1997).

8   Compiled using comparable trade figures available from FAO, WRI and WTO data sets.

9   However, it should be noted that the World Bank/ WWF Alliance does not recognize all certification schemes.

10 Proximate causes are human activities or immediate actions at the local level, such as agricultural expansion, that originate from intended land use and directly impact forest cover. Underlying driving forces are fundamental social processes, such as human population dynamics or government policies that underpin the proximate causes and either operate at the local level or have an indirect impact from the national or global level (Geist and Lambin, 2002).

11 As opposed to 15 per cent on average in other developing countries (WRI, 2000a).

12 This section is drawn from FAO, 2003.

13 According to an Environmental Kuznets Curve (an inverted U-shaped curve), deforestation increases with growth in per capita income up to a point, and then decreases.

14 FAO forum on the Role of Forestry in Poverty Alleviation, Tuscany, Italy, 4–7 September 2001; ECTF/IIED Forestry and Poverty Reduction Workshop, Edinburgh, UK, 13 June 2002; Workshop on Forests in Poverty Reduction Strategies: Capturing the Potential, Ministry of Foreign Affairs of Finland, Tuusala, Finland, October 2002; CIFOR conference on Rural Livelihoods, Forests and Biodiversity, Bonn, Germany, May 2003.

15 However, according to some the actual income contribution of forests is only about 20 per cent (Campbell, 2002 cited in Angelsen and Wunder, 2003).

16 TOF are defined as trees on land not classified as forest or other wooded land (FAO, 2001).

17 CBD and UNFF are two separate but parallel processes dealing with forests and forest biological diversity. The former addresses the conservation and sustainable use of biological diversity and the fair and equitable sharing of benefits arising from the use of genetic resources, including those from forest ecosystems, while the latter looks at the management, conservation and sustainable development of forests on the basis of the outcomes of UNCED, IPF and IFF (FAO, 2003).

References

1   Alves, D. S., Pereira, J., de Sousa, C., Soares, J. and Yamaguchi, F. (1999) 'Characterizing landscape changes in central Rondônia using Landsat TM imagery'. *International Journal of Remote Sensing*, vol 20, pp2877–2882.

2   Angelsen, A. and Wunder, S. (2003) 'Exploring the forest–poverty link: Key concepts, issues and research implications'. *CIFOR Occasional Paper*, no 40. CIFOR, Bogor, Indonesia.

3   Arnold, M., Köhlin, G., Persson, R. and Shepherd, G. (2003) 'Fuelwood revisited: What has changed in the last decade?' *CIFOR Occasional Paper*, no 39. CIFOR, Bogor, Indonesia.

4   Baird, M. (2001) 'Forest crime as a constraint on development. lecture by country director', The World Bank, 13 September 2001, Indonesia. http://wbln0018.worldbank.org/eap/eap. nsf/Attachments/FLEG_S8–2/$File/8+2+Mark+Baird+-+Indonesia,+WB.pdf

5   Barraclough, S. L. and Ghimere, K. B. (2000). *Agricultural Expansion and Tropical Deforestation. Poverty, International Trade and Land Use.* Earthscan, London.

6   Brack, D., Gray, K. and Hayman, G. (2002) *Controlling the International Trade in Illegally Logged Timber and Wood Products.* Royal Institute of International Affairs (RIIA), London. www.riia.org/pdf/research/sdp/tradeinillegaltimber.pdf

7   Campbell, B. M., Sayer, J., Kozanayi, Z., Luckert, M., Matumba, M. and Zindi, C. (2002) *Household Livelihoods in Semi-Arid Regions: Options and Constraints.* CIFOR. Bogor, Indonesia.

8   Carvalho, G., Barros, A. C., Moutinho, P. and Nepstad, D. C. (2001) 'Sensitive development could protect the Amazon instead of destroying it'. *Nature*, vol 409, p131.

9   Cattaneo, A. (2002) *Balancing Agricultural Development and Deforestation in the Brazilian Amazon.* Research Report 129. International Food Policy Research Institute, Washington, DC.

10 Cavendish, W. (1997) *The Economics of Natural Resource Utilisation by Communal Area Farmers of Zimbabwe.* Ph.D. thesis, University of Oxford, Oxford.

11 CIFOR (2003) Bogor, Indonesia. CIFOR website link: www.cifor.cgiar.org/docs/_ref/findoutabout/fire/. Last updated: Sunday, December 21, 2003.

12 Contreras-Hermosilla, A. (2000) *The Underlying Causes of Forest Decline. CIFOR Occasional Paper*, no 30. CIFOR, Bogor, Indonesia. www.cifor.cgiar.org/publications/

13 Costanza, R., d'Arge, R., de Groot, R., Farber, S., Grasso, M., Hannon, B., Limburg, K. et al (1997) 'The value of the world's ecosystem services and natural capital'. *Nature*, vol 387, pp253–260.

14 Culas, R. and Dutta, D. (2002) *The Underlying Causes of Deforestation and Environmental Kuznets Curve: A Cross-country Analysis.* School of Economics and Political Science, University of Sydney, Australia, paper submitted for the Econometric Society of Australasia Meeting (ESAM02) Queensland University of Technology, Brisbane, Australia, 7–10 July 2002.

15 Dudley, N and Stolton, S. (2003) *Running Pure: The Importance of Forest Protected Areas to Drinking Water.* World Bank/WWF, Washington, DC.

16 FAO/UNEP (1982) *Tropical Forest Resources Assessment 1980.* Forestry Paper no 30, FAO, Rome.

17 FAO (1995) *Forest Resources Assessment 1990 – Global Synthesis.* FAO Forestry Paper no 124. FAO, Rome.

18 FAO (1997) *The State of the World's Forests 1997.* FAO, Rome. www.fao.org/forestry/foris/webview/forestry2/index.jsp?siteId=3321&langId=1

19 FAO (2000) 'Trade restrictions and their impact on international trade in forest products' by Bourke, I. J. and Leitch, J., revised edition, FAO, Rome. www.fao.org/forestry/include/frames/english.asp?

20 FAO (2001) *Global Forest Resources Assessment 2000: Main Report.* FAO Forestry Paper no 140. FAO, Rome. www.fao.org/forestry/fo/fra/main/index.jsp

21 FAO (2001a) *The State of the World's Forests 2001.* FAO, Rome. www.fao.org/forestry/foris/webview/forestry2/index.jsp?siteId=3321&langId=1

22 FAO (2001b) *Global Forest Fire Assessment 1990–2000.* Forest Resources Assessment Programme, Working paper 55. FAO, Rome.

23 FAO (2003) *The State of the World's Forests 2003.* FAO, Rome. www.fao.org/forestry/foris/webview/forestry2/index.jsp?siteId=3321&langId=1

24 FAO (2003a) *Expert Consultation on Trade and Sustainable Forest Management – Impacts and Interactions.* 3–5 February, FAO, Rome.

25 FAO (2003b) 'Wood energy'. *Unasylva*, no 211. Forestry Department, FAO, Rome. www. fao.org/DOCREP/005/Y4450E/Y4450E00.HTM

26 FAO (2004) 'Trade and sustainable forest management – Impacts and interactions', Analytic Study of the Global Project CP/INT/775/JPN: *Impact Assessment of Forest Products Trade in the Promotion of Sustainable Forest Management*. FAO, Rome (unpublished draft on website). www.fao.org/docrep/007/ae017e/ae017e03.htm#P770_25328

27 FSC (2004) Forest Stewardship Council website, Global FSC figures (4 October, 2004). www.certified-forests.org/data/types.htm

28 Geist, H. J. and Lambin, E. F. (2002) 'Proximate causes and underlying driving forces of tropical deforestation'. *BioScience*, vol 52, no 2, pp143–150.

29 GFP (2003) *The Global Fire Partnership*. IUCN, WWF and The Nature Conservancy, Gland and Washington, DC.

30 Gutman, P. (2001) *Forest Conservation and the Rural Poor: A Call to Broaden the Conservation Agenda*. WWF Macroeconomics Program Office, Washington, DC.

31 IISD (1999) 'A summary report of the underlying causes of deforestation and forest degradation'. *Sustainable Developments*, vol 21, no 1, IISD. www.iisd.ca/download/pdf/sd/sdvol-21no1e.pdf

32 IPCC (2001) Third Assessment Report, Working Group II Report, *Climate Change 2001: Impacts, Adaptation and Vulnerability*. IPCC, Bonn.

33 ITTO (2002) 'Forest certification: Pending challenges for tropical timber' by Atyi, R. E. and Simula, M., *Technical Series*, no 19, October 2002, ITTO, Yokohama, Japan.

34 IUCN (2003) *IUCN Red List of Threatened Species*. IUCN Species Survival Commission, Gland. www.iucn.org and www.redlist.org

35 IUFRO Research Series no 8. CABI Publishing and International Union of Forestry Research Organization, New York.

36 Jach, M. E., Ceulemans, R. and Murray, M. B. (2001) 'Impact of greenhouse gases on the phenology of forest trees', in Karnosky, D. F., Ceulemans, R., Scarascia-Mugnozza, G. E. and Inees, J. L. (eds) *The Impact of Carbon Dioxide and other Greenhouse Gases on Forest Ecosystems*, pp193–235.

37 Janz, K. and Persson, R. (2002) *How to Know More about Forests? Supply and Use of Information for Forest Policy*. CIFOR Occasional Paper, no 36. CIFOR, Bogor, Indonesia.

38 Kaimowitz, D. and Angelsen, A. (1998) *Economic Models of Tropical Deforestation: A Review*. CIFOR, Bogor, Indonesia.

39 Kasischke, E. S., Bergen, K., Fennimore, R., Sotelo, F., Stephens, G., Janetos, A. and Shugart, H. H. (1999) 'Satellite imagery gives a clear picture of Russia's boreal forest fires'. *EOS – Transactions of the American Geophysical Union*, vol 80, pp141, 147.

40 Landell-Mills, N. and Porras, I. T. (2002) *Silver Bullet or Fools' Gold? A Global Review of Markets for Forest Environmental Services and their Impact on the Poor*. International Institute for Environment and Development (IIED), London.

41 Larson, A. M. (2004) *Democratic Decentralization in the Forestry Sector: Lessons Learned from Africa, Asia and Latin America*. CIFOR, Managua. www.cifor.cgiar.org/publications/pdf_files/interlaken/Anne_Larson.pdf

42 Laurance, W. F. (1999) 'Habitat fragmentation: introduction and synthesis'. *Biological Conservation*, vol 91, pp101–107.

43 Laurance, W. F., Delamnica, P., Laurance, S. G., Vasconcelos, H. L. and Lovejoy, T. E. (2000) 'Conservation: Rainforest fragmentation kills big trees'. *Nature*, vol 404, p836.

44 Laurance, W. F., Albernaz, A. K. M. and Da Costa, C. (2001) *Is Deforestation Accelerating in the Brazilian Amazon?* The Smithsonian Institute and INPA, Brazil. http://lcluc.gsfc.nasa.gov/products/LBA_page/files_fr_2_25_email/Laur2001_Is_Defor_Eng.pdf

45 Lomborg, B. (2001) *The Skeptical Environmentalist: Measuring the Real State of the World*. Cambridge University Press, Cambridge.

46   MacGregor, J. (2003) *Global Trends in Trade of Forest Products and Services*, presented at the FAO Expert Consultation on Trade and Sustainable Forest Management *Impacts and Interactions*, 3–5 February 2003, Rome. www.iisd.ca/linkages/sd/tsfm/sdvol79num1.html

47   Maginnis, S. and Jackson, W. (2003) 'The role of planted forests in forest landscape restoration', in *The Role of Planted Forests in Sustainable Forest Management*. UNFF Intersessional Meeting, Wellington, New Zealand, 25–27 March 2003. www.maf.govt.nz/mafnet/unff-planted-forestry-meeting/index.htm

48   Manor, J. (1999) *The Political Economy of Democratic Decentralization*. World Bank, Washington, DC.

49   Molnar, A. (2003) *Forest Certification and Communities: Looking Forward to the Next Decade*. Forest Trends, Washington, DC. www.forest-trends.org

50   Molnar, A., Scherr, S. and Khare, A. (2004) *Who Conserves the World's Forests? Community-Driven Strategies to Protect Forests and Respect Rights*. Forest Trends and Ecoagriculture Partners, Washington, DC. www.forest-trends.org/resources/pdf/Who%20Conserves%2008–23.pdf

51   Neumann, R. P. and Hirsch, E. (2000) *Commercialisation of Non Timber Forest Products: Review and Analysis of Research*. CIFOR, Bogor, Indonesia and FAO, Rome.

52   Neumayer, E. (1999) *Weak versus Strong Sustainability. Exploring the Limits of Two Opposing Paradigms*. Edward Elgar Publishing, Cheltenham.

53   Oksanen, T., Pajari, B. and Tuomasjukka, T. (eds) (2003) 'Forests in poverty reduction strategies: Capturing the potential'. *EFI Proceedings*, no 47. European Forest Institute, Joensuu, Finland.

54   Owusu, R. (1999) *GM Technology in the Forest Sector*. A coping study for WWF. www.wwf-uk.org/news/gm.pdf

55   Pagiola, S., Bishop, J. and Landell-Mills, N. (eds) (2002) *Selling Forest Environmental Services: Market-based Mechanisms for Conservation and Development*. Earthscan, London.

56   Pearce, D. (2002) *Environmentally Harmful Subsidies: Barriers to Sustainable Development*. OECD, Paris.

57   PEFC (2004) PEFC Council Information Register, Statistic Figures on PEFC Certification. Updated 31 August 2004, www.pefc.org

58   Perrings, C., Williamson, M. and Dalmazzone, S., (eds) (2000) T*he Economics of Biological Invasions*. Edward Elgar Publishing, Cheltenham.

59   Phillips, H. (2004) Global Forest Certification – 2004 Update, presented at the Future Issues for Forest Industries in Europe conference organized by COFORD, 28 April–1 May, Dublin. http://d96968.u29.jodoshared.com/coford/future_issues/presentations/HenryPhillips.ppt#17

60   Poore, D., Burgess, P., Palmer, J., Rietbergen, R. and Synnott, T. (1989) *No Timber without Tree: Sustainability in the Tropical Forest – A study for ITTO*. Earthscan, London.

61   Poore, D. (2003) *Changing Landscapes: The Development of the International Tropical Timber Organization and its Influence on Tropical Forest Management*. Earthscan, London.

62   Reid, W. and Miller, K. (1989) *Keeping Options Alive: The Scientific Basis for Conserving Biodiversity*. WRI, Washington, DC.

63   Ribot, J. C. (2002) *Democratic Decentralization of Natural Resources: Institutionalizing Popular Participation*. WRI, Washington DC.

64   Richards, M., Del Gatto, F. and López, G. A. (2003) *The Cost of Illegal Logging in Central America. How much are the Honduran and Nicaraguan Governments Losing?* ODI, London. www.odi.org.uk/talailegal/eng_evidence.htm

65   Richardson, D. M. (1999) 'Commercial forestry and agroforestry as sources of invasive alien trees and shrubs', in Sandlund, O. T., Schei, P. J. and Viken, A. (eds) *Invasive Species and Biodiversity Management*, Kluwer Academic Publishers, Dordrecht, pp237–257.

66   Scherr, S., White, A. and Khare, A. (2004) 'For services rendered: The current status and future potential of markets for the ecosystem services provided by tropical forests'. *ITTO Technical Series*, no 21. International Tropical Timber Organization, Yokohama, Japan.

67 Scotland, N. and Ludwig, S. (2002) *Deforestation, the Timber Trade and Illegal Logging*. Background paper for EC Workshop on Forest Law Enforcement, Governance and Trade. Brussels, 22–24 April.

68 Sun, X., Katsigiris, E. and White, A. (2004) *Meeting China's Demand for Forest Products: An Overview of Import Trends, Ports of Entry, and Supplying Countries, with Emphasis on the Asia – Pacific Region*. Forest Trends, Chinese Center for Agricultural Policy, and CIFOR, Washington, DC.

69 Sunderlin, W. D., Angelsen, A., Resosudarmo, D. P., Dermawan, A. and Rianto, E. (2001) 'Economic crisis, small farmer well-being and forest cover change in Indonesia'. *World Development*, vol 29, no 5, pp767–782.

70 Sunderlin, W. D., Angelsen, A. and Wunder, S. (2003) 'Forests and poverty alleviation', in FAO. *State of the World's Forests 2003*, pp62–73, FAO, Rome.

71 Sylt, C. (2003) *Wonder Wood* Article in Magazine www.indurite.com/

72 Thomas, C. D., Cameron, A., Green, R. E., Bakkenes, M., Beaumont, L. J., Collingham, Y. C., Erasmus, B. F. et al (2004) 'Extinction risk from climate change'. *Nature*, vol 427, pp145–148.

73 UNEP (1997c)UNEP IUG 1997, Vital Climate Graphics: UNEP-GRID Arendal website www.grida.no/climate/vital/10.htm

74 UNEP/US EPA (1998) Vital Climate Graphics. UNEP-GRID Arendal website www.grida. no/climate/vital/27.htm

75 UNEP (2001) *An Assessment of the Status of the World's Remaining Closed Forests*. UNEP/ DEWA/TR 01–2. Nairobi.

76 UNEP (2001a) *The Global Biodiversity Outlook Report*. CBD/UNEP, Nairobi, Kenya, November 2001. www.biodiv.org/gbo/

77 UNEP (2002) *Global Environmental Outlook 3: Past, Present and Future Perspectives*. Earthscan, London.

78 UNFF (2003) *The Role of Planted Forests in Sustainable Forest Management*. Report of the UNFF Intersessional Experts Meeting, 25–27 March, Wellington, New Zealand.

79 Van Minh (2003) *Interview with the Director of the Ministry of Agriculture and Rural Development*, Viet Nam News, 15 March. http://vietnamnews.vnagency.com.vn/2003–03/14/Comment.htm

80 Victor, D. G. and Ausubel, J. H. (2000) 'Restoring the forests'. *Foreign Affairs*, vol 76, no 6, November/December, pp127–144.

81 WCD (2000) *Dams and Development: A New Framework for Decision Making*. The Report of the World Commission on Dams, WCD, Cape Town, www.dams.org/report/contents.htm

82 White, A. and Martin, A. (2002) *Who Own the World's Forests? Forest Tenure and Public Forests in Transition*. Forest Trends/Center for International Environmental Law, Washington, DC.

83 White, A., Khare, A. and Molnar, A. (2004) 'Who owns, who conserves and why it matters. arborvitae – the IUCN/WWF Forest Conservation Newsletter'. *Issue*, vol 26, Gland. www.iucn.org

84 Wilkie, M. L., Abdel-Nour, H., Carneiro, C. M., Durst, P., Kneeland, D., Kone, P. D., Prins, C. F. L. et al (2003) 'Forest area covered by management plans: global status and trends', presented at the XII World Forestry Congress, *Unasylva*, vol 54, no 214/215, Forest management at the XII World Forestry Congress, FAO, Rome. www.fao.org/DOCREP/006/y5189E/y5189e20.htm

85 Woodmetrics (2003) *The Wall of Wood*. Media release. Auckland. www.woodmetrics.co.nz/FAQs/FAQ3.asp?PAGE=FAQ

86 World Bank (2000) *The World Bank Forest Strategy – Striking the Right Balance* by Lele U., Kumar, N., Husain, S. A., Zazueta, A. and Kelly, L. World Bank, Washington, DC. www.worldbank.org/oed

87 World Bank (2002) *A Revised Forest Strategy for the World Bank Group*. World Bank, Washington, D.C. http://lnweb18.worldbank.org/ESSD/essdext.nsf/14DocByUnid/CB45BCF9 17EA1EE785256BD10068FC8D?Opendocument

88 WRI (1997) *The Last Frontier Forests: Ecosystems and Economies on the Edge*, by Bryant, D., Nielsen, D. and Tangley, L., WRI, Washington, DC.

89 WRI (2000) *People and Ecosystems: The Fraying Web of Life*. WRI, Washington, DC.

90 WRI (2000a) *Pilot Analysis of Global Ecosystems – Forest Ecosystems*. WRI, Washington, DC.

91 Wunder, S. (2001) 'Poverty alleviation and tropical forests – What scope for synergies'. *World Development*, vol 29, no 11, pp1817–1833.

92 Wunder, S. (2003) *Oil Wealth and the Fate of the Forest. A Comparative study of Eight Developing Countries*. Routledge, London.

93 Wunder, S. and Verbist, B. (2003) *The Impact of Trade and Macroeconomic Policies on Frontier Deforestation*. ASB Lecture Note 13. World Agroforestry Center, ICRAF – South East Asia, Bogor.   www.worldagroforestry.org/sea/Products/Training/Materials/lecture%20notes/ASB-LecNotes/ASBLecNote13.pdf

94 WWF (2004) *Quick Overview of Facts on Illegal Logging in Russia*. Compiled by the WWF European Forest Programme from existing studies March 2004. www.panda.org/about_wwf/where_we_work/europe/problems/illegal_logging/Downloads/ILLEGAL%20LOGGI NG%20RUSSIA.pdf

95 Zbicz, D. C. (1999) *Transfrontier Ecosystems and Internationally Adjoining Protected Areas*. Duke University, Durham, North Carolina.

3

Proximate Causes and Underlying Driving Forces of Tropical Deforestation

Helmut J. Geist and Eric F. Lambin

One of the primary causes of global environmental change is tropical deforestation, but the question of what factors drive deforestation remains largely unanswered (NRC, 1999). Various hypotheses have produced rich arguments, but empirical evidence on the causes of deforestation continues to be largely based on cross-national statistical analyses (Bilsborrow, 1994; Brown and Pearce, 1994; Williams, 1994; Painter and Durham, 1995; Sponsel et al, 1996; Murali and Hedge, 1997; Rudel and Roper, 1997; Fairhead and Leach, 1998). In some cases, these analyses are based on debatable data on rates of forest cover change (Palo, 1999). The two major, mutually exclusive – and still unsatisfactory – explanations for tropical deforestation are single-factor causation and irreducible complexity. On the one hand, proponents of single-factor causation suggest various primary causes, such as shifting cultivation (Amelung and Diehl, 1992; Myers, 1993; Rerkasem, 1996; Ranjan and Upadhyay, 1999) and population growth (Allen and Barnes, 1985; Amelung and Diehl, 1992; Cropper and Griffiths, 1994; Ehrhardt-Martinez, 1998; Mather and Needle, 2000). On the other hand, correlations between deforestation and multiple causative factors are many and varied, revealing no distinct pattern (Rudel and Roper, 1996; Bawa and Dayanandan, 1997; Mather et al, 1998; Angelsen and Kaimowitz, 1999).

In addition to chronicling these attempts to identify general causes of deforestation through global-scale statistical analyses, the literature is rich in local-scale case studies investigating the causes and processes of forest cover change in specific localities. Our aim with this study is to generate from local-scale case studies a general understanding of the proximate causes and underlying driving forces of tropical deforestation while preserving the descriptive richness of these studies. Proximate causes are human activities or immediate actions at the local level, such as agricultural expansion, that originate from intended land use and directly impact forest cover. Underlying driving forces are fundamental social processes, such as human population dynamics or agricultural policies, that underpin the proximate causes and either operate at the local level or have an indirect impact from the national or global level.

We analysed the frequency of proximate causes and underlying driving forces of deforestation, including their interactions, as reported in 152 subnational case studies.

Note: Reprinted from *BioScience*, vol 52, Geist, H. J. and Lambin, E. F., 'Proximate causes and underlying driving forces of tropical deforestation', pp143–149, Copyright © (2002), with permission from AIBS

We show that tropical deforestation is driven by identifiable regional patterns of causal factor synergies, of which the most prominent are economic factors, institutions, national policies and remote influences (at the underlying level) driving agricultural expansion, wood extraction and infrastructure extension (at the proximate level). Our findings reveal that prior studies have given too much emphasis to population growth and shifting cultivation as primary causes of deforestation.

Data Analysis

Case studies of net losses of tropical forest cover (n = 152) were analysed to determine whether the proximate causes and underlying driving forces of tropical deforestation fall into any patterns. Study areas range from a community to a multi-province area, and cases span time periods from 1880 to 1996, with 1940 to 1990 being the most frequently covered period. The 152 cases of tropical deforestation were taken from 95 articles published in 40 journals covered by the citation index of the Institute for Scientific Information (Geist and Lambin, 2001). The criteria for selecting studies were the following: quantification of the rates of forest-cover change; net loss of forest cover during at least part of the study period; investigation method based on quantitative data or in-depth field investigations; consideration of clearly named factors as potential causes of deforestation, including basic features of the socioeconomic setting and the natural resource endowment; and absence of obvious disciplinary bias. We assumed that each study revealed the actual causes of deforestation in the study area. Therefore, our comparative analysis of case studies evaluates which causal patterns leading to deforestation are most often found in different tropical regions.

Four broad clusters of proximate causes were identified: agricultural expansion, wood extraction, infrastructure extension, and other factors. Each land-use category was further subdivided; for example, agricultural expansion was divided into permanent cultivation, shifting cultivation, cattle ranching, and colonization (Figure 3.1). Underlying driving forces were categorized into five broad clusters: demographic, economic, technological, policy and institutional, and cultural factors. Each was further subdivided into specific factors; for example, cultural or sociopolitical factors were partitioned into public attitudes, values and beliefs, and individual or household behaviour (Figure 3.1; Ledec, 1985; Lambin, 1994; Ojima et al, 1994; Turner et al, 1995; Lambin, 1997; Contreras-Hermosilla, 2000).

Causal factors were quantified by determining the most frequent proximate and underlying factors in each case. The major interactions and feedback processes between these factors were also identified to reveal the systems dynamics that commonly lead to deforestation. Three modes of causation were distinguished: single-factor causation (ie one individual underlying factor driving one or more proximate factors), chain-logical causation (ie several interlinked factors in combination leading to deforestation), and concomitant occurrence (ie independent, separate operation of factors causing deforestation). Results were broken down by broad geographical regions (Asia, n = 55; Africa, n = 19; Latin America, n = 78). They are given in order of decreasing importance, with factors occurring in less than 25 per cent of the cases not reported.

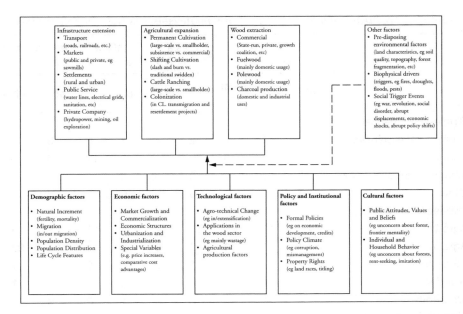

Note: Five broad clusters of underlying driving forces (or fundamental social processes) underpin the proximate causes of tropical deforestation, which are immediate human actions directly impacting forest cover

Figure 3.1 *Causes of forest decline .*

Proximate Causes

At the proximate level, tropical deforestation is best explained by multiple factors rather than by single variables. Dominating the broad clusters of proximate causes is the combination of agricultural expansion, wood extraction and infrastructure expansion, with clear regional variations (Table 3.1). The tables presented here provide a breakdown of proximate causes and underlying driving forces by broad geographical regions (or continents). They show the absolute number as well as the relative percentage of the frequency of causative variables reported in the case studies. Tables 3.1 and 3.3 give only the broad clusters of proximate causes and underlying driving forces, and Tables 3.2 and 3.4 provide a detailed breakdown of the broad clusters by specific factors. Only the frequency of modes of causation (single or multifactorial) by broad clusters of proximate and underlying variables (Tables 3.1 and 3.3) shows cumulative percentages of cases, adding up to 100 per cent. The relative percentages of the frequency of occurrence of specific factors (Tables 3.2 and 3.4) do not add up to 100 per cent, as multiple counts exist because of causal factor synergies (discussed later).

Agricultural expansion is, by far, the leading land-use change associated with nearly all deforestation cases (96 per cent). It includes, with more or less equal frequencies, forest conversion for permanent cropping, cattle ranching, shifting cultivation, and colonization

Table 3.1 *Frequency of broad clusters of proximate causes in tropical deforestation*

	All cases (n = 152)			Asia (n = 55)		Africa (n = 19)		Latin America (n = 78)	
	abs	rel (%)	cum (%)	abs	rel (%)	abs	rel (%)	abs	rel (%)
Single-factor causation									
Agricultural expansion	6	4	4	2	4	1	5	3	4
Wood extraction	2	1	5	0	–	2	11	0	–
Infrastructure expansion	1	1	6	0	–	0	–	1	1
Other[a]	0	–	–	0	–	0	–	0	–
Two-factor causation									
Agro-wood[b]	22	15	20	12	22	2	11	8	10
Agro-infra[c]	30	20	40	3	6	2	11	25	32
Agro-other	5	3	43	1	2	3	16	1	1
Wood-infra	1	1	44	0	–	0	–	1	1
Wood-other	1	1	45	0	–	1	6	0	–
Three-factor causation									
Agro-wood-infra	38	25	70	21	38	2	11	15	19
Agro-wood-other	6	4	74	4	7	1	5	1	1
Agro-infra-other	8	5	79	0	–	0	–	8	10
Wood-infra-other	1	1	80	0	–	0	–	1	1
Four-factor causation									
All	31	20	100	12	22	5	26	14	18
Total	152	100	–	55	100	19	100	78	100

Note: abs, absolute number; rel, relative percentages; cum, cumulative percentages. Relative percentages may not total 100 because of rounding.
a 'Other' refers to predisposing environmental factors, such as land characteristics and social as well as biophysical trigger events.
b Agro, agricultural expansion; wood, wood extraction.
c Infra, infrastructure expansion.

agriculture (Table 3.2). Only permanent agriculture and shifting cultivation display low geographical variation; that is, regional values for permanent cultivation in Asia, Africa and Latin America are close to the 'global' value (ie 44 per cent, 53 per cent and 50 per cent, respectively, compared with a global 48 per cent), similarly for shifting cultivation (ie 44 per cent, 42 per cent and 40 per cent, respectively, compared with a global 41 per cent). Further subdivisions reveal striking regional differences, however. In permanent cultivation, the expansion of food-crop cultivation for subsistence is three times more frequently reported than the expansion of commercial farming (less than 25 per cent for all regions). In shifting cultivation, cases of deforestation driven by slash-and-burn agriculture are more widespread in upland and foothill zones of Asia than elsewhere, whereas when

Table 3.2 *Frequency of specific proximate causes in tropical deforestation.*

	All cases (n = 152)		Asia (n = 55)		Africa (n = 19)		Latin America (n = 78)	
	abs	rel (%)	abs	rel (%)	abs	rel (%)	abs	rel (%)
Agricultural expansion	146	96	55	100	16	84	75	96
Permanent cultivation	73	48	24	44	10	53	39	50
Subsistence agriculture	61	40	20	36	10	53	31	40
Cattle ranching	70	46	3	6	3	16	64	82
Shifting cultivation	63	41	24	44	8	42	31	40
Swidden agriculture	46	30	24	44	7	37	15	19
Colonization[a]	61	40	23	42	4	21	34	44
Infrastructure expansion	110	72	36	66	9	47	65	83
Transport extension	97	64	26	47	9	47	62	80
Roads	93	61	25	46	9	47	59	76
Settlement/market extension	41	27	12	22	3	16	26	33
Wood extraction	102	67	49	89	13	68	40	51
Commercial (for trade)	79	52	43	78	5	26	31	40
Fuelwood (for domestic uses)	42	28	18	33	10	53	14	18
Other factors[b]	52	34	17	31	10	53	25	32

Note: Multiple counts possible; percentages relate to the total of all cases for each category; abs, absolute number; rel, relative percentages. Relative percentages may not total 100 because of rounding.
a Including transmigration and resettlement.
b Predisposing environmental factors such as land characteristics and social or biophysical trigger events.

practised by colonizing migrant settlers in Latin America, it is mainly limited to lowland areas. Pasture creation for cattle ranching is a striking cause of deforestation reported almost exclusively for humid lowland cases from mainland South America.

Commercial wood extraction is frequent in both mainland and insular Asia, whereas in Africa the harvesting of fuelwood and poles by individuals for domestic uses dominates cases of deforestation associated with wood extraction. Among all forms of infrastructure expansion, road construction is by far most frequently reported, mainly in both lowland and mountain cases of Latin America. Predisposing environmental factors such as land characteristics (soil quality, topography) or trigger events, whether biophysical (droughts, floods) or social (mainly wars), are reported to influence deforestation in one-third of the cases.

Among the detailed categories of proximate causes for all regions, the extension of overland transport infrastructure, followed by commercial wood extraction, permanent cultivation and cattle ranching, are the leading proximate causes of deforestation. Contrary to widely held views, case-study evidence suggests that shifting cultivation is not the primary cause of deforestation.

Table 3.3 *Frequency of broad underlying driving forces in tropical deforestation*

	All cases (n = 152)			Asia (n = 55)		Africa (n = 19)		Latin America (n = 78)	
	abs	rel (%)	cum (%)	abs	rel (%)	abs	rel (%)	abs	rel (%)
Single-factor causation									
Economic (econ)	13	9	9	0	–	0	–	13	17
Institutional (inst)	4	3	12	0	–	1	5	3	4
Technological (tech)	0	–	12	0	–	0	–	0	–
Cultural (cult)	0	–	12	0	–	0	–	0	–
Demographic (pop)	0	–	12	0	–	0	–	0	–
Two-factor causation									
Pop-econ	5	3	15	0	–	3	16	2	3
Pop-tech	4	3	17	2	4	1	6	1	1
Pop-inst	1	1	18	0	–	0	–	1	1
Pop-cult	1	1	18	0	–	0	–	1	1
Econ-tech	1	1	19	0	–	0	–	1	1
Econ-inst	5	3	22	0	–	0	–	5	6
Inst-cult	5	3	26	4	7	0	–	1	1
Three-factor causation									
Pop-econ-tech	5	3	29	0	–	4	21	1	1
Pop-econ-inst	1	1	30	1	2	0	–	0	–
Pop-econ-cult	2	1	31	0	–	1	5	1	1
Pop-tech-inst	4	3	34	1	2	1	5	2	3
Econ-tech-cult	1	1	34	0	–	0	–	1	1
Econ-inst-cult	6	4	38	0	–	0	–	6	8
Tech-inst-cult	5	3	42	5	9	0	–	0	–
Four-factor causation									
Pop-econ-tech-inst	8	5	47	5	9	2	11	1	1
Pop-econ-tech-cult	1	1	47	0	–	1	5	0	–
Pop-econ-inst-cult	2	1	49	1	2	0	–	1	1
Pop-tech-inst-cult	5	3	52	4	7	0	–	1	1
Econ-tech-inst-cult	19	13	64	12	22	0	–	7	9
Five-factor causation									
All	54	36	100	20	36	5	26	29	37
Total	152	100	–	55	100	19	100	78	100

Note: abs, absolute number; rel, relative percentages; cum, cumulative percentages. Relative percentages may not total 100 because of rounding.

Table 3.4 *Frequency of specific underlying driving forces in tropical deforestation*

	All cases (n = 152)		Asia (n = 55)		Africa (n = 19)		Latin America (n = 78)	
	abs	rel (%)	abs	rel (%)	abs	rel (%)	abs	rel (%)
Economic factors	123	81	39	71	16	84	68	87
Market growth/ commercialization	103	68	30	55	15	79	58	74
Sectoral market growth[a]	78	51	23	42	13	68	42	54
Demand/consumption[b]	69	45	24	44	13	68	32	41
Market failures	52	34	22	40	6	32	24	31
Urban-industrial growth	58	38	23	42	5	26	30	39
Industrialization	43	28	21	38	1	5	21	27
Foreign exchange[c]	38	25	16	29	5	26	17	22
Special variables[d]	48	32	9	16	5	26	34	44
Institutional/policy factors	119	78	53	96	9	47	57	73
Formal policies	105	69	46	84	7	37	52	67
On land development	60	40	28	51	5	26	27	35
On economic growth[e]	51	34	22	40	5	26	24	31
On credits/subsidies	39	26	11	20	1	5	27	35
Property rights issues[f]	67	44	33	60	5	26	29	37
Policy failures[g]	64	42	31	56	1	5	32	41
Mismanagement	38	25	13	24	1	5	24	31
Technological factors	107	70	49	89	14	74	44	56
Agrotechnological change[h]	70	46	28	51	8	42	34	44
Production changes	50	33	17	31	5	26	28	36
Wood sector related[i]	69	45	39	71	8	42	22	28
Agriculture related	42	28	22	40	4	21	16	21
Cultural/sociopolitical factors	101	66	46	84	7	37	48	62
Public attitudes, values, beliefs	96	63	45	82	5	26	46	59
Public unconcern[j]	66	43	25	46	3	16	38	49
Missing basic values	55	36	33	60	2	11	20	26
Individual/household behaviour	80	53	38	69	6	32	36	46
Situation specific[k]	74	49	36	66	5	26	33	42
Unconcern by individuals[l]	48	32	20	36	4	21	24	31

Table 3.4 *Frequency of specific underlying driving forces in tropical deforestation*

	All cases (n = 152)		Asia (n = 55)		Africa (n = 19)		Latin America (n = 78)	
	abs	rel (%)	abs	rel (%)	abs	rel (%)	abs	rel (%)
Demographic factors[m]	93	61	34	62	18	95	41	53
In-migration	58	38	12	22	9	47	37	47
Growing population density	38	25	12	22	6	32	20	26

Note: Multiple counts possible; percentages relate to the total of all cases in each category; abs, absolute number; rel, relative percentages. Relative percentages may not total 100 because of rounding.
a Growth of markets for wood (eg timber products) 29 per cent, agricultural products (eg food) 29 per cent and minerals 15 per cent (eg oil energy).
b Demand for wood (eg processed timber) 32 per cent and agricultural products (eg food) 18 per cent.
c Generation of foreign exchange earnings.
d Low cost conditions (production factors) and price changes (increases and decreases).
e Especially agricultural and infrastructure development policies.
f 'Land races', land tenure insecurity, quasi-open access conditions, maladjusted customary rights, titling/legalization, low empowerment of local user groups.
g Corruption, lawlessness, clientelism, and the operation of vested interests and 'growth coalitions', besides mismanagement or poor performance.
h Land-use intensification and extension, besides changes in market vs subsistence orientation, in intensity of labour vs capital used, and in holding size (productional changes).
i Poor logging performance, wastage in timber processing and poor domestic or industrial furnace performance.
j Dominant frontier mentality, prevailing attitudes of nation-building, modernization, development and low (public) morale.
k Mainly rent-seeking behaviour (35 per cent).
l Unconcern about the forest environment as reflected in increasing levels of demand, aspiration and consumption, commonly associated with increased income.
m. Including natural increment spatial distribution and life cycle features.

Underlying Driving Forces

At the underlying level, tropical deforestation is also best explained by multiple factors and drivers acting synergistically rather than by single-factor causation, with more than one-third of the cases being driven by the full interplay of economic, institutional, technological, cultural and demographic variables (Table 3.3).

Economic factors are prominent underlying forces of tropical deforestation (81 per cent). Commercialization and the growth of mainly timber markets (as driven by national and international demands) as well as market failures are frequently reported to drive deforestation (Table 3.4). Economic variables such as low domestic costs (for land, labour, fuel or timber), product price increases (mostly for cash crops) and the ecological footprint of remote urban-industrial centers underpin about one-third of the cases each, whereas the requirement to generate foreign exchange earnings at a national level intervenes in a quarter of the cases. With few exceptions, factors related to economic development through a growing cash economy show little regional variation and, thus, constitute a robust underlying force of deforestation. A number of case studies describe a process of frontier colonization with a sequence of poverty- and capital-driven deforestation (Rudel and Roper, 1997). Poverty-driven deforestation refers to the ecological marginalization of farmers who have lost their resource entitlements, and capital-driven deforestation refers to public or private investments to develop the frontier for political, economic or social reasons. Underlying 42 per cent of the cases each, both processes overlap considerably.

Institutional factors also drive many cases of deforestation (78 per cent). These factors mainly include formal pro-deforestation measures such as policies on land use and economic development as related to colonization, transportation or subsidies for land-based activities. Land tenure arrangements and policy failures (such as corruption or mismanagement in the forestry sector) are also important drivers of deforestation. Though much discussed as a general cause of deforestation (eg Deacon, 1994; Mendelsohn and Balick, 1995), property rights issues are mainly a characteristic of Asian cases and tend to have ambiguous effects upon forest cover: insecure ownership, quasi-open access conditions, maladjusted customary rights, as well as the legalization of land titles, are all reported to influence deforestation in a similar manner.

Among technological factors (70 per cent), important processes affecting deforestation are agrotechnological change, with agricultural intensification having no distinct impact separate from agricultural expansion and poor technological applications in the wood sector (leading to wasteful logging practices).

Cultural or sociopolitical factors (66 per cent) are reported to underlie mainly economic and policy forces in the form of attitudes of public unconcern towards forest environments. These factors also shape the rent-seeking behaviour of individual agents causing deforestation.

Among demographic factors (61 per cent), only in-migration of colonizing settlers into sparsely populated forest areas, with the consequence of increasing population density there, shows a notable influence on deforestation. It tends to feature African and Latin American rather than Asian cases. Contrary to a common misconception, population increase due to high fertility rates is not a primary driver of deforestation at a local scale over a time period of a few decades, as it intervenes in 8 per cent of the cases only and is always combined with other factors.

Interactions and Feedbacks

Not only are multiple causal factors at work, but their interactions also lead to deforestation, which is why it is important to understand the systems dynamics (Figure 3.2). Our analysis reveals that, in most cases, three to four underlying causes are driving two to three proximate causes. A frequent pattern of causal interaction stems from the necessity for road construction that is associated with wood extraction or agricultural expansion, which is mostly driven by policy and institutional factors but also by economic and

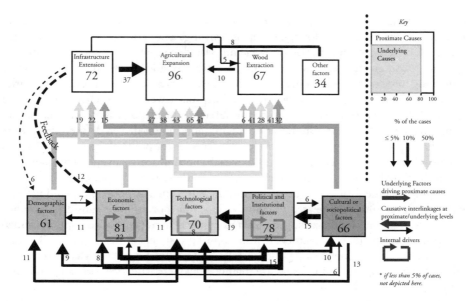

Note: Systems dynamics commonly lead to tropical deforestation. No single or key variable, such as population growth or shifting cultivation, unilaterally impacts forest cover change; synergies between proximate causes and underlying (social) driving forces best explain tropical forest cover losses. A recurrent set of mainly economic, political and institutional driving forces underpins proximate causes, such as agricultural expansion, infrastructure extension and wood extraction, leading to deforestation. Though some investigators have claimed irreducible complexity is the explanation, distinct regional patterns exist.

Figure 3.2 *Causative pattern of tropical deforestation (n = 152 cases)**

cultural factors. Pro-deforestation state policies aimed at land use and economic development (eg credits, low taxation, incentives for cash cropping, legal land titling) lead to the expansion of commercial crops and pastures in combination with an extension of the road network. Another pattern, seen mostly in Africa, comes from insecure ownership related to uncertainties of land tenure, which drives the shift from communal to private property and underlies many cases in which traditional shifting cultivation is a direct cause of deforestation. Policies facilitating the establishment of state agricultural and forestry plantations lead to deforestation in both insular and continental Asia. Agricultural colonization in Latin America is often favoured by land policies that are directed towards deregulation of land access, transfer of public forest land to private holdings, and state regulations in favour of large individual landholdings.

Policy, institutional and economic factors are also driving wood extraction. Cases of deforestation related to both private and state-run timber logging, especially in Asia, are almost exclusively driven by the liberal granting of concessions, development projects and state claims for logging areas, in conjunction with corruption and poor implementation of forestry rules.

In-migration and, to a much lesser degree, natural population growth drive the expansion of cropped land and pasture in 47 per cent of the cases in Africa and Latin

America (22 per cent in Asia), concomitantly with other underlying drivers. The extension of permanently cropped land for subsistence farming to meet the needs of a growing population is reported particularly for African cases. Expansion of pastures emerges exclusively from mainland South American cases, in association with processes of both planned colonization and spontaneous settlement by colonist agriculturalists.

Some feedbacks amplify the process of deforestation. The most frequent feedback identified is that from road construction and the creation of new settlements in a frontier area, which work upon economic factors such as the growth of wood and food markets. The development of commercialization induces further deforestation and agricultural modernization, mainly in frontier regions of the Amazon lowlands and in South-East Asia. Shifting cultivators turn into sedentary cash croppers and permanently settled subsistence farmers who respond to market signals.

Conclusion

Evidence from empirical case studies that identify both proximate causes and underlying forces of tropical deforestation suggests that no universal link between cause and effect exists. Rather than providing support for dominant theories of global deforestation (neoclassical, impoverishment, political ecology), analysis of these studies shows that tropical forest decline is determined by different combinations of various proximate causes and underlying driving forces in varying geographical and historical contexts. Some of these combinations are robust geographically (such as the development of market economies and the expansion of permanently cropped land for food), whereas most of them are region specific. The observed causal factor synergies challenge single-factor explanations that put most of the blame of deforestation upon shifting cultivators and population growth (caused by natural increment). Rather, our analysis reveals that, at the underlying level, public and individual decisions largely respond to changing, national- to global-scale economic opportunities and/or policies, as mediated by local-scale institutional factors, and that, at the proximate level, regionally distinct modes of agricultural expansion, wood extraction and infrastructure extension prevail in causing deforestation. As a major implication, case study-based evidence reveals that no universal policy for controlling tropical deforestation can be conceived. Rather, a detailed understanding of the complex set of proximate causes and underlying driving forces affecting forest cover changes in a given location is required prior to any policy intervention.

Acknowledgements

Support is acknowledged from the services of the Prime Minister of Belgium, Office for Scientific, Technical and Cultural Affairs. Thanks go to participants of the international IGBP/IHDP-LUCC workshop, 'Human modification of the biosphere: Key drivers of land use/cover change processes', for their valuable comments (Royal Academy of Sciences; 13–15 March 2000; Stockholm, Sweden).

References

1   Allen, J. C. and Barnes, D. F. (1985) 'The causes of deforestation in developing countries'. *Annals of the Association of American Geographers*, vol 75, pp163–184.
2   Amelung, T. and Diehl, M. (1992) *Deforestation of Tropical Rain Forests: Economic Causes and Impact on Development*. J. C. B. Mohr, Tübingen, Germany.
3   Angelsen, A. and Kaimowitz, D. (1999) 'Rethinking the causes of deforestation: Lessons from economic models'. *World Bank Research Observer*, vol 14, pp73–98.
4   Bawa, K. S. and Dayanandan, S. (1997) 'Socioeconomic factors and tropical deforestation'. *Nature*, vol 386, pp562–563.
5   Bilsborrow, R. E. (1994) 'Population, development and deforestation: Some recent evidence', in *United Nations, Department of Economic and Social Information and Policy Analysis*, (ed) *Population, Environment and Development*. United Nations Press, New York, pp117–134.
6   Brown, K. and Pearce, D. W. (eds) (1994) *The Causes of Tropical Deforestation: The Economic and Statistical Analysis of Factors Giving Rise to the Loss of the Tropical Forests*. University College London Press, London.
7   Contreras-Hermosilla, A. (2000) *The Underlying Causes of Forest Decline*. Centre for International Forestry Research, Bogor, Indonesia.
8   Cropper, M. and Griffiths, C. (1994) 'The interaction of population growth and environmental quality'. *American Economic Review*, vol 84, pp250–254.
9   Deacon, R. T. (1994) 'Deforestation and the rule of law in a cross-section of countries'. *Land Economics*, vol 70, pp414–30.
10   Ehrhardt-Martinez, K. (1998) 'Social determinants of deforestation in developing countries: A cross-national study'. *Social Forces*, vol 77, pp567–586.
11   Fairhead, J. and Leach, M. (1998) *Reframing Deforestation: Global Analyses and Local Realities – Studies in West Africa*. Routledge, London.
12   Geist, H. J. and Lambin, E. F. (2001) *What Drives Tropical Deforestation? A Meta-Analysis of Proximate and Underlying Causes of Deforestation Based on Subnational Case Study Evidence*. LUCC International Project Office, Louvain-la-Neuve, Belgium, LUCC Report Series no 4.
13   Lambin, E. F. (1994) *Modelling Deforestation Processes: A Review*. European Commission, Luxembourg, Directorate-General XIII. Report no. EUR-15744-EN. Available from Office of Official Publications of the European Community, Luxembourg.
14   Lambin, E. F. (1997) 'Modelling and monitoring land-cover change processes in tropical regions'. *Progress in Physical Geography*, vol 21, pp375–393.
15   Ledec, G. (1985) 'The political economy of tropical deforestation', in Leonard, J. H. (ed) *Diverting Nature's Capital: The Political Economy of Environmental Abuse in the Third World*. Holmes and Meier, New York, pp179–226.
16   Mather, A. S. and Needle, C. L. (2000) 'The relationships of population and forest trends'. *Geographical Journal*, vol 166, pp2–13.
17   Mather, A. S., Needle, C. L. and Fairbairn, J. (1998) 'The human drivers of global land cover change: The case of forests'. *Hydrological Processes*, vol 12, pp1983–1994.
18   Mendelsohn, R. and Balick, M. (1995) 'Private property and rainforest conservation'. *Conservation Biology*, vol 9, pp1322–1323.
19   Murali, K. S. and Hedge, R. (1997) 'Patterns of tropical deforestation'. *Journal of Tropical Forest Science*, vol 9, pp465–476.
20   Myers, N. (1993) 'Tropical forests: The main deforestation fronts'. *Environmental Conservation*, vol 20, pp9–16.
21   [NRC] National Research Council, Board on Sustainable Development, Policy Division, Committee on Global Change Research (1999) *Global Environmental Change: Research Pathways for the Next Decade*. National Academy Press, Washington, DC.

22 Ojima, D. S., Galvin, K. A. and Turner, B. L. II. (1994) 'The global impact of land-use change'. *BioScience*, vol 44, pp300–304.

23 Painter, M. and Durham, W. H. (eds) (1995) *The Social Causes of Environmental Destruction in Latin America*. University of Michigan Press, Ann Arbor, MI.

24 Palo, M. (1999) 'No end to deforestation?' in Palo, M. and Uusivuori, J. (eds) *World Forests, Society and Environment*. Kluwer Academic, Boston, pp65–77.

25 Ranjan, R. and Upadhyay, V. P. (1999) 'Ecological problems due to shifting cultivation'. *Current Science*, vol 77, pp1246–1250.

26 Rerkasem, B. (ed) (1996) *Montane Mainland Southeast Asia in Transition*. Chiang Mai University Consortium, Chiang Mai, Thailand.

27 Rudel, T. and Roper, J. (1996) 'Regional patterns and historical trends in tropical deforestation, 1976–1990: A qualitative comparative analysis'. *Ambio*, vol 25, pp160–166.

28 Rudel, T. and Roper, J. (1997) 'The paths to rain forest destruction: Crossnational patterns of tropical deforestation'. *World Development*, vol 25, pp53–65.

29 Sponsel, L. E., Headland, T. N. and Bailey, R. (eds) (1996) *Tropical Deforestation: The Human Dimension*. Columbia University Press, New York.

30 Turner, B. L., Skole, D., Sanderson, S., Fischer, G., Fresco, L. and Leemans, R. (1995) *Land-Use and Land-Cover Change: Science/Research Plan*. Royal Swedish Academy of Sciences, Stockholm, Report no 35/7. Available from International Geosphere-Biosphere Program Secretariat, Stockholm.

31 Williams, M. (1994) 'Forests and tree cover', in Meyer, W. B. and Turner, B. L. II. (eds) *Changes in Land Use and Land Cover: A Global Perspective*. Press Syndicate of the University of Cambridge, New York, pp97–124.

4

Who Owns the World's Forests? Forest Tenure and Public Forests in Transition

Andy White and Alejandra Martin

Introduction and Rationale

Much attention and effort – on national and international levels – has been devoted over past decades to global problems of deforestation and forest degradation, and to improving forest management and conservation. The number of protected areas has increased dramatically, new funds have been established to finance protection and many international and non-governmental organizations (NGOs) have moved to develop markets for sustainably produced forest products. Despite these efforts, forest degradation has steadily increased throughout much of the world. At the same time there is growing realization that insecure property rights are a key underlying problem and cause of degradation. Property rights to forest-lands and resources are often, if not usually, contested, overlapping or simply not enforced. Much of the global forest estate is characterized by confusion and insecurity over prop-erty rights. This insecurity undermines sound forest management, for without secure rights forest holders have few incentives – and often lack legal status – to invest in man-aging and protecting their forest resources. While secure property rights cannot ensure sustained protection and investments in an asset, they are often a necessary condition.[1]

This growing global recognition of the importance of property rights is mirrored by a longstanding preoccupation with rights issues at local levels. The questions of who owns the forests, who claims them, who has access to them and further, who *should* own them, are hotly contested in many forest regions of the world. These are often the pri-mary concerns of local people most directly dependent on forest resources.

Growing interest in developing markets for the environmental services that forests provide has also brought new attention to property rights issues. Many governments, local organizations and private-sector actors are beginning to consider questions of ownership regarding the services provided by forests, such as carbon sequestration, biodiversity hab-itat and watershed protection. They are considering who should pay for the production and maintenance of these services.[2] The ways in which existing cultural, legal and regu-latory mechanisms should be applied to these potentially marketable services is likewise becoming a source of considerable debate.

Until recently, the answer to the question of who owns the world's forests was fairly straightforward. For most of modern history, governments have legally owned most forests. In the Western world, this tradition of government ownership began in medieval Europe, where royalty excluded commoners and laid claim to forests to serve the interests of the manor. It was there, too, that the practice of modern forest management had its roots. The profession of forestry grew from the initial tasks of policing the grounds and ensuring a steady supply of forest products and wildlife for the crown (Perlin, 1989; Bloch, 1966; Rampey, 2001). This tradition of government ownership and government-led forest management was transported to many colonies and adopted by imperial states in the 16th and 17th centuries. Throughout Africa, the Americas and South and East Asia, new governments took rights from native peoples and gave public forest agencies authority over essentially all natural forests – and indirectly, over large numbers of native inhabitants (Wilmsen, 1989; Poffenberger, 1990; Peluso, 1992; Lynch et al, 1995; Seymour, 1995. For Africa, see CPRD, 2001). A number of countries, including the United States, Mexico, China and Papua New Guinea, did not follow this path of government-dominated ownership. Still, since the Chinese revolution of the 1940s, global forest ownership has been largely static, and it is still largely dominated by government ownership. Most forested countries have not rethought forest ownership and the legal framework of ownership since the colonial and imperial eras.

But this picture of public ownership is beginning to change. Since the late 1980s, some governments of major forested countries have begun to reconsider and reform public forest ownership policies. These transitions are driven by three primary considerations. First, governments are increasingly aware of the fact that official forest tenure systems in many countries discriminate against the rights and claims of indigenous people and other local communities. Although the data are incomplete, it is estimated that some 60 million highly forest-dependent indigenous forest people live in the rainforests of Latin America, West Africa and South-East Asia (World Bank, 2001). An additional 400–500 million people are estimated to be directly dependent on forest resources for their livelihoods.[3] Around the world, indigenous people have legitimate claims to more forest areas than governments currently acknowledge. In South and South-East Asia alone, several hundred million people live on land classified as public forest (Lynch and Talbot, 1995). International conventions and national political movements are driving governments to recognize the traditional ownership claims of indigenous peoples and recognize legal ownership and land-use rights held by them and other local communities (Colchester, 2001).

This growing recognition of rights for indigenous and other local communities is not simply an issue of justice. There is also a growing convergence of economic development and environmental protection agendas. Without secure rights, indigenous and other local community groups lack long-term financial incentives for converting their forest resources into economically productive assets for their own development. There is growing evidence that local community-based entities are as good, and often better, managers of forests than federal, regional and local governments (see eg Lynch and Alcorn, 1994 and Thrupp et al, 1997). In addition, biologists and protected area specialists are beginning to change perspectives on human interactions with nature, acknowledging that the traditional management practices of indigenous peoples can be positive for biodiversity conservation and ecosystem maintenance. This positive outcome is best gained by devolving control of forest land to communities (Tuxhall and Nabhan, 2001). A recent review of property rights and deforestation in Ecuador, for example, found that community ownership often provides a disincentive to forest conversion (Wunder, 2001).

A third reason for this transition is the growing recognition that governments and public forest management agencies often have not been good stewards of public forests. While many countries have proven that public ownership can be effective in protecting and managing forests, others have not developed the governance structures and management capacities necessary to ensure effectiveness. While exploitation is a legitimate use of public forests, in many places forests have been abused to finance political elites and curry political favours. The findings from a number of recent studies on illegal logging and corruption are staggering; illegal logging on public forestlands is estimated to cost forest country governments at least US$10–15 billion a year – an amount greater than total World Bank lending to client countries and greater than total annual development assistance in public education and health.[4]

Understanding tenure issues and trends is essential for all actors concerned with forests – for governments that seek to promote sustainable use, combat illegal logging and quell local unrest; for indigenous and other local communities and their supporters who want legal recognition of community rights and broader political participation; for environmental organizations that seek conservation without undermining local tenure rights; for sustainable forest management certifying bodies that want to ensure they do not unwittingly reinforce unjust claims; for private industries requiring reliable sources of timber and fibre; and for investors seeking low-risk investment opportunities.

Unfortunately, progress on studying, debating and addressing tenure issues is constrained by a lack of data and information. Despite the importance of forest tenure, there is no comprehensive overview of the global situation. Few countries keep accurate tenure data and little has been done to document and describe tenure distribution at the regional or global levels.[5] This lack of information is partly due to the complexity of the issue. To start, tenure definitions vary, and the spectrum of tenure types is wide and diverse. Ownership also is often in dispute, with a great difference between official government claims and those of local communities, especially indigenous ones. Finally, when they are available, tenure data are frequently outdated and often contradictory.[6]

This report is an initial attempt to capture the pieces of this global picture using the available information. It presents a newly collected and aggregated set of official tenure data for 24 of the 30 most forested countries of the world and summarizes the findings from those data.[7] These 24 countries represent approximately 93 per cent of the world's remaining natural forest of approximately 3.9 billion hectares. These data are presented in a novel manner that more clearly distinguishes between community lands that are legally regarded as private property and those that are ostensibly in state ownership. This is a central difference, as private property rights are typically much stronger than rights to public forests. Given the predominance of public forest ownership, this report then describes the status of logging concessions in public forests as a primary mode of government exploitation of forestlands.

Next, changes in forest tenure are illustrated by three major global trends in public forest ownership, all of which entail shifting ownership and access rights from governments to indigenous and other local communities. Given the growing importance of community ownership, this report also reviews the emerging evidence of the possibilities created by community forest ownership and management. The report concludes with a description of the data's key implications and opportunities for key actors to address forest tenure issues.

Official Forest Ownership: The Government Perspective

The data

Tenure data collected for 24 of the 30 most forested countries appears in Table 4.1. The 30 most forested countries include 14 of the 17 megadiversity countries in the world (Bowles et al, 1998)[8] as well as the top seven of the ten leading industrial roundwood-producing countries – the United States, China, Brazil, Canada, Indonesia, the Russian Federation and Sweden (the other three major roundwood producers are France, Malaysia and Germany). Reliable tenure distribution data for six of the top 30 countries (Angola, Venezuela, Zambia, Mozambique, Paraguay and the Democratic Republic of Congo) were unavailable. For Tanzania, Indonesia, Peru and Colombia data are incomplete.

This report presents forest ownership data as identified by governments, recognizing that these statistics do not acknowledge the unrecognized claims of indigenous and other forest-dependent local communities, including areas where communities, in fact, exercise authority over forest resources without official sanction. The administration of forest resources by indigenous and other local communities is thus greater, perhaps much greater, than reflected in the official data. This difference between official ownership and effective authority is perhaps most pronounced in Africa and Asia, where governments own most forest land, but often appear to have authority over very little of it (Lynch and Talbott, 1995; Wily and Mbaya, 2001). A further complication is the fact that governments often adopt quite different legal definitions of the term 'forest'; these definitions can have a profound impact on the ways in which forest tenure issues are perceived and addressed.[9] Another complication in presenting the data derives from the complex relationship between legal 'ownership' of land and forest resources and rights to access, use and exploit those resources. Governments universally retain some rights to control the use of land and resources, regardless of ownership. For this reason, ownership often does not equate to effective control over the land.

In order to better understand and depict tenure status, this report adopts a set of categories of ownership, starting with the predominant legal categories of public and private property – recognizing that in reality there is a broad spectrum of tenure arrangements between these two.[10] These categories, and the particular placement of country data within these categories, will need to be refined and improved over time.

This report defines public ownership as all lands owned by central, regional or local governments. The public category is further divided into two subcategories: lands administered by government entities and lands set aside – or 'reserved' – for local communities, including indigenous groups, on a semi-permanent, but conditional, basis. In this latter category, governments retain ownership and the entitlement to unilaterally extinguish local groups' rights over entire areas. Under this arrangement, local groups typically lack rights to sell or otherwise alienate land through mortgages or other financial instruments.[11] Although the distribution of rights between government and community in this category is different in almost every country, invariably governments retain strong authority to extract and manage forest resources. Examples of this category include tracts of government lands 'reserved' for indigenous peoples in Brazil and the United States, the joint-forest management schemes of India, areas covered by social forestry instruments in Thailand, the Philippines and Indonesia, and areas included in

Table 4.1 *Official forest ownership in 24 of the 30 most forested countries*

Country (by descending area of forest cover as identified by FAO, 2001)	Area in million hectares (% of country total)			
	Public		Private	
	Administered by government	Reserved for community and indigenous groups	Community/ indigenous	Individual/ firm
Russian Federation[12]	886.5 (100)	0 (0)	0 (0)	0 (0)
Brazil[13]	423.7 (77.0)	74.5 (13.0)	0 (0)	57.3 (10)
Canada[14]	388.9 (93.2)	1.4 (0.3)	0 (0)	27.2 (6.5)
United States[15]	110.0 (37.8)	17.1 (5.9)	0 (0)	164.1 (56.3)
China[16]	58.2 (45.0)	0 (0)	70.3 (55.0)	0 (0)
Australia[17]	410.3 (70.9)	0 (0)	53.5 (9.3)	114.6 (19.8)
Democratic Republic of Congo[18]	109.2 (100)	0 (0)	0 (0)	0 (0)
Indonesia[19]	104.0 (99.4)	0.6 (0.6)	0 (0)	0 (0)
Peru[20]	–	8.4 (1.2)	22.5 (33.0)	–
India[21]	53.6 (76.1)	11.6 (16.5)	0 (0)	5.2 (7.4)
Sudan[22]	40.6 (98.0)	0.8 (2.0)	0(0)	0(0)
Mexico[23]	2.75 (5.0)	0 (0)	44.0 (80.0)	8.3 (15.0)
Bolivia[24]	28.2 (53.2)[25]	16.6 (31.3)[26]	2.8 (5.3)[27]	5.4 (10.2)[28]
Colombia[29]	–	–	24.5 (46.0)	–
Tanzania[30]	38.5 (99.1)	0.4 (0.9)[31]	0 (0)	0 (0)
Argentina[32]	5.7 (20.5)	0 (0)	0 (0)	22.2 (79.5)
Myanmar[33]	27.1 (100)	0 (0)	0 (0)	0 (0)
Papua New Guinea[34]	0.8 (3.0)	0 (0)	25.9 (97.0)	0 (0)
Sweden[35]	6.1 (20.2)	0 (0)	0 (0)	24.1 (79.8)
Japan[36]	10.5 (41.8)	0 (0)	0 (0)	14.6 (58.2)
Cameroon[37]	22.8 (100)	0 (0)	0 (0)	0 (0)
Central African Republic[38]	22.9 (100)	0 (0)	0 (0)	0 (0)
Gabon[39]	21.0 (100)	0 (0)	0 (0)	0 (0)
Guyana[40]	30.9 (91.7)	0 (0)	2.8 (8.3)	0 (0)
Total	2803.2	131.4	246.3	443.0

Zimbabwe's pioneering Communal Areas Management Programme for Indigenous Resources (CAMPFIRE) programme.

The private ownership category is also subdivided into two categories: forest areas owned by indigenous and other local community groups and those owned by private individuals and firms. Private ownership is defined as rights over a specific area that cannot unilaterally be terminated by a government without some form of due process and compensation. Owners of private property typically have rights to access, sell or otherwise alienate, manage and withdraw resources and exclude outsiders. This situation is described as 'fee-simple' ownership in countries with Anglo-American property traditions. Private group ownership, by contrast, simply refers to an area owned by a group as private property.

These categories may appear academic – particularly the distinction between private group rights and public 'reserves' – but the distinctions are important. Private rights are more secure because they are not as easily controlled or expropriated by government. Communities that hold private rights have more leverage than communities with long-term public use rights when negotiating with governments and external commercial interests. The importance of this distinction may become more apparent as the importance of ecosystem services provided by forests grows. Communities with private rights generally have much stronger claims to the benefits of ecosystem services and opportunities generated by their forests – as well as control over more traditional uses – than do communities on public land. Communities with corporate governance systems recognized in law can also be eligible in some countries to take on collective debt and thereby attract private investment for their forestry activities.

Summary analysis of ownership

The aggregate tenure distribution for the 24 countries reported in Table 4.1 is: about 2.8 billion hectares owned and administered by governments; 131 million hectares reserved for communities; 246 million hectares owned by indigenous and community groups; and 443 million hectares privately owned by individuals and firms. These aggregate statistics should be used with caution given that they are derived from only 24 of the many dozens of countries with forests, include only official data for lands that are known to be in each of the categories, and do not reflect the amount of forest land actively claimed by indigenous and other local communities. For these reasons the aggregate numbers should be understood as low-end estimates, particularly for the three non-government-administered categories. The percentages of forest in each tenure category are perhaps more reliable. Extrapolated to the global forest estate of 3.9 billion hectares, these data suggest that approximately 77 per cent of the world's forest is – according to national laws – owned and administered by governments, at least 4 per cent is reserved for communities, at least 7 per cent is owned by local communities, and approximately 12 per cent is owned by individuals (Table 4.2).

When aggregated for developing countries only – which excludes statistics for Canada, the United States, Russia, Australia, Sweden and Japan – the importance of community reserves and ownership increases. Community reserves represent at least 8 per cent of all developing country forests, community ownership represents at least 14 per cent and individual ownership represents only 7 per cent. This suggests that in developing countries, community reserves and ownership total at least 22 per cent of all forests – approximately three times the amount held by private individuals and firms.

Table 4.2 *Estimated distribution of forest ownership for selected categories (expressed in percentage of total)*

Categories	Public		Private	
	Administered by government	Reserved for community and indigenous groups	Community/ indigenous	Individual/ firm
Global forest estate	77	4	7	12
Developing countries	71	8	14	7
Developed countries	81	1	2	16
Countries with tropical forests	71	6	13	10
Top 17 megadiverse countries	65	6	12	17
Top 5 roundwood producers	80	7	6	7

Interestingly, the percentage of government-owned forest is higher in developed countries as is the percentage of privately held forest. To date, community ownership and access is largely concentrated in developing countries.

There are some important exceptions to these generalities within certain countries. First, in the United States private individuals and firms own more than half of the forests, or 55 per cent. The United States is joined by two other commercially important northern forested countries, Sweden and Finland, at 70 and 80 per cent, respectively,[41] and Argentina, where some 80 per cent of forests are also privately owned by individuals and firms. Other important exceptions are Mexico and Papua New Guinea, where indigenous and other local communities own some 80 per cent and 90 per cent of forests respectively.

African countries with available data have essentially no land areas officially reserved for communities and no privately held forest – community or individual. These data do not accurately represent all of Africa, as some countries are beginning to reform their land laws and recognize customary use of forest resources.[42] Further research will be needed to document this trend as soon as data become available.

Public logging concessions

While identifying whether forest areas are publicly or privately owned is critical, it is also important to ascertain how owners exercise their property rights. Forest owners, for example, frequently grant access and use rights to other parties. Given that governments own and administer the vast majority of the world's forests, the question of who has access and rights to those public forests becomes important. Although there are exceptions, governments in countries with large amounts of forest have traditionally chosen to transfer access rights and management authority to large-scale private forest industry through logging concessions. In most concessions, companies have long-term rights to access, manage the land, harvest timber and exclude the public. In return, firms typically promise to pay royalties and other fees to the government. One author estimated that in 1980 some 90 per cent of all industrial roundwood derived from logging concessions.[43]

Data for the 16 countries for which concession information is available are described in Table 4.3. These countries constitute approximately 23 per cent of the global forest estate. Reporting concession data is very difficult for a number of reasons. First of all, most governments do not publish data regarding concessions, rendering difficult the task of locating and verifying information. Secondly, concession area and access can change quickly, making information quickly out of date. Nonetheless, given existing data sources, for these 16 countries alone, the total number of hectares of public forest currently allocated to private industry is about 396 million hectares. This figure is most probably low, given, for example, that some 81 per cent of Cameroon's forests have been allocated to concessions, but only 37 per cent is currently under concession.[44]

The total area of public forest currently allocated to large industry in these 16 countries is about 150 million hectares greater than the amount owned by local communities, about 265 million hectares greater than the amount of public forest reserved for communities, and only some 47 million hectares less than the amount of all forest area owned by private individuals and firms. Concessions occupy the majority of public forest land in eight of the 16 countries listed above: Canada, Democratic Republic of Congo, Central African Republic, Gabon, Equatorial Guinea, Malaysia, Cambodia and Indonesia. In addition, logging concessions are known to predominate in Russia, Papua New Guinea, Ivory Coast and Ghana.

The literature reviewed in Table 4.3 also indicates that the governments of many countries grant logging rights to a relatively small number of companies, and that corruption and illegal logging are commonplace. According to a recent global review of illegal logging prepared by Britain's Royal Institute for International Affairs, the

> allocation of timber concessions has often been used as a mechanism of mobilizing wealth to reward allies and engender patronage. Protected by powerful patrons, timber companies may evade national regulations with relative impunity. State forestry institutions may be subject to regulatory capture, becoming clients of concession-holding industrial interests of the ruling elite, exercising their powers as a form of private property rather than a public service.[62]

As has been reported in Indonesia, government-required forest management plans often have been little more than fig leaves, covering the corruption and illegal logging common on public lands.[63]

Unfortunately, in many countries few profits from concessions are reinvested in forest communities or social programmes that aim to provide sustainable livelihoods to local people. Harvest levels are often unsustainably high and lead to boom-bust cycles in local development.[64] This approach to public forest land management has often led to environmental degradation, social instability and insecurity, and additional financial burdens on cash-starved governments.

Logging concession policy has been studied by a number of authors who have generally focused on the technical mechanics of concessions and measures to improve performance.[65] However, new evidence suggests that the fundamental assumptions underlying

Table 4.3 *Public forest concessions in 16 forest countries*

Country	Total forest (million hectares)	Area under concession (% of total)	Number of companies operating concessions	Comments
Africa				
Central African Republic[45]	4.9	3.5. (71)	10[46]	This 10 is composed of 4 French companies, 2 from Romania, 1 Former Yugoslavia and 3 domestic.
Cameroon	19.4	7.2 (37)	177[47]	Only 84 of the 177 companies have valid documented logging rights Half of the total area under concessions is abandoned
Republic of Congo	21.6	17.1 (79)	17[48]	
Democratic Republic of Congo	112.5	40.9 (36)	–	
Equatorial Guinea	2.1	1.5 (71)	–	
Gabon	21.0	11.9 (56)	221[49]	13 companies hold 50% of the concession area. This equals almost 21% of Gabon's total forest cover
Americas				
Bolivia[50]	53.0[51]	5.4 (10.2)	86	In addition to regular concessions, the government awards different types of 'land leases' for long-term contracts (400,000ha) and scientific research (200,000ha)
Canada[52]	388.9	220 (56.6)	–	Thirteen companies hold 48% of the area of commercial forestry operations
Guatemala[53]	4.2	0.2 (4.8)	–	
Peru[54]	68.1	1.2 (1.7)	–	
Suriname[55]	14.3	3.2 (22.4)	32	
Venezuela[56]	45.8	2.7 (5.9)	–	
South-East Asia				
Cambodia[57]	9.8	6.3 (64.3)	33	Legally reported log volume is 220, 894m3, export volume of 'unreported' logs is 902,500m^3 [58]
Indonesia[59]	104.6	62.5 (60.0)	–	There were 585 concessions in 1994–1995. The top ten conglomerates held 228 of these concessions, 45% of the concession area

Table 4.3 *Public forest concessions in 16 forest countries*

Country	Total forest (million hectares)	Area under concession (% of total)	Number of companies operating concessions	Comments
Malaysia[60]	18.9	10.9 (57.7)	–	Figure refers to area designated for production, not necessarily currently under concession
Philippines[61]	6.6	1.5 (22.7)	–	Area under concession refers to active and inactive zones
Total	895.7	396.0		

concession policy merit reconsideration, including the assumptions that governments have the capacity to effectively administer these complicated systems and that large private firms are uniquely qualified to administer and exploit public forest resources.[66] Most fundamentally, concession policy is founded on the notion that government ownership is the optimal – or at least the most adequate – form of forest tenure. The weight of evidence on illegal logging and corruption, along with the social consequences of the quick liquidation strategies practised by many industries, are driving some countries to rethink and reform concession policies.

Public Forests in Transition

In some countries governments have begun to shift dramatically toward community access and ownership in the last decade, partly in recognition of the legitimate claims of indigenous and other local communities and the limits of public forest governance. At least ten forest countries have enacted new legislation to strengthen indigenous ownership in this period (Table 4.4). Approximately 57 per cent of the legal rights over the some 380 million hectares now owned by, or reserved for, communities have been transferred in the last 15 years (Table 4.5). In the eight Amazon basin countries alone, the total area of public land reserved for communities or legally recognized as being under community ownership now exceeds 1 million km², roughly the area of Bolivia, and all this land has been transferred since 1985.[67] These legal reforms represent a dramatic shift in forest ownership. This shift, which began in Latin America in the late 1970s, appears to have gained momentum in Africa by the late 1990s and then spread more recently to Asia. The bulk of the some 215 million hectares recognized or legally transferred in the last decade have been in Latin America. It is likely that once the African governments begin to delimit forests and implement new legislation, a second wave of community ownership will occur.

Within these many shifts and changes in public forest ownership in the major forest countries of the world, three new trends in public forest ownership are most pronounced – each reflecting a different degree of governments' legal transfer of rights to communities or recognition of pre-existing community-based rights.[68] Some governments have begun

Table 4.4 *Recent legal reforms strengthening community forest tenure in forest countries*

Country	Year enacted	Key feature of legal reform
Australia	1996	The federal government returned ownership rights to traditional Aboriginal groups and portions are leased back to the National Parks & Wildlife Service for national parks
Bolivia	1996	Ancestral rights of community groups have precedence over forest concession holders where these rights overlap. Subsequent laws have strengthened community rights
Brazil	1988	The Constitution recognizes ancestral rights over land areas that indigenous groups and former slave communities traditionally occupied. Federal government is responsible for demarcating indigenous reserves on public lands and ensuring that land rights of indigenous groups are protected
Colombia	1991	Constitution of 1991 recognizes and outlines a framework for collective territorial rights for indigenous groups and Afro-Colombian traditional communities
Indonesia	2000	New regulatory process recently established by which customary ownership can be recognized
Mozambique	1997	Titles for customary rights are available
Philippines	1997	Constitution of 1987 protects ancestral domain rights. Indigenous Peoples Rights Act of 1997 provides legal recognition of ancestral domain rights pursuant to indigenous concepts of ownership
Tanzania	1999	Customary tenure is given statutory protection whether registered or not. Titles for customary rights are available
Uganda	2000	2000 draft currently under revision. Government is embarking upon an ambitious programme of devolution to district and local councils
Zambia	1995	Recognizes customary tenure but with strong encouragement to convert to leaseholds; titles for customary rights are not available

to recognize community ownership and reform their legal frameworks accordingly, others are devolving responsibility for managing public forestlands to communities, and a third group is beginning to reform public logging concessions to support greater local community access. Each of these three trends is discussed below.

Legal reforms to recognize community-based property rights

The legal recognition of indigenous and other local community property rights is the subject of major national debate and conflict in many nations – including Bolivia, Peru

Table 4.5 *Amount of forest recognized as community ownership or reserved for communities in selected countries since 1985*

Country	Recognized community ownership (millions of hectares)	Reserved for community administration (millions of hectares)
Australia	53.5	–
Bolivia	2.8	16.6
Brazil	–	74.5
Colombia	24.5	–
India	–	11.6
Indonesia	–	0.6
Peru	22.5	8.4
Sudan	–	0.8
Tanzania	–	0.4
TOTAL	103.3	112.9

and Colombia in South America, and Uganda, Tanzania, South Africa and Zimbabwe in Africa. Some nations have reformed land laws to recognize private community-based property rights (CBPRs) of forest-dependent communities. In these countries the process of recognizing of community property rights has developed in tandem with demands for rights to self-determination and cultural differentiation. In Colombia, for example, legal changes in 1995 allowed indigenous groups and Afro-Colombian communities to register their rights to territories they have historically occupied.[69] Titles to land have been granted to 404 communities, protecting them against government expropriation.[70] The Philippine Supreme Court recently upheld the constitutionality of the Indigenous Peoples Rights Act of 1997, providing legal recognition of ancestral domain rights pursuant to indigenous concepts of ownership over potentially as much as 20 per cent of the nation's total land mass, including well over a third of the previously public forest estate.[71]

In Canada, Indonesia, Malaysia and Nicaragua, recognition of indigenous and community rights is the subject of major national debate and conflict. A 1997 decision by Canada's Supreme Court on the Delgamuukw case recognized the sovereign land rights of First Nations over land they can document as traditional territory.[72] Some provincial governments have yet to fully accept this decision, sustaining the conflict between First Nations and governments over forests.[73] In Indonesia, the government recently established a process by which customary ownership can be recognized, although the process is so cumbersome and vulnerable to political manipulation that community advocates question its utility.[74] In October 1998 a district court in East Kalimantan issued precedent-setting statements recognizing the existence of indigenous peoples in Indonesia and their right to protect their territories.[75] In a recent landmark case in Malaysia, the High Court in the state of Sarawak upheld the customary rights of the Iban village Rumah Nor against a major forest company operating a government concession. The court faulted the Borneo Paper and Pulp Company for unlawfully accessing and exploiting the group's forest. The ruling expanded the definition of customary law to include rivers, streams and communal forests.[76]

In 1995 the Supreme Court of Nicaragua ruled that a 30-year logging concession granted to a company on Mayagna ancestral lands was unconstitutional. This was the first case before the Inter-American Court on Human Rights to directly address the territorial rights of indigenous communities. The government of Nicaragua refused to revoke the logging permit and, in a landmark decision in September 2001, the Inter-American Court affirmed the Mayagna peoples' collective right to their land, resources and environment by 'declaring that the community's rights to property and judicial protection were violated by the government of Nicaragua when it granted concessions to a foreign company without either consulting with the community or obtaining their consent'. The court found that the government discriminated against the community by denying equal protection under the laws of the state and violated its obligations under international law.[77] While little action has yet taken place, the government has pledged to carry out the court's decision.

Government devolution of limited rights to indigenous and other communities

Over the last 15 years several governments, such as that of Brazil, have begun to set aside public lands for indigenous communities. Other countries, including India and Nepal, have devolved limited rights to local communities to manage and benefit from forests that are still officially considered public land. This process is actively underway in most of the African subcontinent, with more complete transfer of rights present only in Tanzania, Gambia and Cameroon.[78] These arrangements, known by such terms as 'joint management' and 'co-management', do not alter state ownership. They represent a much weaker form of property rights than those provided by private community-based ownership.

These new arrangements involving various types of collaborative management involving governments and local communities are increasingly common in areas where governments recognize their limited capacity to manage public forests lands effectively. These are also increasing in forest areas that have already been severely degraded and are no longer of interest to private industry.

In India, out of the 65.2 million hectares recognized by the government as public forests, over 10 million are co-managed with forest user groups. Nepal uses a similar type of joint forest management, and the percentage of benefits going to communities increased from 40 per cent in 1978 to 100 per cent in 1995.[79] However, the Nepalese government is now attempting to take back many of the rights and benefits that it previously devolved. In Indonesia, forestry laws have been modified to allow forest communities to form cooperatives and extract timber from public forests, but this has also served as a screen for industries to act as cooperatives. These rights can be unilaterally revoked by the state at any time.[80]

Mounting evidence suggests these efforts to devolve government authority to communities can increase local benefits and incomes.[81] Responsibilities for resource management, however, are often being transferred to local users without transferring commensurate rights and access to benefits. In some cases these arrangements can actually reduce and erode pre-existing community rights.[82] In Brazil, for example, where some 75 million hectares have been set aside for indigenous communities, these communities have no right to harvest their timber, even under sustainable management regimes. The evidence from India indicates that forest cover and government forest departments have benefited from joint forest management, but the impact on communities has been mixed. In many countries, local groups see joint management as an inadequate but politically acceptable step toward increasing local community management authority and benefits.

Reforming public forest concession policy

Some countries, including Canada, Laos and Guatemala, are beginning to adjust traditional industrial logging concession arrangements to encompass indigenous and other local communities. In British Columbia, the provincial government recently agreed to allow Weyerhaeuser Limited to transfer its concession rights to a new business venture with a coalition of indigenous groups as the lead partner.[83] This venture, named Iisaak Forest Resources, is composed of Weyerhaeuser Limited and a coalition of First Nations that holds majority ownership. The coalition now has majority ownership of access and use rights to a portion of its ancestral homelands – but not to the land itself. More transfers of use rights between companies and communities are underway and more joint ventures are being explored in British Columbia. These agreements are often seen as 'interim measures' by First Nations, helpful in their ongoing treaty negotiations with governments.

The Guatemalan government has piloted the granting of timber concessions to local communities rather than large industries, and the early experience is positive (Box 4.1). In Laos, the government has launched a similar participatory management pilot programme involving 60 villages through 50-year management contracts. Preliminary evaluations indicate that the quality of management has improved, the amount of illegal activity has declined and the royalty payments to the government have increased, although the share of income from forests destined for local people remains low.[84]

Box 4.1 *Community concessions in Guatemala*

The Mayan Biosphere Reserve is the largest area of natural forest in Guatemala. Encroachment and illegal logging have long been major threats to the Reserve. In 1998 the National Council of Protected Areas (CONAP) issued at least four forest management concessions to local communities that were supported by partner NGOs that provide technical, administrative and community organizing expertise. The concessions range from 7,000 to 55, 000 hectares. Timber and non-timber resources are managed under a single plan. Timber is sawn on site to increase local employment. Experience has been varied but generally positive. One community's operation produced a net profit of US$89,500 for the first year in operation, roughly translated to US$318 per hectare or US$4,400 per family. Satellite images recently revealed that illegal logging and the agricultural frontier have continued to expand in protected areas, while in the community concession areas, logging has decreased.

International Tropical Timber Organization Newsletter. 2000. Spencer Ortiz. Community Forestry for Profit and Conservation: A successful community management experience in timber production and marketing in Guatemala.

The Possibilities and Potential of Community Forest Ownership

Indigenous and other communities are increasingly acknowledged for being important stewards of the global forest estate. This relatively new development provides an historic opportunity for sustainable forest conservation and the economic development of some of the world's poorest regions. In practice – as is the case with governments, individuals and firms – some communities have translated ownership into effective forest management, and others have not. Unfortunately, in addition to contending with historic political discrimination, community management is often doubly disadvantaged from a policy perspective: first because the policy frameworks of most governments privilege agriculture over forestry, and second because most forest policies privilege large producers over small.[85] Performing in these 'unlevel playing fields' is very difficult, making it hard for communities to compete with other enterprises.

In parallel with policy and market discrimination against communities, community groups face their own internal challenges of building on traditional governance systems to manage their forests.[86] Papua New Guinea is a frequently cited case where community ownership has not led to effective management (Box 4.2). There, and in other countries across the world, local communities face many of the same temptations as governments and private timber companies.

Box 4.2 *Challenges to effective community ownership in Papua New Guinea*

Papua New Guinea contains the largest intact tropical rainforest wilderness in the Asia-Pacific region and the third largest on the planet. Some 90–97 per cent of its forests are owned by some 8000 traditionally autonomous tribal groups. Despite state-recognized local community ownership, government and industry continue to wield tremendous influence, often taking advantage of limited community capacities to defend local interests against outside entrepreneurs. Almost half the country's accessible forests are already committed to industrial logging, and over 30 proposed timber projects threaten the rest. Some community leaders participate in corrupt deals that advance their own interests at the expense of their communities. Rapid deforestation, widespread corruption and illegal logging have led to a moratorium on all new logging concessions in Papua New Guinea and new support for community-based forestry management initiatives.

International Institute for Environment and Development (IIED). Papua New Guinea Country Study. Policy that Works for Forests and People Series No: 2. 1999. Forestry and Land Use Program. London.

Nonetheless, there are many examples of sound community management where harvest levels appear sustainable and benefits are distributed more equitably to local community members.[87] In Mexico, for example, community-owned forests contribute substantially to local livelihoods (Box 4.3). As with private individually owned property, this occurs when communities have clear rights and there are mechanisms in place to monitor and regulate use and exclude outsiders.[88]

Box 4.3 *Mexico's community forests*

Mexico is one of the few countries in the world where the vast majority of forests are privately owned by indigenous and other communities. Even so, there are important barriers to effective community forestry stemming from inadequate policies, technical support and markets. There are approximately 8000 communities who own 44 million hectares of forest or an average of 5000 hectares per community. The legal status of these lands derives from the creation of *ejidos* as land reform blocks transferred to producers from large landowners after the Revolution, recognition of traditional claims of indigenous groups to ancestral territories, or lands to which they fled after conquest. Many of the social systems governing these communities predate the Spanish conquest although they have been strongly modified by cultural contact. A constitutional amendment proposed in early 2001 attempts to regulate indigenous self-governance rights, including land and forest use, but this amendment has been criticized as not ratifying basic principles of indigenous rights.

Despite official community ownership, until 1986 the government unilaterally granted access to commercial value community forests to private concessionaires and then to parastatals, giving communities little voice over management decisions and transferring limited benefits from logging. This approach, combined with historical agricultural and forest policies that were biased in favour of large private industry and urban dwellers, created a situation whereby indigenous and *ejido* communities remain among the poorest people in Mexico and community-based forest industries account for less than 18 per cent of the registered national forest industry capacity. The 1986 Forest Law and subsequent modifications in 1992 and 1997 suspended the concessions system and increased opportunities for communities to direct their own forest enterprises – including extraction, provision of services and processing, as long as they prepared legal management plans for their forests. Less than a quarter of all communities and *ejidos* have active management plans – related to the cost of preparing such plans and encouraging illegal logging. The poor quality of original territorial surveys of community boundaries leaves many inter-community conflicts over boundaries, and encourages substantial investment of community funds to police their lands and advance their claims in court.

Despite these limitations, some 500 communities and *ejidos* have developed quite successful integrated forest enterprises, generating local employment and technical expertise and providing an alternative to labour migration and deforestation. Some communities hold strong cultural values that lead them to invest profits in social services and infrastructure, and conservation of biodiverse areas. The forest sector has an enormous potential to provide economic, environmental and social services, with opportunities in timber and non-timber forest products and small-scale tourism, but the sector has yet to receive equal treatment as agriculture or cattle raising. It is clearly time to reconsider the role of social forestry in Mexico as a development strategy that addresses poverty alleviation, economic development and environmental protection.

A. Molnar and A. White, 2001. 'Forestry and Land Management' in Mexico: A Comprehensive Development Agenda for the New Era. Giugale, Lafourcade and Nguyen (eds) World Bank. Washington, DC.

Segura, Gerardo. The State of Mexico's Forests Resources Management and Conservation. UNAM. Mexico.

www.semarnap.gob.mx/ssrn/conaf/statfor.htm

A key lesson from these experiences is that, just as in the case of government and private land owned by individuals, there must be a legal and policy environment to support community ownership in order for it to be effective. Official legal recognition of community-based rights to forest resources must be feasible to acquire, and must be defensible in both political and judicial arenas.[89] Unfortunately, progress on these fronts is lagging in many countries.

For example, in 1996 the government of Bolivia enacted legislation recognizing indigenous territorial ownership, but the conditions for achieving formal documentation of communal legal rights is complicated and costly. By 1999, only 10 per cent of the designated territories had received titles.[90] Even when native tenure rights are officially recognized, timber concessions, other extractive commercial enterprises and protected areas are often 'legally' overlaid on ancestral domains without any notice, participation or benefits accorded to indigenous and other local peoples.[91] In China, although the law states that local communities have clear authority over collective forests, the government has frequently reallocated rights to forest resources without any local consultation or compensation.[92] A recent example is the so-called 'logging ban' of 1998 that originally was intended to address over-harvesting on public forests. However, in many provinces it was arbitrarily extended to community forests – which by some accounts are better managed than public forests.[93]

Recent experiences in Africa have been similar. Although Uganda and Tanzania now legally recognize community ownership, as in most other countries that have enacted similar reforms, there is no legal guidance on the principles and rules for formally recognizing and governing these areas.[94] Tanzania has gone farthest, in devolving formal registration of Village Land Forest Reserves to local communities with or without the explicit support of the central forestry administration.[95]

Relationships between community-based property rights and environmental services that benefit the general public are drawing new attention. Governments' traditional approach to preserving natural areas has been the creation of official protected areas – either by delineating an area of existing public property or by using the right of eminent domain to expropriate land.[96] Not only has this approach often abused legitimate community rights, but it has often led to mixed results, largely because governments often lack the incentives, resources and political will needed to protect and sustainably manage natural resources that are public property. At the same time, valuable public services provided by the private forests of indigenous and other communities have often been taken for granted by the wider public that benefits. Additionally, it is still widely believed by political and economic elites in many nations that local communities should not or will not use their forest assets for economic development. Official efforts at national levels, therefore, to protect and maintain publicly valuable environmental services often have not been harmonized with local communities' rights and incentives to manage their private property. Nor do these strategies acknowledge the role of indigenous peoples as a living repository of cultural norms and technical knowledge that shapes the biodiversity of ecosystems, which can be an important element in biodiversity conservation and in carrying out ex situ conservation.

Australia provides an innovative exception to this long-standing disconnect. In 1998 the government began recognizing the community-based private ownership of national park lands covered by Aboriginal titles. Local Aboriginal communities are, in turn, now leasing some areas back to the government.[97] A related approach is under development in Mexico, where, rather than investing in the establishment of new protected areas, local communities would be compensated for continuing to conserve their privately-held forest resources.[98] This approach is akin to the conservation easement approach used in some developed countries – a proven approach that is based on secure private property rights and incentives for local landowners to enhance the value of their assets.

Identifying and promoting innovative legal processes and mechanisms for achieving forest conservation merits much more attention and support. New tools can contribute to low-cost, poverty-alleviating strategies for improving forest management and conservation with fewer complications than traditional state-centric approaches. In short, clearly identifying and recognizing private property rights held by indigenous and other qualified local community groups can lead to fairer, more effective and more efficient compensation schemes for environmental services. It can lead to more sustainable management and conservation of forest resources.

Managing the Transition

Realizing the potential of private community-based ownership and avoiding further forest degradation will require removing existing legal, market and policy barriers. Just as in the case of individual and government ownership, there must be clearer legal protections accorded to community ownership appropriate to the particular situation. In some situations, community ownership should be equal to individual ownership, but allow the community to define rights, including private property rights within its ownership; in others the state might have some mediating or protectorate role. But not providing clear, specific and adequate legal rights results in a non-viable arrangement.

In many countries, reforming ownership inevitably has implications for restructuring industry, as well as the existing policy and subsidy frameworks. Effectively guiding these transitions is a daunting political, technical and fiscal challenge. The experiences of some countries undergoing similar forest-sector transitions are instructive. For example, Indonesia's central government recently decentralized management responsibility to local governments in a very short period of time, with limited preparation and guidance to the local entities. The limited management capacity of local governments threatens to enhance degradation and deforestation – the very problems that drove the reforms in the first place.[99]

Russia provides a similar example. Sudden dissolution of that country's Forest Service in 2000, collapse of the budget for basic public forest management and confusion over rights to forest resources have fuelled a rapid rise in illegal logging. Unfortunately, this is occurring at a time when indigenous people are beginning to gain additional rights to publicly held forests (Box 4.4). Recent experience in the newly independent states of Eastern Europe shows that returning forests to private ownership after the collapse of communist rule has sometimes led to rapid deforestation.[100] A more promising example comes from Bolivia, where a more structured reform, accompanied by long-term technical assistance, has led to much better outcomes (Box 4.5). As the examples given here illustrate, governments need to carefully plan and manage transitions in public forest ownership. The findings of the Treaty Commission in British Columbia are also instructive: 'A treaty is a process, not an event. Our expectations for comprehensive treaties were unrealistic. We tried to accomplish too much too soon.'[101]

Box 4.4 *The status of indigenous land rights in the Russian Far East*

Siberia and the Russian Far East are home to some of the world's largest forest and woodlands ecosystems. Covering an area of 2.3 million square miles, these forests represent 21 per cent of the earth's remaining forests – an area equivalent to the continental United States and almost twice as large as the Amazon rainforests. All Russia's forests are currently owned by the government. There is no legal recognition of indigenous ownership, forest or otherwise. Although indigenous ownership is not recognized, there are Guarantees in the Russian Constitution, and in a federal law on Guarantees, that specify the rights of indigenous groups to traditional lands, including forests. During the 1990s a number of different regional governments began to pass legislation that enabled the designation of 'Territories of Traditional Natural Resource Use' (TTPs) – areas that are designed to support native peoples in their traditional activities and where native peoples have a voice in the development activities that occur on the land.

However, the exact set of rights within this designation and enforcement has varied tremendously across the region and not all areas where indigenous peoples live passed necessary legislation, and thus, not every native group with a claim has been able to establish a TTP. In general though, most legislation provided indigenous groups residing within a given TTP the capacity to 'veto' development activities (eg timber harvest, mining, etc) not in keeping with nor supporting a traditional way of life. In practice, this is a 'fuzzy' right and it has been abused, terribly at times, but nevertheless this capacity is as close as native peoples have come to ownership or even co-management. The Khabarovsk Region, the home of some 70 per cent of all forest processing operations in the Russian Far East, for example, has been relatively aggressive in establishing TTPs. There are 43 TTPs – covering 29,819,200 hectares (more than half of all the forestlands in the region). Despite this enthusiasm for the legal designation, the new rights of indigenous peoples were routinely ignored until the Khabarovsk Regional Association of Indigenous Peoples of the North (RAIPON) successfully contested the application of the law to ensure that timber companies negotiate with indigenous peoples. In the neighbouring Primorsky Region, however, only one territory has been designated for traditional use. There, regional native peoples and environmental groups have been advocating that more forests, including forests currently tendered for logging, should be designated as TTPs.

Given the mixed passage and application of TTP legislation by regional governments, a new law on TTPs was passed by the Federal Duma and signed into law by Russian President Vladimir Putin on 7 May 2001. This legislation establishes a framework though which native peoples can pursue their claim to the use of traditional lands. However, some native groups worry that the new law appears to provide regional governments with a means to remove some lands from this designation, and there is concern that the Khabarovsk Regional Government, for example, might seek to roll back the number and extent of TTPs.

A new national Land Code was signed by President Putin on 26 October 2001. This code provides clear rights for private parties to own urban and industrial land, and reconfirms private rights to small plots, notably dacha and household plots. It also provides clear authority for private owners to decide how to use and transfer their land. The law theoretically does not affect indigenous claims and forests; however, forest conservationists in Moscow worry that the lack of clarity in the new Land Code could allow the development of a 'black market' in forestlands. These worries may be justified, considering that Russian governmental control over forests has been severely weakened under the Putin administration. In May 2000, President Putin liquidated the 200-year-old Russian Forest Service, transferring its functions to the Ministry for Natural Resources – an agency driven by a resource-exploitation mission. As government oversight over forest use has weakened and confusion over access rights has increased, illegal logging and trade have grown in Siberia and the Russian Far East, including a sharp increase in illegal trade with China. Ministry officials now calculate the total amount of illegal exports to be approximately 50 million m³/year, equal to total legal exports.

Personal communication with David Gordon and Misha Jones, Pacific Environment.

Box 4.5 *Guiding the transition in Bolivia*

In 1990 Bolivia, a country of some 53 million hectares of forests launched comprehensive and ambitious reforms of its forestry sector. Progress has been remarkable – the difficult process of confirming land ownership rights to indigenous communities is well underway, with some 1.4 million hectares with clear property boundaries and ownership rights. A professional and transparent public forest administration has replaced corruption and inefficiency. And there has been significant progress toward decentralizing and devolving responsibilities and decisions regarding forest resources management to rural communities. In addition, 7 million hectares of forest are now under sustainable forest management plans and the country is a world leader in tropical forest certification with some 800,000 hectares of forest resources certified. The forest industry is restructuring towards more vertical integration, creating greater efficiencies in processing, diversification of species exploitation and raising the value of export composition. These reforms have been possible, in part, due to the strong and long-term support of international donors, particularly USAID through its BOLFOR project. However, the reform process has faced many obstacles and offers the following main lessons for policy makers who are considering reforming their forestry sectors:

* **Radical reforms require strong political will.** These sectoral reforms were carried out in an environment of innovation at the federal level, and under political conditions that facilitated public participation.

* **Provide clear ownership and use rights.** The absence of clear land ownership rights has proven a major obstacle in promoting sustainable forest management in Bolivia. Reformers must keep in mind that this means more than the simple division of space in neat parcels of land. Use rights by different groups

tend to overlap; thus, the distribution of land may not be the primary policy consideration. Rather, the much more complex legal establishment of use rights and responsibilities may be a key element in organizing forest resources management.

- **Consider establishing participatory consultative strategies in designing policy reforms.** In countries where the forestry sector has potential economic and environmental importance, and provides support and livelihoods for a large number of poor families, the benefits of consultation can justify its costs.

- **Establish a sectoral policy vision with quantitative goals.** Several of the problems encountered in preparing and executing Bolivia's national forestry system stemmed from the lack of an articulated sectoral development policy with clear and quantitative goals. The discussion of quantitative goals forces a more detailed analysis of the feasibility of measures proposed in policy reforms, and exposes those that are simply unrealizable.

- **Pay attention to the financial impact on principal stakeholders and provide economic incentives to guide private action in the desired direction.** But these incentives need to be paired with adequate penalties for not complying with the law. Economic incentives are not a sufficient condition if illegal activities can generate better commercial results.

- **Obtain an adequate understanding of the capacity of, and obstacles faced by, main actors in following the policy and law.** Policy reforms must strive to be realistic. Some have argued that Bolivian entrepreneurs have not been able to adapt to the changing economic conditions created by the forestry regime. This is because of an acute scarcity of managerial and technical expertise and a dearth of capital markets that would have facilitated the adoption of more advanced, as well as more expensive, technologies.

- **Promote vertical as well as horizontal integration of forest-based operations** (not necessarily enterprises or ownership).

- **Establish transparent procedures open to the public and administrative strategies aimed at reducing incentives and opportunities for corruption and manipulation of the public forestry administration. Harmonize legislation of related sectors that could affect the progress of the forestry sector,** such as agriculture, industry, transportation infrastructure and international commerce.

- **Strive to make sustainable forest management more profitable, rather than more complex and costly.**

Extracted from Conteras-Hermosilla, Arnoldo and Maria Teresa Vargas. Social, Environmental and Economic Impacts of Forest Policy Reforms in Bolivia. Forest Trends and CIFOR. 2002.

Looking Forward: Opportunities to Reform Tenure and Improve Forest Management

There is a major, unprecedented transition in forest ownership underway. This transition presents both opportunities and challenges to the global forest community. The recognition of indigenous rights and community ownership – and the broader rationalization of public forest tenure – present an historic opportunity for countries to dramatically improve the livelihoods of millions of forest inhabitants. Governments have an opportunity to make real headway in establishing the conditions for effective forest conservation. But seizing this opportunity and preventing further forest degradation will require ambitious and concerted action by the global forest community. Some of the more important opportunities are listed below:

- **Better knowledge on actual forest tenure claims, disputes and ownership is needed.** As evidenced by the difficulty in collecting information for this report, it is clear that data and information regarding who owns, and who has access to, the world's forests are incomplete. Where they exist, they are often questionable in quality and difficult to compare. New mapping technology presents a unique opportunity to collect, describe and monitor tenure distribution. The Global Forest Watch project, for example, is making a great contribution to knowledge at the local, national and global levels by mapping forest cover. Work of this type provides a platform for mapping tenure and access. With this information, all players will be able to better conduct informed debates on reforming forest tenure.
- **Greater awareness of transition strategies, lessons and best practices is needed.** Many governments and supporting actors are reforming tenure systems, but the knowledge generated from these experiences is often difficult for innovators in other countries to find. Disseminating information on the most effective uses of these strategies would be very valuable.
- **Major investments will be required to facilitate this transition.** Assessing community claims, mapping tenure, delimiting property, reforming legal frameworks, devising regulations and establishing new enforcement mechanisms are not inexpensive. Developed countries and multilateral and bilateral organizations need to dedicate their technology and financial resources to the monumental task of reforming tenure.
- **Property rights to ecosystem services should be identified and clarified.** Growing recognition of the linkages between forest cover and water quality, carbon absorption and biodiversity habitat has led policy makers and practitioners to design market mechanisms for capturing those benefits. Further legal development is needed to fill the void in property rights to environmental services in all countries. This would help compensate those whose lands and resources provide these services. As indigenous and other communities own and have long-term rights to a large and increasing percentage of the world's forests, it is important to ensure that their rights and economic interests are protected. Recognizing their private property rights will also serve as a critical step in achieving effective and efficient forest conservation. It is important to better understand the linkage between traditional community

management systems and the corridor of ecosystems to be managed. In many cases, the human-managed part of the system is integral to the long-term maintenance of the whole.

- **Markets and global finance must be creatively leveraged to support tenure reforms.** Without secure property rights, the incentives to manage forests responsibly are limited. Advancing the tenure reform agenda is in the interest of socially responsible industries that rely on forest fibre, environmental organizations interested in habitat protection and governments that truly seek economic development for their citizens.
- **All stakeholders will play a critical role in advancing tenure reforms and ensuring that this transition serves the interests of people and forests:**

  - **Forest-dependent peoples** can learn more about their rights and the experiences of other forest groups, building on proven strategies to assert their rights to forest resources.
  - **Socially responsible forest industries** know and appreciate the importance of secure property rights. They can support the resolution of indigenous and community tenure conflicts, develop partnerships that serve the mutual interests of communities and industry, and creatively leverage market forces to support tenure reforms.
  - **Forest management certifying bodies** can be mindful of tenure conflicts and carefully avoid reinforcing any one party's claims where tenure is contested. Certifiers can learn more about the nature of community-managed forests and multiple-use arrangements and combinations of agroforestry, which can in certain conditions help to sustain an ecosystem over a period of time.
  - **Multilateral development banks** can leverage clout and carry the urgency of addressing tenure issues to finance ministers and other influential actors by sponsoring national studies and debates on the effective ways to advance tenure reforms and their associated economic benefits. International investors and NGOs should work with communities and governments to develop capacity in communities for management, governance and strategic planning, and create linkages between sets of communities to share learning.
  - **Investors** can incorporate tenure considerations in their due diligence procedures and aggressively support investments that address community tenure issues. Investors can learn to recognize environmental and social risks and the very real impact of inappropriate action on companies' financial performance. Respecting local access to resources is essential to pleasing shareholders.
  - **Environmental NGOs** can be cognizant of humanity's historic role in shaping forest ecosystems, recognize indigenous and other local community claims and property rights, fully embrace these communities as equal partners in conservation and promote strategies that recognize and compensate communities for the environmental services their forests provide.
  - **Governments** can recognize ancestral domain rights of indigenous groups and provide full rights to forest owners to use and manage their forest assets. Governments also can 'level the playing field' for communities by removing subsidies to large industry and providing community property equal protection under the law.

Acknowledgements

Many people have contributed to this chapter. Forest Trends would like to thank Owen J. Lynch in particular for his support and encouragement throughout its preparation. In addition we would like to thank Augusta Molnar, Lynn Ellsworth, Chetan Agarwal and Jon Lindsay for their substantial contributions, as well as Sara Scherr, David Kaimowitz, Monty Roper, N. C. Saxena, Ian Gill, Gary Bull, Bob Kirmse, Peter Dewees, Misha Jones, Alexander Sheingauz, David Gordon, Joost Polak, Ali Mekouar and Shivani Chaudhry for their comments and suggestions.

Notes

1   This statement does not contradict the fact that people are known to invest in land that is insecure to strengthen their claims. For example, poor farmers are known to plant trees on contested land to enhance the security of their claim. Security is a condition of continued investment over the long term.

2   See Palmisano (2001). One exception is the Australian State of New South Wales, where recent legislative changes enable forest owners to separate the rights to land, forests, carbon and biodiversity. Another is the Dominican Republic, which passed a new forest law in 2000.

3   The World Resources Institute estimates around 400 million people worldwide live and depend on tropical forests alone (www.wri.org/forests/tropical.html), while Lynch and Talbott (1995) estimate that there are 447 million in India, Indonesia, Nepal, the Philippines, Sri Lanka and Thailand alone.

4   See Contreras-Hermosilla (2001) for an impressive compilation of studies on illegal logging and corruption from around the world. The scale of this problem is hard to overestimate. In Indonesia, for example, 84 per cent of concessionaires violated the law during the mid-1990s, including systematic operations in important national parks, an estimated 40 per cent of all wood supplies to the pulp and paper industries came from undocumented sources, and the sum of these illegal acts cost the government some $3.5 billion a year. In Cameroon, over half of all active logging licences were illegal in 1999 and at least three companies held concessions larger than the legal limit of 200,000 hectares. These stories are repeated across Asia, South America, Africa and Russia. The comparison of the costs of illegal logging and World Bank lending is reported in *A Revised Forest Strategy for the World Bank Group* (The World Bank, July 2001).

5   There have been some regional efforts, most notably UN-ECE/FAO (2000), Lynch and Talbott (1995) and Brooks, D. J. (1993). For a survey of indigenous land tenure from the colonial period to the present, see Colchester (2001).

6   Tenure advocates and researchers have sometimes claimed this dearth of information as fully intentional and motivated by government desire to maintain control of natural resources. See Lynch (1990). See also Dove (1983).

7   The list of 30 countries with highest forest cover derives from FAO (2001). FAO's 2001 definition of land with forest cover is: Land with tree crown cover (or equivalent

stocking level) of more than 10 per cent and area of more than 0.5 hectares (ha). The trees should be able to reach a minimum height of 5 metres (m) at maturity in situ.

8   Megadiversity is a concept first proposed in a paper at the Smithsonian's 1988 Biodiversity Conference. This approach looks at biodiversity priorities by political units, in this case sovereign nations, rather than by ecosystems. It recognizes that a very small number of units (17 countries out of a global 200+) are home to an inordinately large share of the world's biodiversity. The only megadiverse countries which are not in the 30 most forested cover list are Madagascar, Ecuador and the Philippines. The other top megadiverse countries are Brazil, Colombia, Indonesia, Peru, Mexico, China, India, Venezuela, Papua New Guinea, the United States, Australia, Democratic Republic of Congo, Cameroon and Tanzania.

9   It has been estimated that almost one-half of the state forest domain in Indonesia actually remains forested, the rest of the domain has either been converted to agriculture or agroforests and much of this land is populated by millions of local and indigenous peoples.

10 This categorization is informed by the typology proposed by Lynch (2000). For a more expansive and recent discussion of CBPRs see Chapter 1 on 'Community-Based Property Rights: A Conceptual Note' in CIEL (2002).

11 Of course, all governments retain the right of 'eminent domain' to unilaterally 'take' lands and rights to select portions of private holdings when deemed in the public interest to do so. The difference here is that in the case of public 'reserves' the government can unilaterally extinguish all rights to the entire parcel of land allocated to local groups.

12 UN-ECE/FAO (2000).

13 Government figures: www.un.org/esa/agenda21/natlinfo/countr/brazil/natur.htm# forests Community figures refer to indigenous lands only, quilombos not included. Proxy figures based on indigenous population in the Amazon Basin by country. See Tresierra (1999).

14 UN-ECE/FAO (2000).

15 Ibid (2000).

16 CAF-IIED (forthcoming).

17 UN-ECE/FAO (2000).

18 Except for Cameroon, the data regarding African countries derives from Egli, N. Annex 1 in Grut et al (1991) and is somewhat dated. Grut et al state that the forests of Africa are predominantly publicly owned, almost wholly by central governments.

19 All natural forests are owned and administered by the State. Community figure refers to 'Community Forest Development Program'. FAO. Country Profile, Indonesia. Forest Management, Status and Trends. Public Participation in Forest Management. www.fao.org/forestry/fo/country/index.jsp?geo_id=82&lang_id=1

20 Out of the 1000 indigenous communities in Peru, 673 have been demarcated and titled. This equals an area of 22,488km^2 for titled land, and 8403 given for use only.

21 State Forest Department. 'Progress of the Joint Forest Management Programme in India' as of 30 April 2001, India (Poffenberger, 2000, p54).

22 Energy Sector Management Assistance Program. *Sudan Activity Completion Report.* April 1988. No. 073/88. Joint World Bank/UNDP. This is the minimum number allocated for village management under supervised protection. The number is believed to be higher in reality.

23 CTCNF (1998).

24 SIRENARE (1999) *Informe Anual de la Superintendencia Forestal.* Sistema de Regulación de los Recursos Naturales Renovables, Bolivia.

25 *Mapa Forestal de Bolivia, Memoria Explicativa,* 2000. Ministerio de Desarrollo Sostenible y Medio Ambiente, BOLFOR, Bolivia.

26 These hectares are legally on hold by any other type of claims since they have been claimed by indigenous groups and they have judicial preference. The government has stopped awarding any type of concessions to those lands. Legalizing devolution requires a lengthy and expensive procedure. The intention is that this territory will be eventually owned and managed by indigenous groups. Conteras-Hermosilla, Arnoldo and Maria Teresa Vargas. *Social, Environmental and Economic Impacts of Forest Policy Reforms in Bolivia.* Forest Trends and CIFOR 2002.

27 These lands are known as TCOs or Indigenous Communal Lands (Tierras Comunitarias de Origen).

28 *Mapa Forestal de Bolivia* (2000) SIRENARE. La Paz, Bolivia.

29 Community figures refer to indigenous lands as well as Afro-Colombian territories. This reflects 72 per cent of the territories claimed by Afro-Colombian and indigenous groups, as the process of legal demarcation has not been completed. Bettina Ng'weno. *On Titling Collective Property, Participation, and Natural Resource Management: Implementing Indigenous and Afro-Colombian Demands. A Review of Bank Experience in Colombia.* World Bank. September 2000. www.cidh.org/countryrep/Colom99sp/capitulo-10.htm. This process focuses on giving property titles to communities.

30 FAO Country Profile. Tanzania. Forest Cover 2000. www.fao.org/forestry/fo/country/index

31 This figure continues to grow monthly; joint management is also on the rise, but numbers are not available. There are many types of community management schemes such as community-owned and managed forest reserves and Joint Management Agreements in at least 4 of the 500 government-owned forest reserves. *From Users to Custodians: Changing Relations Between People and the State in Forest Management in Tanzania.* Liz Alden Wily and Peter A. Dewees. Working Paper no 2569. World Bank, Washington, DC. 28 March 2001.

32 *Argentina Forestry Sector Review.* April 1993. Report 11833-AR. World Bank, Washington, DC.

33 Asia-Pacific Centre for Environmental Law Faculty of Law National University of Singapore. Myanmar Primary Legislation: FOREST LAW State Law and Order Restoration Council Law No 8/92. 3 November 1992. http://sunsite.nus.edu.sg/apcel/dbase/myanmar/primary/myafor.html#ch3

34 World Bank. *Papua New Guinea and Conservation Project.* Project Information Document. 1 April 1998. www.worldbank.org/pics/gef/pg4398ge.txt

35 *Forest Resources of Europe, CIS, North America, Australia, Japan and New Zealand, 2000.*

36 Ibid (2000).

37 Egbe, Samuel (1997) *Forest Tenure and Access to Forest Resources in Cameroon: An Overview.* IIED, London

38 Egli, N. (1991) Summaries of Country Case Studies of Selected West and Central African Countries, Annex 1 in Grut, M., J. Gray, and N. Egli, 1991, *Forest Pricing and Concession Policies: Managing the High Forests of West and Central Africa.* World Bank Technical Paper no 143, World Bank, Washington, DC.

39 Ibid (1991) and Global Forest Watch (2000). *A First Look at Logging in Gabon.* WRI, Washington, DC.

40 Guyana Forestry Commission. 29 April 1996. National Development Strategy. Chapter 30. www.guyana.org/NDS/chap30.htm

41 United Nations Economic Commission for Europe, Food and Agriculture Organization of the United Nations. *Forest Resources of Europe, CIS, North America, Australia, Japan and New Zealand (industrialized temperate/boreal countries).* UN-ECE/FAO Contribution to the Global Forest Resources Assessment 2000. Geneva Timber and Forest Study Papers, no 17, New York and Geneva (2000).

42 H. W. O. Okoth-Ogendo (2001) 'Legislating the Commons: The Opportunity and the Challenge'. The International Association for the Study of Common Property, www.iascp.org

43 Gillis, M. (1992) 'Forest Concession Management and Revenue Policies', Chapter 7 in Sharma, N. P. (ed) *Managing the World's Forests: Looking for Balance Between Conservation and Development.* Kendall/Hunt, Iowa.

44 Global Forest Watch (2000). *An overview of logging in Cameroon.* WRI, Washington, DC.

45 Data on forest cover, concession area and per cent of total forest area in concession for all African countries described here (Central African Republic, Cameroon, Democratic Republic of Congo, Equitorial Guinea, Gabon and the Republic of Congo) all derive from Table 2 of Global Forest Watch (2002). *An analysis of access into Central Africa's Rainforests.*

46 www.iucn.org/themes/forests/2/globalmeg.html

47 Global Forest Watch (2000). *An overview of logging in Cameroon.* WRI, Washington, DC.

48 Grut, M, J. Gray, and N. Egli (1991) *Forest Pricing and Concession Policies: Managing the High Forests of West and Central Africa.* World Bank Technical Paper no 143, World Bank, Washington, DC.

49 Global Forest Watch (2000) *A First Look at Logging in Gabon.* WRI, Washington, DC.

50 SIRENARE (1999) *Informe Anual de la Superintendencia Forestal.* Sistema de Regulación de los Recursos Naturales Renovables, Bolivia.

51 Only 28.8 million hectares are considered as natural production forest area.

52 Global Forest Watch (2000) *Canada's Forests at a Crossroads: An Assessment in the Year 2000.* WRI, Washington, DC.

53 Hardner, J. J. and Rice, R. (1999) 'Rethinking Forest Concession Policies', in Keipi, K. (ed) *Forest Resource Policy in Latin America,* IDB, Washington, DC.

54 Ibid (1999).

55 Sizer, N. and Rice, R. (1995) *Backs to the wall in Suriname: Forest policy in a country in crisis.* WRI, Washington. Mitchell, A. (1998) *Report on Forest Concessions, Suriname.* Fortech, London, UK.

56 *Forest Concession Policy in Venezuela.* Julio Cesar Centeno. WRI, April 1995. www. ciens.ula.ve/~jcenteno/concess.html

57 Development Alternatives Inc. *Findings and Recommendations of the Log Monitoring and Logging Control Project.* Main Report. Final Edition. Bethesda, MD (September 1998) www.oneworld.org/glboalwitness/reports/deserts/concess.htm

58 *Findings and Recommendations of the Log Monitoring Control Project* (September 1998 Main Report. Development Alternatives Inc, Washington, DC.)

59 Brown, David (1999) 'Addicted to Rent: Corporate and Spatial Distribution of Forest Resources in Indonesia; Implications for Forest Sustainability and Government Policy'. DFID/Indonesia-UK Tropical Forest Management Programme, Report no PFM/EC/99/06.

60 Sustainable Forest Management in Malaysia – The Way Forward. Dato Ismail Awang. www.mtc.com.my/publication/speech/sustainable.htm

61 Cruz vs Secretary of Environment and Natural Resources, Philippine Supreme Court Reports Annotated. See also Vicente, Paolo B. Yu III, *Selling off the National Patrimony: Assessing the Economic and Natural Resources Policies in the Philippines (1986–1999).* Draft. 24 May 1999. Legal Rights and Natural Resources Center-Kasama sa Kalikasan/Friends of the Earth Philippines.

62 Brack, D. and Hayman, G. (2001) *Intergovernmental Actions on Illegal Logging: Options for intergovernmental action to combat illegal logging and illegal trade in timber and forest products.* The Royal Institute of International Affairs, London.

63 ELSAM (Lembaga Studi dan Advokasi Masyarakat, The Institute for Policy Research and Advocacy) and CIEL (Center for International Environmental Law). *Whose Resources? Whose Common Good? A New Paradigm of Environmental Justice and the National Interest in Indonesia.* June 2001. ELSAM and CIEL. Jakarta, Indonesia and Washington, DC.

64 Robert Schneider, E. Arima, A. Verissimo, P. Barreto, and C. Souza Jr. (2000) *Sustainable Amazon: Limitations and Opportunities for Rural Development.* IMAZON and World Bank. Michael Dove (1983) 'Theories of Swidden Agriculture and the Political Economy of Ignorance'. *Agroforestry Systems,* vol 1, pp85–99; and Thomas M. Power (1996) *Lost Landscapes and Failed Economies: The Search for a Value of Place.* Island Press.

65 Gillis, G. et al (1991). Repetto, R. and Gillis, M. (1988) *Deforestation and Government Policy.* International Center for Economic Growth, San Francisco.

66 Barr, C. (2001) 'Will HPH Reform Lead to Sustainable Forest Management?: Questioning the Assumptions of the "Sustainable Logging" Paradigm in Indonesia', in *Which way forward? People, Forests and Policymaking in Indonesia,* Carol J. Pierce Colfer and Ida Aju Pradnja Resosudarmo (eds) CIFOR, Bogor, Indonesia.

67 The total area of land in community reserves or ownership in the Amazon basin is 1,007,678km². Bolivia is 1,098,581km² in size.

68 There is a separate, and important, shift in forest tenure taking place in Eastern Europe in countries that are not among the top forest covered countries of the world. Ten of the 11 Eastern European countries – Poland is the exception – formerly under Soviet rule have begun returning forest lands seized under their communist regimes to their previous owners. When the process is finished, they expect an average of about a third of those lands to be privately owned. *Implications of Land Restitution*

for Achieving World Bank/WWF Alliance Targets in Eastern Europe and the Central Asian Region. INDUFOR OY & ECO, Helsinki, 2001.

69 Third Report on the State of Human Rights in Colombia. Chapter 10 Indigenous Rights. Organization of the American States. www.cidh.oas.org/countryrep/Colom99sp/capitulo-10.htm

70 http://cidh.org/countryrep/Colom99sp/capitulo-10.htm

71 CIEL Press Release. *CIEL helps score major victory for indigenous rights in Philippine Supreme Court.* 29 January 2001. www.ciel.org

72 Interior Alliance Update – no date, no issue number. International edition. 'The Interior – A Stronghold for Aboriginal Title'. Canada.

73 Efforts are underway by the Canadian government and British Columbia's largest First Nation group to draft a permanent treaty. 'Canada and British Columbia's Largest Indian Group Taking Steps to First Permanent Treaty'. Sunday, 11 March 2001. p10 NE version. *New York Times.*

74 ELSAM (Lembaga Studi dan Advokasi Masyarakat, The Institute for Policy Research and Advocacy) and CIEL (Center for International Environmental Law). *Whose Resources? Whose Common Good? A New Paradigm of Environmental Justice and the National Interest in Indonesia.* June 2001. ELSAM and CIEL. Jakarta, Indonesia and Washington, DC.

75 See Fried, S. G. (2000) 'Tropical Forest Forever? A Contextual Ecology of Bentian Rattan Agroforestry Systems', in *Peoples, Plants and Justice: The Politics of Nature Conservation,* Charles Zerner (ed). New York, Columbia University Press.

76 The Borneo Project. 'Landmark Ruling Secures Native Land Rights'. www.earthisland.org/borneo/news/articles/010509article.html. Meanwhile, in Sarawak a draft Land Surveyors Bill 2001 was introduced in the Dewan Undangan Negeri (DUN) (former General Council) on 31 October. According to the draft bill, any map that shows 'the delimitation of the boundaries of any land, including State land and any land lawfully held under native customary rights' can be done only by a licensed surveyor. Community mappers do not possess the necessary professional degrees that will be recognized by the Board of Surveyors.

77 'Case of the Indigenous Mayagna Community of Awas Tingni (Nicaragua) Before the Inter-American Court on Human Rights'. September 2001. www.indianlaw.org/body_awas_tingni_summary.htm

78 Wily, L. A. (2002) 'Participatory Forest Management in Africa Today: An overview of progress and trends'. Key Note Address, Second International Workshop of Participatory Forestry in Africa, Tanzania, 18–22 February 2002.

79 Poffenberger, M. (ed) (2000) 'Communities and Forest Management in South Asia'. Working Group on Community Involvement in Forest Management, IUCN, September 2000, p63.

80 ELSAM (Lembaga Studi dan Advokasi Masyarakat, The Institute for Policy Research and Advocacy) and CIEL (Center for International Environmental Law). *Whose Resources? Whose Common Good? A New Paradigm of Environmental Justice and the National Interest in Indonesia.* June 2001. ELSAM and CIEL. Jakarta, Indonesia and Washington, DC.

81 World Bank, Implementation Completion Report for Madhya Pradesh Forestry Project, Internal Evaluation Document, Washington, DC.

82 Wollenberg, E., Edmunds, D. and Buck, L. (2000) *Scenarios As a Tool for Adaptive Forest Management: A Guide.* CIFOR. Bogor, Indonesia. Katon, B., Knox, A. and Meinzen-Dick, R. (January 2001) 'Collective Action, Property Rights, and Devolution of Natural Resource Management'. Policy Brief #2. CAPRi: CGIAR Systemwide program on Collective Action and Property Rights. International Food Policy Institute. Washington, DC.

83 Iisaak Forest Resources (May 2000) Submission to the Green Economy Secretariat, Government of British Columbia.

84 Williams, P. (2000) *Draft Evaluation Summary, Evaluation of 3 Pilot Models for Participatory Forest Management; Villager Involvement in Production Forestry in Lao PDR.* 6 September 2000. World Bank.

85 Scherr, S., White, A. and Kaimowitz, D. (2002) *Strategies to Improve Rural Livelihoods Through Markets for Forest Products and Services.* Forest Trends.

86 Richards, M. (September 1997) 'Tragedy of the Commons for Community-Based Forest Management in Latin America?' no 22. *Natural Resource Perspectives.* Overseas Development Institute. London, UK; and Ribot, J. C. 'Participation Without Representation: Chiefs, Councils and Forestry Law in the West African Sahel'. Fall 1996. *Cultural Survival Quarterly.* Cambridge, MA, US.

87 Agrawal, A. and Gibson, C. (eds) (2001) *Communities and the Environment: Ethnicity, Gender, and the State in Community-Based Conservation.* Rutgers University Press. New Jersey, US.

88 Pye-Smith, C. and Feyerabend, B. (1994) *The Wealth of Communities.* Earthscan, London.

89 Forest Trends (2001) *Strategies for Strengthening Community Property Rights Over Forests: Lessons and Opportunities for Practitioners.* Report Developed for Ford Foundation.

90 Becker, C. D. and León, R. (2001) 'Indigenous Forest Management in the Bolivian Amazon: Lessons from the Yucaré People', in Gibson, C. C., McKean, M. and Ostrom, E. (eds) *People and Forests: Communities, Institutions and Governance.* Cambridge, MA and London, The MIT Press.

91 In Chile, at least 23 indigenous communities have contested legal claims against seven forest companies and private owners for 48,554 hectares of forests. Comunidades Mapuche en Conflicto. http://members.aol.com/mapulink/espanol/conflicto.html

92 Dachang, L. (2001) 'Tenure and Management of Non-State Forests in China since 1950: A Historical Review'. *Environmental History*, vol 6, no 2.

93 CCICED (2001) *Forest /Grassland Taskforce Report to the CCICED 2001 Annual Meeting.* The Fifth Meeting of the 2nd Phase of CCICED. August.

94 Okoth-Ogendo, H. W. O. 'Legislating the Commons: The Opportunity and the Challenge', in *The Common Property Resource Digest.* June 2001, no 57. Quarterly publication of the International Association for the Study of Common Property. Gary, Indiana.

95 Wily (2002), p15.

96 See for example, Burnham, P. (2000) *Indian country, God's country: Native Americans and the National Parks.* Island Press, Washington, DC. for an account of how American Indians were removed from lands that became national parks.

97 The 76,000 hectare Mutawintji National Park in the far west of New South Wales (NSW), Australia, became the first park returned to its traditional Aboriginal owners

and then leased back to the NSW National Parks & Wildlife Service (NPWS).
Other aboriginal lands are also currently leased to state governments for state and
national park use, such as the 1325 hectare Uluru/Kata Tjuta National Park where
Ayers Rock is located. Australian legislation provides for an annual rental fee paid
by the government, which provides extra funding to ensure the lands are managed
to meet Aboriginal community interests. Mark Sutton. Aboriginal Ownership of
National Parks and Tourism. www.cs.org/newdirection/voices/sutton.htm and Northern Territory Visitors Centre. www.northernterritory.com/3–2a.html

98 Molnar, A. and White, A. (2001) 'Forestry and Land Management', in Giugale,
Lafourcade and Nguyen (eds) *Mexico: A Comprehensive Development Agenda for the
New Era.* World Bank, Washington, DC.

99 Barr, C. (2001) *Banking on Sustainability: Structural Adjustment and Forestry Reform
in Post-Suharto Indonesia.* Macroeconomics for Sustainable Development program
Office, WWF and CIFOR, Bogor, Indonesia.

100 INDUFOR OY/ECO for The World Bank and World Wildlife Fund Alliance
(2001) *Implications of Land Restitution for Achieving World Bank/WWF Alliance Targets in Eastern Europe and the Central Asian Region.*

101 British Columbia Treaty Commission (2001) *Looking Back Looking Forward: A
Review of the BC Treaty Process.* British Columbia Treaty Commission, Vancouver,
Canada.

References

1 Bloch, M. (1966) *French Rural History.* University of California Press, Davis.
2 Bowles, I., Rice, R., Mittermeier, R. and da Fonseca, A. (1998) 'Logging and tropical forest
conservation'. *Science*, vol 280, pp1899–1900.
3 Brooks, D. J. (1993) *U.S. Forests in a Global Context.* General Technical Report RM-228. US
Department of Agriculture Forest Service. Rocky Mountain Forest and Range Experiment
Station. Fort Collins, Colorado for United States, Canada, Nordic Countries, Europe, Japan,
New Zealand, Australia, Tropical forested countries.
4 CAF-IIED (forthcoming) *Instruments for Sustainable Private Sector Forestry.* Chinese Academy of Forestry, International Institute for Environment and Development, Overseas Development Institute, China Country Study, Work Plan: December 1999 to December 2000.
5 CIEL (2002) *Whose Nation? Whose Resources? Towards a New Paradigm of Environmental Justice and the National Interest in Indonesia.* Center for International Environmental Law,
Washington, DC.
6 Colchester, M. (ed) (2001) *Survey of Indigenous Land Tenure: A report for the land tenure
service of the Food and Agricultural Service of the Food and Agriculture Organization.* unpublished draft, December 2001.
7 Contreras-Hermosilla, A. (2001) *Forest law enforcement.* World Bank, Washington, DC.
8 CPRD (2001) 'The tragic african commons: A century of expropriation, suppression and
subversion' in *The Common Property Resource Digest.* June 2001, no 57. Quarterly publication of the International Association for the Study of Common Property, Gary, Indiana.
9 CTCNF (1998) *El Subsector Forestal en México.* Consejo Técnico Consultivo Nacional Forestal. www.semarnap.gob.mx/ssrn/conar/subsecto.htm
10 Dove, M. R. (1983) 'Theories of Swidden agriculture and the political economy of ignorance'. *Agroforestry Systems*, vol 1, pp85–99.

11 FAO (2001) *Forest Resource Assessment – Forest Cover 2000*. Rome, Italy. www.fao.org/forestry/fo/fra/index.jsp

12 Grut, M., Gray, J.and Egli, N. (1991) *Forest Pricing and Concession Policies: Managing the High Forests of West and Central Africa*. World Bank Technical Paper no 143, World Bank, Washington, DC.

13 Lynch, O. J. (1990) *Whither the People? Demographic, Tenurial and Agricultural Aspects of the Tropical Forestry Action Plan*. World Resources Institute, Washington, DC.

14 Lynch, O. J. and Alcorn, J. (1994) 'Tenurial rights and community-based conservation' in Western, D., Wright, M. and Strum, S (eds) *Natural Connections: Perspectives on Community-Based Conservation*. Island Press, Washington, DC.

15 Lynch, O. J. and Talbott, K. (1995) *Balancing Acts: Community-Based Forest Management and National Law in Asia and the Pacific for Indonesia, India, Nepal, Philippines, Sri Lanka and Thailand*. World Resources Institute, Washington, DC.

16 Lynch, O. J. (2000) *Promoting Legal Recognition of Community-Based Property Rights, Including the Commons: Some Theoretical Considerations*, paper presented at the Symposium of the International Association for the Study of Common Property, Indiana University, Bloomington, Indiana.

17 Palmisano, J. (2001) 'Property rights and risk management'. *Environmental Finance*, March 2001.

18 Peluso, N. L. (1992) *Rich Forests, Poor People: Resource Control and Resistance in Java*. University of California Press, Berkeley and Los Angeles.

19 Perlin, J. (1989) *A Forest Journey: The Role of Wood in the Development of Civilization*. Harvard University Press, Cambridge, MA and London, UK.

20 Poffenberger, M. (ed) (1990) *Keepers of the Forest: Land Management Alternatives in Southeast Asia*. Ateneo de Manila University Press, Manila.

21 Poffenberger, M. (ed) (2000) *Communities and Forest Management in South Asia*. Working Group on Community Involvement in Forest Management, Forests, People and Policies. IUCN, September.

22 Rampey, L. (2001) *The English Royal Forest and the Tale of Gamelyn* at www.geocities.com/lrampey/erf.htm

23 Seymour, F. (1995) *A Preliminary Review of Experience with Resettlement to Achieve Conservation Objectives*. World Wildlife Fund, Washington, DC.

24 Thrupp, L., Hecht, S. B. and Browder, J. O. (1997) *Diversity and Dynamics of Shifting Cultivation: Myths, Realities and Policy Implications*. World Resources Institute, Washington, DC.

25 Tresierra, J. (1999) 'Rights of indigenous peoples over tropical forest resources' in Keipi, K. (ed) *Forest Resource Policy in Latin America*. Inter-American Development Bank, Washington, DC.

26 Tuxall, J. and Nabhan, G. P. (2001) *People, Plants and Protected Areas: A Guide to In Situ Management, People and Plants Conservation Manual*. Earthscan, London.

27 UN-ECE/FAO (2000) *Forest Resources of Europe, CIS, North America, Australia, Japan and New Zealand (industrialized temperate/boreal countries)*. United Nations Economic Commission for Europe/Food and Agriculture Organization of the United Nations, Contribution to the Global Forest Resources Assessment 2000, Geneva Timber and Forest Study Papers no 17, New York and Geneva.

28 Wilmsen, E. N. (ed) (1989) *We Are Here: Politics of Aboriginal Land Tenure*. University of California Press, Berkeley.

29 Wily, L. A. and Mbaya, S. (2001) *Land, People and Forests in Eastern and Southern Africa at the Beginning of the 21st Century*. Natural Resources International, The World Conservation Union.

30 World Bank (2001) *Draft, Revised Forest Strategy for the World Bank Group*. July.

31 Wunder, S. (2001) *The Economics of Deforestation: The Example of Ecuador, St Anthony's Series*. Macmillan & St Martin's Press, London and New York.

Part 2

Forests and Livelihoods

There are a number of older books that provide fascinating accounts of the complexity of local management of forests in historical times. More recent books, for instance those by Watkins (1998) and Rackham (1976), provide valuable accounts of the sophisticated forest management systems that have developed in Western Europe over the last 1000 years or so. They also give guidance in accessing the literature on forest history. However, the period of rapid economic expansion following the Second World War led forest institutions in both tropical and temperate zones to focus their attention much more on the emerging phenomenon of industrial forestry. Forests went rapidly from being mainly managed as local resources to being the source of internationally traded commodities. In the closing years of the 20th century the pendulum began to swing back and a lot more interest was shown in forests as the sources of the livelihoods of very large numbers of people, especially in the developing world. The pioneering work by FAO on social forestry and community forestry opened people's eyes to the strong dependence that many people throughout the world still had on the multiple products and services that were provided by forests.

This interest in local forest values led to some sweeping assumptions on the potential roles of forests as the resource upon which sustainable development of poor countries could be based. A widely quoted paper by Peters et al (1989, published in Nature) on valuation of an Amazonian rainforest was especially influential. Conservation organizations and development assistance agencies began to promote the maintenance of traditional forest management systems as a means of achieving both conservation and development objectives.

Chapter 5 by Wunder and Chapter 6 by Arnold and Ruiz-Perez set some of these claims in perspective. They provide a more realistic and rigorous evaluation of the ways in which forest conservation and poverty alleviation objectives can be complementary, but can also be in conflict. Forests are indeed important for poor people, but forest-based activities do not often provide the shortest route out of poverty. Non-timber forest products are surprisingly important economically and many poor people profit from them. However, trying to maintain traditional systems of non-timber product management in their present state is often unrealistic. Markets change, products previously gathered from the wild are domesticated and a host of other factors complicate the story.

This raised the difficult issue of what future one should envisage for the people living in and off forests. Byron and Arnold examine this in Chapter 7 and, again, it emerges that change is inevitable. Trying to tie these people to existing ways of life may be condemning them to a 'poverty trap'. These people must be empowered to determine their own futures – forests may be, but more often will not be, the route that they choose.

Finding the best ways for people to benefit sustainably from their forests is important. Local participation in forest management has been a dominant theme in forestry in recent decades. Various forms of local forest management are likely to be even more widespread in the future and a lot has been learned in recent years on how this can be achieved successfully. We will return to these themes in Chapters 19 and 20.

References

1   Peters, C. M., Gentry, A. H., and Mendelsohn, R. O. (1989) 'Valuation of an Amazonian Rainforest'. *Nature*, vol 339, pp655–656.
2   Rackham, O. (1976) *Trees and Woodlands in the British Landscape: The Complete History of Britain's Trees, Woods and Hedgerows*. Phoenix Press, London.
3   Watkins, C. (1998) *European Woods and Forests: Studies in Cultural History*. CAB International, Wallingford.

5

Poverty Alleviation and Tropical Forests: What Scope for Synergies?

Sven Wunder

Introduction

How much can forest research and development (R&D) – the generation of new knowledge on both forest commodity production and on the broader principles of natural resource management (NRM) – contribute to the simultaneous goals of poverty alleviation and biodiversity conservation? This chapter gives an overview of the literature on this topic, and draws on ongoing research at the Center for International Forestry Research (CIFOR), particularly in Latin America. It also puts forward some hypotheses on expected synergies and contradictions between environmental and developmental forest objectives. I argue that, for the case of tropical forests, the very optimistic outlook on 'win-win' potentials in the Brundtland report and from RIO 1992 was based on the inadequate implicit diagnosis that poverty is the cause of forest destruction. This, in turn, has raised overly optimistic expectations on the scope for integrating forest conservation and development objectives. Forests may sustain poor people and help them survive, but degrading and converting forests may also be an important but not always 'unsustainable', pathway out of their poverty.

In several respects, natural forests may have a poor comparative advantage for alleviating human poverty. The contradictions tend to outweigh the synergies.

The structure of the chapter is the following. The section on key definitions and concepts provides basic definitions in regard to both forests and poverty. 'The forest-dependent poor' reviews some of the thinking on the poverty-forest interaction over the last decades. The sections on macro- and micro-economic poverty examine the impact of poverty on forests, at the macro- and microlevel, respectively. In 'Poverty profiles' and 'Natural forests for poverty alleviation', we are interested in the reverse causality – identifying recent poverty trends in tropical developing countries and the potential of tropical forests for poverty alleviation. The conclusion reviews the main arguments of this chapter, while the discussion looks at main policy lessons for both conservation and poverty-alleviation strategies.

Note: Reprinted from *World Development*, vol 29, Wunder, S., 'Poverty alleviation and tropical Forests – What scope for synergies?', pp1617–1833, Copyright © (2001), with permission from Elsevier

Key Definitions and Concepts

This chapter focuses on natural tropical forests, rather than on modified systems such as plantations or agroforestry, due to the prime importance of natural forests for biodiversity. Following FAO, 'deforestation' is defined as a radical removal of vegetation, to less than 10 per cent crown cover. Deforestation implies costs, and it is basically an investment in future alternative land uses. In turn, 'forest degradation' refers to all other interventions with biodiversity-loss impacts (logging, fires, overgrazing, overhunting etc); rather than an investment, degradation is mostly a rent cashing-in of the 'subsidy from nature'.

Poverty is by definition extremely complex, multidimensional and linked to many variables. The links to land use and deforestation are thus in many cases indirect, that is, connected to a third set of intermediate variables. Following Reardon and Vosti (1995), I distinguish between two interrelated concepts of poverty. 'Asset poverty' is related to producers, which entails binding restrictions on the choice of economic activities, crops, investments, technologies, and so on.[1] 'Welfare poverty' is related to absolute restrictions in (monetary and nonmonetary) household consumption, for example vis-à-vis the simple poverty line of US$1 per capita suggested by the World Bank. 'Poverty' to me thus retains a strong economic component. I also define it as an absolute concept, neither a matter of households' 'self-perception' nor of their welfare relative to other groups; the latter phenomena are rather related to income distribution and inequality.[2]

The Forest-Dependent Poor

In the tropics, extensive forest areas often coincide geographically with a large number of poor people that depend on forests for their livelihoods.[3] They may, on the one extreme, be native forest dwellers that have a long, culturally rooted tradition of extracting a broad range of commodities from vast forest areas. At the other end of the spectrum, one finds immigrant newcomers that opportunistically take advantage of selected forest resources, but are most interested in the soil under the trees to cultivate crops and raise livestock. Some of the forest-derived benefits are converted to monetary income, for example by selling logs, charcoal or resins. Others, such as firewood, vines or fruits collected for household consumption remain a 'hidden harvest', the value of which is often difficult for economists to quantify (IIED, 1995). Finally, some poor use the forest as production input, such as slash-and-burn swiddeners for long fallow systems or herders in the Sahel for grazing (Warner, 2000). There is by now a large literature that documents how natural forests worldwide serve as 'the poor man's overcoat' (Westoby, 1989, p58), providing vital safety-net functions for rural livelihoods in terms of risk safeguarding ('famine foods'), health (medicinal plants), filling income gaps and balancing nutrition.[4] There is also evidence that poor households derive a relatively larger share of their income from forests and wildlands than better-off households within the same community (eg Cavendish, 1997).

Access rights of poor people to the forest tend to be open or informal, difficult to protect against external interests, and sometimes conflictive between on-site users. When powerful actors from outside, such as logging or mining firms, commercial farmers or ranchers, find it profitable to exploit the forest for their own purposes, conflicts of interest arise and poor forest-dwellers' benefits are endangered. The forest-dependent poor become poorer; protecting their access is thus vital to their current wellbeing. This is a story that we recognize from many forest areas where externally induced development processes are advancing. From this perspective, it would seem straightforward that 'the fate of the forest' and 'the fate of the poor' are linked in an interdependent and harmonious manner.

The static argument that the forest-dependent poor are made worse off when they lose forest access is thus very important for 'poverty prevention'. But it says little about the *dynamic* potential of natural forests to actively reduce poverty over time, that is, to produce more benefits for the poor. In some cases, the poor may only have been 'allowed' to use forests without interference from powerful groups because of these forests' poor financial returns. The same vital time distinction has to be made for the impact of poverty on forests. The simplistic view states that forest degradation is 'a short-term solution' to poverty (Schmidt et al, 1999, p18). But to be poor is in itself a static condition that cannot explain the dynamic processes of forest loss and degradation: many welfare-poor indigenous peoples have conserved their forests for centuries.[5] Poverty combined with internal and/or external change factors – population growth, land races and external interventions – can possibly set off self-perpetrating processes of forest degradation. As expressed by Eckholm et al (1984, p6): 'The poor are not ignorant of the process of deforestation nor blind to its effect. They cut because they must.'

The idea behind these 'vicious circles' is that poor resource access triggers environmental degradation, which aggravates poverty and promotes reinforced degradation and so on. The concept goes back at least to Myrdal (1957), but in the late 1980s it was broadly applied in the UN-commissioned and influential Brundtlandt report, including to the problem of forest loss (WCED, 1987, p28). A policy conclusion was that a general alleviation of poor forest dwellers' resource constraints would be an important pathway to protect the resource. By making people better off, one would enable them to make a rational and 'sustainable' (nondegrading) use of their forests, and help to protect these against external pressures. This philosophy also became mainstream thinking in major international organizations, such as in UNEP, the WWF and IUCN World Conservation Strategy from 1980 (Sayer, 1995).

By the time the RIO 1992 conference was convened, some of the difficulties of 'sustainable development' had already been recognized. But the diagnosis of poverty-led natural resource degradation proved to be a convenient point of consensus between, on the one hand, developed countries looking for means of protecting the global commons and, on the other, developing countries striving for more economic progress. Environmentally speaking, there was an idealistic notion that the 'sustainable-development' stone could kill both the poverty-alleviation and the conservation birds, yielding an efficient 'win-win' strategy to satisfy both parties' needs. During the late 1980s and 1990s, many investments in 'buffer zone' projects and in Integrated Conservation and Development Projects (ICDPs) sought to alleviate resource constraints for local stakeholders. The hope was that this would empower them to stop cutting down their forests. People moved to the centre of the conservation ethics and practice – indeed, who honestly wants to argue against people? The Green agenda moved significantly from the political right to the left, it won new allies, and 'conservation with a human face' had come to stay.

The overriding question is whether the poverty-degradation causal link reflects a correct diagnosis of the environmental problem at stake. Are tropical forest degradation and deforestation caused by a high level of (or rises in) poverty? Can forests provide for a stream of benefits to reduce poverty at the national scale? Do they thus lend themselves easily to 'win-win' solutions? Or are we, on the contrary, confronted with a case of wishful thinking by politicians and multilateral negotiators, readily taken up by environmental agencies and opportunistic non-governmental organizations (NGOs)? The answer is bound to lie in between the two extremes, but I will argue that the latter position may be much closer to the truth than the former.

Macroeconomic Poverty – Ambiguous Impacts on Forests

A basic point in question is that, at the global level, the shift toward a people-centred conservation paradigm has not succeeded in slowing down the loss of tropical forests and their biological diversity – let alone halting and eventually reversing these trends. Our quantitative knowledge of both phenomena remains insecure, though. The latest update of FAO's Forest Resources Assessment (FAO, 1997, p18) claims that there has been a slowdown in tropical deforestation, from an annual 14.6 million ha in the 1980s to 12.8 million ha during 1990–1995, but some of this may reflect a modelling mirage.[6] Some sources have also tried to assess logging and forest-degradation trends over time for different countries (WRI, 1994, pp306–309), but these estimates are still more uncertain than the deforestation figures.

At the macroeconomic level, do growth and poverty reduction reduce deforestation and, vice versa, does economic crisis accelerate forest loss? Let us first look at the long-term trends. The empirical evidence denies this: for 1981–1998, Thomas et al, (2000, p3) find that forest loss is positively correlated with both economic growth (81 countries) and with poverty reduction (26 countries) – though neither coefficient is significant. Many cross-country studies test the hypothesis of an 'environmental Kuznets curve' – an inverted U-curve where rising income increases forest loss but forest cover increases again at higher-income levels. The literature review by Kaimowitz and Angelsen (1998, pp72–88) finds only weak empirical support for a U-curve. Many developed countries exhibit a 'forest transition' U-curve over time and income, with rapid land clearing followed by some net reforestation later on. Important causal factors for this turnaround are the high-income demand for forest services (hydrological, recreation etc), forest regrowth on abandoned low-return crop lands, and the creation of forest plantations. The transition has only been found in a few tropical developing countries (such as Costa Rica, Cuba and Puerto Rico); for others, the pace of forest loss has levelled off (eg, Thailand and the Philippines) – but at quite low forest-cover levels.[7]

Over the last three decades, South-East Asia has been one of the most successful regions in regard to poverty alleviation, mainly due to improvements in agriculture, education and the rise of labour-intensive industries. It has also been one of the tropical regions with the highest deforestation rates. Indeed, as noted for the case of Thailand, deforestation for extensive crops may have been an important avenue of rural poverty alleviation, by allowing the landless poor to acquire agricultural land (Reed, 1992, pp122–123), increasing the supply and reducing the price of staple crops for urban consumers.

How is deforestation related to the short-term year-to-year fluctuations in economic welfare? Few developing countries monitor their forest resources yearly so as to allow a comparison of deforestation directly with economic growth rates over time. Brazil is an exception; after 1988, annual forest-loss data from the Amazon region have been produced by INPE, the Brazilian Space Research Institute. Comparing GDP growth of the Brazilian economy with forest loss, Figure 5.1 shows signs of a pro-cyclical pattern of deforestation: the high-growth periods 1977–1988 and 1993–1994 coincide with high forest loss. Conversely, deforestation slowed down during the years of severe economic crisis, mainly because public projects (roads, dams etc) had to be postponed, agricultural subsidies were cut, and private investment in agriculture receded (Young, 1995). Moreover, the economic growth rates of the high-deforesting Legal Amazon federal states has during the last three decades, that is, since the deforestation process took off, been higher for the whole of Brazil (Andersen et al, 1996, Chapter 2). This trend links deforestation to neither increased asset nor welfare poverty, but rather to unconstrained public and private investments in agricultural expansion.

What if we look at a group of wealthy or rapidly growing developing countries, for instance, at those specialized in oil exports?[8] Their wealth should imply that, similar to the developed countries, less-productive crops are abandoned, thus reducing deforestation. An affirmative example from Central Africa is Gabon, where the combination of low population density and oil wealth wiped out a small-scale agricultural sector that could not compete with imported foodstuff. Urbanization was dramatic, and new forest clearing was restricted to 'point impacts', near the cities and some agro-industrial plantations. Oil also financed the costly construction of the Transgabonese Railway, which opened up large areas for selective timber extraction, thus increasing forest degradation.[9]

In Venezuela, from the 1920s onward oil basically changed a resource-poor agrarian export economy (coffee, cocoa) to a highly urbanized consumer society. This meant that the region south of the Orinoco was basically left untouched until recently. In the areas close to the cities, growing urban food demand continued to cause forest loss, in particular because of a pronounced Latin American phenomenon: a richer population consumes more meat, and extensive cattle ranching advances. Severe economic and political crises since the mid-1980s have, however, led a new quest for economic diversification, implying that agriculture, timber extraction and mining have expanded, generally to the detriment of forest size and quality.[10]

Ecuador only became a net oil exporter in the 1970s, and its limited reserves may soon be exhausted. Being traditionally divided between the coastal and the highland region, Ecuador used much of its oil wealth to increase market integration by massive road construction. While this strategy indeed helped to alleviate poverty and increase the rural poor's off-farm employment, it also took a heavy toll on forests by opening up new forested areas for colonization, with most areas eventually being converted to pasture. During the crisis of the 1980s, the new road infrastructure and currency devaluations helped to boost growth in the agricultural sector and shrimp farming, both of which caused additional deforestation (Wunder, 1997, 2000a).

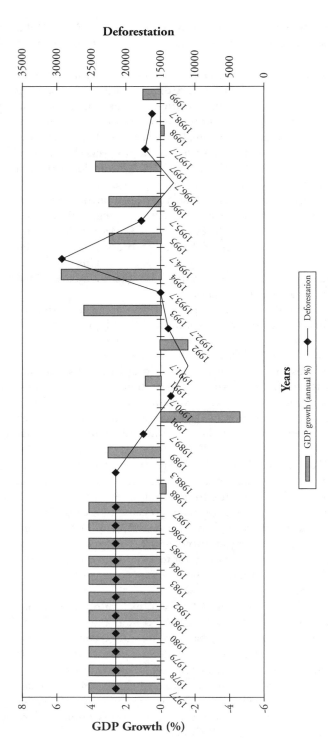

Figure 5.1 *Brazil: Economic growth and Amazon deforestation (in km²) from 1977–1988 to 1998*

Source: INPE data at www.inpe.brlinformacoes_Eventoslamz1998_1999lpagina7.htm (May 20); *Financial Times* (12 April 2000); Brazilian forest logging escalates; World Bank (1998). World Development Indicators; IPEA (2000). *Boletim Conjuntural*, 48.

Note: The timescale of the deforestation series has been adjusted to the fact that measurements, unlike economic growth rates, do not follow the calendar year (most images are from August of the respective year).

The combination of macroeconomic crisis and structural adjustment – including currency devaluation and price liberalization – can often put an additional strain on forests. In the case of Bolivia, devaluation clearly increased timber exports (Kaimowitz et al, 1996). For Cameroon, a marked economic crisis from 1986 onward reversed the rapid urbanization trend during the oil boom; more people settled in the countryside, including the humid forest zone in the south. They also increasingly switched back from commercial crops to slash-and-burn foodcrops such as plantain and tubers. As documented by satellite imagery,[11] crisis-hit Cameroon became an example of the case analysed by Döös (2000) where increased food demand leads to extensification and continuous forest loss.

Hence, macroeconomic wealth may or may not reduce forest loss and degradation, depending on whether price-incentive-reduction effects dominate over capital-endowment expansion effects. One possibility is that the forest outcome is inversely proportional to the fate of land-extensive agricultural sectors. If extensive agriculture or cattle ranching is promoted by an economic crisis and by accompanying government policies, then forest loss will rise. Agriculture often benefits from currency devaluation, price liberalization and increased rural labour supply; it loses typically from cuts in road construction, agricultural subsidies and private investment. So, rising poverty is usually accompanied by price incentives for crop expansion into forested areas, but it also decreases the capital available to implement it. The outcome depends on which effect is strongest.

Microeconomic Poverty – Ambiguous Impacts on Forests

At the microeconomic level of forested areas, similar ambiguities apply, and may be analysed in greater detail. Welfare poverty tends to make cheap labour available, which in turn is useful both to cash in low-return forest rents (degradation) and to undertake labour-intensive investment in forest clearing (deforestation). Reducing welfare poverty makes labour more expensive and, other things being equal, deforestation less attractive. Yet, reduced poverty may go hand-in-hand with higher rural investments that require land clearing. The alleviation of asset-poverty constraints can thus either reduce or augment deforestation.

A classic acceleration example is when asset-poverty alleviation introduces chain-saws – the prime technology to accelerate forest clearing. A counterexample refers to poor farmers in the Brazilian Amazon who, when endowed with the necessary capital, switch from land-extensive annual crops to land-intensive perennials that need far less land clearing (Andersen et al, 1996, Chapter 7). But capital constraints may in general be the single most important impediment to forest clearing at the frontier. Carpentier et al (1999) find in their model for settlers in the western Amazon that a tripling of the farm-gate price in Brazil nuts, the main pro-conservation extractive product, leads to more, rather than less forest clearing. This counterintuitive result occurs because this price change is insufficient to make forest-conservation-cum-nut-extraction a viable alternative to deforestation for cattle ranching. Households therefore welcome extra forest-extraction earnings and reinvest them in cattle – capital-endowment impacts dominate over the relative price effect.

Evaluations of the aforementioned ICDPs and buffer zone projects show that these have had some success in poverty alleviation, but much less so in changing local forest use and promoting conservation.[12] This trend suggests that ICDPs have often been built on a wrong diagnosis of why local communities and households act as they do. It may also indicate that asset constraints dominate, and that investments in frontier areas often attract additional resources (investors, migrants) that end up increasing pressures on forests.

Is it the poorest that make their way out to the forest frontier to colonize new territories in forest-rich areas? For Latin America, the answer mostly seems to be 'no'. The landless without savings or credit access do not have the resources to indulge unassisted in the risky project of frontier deforestation (Rudel, 1993). In many cases, they may actually physically implement the clearing, but are being paid by a capital-endowed landowner or squatter who decides to cut down forested areas. As part of this arrangement, benefit-sharing schemes may give the poor access to harvest wood products, such as logs, charcoal and firewood. Yet, their role in the clearing is instrumental; they are seldom decision makers. Opportunities tend to drive tropical frontier deforestation more than poverty – this holds true in most of Latin America and probably in much of Asia; in Africa, the large role of shifting cultivators in forest clearing may imply a slightly different picture.

On the other hand, many Latin American case studies confirm that forest degradation and deforestation raise income for the responsible agents. Jones (1990) reaches this conclusion by comparing Central American colonization cases. Sunderlin and Rodríguez (1996, p13) found that 90 per cent of immigrant forest-clearing farmers in the broadleaf forest region of Olancho, Honduras, had improved their material living standards, compared to their place of origin, cattle ranching being their prime source of income. Wunder (2000a, Chapter 6) finds pull factors to be decisive for the loss of forest fragments in the Ecuadorian highlands. In Latin American frontier deforestation, most commonly the 'deforestation decision maker' is a small entrepreneur with some financial capital. Although poor by developed country standards, he belongs to a middle class within the hierarchy of his rural society. Besides seeing profits in his risky, commercially orientated enterprise, he often seeks to obtain land rights ('homesteading') through the active use (ie deforestation) of his plot.

A link between forest 'mining' and conversion to micropoverty alleviation may also be observed elsewhere (Lipper, 2000), but the situation may differ slightly in forest-scarce environments, such as the Sahel or the Himalayas. Dry forest areas are generally characterized by a higher population density and by the extensive use of non-timber forest products (NTFPs, firewood etc). Poverty and population growth may here combine in 'immizeration' scenarios. Short-run poverty alleviation is often linked to gradual degradation processes, which negatively affect biodiversity (see also 'discussion section'). Collective degradation of open-access forest resources ('tragedy of the commons') is a frequent scenario here, and it may take special preconditions for collective forest management to work effectively (Ostrom, 1999).

Normally, forest degradation and deforestation reflect two quite different types of land-use decisions, which have a distinct link to poverty. Forest degradation may occur through selective logging, mining, overhunting, overgrazing, non-wood overharvesting, fuelwood collection etc. If the corresponding agent were better off in terms of fewer resource constraints, would (s)he reduce degradation? This may be the case. Often, forest degradation is a rent-collection activity, cashing in a once-and-for-all gain and then moving on in a 'cowboy-economics' logic. If this activity is economically marginal, for example, firewood collection, lower welfare poverty reduces this type of degradation.

Deforestation, in turn, is basically an investment in future land uses. It may actually be an 'overinvestment' with speculative 'homesteading' motives – hoping to gain land titles from forest clearing (Angelsen, 1997, p149). It may contain the rent collection element of nutrient mining for enhancement of soil fertility; asset poverty may impede the use of fertilizers and other inputs, thus favouring forest burning. Again, this connection supports the link observed between poverty and increased slash-and-burn deforestation in Africa. But normally deforestation is led by expected future pay-offs, in terms of income and formal land tenure. Investment involves a decision about the alternative use of financial resources. Poverty, and changes in its level, may be part of the investment criteria (eg for the choice of crops or technologies), but asset poverty's impact on the deforestation decision is ambiguous. It depends on the type of poor people, their changed incentives, their asset constraints and how these assets are linked to forests. This set of factors may entail a far more complicated equation than the Brundtland report would have us believe.

Poverty Profiles and Current Poverty Trends

In the first part of this chapter, the implicit problem was principally a 'Green' one – the impact of poverty on forests, that is, a concern for the environment with the poor in an instrumental position. During the recent decade, however, many international and bilateral donors have explicitly adopted poverty alleviation as their prime overall objective. The question to them is what natural forests, and organizations working with them, can contribute in delivering benefits to poor people and in the general quest to reduce a country's poverty. This section will prepare the ground for a discussion of this issue by identifying broader poverty trends and pathways, and their relation to NRM, including agriculture. Who are the poor in tropical countries, and where do they live? What do they do for a living? What pathways do they use to improve their livelihoods?[13]

The incidence of absolute poverty has over the last three decades been reduced markedly in Asia, slightly in Latin America, but much less so in sub-Saharan Africa. Previously, the poor were concentrated in rural areas, but strong urbanization meant that almost half of the poor now live in cities. Even for rural livelihoods, off-farm income is increasingly important. Diversification of activities and income sources is essential. The controversial 'trickle-down' effects from economic growth to poverty alleviation have actually worked, except for those countries with an extremely unequal asset distribution (Ravallion, 1997; World Bank, 2001).

Agricultural growth and development have been key to poverty alleviation, and so has access to trade and geographical mobility, being vital elements in an overall trend of increasing market orientation (Mellor, 1999). However, people who do not fulfil the requirements to participate in the market economy, for example those living in isolated areas with poor infrastructure and education, are likely to remain poor. The distribution of physical assets, such as land distribution, has often proved less important for poverty alleviation than the quality of these assets, for example market access is more important than the sheer size of a certain piece of land. Access to 'soft' assets (human and social capital) is vital. Education is perhaps the single most cost-effective tool for poverty reduction. Low population growth and smaller family sizes also go hand-in-hand with less poverty. What types of benefits can NRM and agricultural R&D deliver to achieve reductions in poverty? Three different impact types can be distinguished:

- producer benefits (direct impacts);
- consumer benefits (indirect impacts);
- economy-wide labour absorption (derived impacts).

The Green Revolution can be used as an example of all three benefit types. The adoption of high-yield crops and other productivity-enhancing R&D initially produced large producer benefits (direct impact). Nevertheless, with the general diffusion of new technologies among farmers, a large increase in overall supplies to food markets occurred. Falling food prices reduce producer benefits, so most benefits from technological innovation were eventually captured by urban consumers (indirect benefits). Price reductions were not equal across products, so staple-crop substitution, such as the dramatic rise in Latin American rice consumption, also made food more affordable to the poor, and thus alleviated urban poverty significantly. Finally, making the basic food basket cheaper will also put downward pressure on nominal salaries. This reduces labour costs and increases labour absorption. An economy-wide reduction in unemployment can reduce poverty significantly,[14] as has been observed in Asia. Based on these effects, four predominant pathways out of rural poverty can be identified:

1   the multiple-activity path (rural diversification into non-farm activities);
2   the rural-urban migration path (labour movement and reallocation);
3   the agricultural path (the traditional rural development model);
4   the assistance path (aid, transfers and public investments).

Pathway 4 (aid) is a policy option complementary to the remaining three development options; pathway 1 (off-farm diversification) lies beyond the scope of NRM and agricultural R&D. This leaves pathways 2 (migration) and 3 (rural development) as the fields of action that have our interest. The main poverty-alleviation impact of agricultural R&D has been to support pathway 2 – making the urban poor markedly better off and increasing labour absorption. Indirect and derived benefits have thus been more important than the gains reaped by the poor remaining in rural areas.

Natural Forests for Poverty Alleviation – A Limited Potential

In this section, I use both the poverty profiles/trends, R&D benefit types and poverty-reduction pathways to provide some ideas about forests' potential to alleviate poverty, compared to other land uses. What are the main forest benefits valued by the poor? How much do they matter, and what poverty pathways can forest products contribute to?

In terms of poverty trends and profiles, increasing urbanization and off-farm diversification have further detached most forest-rich areas and their poor people from poverty reduction. Less and less tropical forest cover remains around cities and densely populated areas in the developing world, because of forest degradation and conversion to other uses. Forest-rich areas are thus concentrated in remote areas, often with poor transport access. But this implies that people living there have fewer off-farm income opportunities (except for logging and mining); they cannot commute to the cities. Their options of participating in a booming urban economy are limited to the decision to migrate. Moreover, low presence of the state in remote areas implies that access to health and education is very restricted (low social and human capital). Thus, geographical and economic marginalization become mutually reinforcing. Finally, when trade creates agricultural 'trickle-down' in remote areas, expansion in cropped areas will often fuel forest loss. In fact, some of the agricultural technologies particularly targeted to the poor in marginal production areas (restricted soils, rain-fed hillsides etc) may also be those that are most likely to promote strong 'win-lose' deforestation trade-offs.[15]

What benefits can forest R&D and innovation generate? The categories from above are:

- producer benefits in forested areas (direct impacts);
- forest product consumer benefits (indirect impacts);
- economy-wide labour absorption (derived impacts).

How large are the potentials for increasing overall benefits from natural forests by raising productivity and improving NRM? This is difficult to answer in depth because tropical forests continue to provide new products and genetic information to the market economy. Economists have extreme difficulties in converting these 'option values' to monetary figures. Concentrating on currently marketed products, logging is the most important commercial activity in tropical forests. The GDP share of forestry only becomes significant, however, in exceptional cases (Whiteman, 2000); by the mid-1990s, it only exceeded 15 per cent in nine developing countries (FAO, 1997, p36). Timber extraction, transport and processing offer large options for increasing efficiency that are presently forgone because continued open-access and low-priced resources do not favour technological innovation. Many secondary timber species are only accepted in the marketplace when traditional species become scarce. Improved NRM could eventually increase efficiency and benefits in a variety of ways. But one should not expect a future Green Revolution in tropical forests; value increases will most likely be inferior to the ones observed for agriculture in the past.

What about timber and the poor? Most tropical timber is produced by (non-poor) companies, but some is also produced by small-scale extractors, many of which use timber as an income source supplementary to farming. Their logging is often simple, adhoc, geographically dispersed and only slightly favourable to the adoption of new technologies. Furthermore, logging and wood-energy extraction (commercialization of firewood and charcoal) tend to cause forest degradation, especially in forest-scarce environments, or are direct by-products of forest conversion. Small-scale wood producers are either reluctant to embrace technological change,[16] or the technologies they adopted are poverty reducing but forest-loss accelerating (chainsaws). In the aggregate, the prospects for generating significant pro-poor growth from logging seem dim. Indicative for this view are the words of Jack Westoby, one of the fathers of social forestry and originally an optimistic advocate of forest-led development in the 1960s. He later had to revise his view:

> as yet, forest industries have made little or no contribution to socioeconomic develop-
> ment in the under-developed world – certainly not the significant contribution that was
> envisaged for them a couple of decades ago (Westoby, 1987, pp246–247).[17]

Perhaps redistribution of existing revenues from logging companies to poor forest dwellers is more promising for poverty alleviation than growth. Payments to local communities already occur in many logged-over tropical forests, formally or informally, by law or negotiation, but their size is generally small. Rents in the timber business can be high, so if decentralization and devolution of property rights to communities succeed in redistributing just a minor amount, poverty-alleviation potentials could be significant. High timber rents are often associated with corruption. Preliminary evidence suggests that corruption is particularly harmful to the poor (Thomas et al, 2000, pp146–150), so improved governance in forestry may help reduce poverty.

The presence or emergence of supra-normal forest rents tend, however, to attract powerful outsiders that seek control over these rents (Dove, 1993; Fisher, 2001, p6). Large rents thus set in motion a political economy that generates a skewed distribution. A recent book by Ross (2001) describes how the timber booms in Malaysia, the Philippines and Indonesia undermined forestry institutions through both rent-seeking by logging firms and rent-seizing by politicians who gained the right to allocate rents. Natural resource-rich countries thus tend to have a high inequality. This is not meant to discourage those that fight for a more equitable distribution of forestry rents, but it is an uphill political struggle against the vested interests of the powerful.

Nonwood forest products (NWFPs) offer more examples of targeted benefits for poor producers, and their extraction tends to have less ecological impact than that of wood. The literature on community forestry holds some references to multiple-product forest management particularly benefiting the poor. But typically the Achilles heel of NWFPs as alleged cornerstones of livelihood strategies is their poor economic potential, especially in biologically diverse tropical forests. The global overview by Neumann and Hirsch (2000) shows that specialized NTFP extractors in the tropics are among the poorest people in the world, but that they mostly practise extraction for the lack of better alternatives. The following is a brief sketch of the situation for three prototypes of potentially poor forest-user groups.

The first prototype is native indigenous populations using large, remotely located areas for the extensive extraction of forest products, mainly for autoconsumption. Often, their forest use is sustainable as long as their territories are not encroached upon, if population density remains very low, and if they have few incentives to scale up their extraction activities, due to limited market linkages. But it is difficult to see how forest research can help them, other than by aiding to secure their land rights – a 'defensive' microintervention which can also be controversial vis-à-vis society's overall poverty-alleviation objectives.[18]

Another group of poor NWFP extractors refers to smallholders living in forest margins that derive their livelihoods from agriculture, with forest extraction as a supplementary (monetary and nonmonetary) income source. To them, forests are most important as sources of land and nutrients for slash-and-burn agriculture. These systems often become unsustainable when population densities grow. NRM practices applied often imply gradual forest degradation (eg overharvesting of game) and conversion (crop expansion). The importance of NWFPs in this setting may reach around 10–25 per cent of household income. It has been widely recognized that NWFPs are important for food security, for example in the case of failed crop harvests and seasonal gaps. Nevertheless, as one review concludes:

> NTFP extraction activities seem primarily to make up shortfalls in income rather than provide a path to socioeconomic advancement. NTFP extraction is often less about 'poverty alleviation' ... than about basic survival for the poor (Neumann and Hirsch, 2000, p34).

Finally, more specialized NWFP producers extract a limited range of products for commercialization in urban markets (rubber, honey, resins, nuts etc). In some cases, they find a profitable market niche, which allows them to derive exceptionally competitive income from their activity.[19] Helping them to manage and commercialize their products may be a good target for poverty alleviation for forest R&D. But as the example of the Brazilian extractive reserves shows, profitability may depend on implicit price subsidies, and can be highly fluctuating over time. Demand for extracted products may follow certain stages of commercial development. Forest supply often cannot respond adequately to rising market demand, so extractors may gradually either overharvest forest resources, remove it from natural forests by domestication, or synthetic substitutes may emerge (Homma, 1996). Forest-dependent extractors often respond by diversifying into agriculture to escape poverty.

One basic problem of producers is poor returns to labour in remote areas with large transport costs. Low frequency of commercial species in biodiversity-rich tropical forests is another drawback. For instance, Brazil nuts are harvested from 2–3 million ha in the Amazon, but only at an average of one tree per ha, implying that the returns from nut extraction in Southern Pará (Brazil) in 1995 were only US$7/ha/yr (Homma et al, 1996, p529, Clay, 1997). In the entire Brazilian Legal Amazon region, NWFP extraction values registered in the Brazilian agricultural census are extremely concentrated on few products and regions. Most high-value products (assai palm hearts and fruits, babassu kernels, piassava fibres) come from market-near areas that are often ecologically disturbed, for example due to previous burning and/or deforestation. In comparison, the market-remote but biologically diverse terra firma rainforests generate negligible NWFP income (Wunder, 1999).

So far, direct producer benefits have been addressed. The last section actually pointed to indirect benefits from R&D and technological innovation to urban consumers as a more important tool for poverty alleviation, especially when a country's urbanization has reached an advanced stage. What role are forest products likely to play for the urban poor? Can cheaper timber, wood-energy and NWFP supply help poor urban consumers to overcome welfare constraints?

In many tropical developing countries, national markets for timber and wood-based products are evolving rapidly. But this rise occurs more in the transition to middle-income markets than in the poorest countries (FAO, 1997, p6). This is also true at the cross-section level. Timber is bound to have a negligible weight in the consumption basket of those earning less than US$1 per day. Wood uses for furniture, construction, pulp and paper and so on are mostly appealing to middle- and high-income groups. They would hence reap the potential consumer benefits from technological advances in forest production leading to cheaper supplies.

The reverse is obviously true for firewood, 'the poor man's energy', which is gradually replaced by fossil fuels when income rises. In urban areas, however, wood substitution even occurs in poor environments, due to high transport costs and the inconvenience of storing and burning a large amount of biomass in an urban space. This generally seems to be the case in Latin America and Asia, although less so in Africa. Firewood's close relative, charcoal, shares some of these features but is often used by higher-income groups, for example in barbecues or in broiler restaurants. For both woodfuels, the following general strategy would seem economically rational: First, harvest cheap wood from peri-urban natural forests in an 'unsustainable' way (forest degradation); when transport costs become prohibitive, convert gradually to fossil fuels and/or to fast-growing plantation supplies established in peri-urban areas. Both processes can frequently be observed in developing countries. They imply that woodfuel consumption does help the urban poor, but only in the transition from forest abundance to scarcity, and eventually with little scope for technological progress.

Urban consumption of NWFPs is heterogeneous, involving products favoured both by high- and low-income groups. Many NWFPs are, however, economically 'inferior', in the sense that they are replaced by synthetic or agricultural substitutes when consumers get richer, and vice versa during crisis periods. For instance, in a Latin American consultation on NWFPs, it was noted that the economic crisis of the 1980s, by imposing strict foreign exchange limits on import capacity, had caused a revival of many of those NWFPs that constituted import substitutes (FAO, 1996). Some products have region-specific importance for the consumption basket of the urban poor, for instance bush-meat in Central and West Africa. In the aggregate, the urban poor may reap special gains from certain NWFPs, but in most cases, the budget share of these products is too limited to make a marked difference for their livelihoods: the bulk of the natural resources they consume tends to come from cultivated production systems.

Finally, R&D-generated benefits and new technologies could also have an economy-wide impact, for example by promoting labour-intensive growth paths that ease unemployment and reduce poverty. One prerequisite is that the economic activity actually is sufficiently large to make an economy-wide impact. We have seen above that this condition is seldom met. A second prerequisite would be that forestry was more labour-intensive than other commodity-producing sectors. This assumption is even more dubious. Forestry tends to be a rather capital-intensive activity, in particular compared to agriculture.[20] Whiteman (2000) offers some figures and concludes that 'formal forestry employment is probably not a major contributor to rural incomes except, perhaps, in a few small areas' (p15). For economy-wide employment generation, forestry is less attractive than agriculture and urban-based sectors, but more important than oil, mining and other enclave sectors. Most important for the poor is probably the reinvestment of timber rents in other sectors, an area which has been underresearched.

Conclusion

This paper attempts to provide inputs for the analysis of the two-way causal links between poverty alleviation and natural tropical forests – that is the options relating to their conservation, management, degradation or conversion. Experiences differ according to factors such as geographical regions, forest types, the degree of forest abundance and the wealth of forest resources. The general picture indicates, however, that there may be fewer 'win-win' synergies between natural forest conservation and poverty alleviation than those of us who are sympathetic to forest conservation would prefer.

Poverty can be defined in both welfare and asset terms, but the latter is likely to have more explanatory power vis-à-vis the fate of forests. Deforestation, that is a radical removal of tree cover, is mostly an investment in future land uses, and as such is quite different from forest degradation (logging, overharvesting, etc). At the microlevel of forested areas, reduced poverty can cause either more or less deforestation. At the macrolevel, less poverty also has an ambiguous effect on forests, but in the initial stages of the forest transition, higher income is likely to fuel crop-land demand and forest conversion. For the reverse causality, the potential for forest-led development to alleviate a country's poverty tends to be limited. Current benefits from natural forests to poor rural producers may be safeguarded, as a 'defensive' strategy, but it is difficult to raise benefits in a sustained manner, except perhaps redistributing timber-profits and promoting selected commercial NWFPs. Urban consumer benefits, an area of growing poverty relevance, are likely to be even more restricted from forests, and will seldom favour the poor. Likewise, the economy-wide absorption of (poor) unskilled labour is not helped much by forestry, which is a relatively capital-intensive sector.

A general conclusion is that, in most settings, natural forests tend to have little comparative advantage for the large-scale alleviation of poverty, especially compared to their great land-use competitor, agriculture. Like other 'wildlands', their advantage is that nature provides for multiple commodities that can be harvested without capital inputs and human production efforts. Their disadvantage is that extraction is usually labour-intensive, land-extensive and supply is inflexible vis-à-vis demand changes. This becomes important in areas where 'development' is advancing, in the sense that more production factors and infrastructure per land unit are becoming available. It can be synthesized in an overall, rather pessimistic hypothesis:

> In a landscape designed exclusively for the objective of poverty alleviation, where a flexible supply of labour, capital and access to physical infrastructure (roads, rail, ports) and alternative land use options exist, there will be little place for natural forests.

Two underlying assumptions accompany this hypothesis on the transition towards a 'full world' scenario. One is that alternative land uses actually exist, such as agriculture, cattle ranching, but also agroforestry or plantation forests, where the opportunity costs of conservation are greater than zero. This probably applies to most tropical areas.

The second premise is that environmental services at the global (carbon, biodiversity conservation etc) and national/regional level (ecotourism, hydrological benefits etc) are uncompensated, so that they generate no income to local land-use decision makers, and thus have no bearing on poverty. Unfortunately, this assumption also reflects current reality quite well. The current 'free-rider' character of forest ecoservices is detrimental both to the forest-dwelling poor (who lose a potential income), to global interests (who irreversibly lose environmental assets), but also to the tropical poor outside of the forest. For instance, global warming, which deforestation contributes to, is expected to hit poor people in tropical countries particularly hard, because of water shortages, declining crop yields and greater disease-proneness (IPCC, 2001). Yet, counterexamples of poverty-alleviating forest services can be built upon in the search for 'win-win' niches. For instance, ecotourism has proved in a number of cases to provide both successful 'trickle-down' to local forest dwellers and direct incentives to protect, if not large frontier forests, than at least valuable forest fragments.[21]

Of course, at present landscapes are not all 'designed exclusively for the objective of poverty alleviation', nor for satisfying human needs in a broader sense. There are places that are quite close to that premise, such as the island of Java or Bangladesh, whereas others are far from that, such as the Guyanas or Gabon. The general point is that future population and consumption growth implies that more and more areas are approaching that stage, so that pressures on forests will continue. Natural tropical forests have not evolved in a way that makes them suitable to serve human needs exclusively. In a static sense, forests may be very important safety nets for poor people's food security and subsistence needs ('defensive' objectives). In a transition phase, they may also be important sources of rent extracted and reinvested in sectors with greater production potentials, allowing for a diversification of income sources. But in cases where people are prevented from diversification, for example by debt peonage, contracts or lack of alternatives, forest products may actually turn from 'safety nets' into 'poverty traps' (Browder, 1992; Neumann and Hirsch, 2000, pp32–43).

To the extent that 'development' empowers humans to dominate landscapes for their own purposes, they will inevitably seek to modify natural ecosystems and domesticate its resources, in order to produce more specialized outputs that meet human demands. These modifications, beyond a certain threshold, are bound to cause declines in the biological diversity of natural ecosystems, until the point at which they cease to be natural and instead become home gardens, tree plantations, crop fields or pastures. Conversely, the conservation of biologically rich but commercially poor natural forests, and of associated modes of their human exploitation, is seldom conducive to the 'assertive' objective of poverty alleviation.

At present, what protects the remaining natural tropical forests from a sad fate is not that they provide an overwhelming flow of products for poverty alleviation, triggering active and robust human conservation efforts. On the contrary, their present state is rather due to features of 'underdevelopment' that have hitherto excluded these areas from (full) integration into national economies, and left them passively behind as areas not (yet) profitable to exploit and convert. These features include low population density (inelastic labour supply), lacking access to credit and to physical infrastructure, rugged topography, limited soil potentials, tropical diseases or tribal warfare – it is these, from a human viewpoint, grim features that tend to be the silent but gradually weakening protectors of tropical forests. Unfortunately, these factors are actually more correlated to sustained human poverty than to its alleviation.

Discussion and Policy Implications

In general, many of the quantitative relations stipulated in this chapter would need to be further verified and documented by future research. It draws primarily on Latin American examples, although Fisher (2001) reaches surprisingly similar conclusions in his overview of Asian experiences. Some observers claim that dry forests offer better poverty-alleviation potentials, especially in India.[22] But these studies equally seem to refer mostly to 'static-defensive' functions of protecting poor people's forest access, rather than options of reducing poverty. They also admit that the poor's current uses tend to degrade dry forests over time.

Nevertheless, this paper is not meant to belittle those efforts that in a specific micro-setting simultaneously achieve forest conservation and poverty alleviation. In some cases, forests can definitely alleviate poverty for specific groups and regions at the subnational level, implying that the scale of analysis is also quite important. Historical cases from the northern hemisphere (Canada, Sweden, Finland) suggest that forestry-led poverty reduction is possible at the national level, though probably conditioned by low rural population densities, poor agricultural soils and biodiversity-reducing forest modifications. The considerations presented in this paper should thus not discourage, but serve to downscale unrealistic expectations on the forest vs poverty-alleviation synergies, and on the prospects of 'sustainable forest development' based on local benefit generation. This also means that compensation schemes for global/national/regional forest services, and the search for ways to implement them successfully, may deserve much more attention.

If we take the main points from this paper at face value, what are the implications for forest conservation strategies and poverty alleviation efforts, respectively?

As a donor interested only in poverty alleviation, and streamlining my activities to serve this single purpose, I would not put much of my money into natural forests. Education, health and institutional capacity-building would be my favourite areas of intervention, but part of the funding would also go into agriculture, thereby implicitly accepting that higher demand for agricultural land might fuel deforestation. I would also allocate a minor share of my funds for projects at the forest margin and in extensive, forest-rich areas. NRM, including stakeholder dialogue, distribution of timber rents and so on, would have top priority on this sublist, together with commercial NWFPs. But being honest with myself, I would not expect to achieve much in reducing sizeable target groups' poverty at a substantial scale. My aim would rather be defensive, that is to avoid a rise in poverty which may otherwise occur for these marginal groups. I would then, under the hard pressure from my worst critics, have to admit that I had not been thinking only about poverty, but also about other nice things, such as environmental ethics, existence values, human rights, democratic principles and the preservation of cultural and ethnic diversity.

As a large NGO with a mandate exclusively in natural forests and biodiversity conservation, I would downscale my expectations and change the design of my ICDPs. I would ensure that my consultants, project managers and associated researchers looked with much more vigour into the factual links between resource constraints and emerging environmental pressures. I would be extremely suspicious if they told me that forests were felled because the local people are poor. In my project portfolio, I would cut back on large-scale planning exercises and people-centred projects, where we recently actually had considered bringing more people into national parks to make our projects look more participatory. I would let the pendulum swing somewhat back to pure conservation activities, including acquiring biologically rich forested land areas in remote regions. I would, however, refresh this revisionist approach with a variety of experiments on compensation schemes and types of conservation contracts with different stakeholders, mostly at the local, but also at the more aggregate levels. I would generally try to make my funded activities contingent upon well-defined concepts of conservation conditionalities, which I would be able to monitor over time. I would support the conservation efforts of private forest owners, even if they should turn out to be much less poor than the squatters that try to encroach on their lands. I would pay special attention to key extra-sectoral policy decisions; for example I would lobby strongly against road building into forested areas, even when poor local dwellers are much in favour of it.

By the end of the day, I would worry much about my home constituency, and discover that I had lost a few friends. I would probably note that my funding from both bilateral donors and civil society sources had declined considerably, and that citizens and politicians alike had found my old but generally unsuccessful 'win-win' projects much more worthy of support.

Acknowledgements

I am grateful for comments on earlier drafts from Arild Angelsen, Bruce Campbell, Mafa Chipeta, Wil de Jong, Robin Grimble, David Kaimowitz, Jeffrey Sayer, Joyotee Smith, Laura Snook and participants of the 'Biodiversity for Poverty Alleviation' workshop at the 15th Global Biodiversity Forum, held on 12–14 May 2000 in Nairobi, Kenya. The ideas expressed are the sole responsibility of the author and do not represent an official view on behalf of CIFOR. Final revision accepted: 6 June 2001.

Notes

1   Reardon and Vosti (1995) use the term 'investment poverty', which has a more restricted functional scope.
2   See Fields (1980) for an extensive discussion and comparison of poverty and inequality measures.
3   Efforts to quantify the number of forest-dependent people (eg Calibre Consultants and SSC, 2000) have been frustrated by the lack of adequate data and by blurred criteria for how much forest use justifies the term 'dependency'. For the time being, it may be preferable to simply keep in mind a typology of poor forest users, as proposed by Byron and Arnold (1999).
4   Townson (1994), Shepherd et al (1999) and Warner (2000) provide overviews on this issue.
5   See, for instance, Balée (1992). Nevertheless, their de facto forest conservation may not so much be due to their being ecologically 'noble savages' (Redford, 1990) as to

low population densities, resource abundance and 'asset poverty' restricting their ability to transform the forest environment.

6 The national deforestation figures published by FAO are for generally model-based extrapolations of one or two national forest assessments. The decisive change factor in the FORIS model is population growth, which globally has slowed down slightly from the 1980s to the 1990s. An actual measurement of deforestation trends over time is currently not available, but may be produced in the future by existing research projects (TREES, NASA-Pathfinder etc), aiming at the interpretation of historical satellite images.

7 On forest transition processes, see eg Mather (1996), Grainger (1999) and Rudel (2001).

8 The following three paragraphs draw on my current comparative research project at CIFOR, 'Oil wealth and the fate of tropical forests'.

9 Pourtier (1989a, b) gives an excellent account of the adjustment to oil wealth in Gabon.

10 Miranda et al (1998) describe forest threats arising from responses to Venezuela's economic crisis.

11 See Ndoye and Kaimowitz (2000), Sunderlin et al (2000) and Mertens et al (2000).

12 See Sayer (1991), Wells and Brandon (1992) and Gilmour (1994).

13 In addition to the referenced literature, this section draws on presentations from the international conference 'Assessing the Impact of Agricultural Research on Poverty Alleviation', 14–16 September 1999, San José, Costa Rica.

14 Of course, in the long run cheaper foodstuff can also fuel population growth, which may counteract the reduction in the absolute number of poor.

15 One example is the development of new cassava varieties, adaptable to many types of marginal areas (CIAT, 1999) which are promoting frontier crop expansion, often at the cost of forest loss.

16 One case in question is the poor success in introducing improved cooking stoves and energy-saving charcoal kilns in rural areas, both basically aimed at reducing wood consumption. Often, the main problem is that producers do not see a rationale for adopting complex technologies to economize on a wood resource they still perceive as abundant.

17 This quote is from Westoby's speech to the 8th World Forestry Congress, Jakarta, 1978, reprinted in Westoby (1987, pp241–254). His early, optimistic essay from 1962, 'The role of forest industries in the attack on underdevelopment' is reprinted in Westoby (1987, pp3–70).

18 For instance, this will be the case if in an area of high population density poor, land-hungry squatters are denied access to the colonization of extensively used indigenous lands; in other words, the argument will apply when agricultural expansion is a main potential pathway of poverty alleviation for migrants.

19 For instance, Monela et al (1999) find for six Tanzanian villages that honey alone made up one third of households' cash income, while all non-timber products (including charcoal) averaged 58 per cent of monetary income.

20 A counterexample is Gabon, where the timber sector in fact is a significant source of employment and labour absorption. This is because the country has an extremely

low population density, its agriculture is neglected and the alternative commodity-producing sectors (mining, oil, industry) are even more capital-intensive.

21 For instance, see Zimbabwe Trust, The Department of National Parks, and CAMP-FIRE Association (1994) on the CAMPFIRE project in Zimbabwe; Gurung and Coursey (1994) on the Annapurna Conservation Area Project in Nepal; and my own research on Brazil (Wunder, 2000b) and Ecuador (Wunder, 2000c).

22 For example, see Reddy and Chakravarty (1999); Bahuguna (2000); Kumar et al (2000) for a national overview.

References

1 Andersen, E. L., Granger, C. W. J., Huang, L. L., Reis, E. and Weinhold, D. (1996) *Report on Amazon Deforestation*. Discussion Paper 9640, Department of Economics, University of California, San Diego.

2 Angelsen, A. (1997) *The Poverty-Environment Thesis: Was Brundtland Wrong?* Forum for Development Studies, no 1.

3 Bahuguna, V. K. (2000) 'Forests in the economy of the rural poor: An estimation of the dependency level'. *Ambio*, vol 29, no 3, p126.

4 Balée, W. (1992) 'Indigenous history and Amazonian biodiversity', in Steen, H. K. and Tucker, R. P. (eds) *Changing Tropical Forests. Historical Perspectives on Today's Challenges in Central and South America*. The Forest History Society, USA.

5 Browder, J. O. (1992). 'The limits of extractivism: Tropical forest strategies beyond extractive reserves'. *BioScience*, vol 42, pp174–182.

6 Byron, N. and Arnold, M. (1999) 'What futures for the people of the tropical forests?' *World Development*, vol 27, no 5, pp789–805.

7 Calibre Consultants and The Statistical Services Centre (SSC) (2000) *Numbers of Forest Dependent People*. A feasibility study for DFID's forestry research programme. University of Reading, Reading, UK.

8 Carpentier, C. L., Vosti, S. A. and Witcover, J. (1999) *Impacts of Subsidized Brazil Nut Prices on Deforestation, Use of Cleared Land and Farm Income*. Technical Note 8.1, University of California at Davis, Davis, CA.

9 Cavendish, W. (1997) *The Economics of Natural Resource Utilisation by Communal Area Farmers of Zimbabwe*. PhD thesis, Economics, Social Studies Faculty, University of Oxford.

10 CIAT (1999) *Pathways Out of Poverty*. CIAT in perspective. CIAT, Palmiras, Colombia.

11 Clay, J. W. (1997) 'Brazil nuts. The use of a keystone species for conservation and development', in Freese, C. H. (ed) *Harvesting Wild Species. Implications for Biodiversity Conservation*. Johns Hopkins University Press, Baltimore.

12 Doos, B. R. (2000) 'Increasing food production at the expense of tropical forests'. *Integrated Assessment*, vol 1, pp189–202.

13 Dove, M. (1993) 'A revisionist view of tropical deforestation and development'. *Environmental Conservation*, vol 20, no 1, pp17–24.

14 Eckholm, E., Foley, G., Barnard, G. and Timberlake, L. (1984) *Fuelwood: The Energy Crisis That Won't Go Away*. Earthscan, London.

15 FAO (1996) *Report from the Meeting of Latin American Experts on Non-timber Forest Products*. Food and Agricultural Organization of the United Nations, Rome.

16 FAO (1997) *State of the World's Forests*. Food and Agricultural Organization Nations, Rome.

17 Fields, G. S. (1980) *Poverty, Inequality and Development*. Cambridge University Press, Cambridge.

18   Fisher, R. J. (2001) *Poverty Alleviation and Forests: Experiences from Asia*. RECOFTC, Kaset-sart University, Bangkok: www.RECOFTC.org/publications_at_recoftc.htm
19   Gilmour, D. A. (1994) *Conservation and Development – Seeking the Linkages*. International Symposium on Management of Rainforests in Asia, University of Oslo, 23–26 March.
20   Grainger, A. (1999) The National Land-Use Transition and Technological Change. CIFOR workshop on 'Technological Change in Agriculture and Deforestation'. Turrialba, Costa Rica, 11–13 March.
21   Gurung, C. P. and Coursey, M. D. (1994) *Nepal, Pioneering Sustainable Tourism. The Annapurna Conservation Area Project: An Applied Experiment in Integrated Conservation and Development*. The Rural Extension Bulletin, no 5, August, University of Reading.
22   Homma, A. K. O. (1996) 'Modernisation and technological dualism in the extractive econ-omy in Amazonia', in Perez, M. R. and Arnold, J. E. M. (eds) *Current Issues in Non-Timber Forest Products Research*. CIFOR, Bogor, Indonesia.
23   Homma, A. K. O., Walker, R. T., Carvalho, R. A., Conto, A. J. and Ferreiras, C. A. P. (1996) 'Razoes de risco e rentabilidade na destruicao de recursos florestais: o caso de castanhais em lotes de colonos no sul do Para'. *Rev. Econ. Nord. Fortaleza*, vol 27, no 3, July/September.
24   IIED (1995) *The Hidden Harvest: The Value of Wild Resources in Agricultural Systems. A Sum-mary*. International Institute for Environment and Development, London.
25   IPCC (2001) *Climate Change 2001: Impacts, Adaption and Vulnerability. Summary for Policy-makers*. Intergovernmental Panel on Climate Change (WGII), available at: www.ipcc.ch/
26   Jones, J. R. (1990) *Colonization and Environment. Land Settlement Projects in Central Amer-ica*. United Nations University, Tokyo.
27   Kaimowitz, D., Thiele, G. and Pacheco, P. (1996) *The Effects of Structural Adjustment on Defor-estation and Forest Degradation in Lowland Bolivia*. CIFOR, Bogor, Indonesia.
28   Kaimowitz, D. and Angelsen, A. (1998) *Economic Models of Tropical Deforestation. A Review*. Center for International Forestry Research, Bogor, Indonesia.
29   Kumar, N., Saxena, N., Alagh, Y. and Mitra, K. (2000) *India Alleviating Poverty through For-est Development. Evaluation Country Case Study*. The World Bank, Washington, DC.
30   Lipper, L. (2000) 'Forest degradation and food security'. *Unasylva*, vol 51, no 202, pp24–31.
31   Mather, A. (1996) *South-North Challenges in Global Forestry*. CIFOR/UNU-WIDER work-shop on Theory and Modeling of Tropical Deforestation, Indonesia.
32   Mellor, J. W. (1999) *Pro-poor Growth. The Relation between Growth in Agriculture and Poverty Reduction*. Paper prepared for USAID/G/EGAD.
33   Mertens, B., Sunderlin, W. D., Ndoye, O. and Lambin, E. F. (2000) 'Impact of macro-economic change on deforestation in South Cameroon: Integration of household survey and remotely-sensed data'. *World Development*, vol 28, no 6, pp983–999.
34   Miranda, M., Blanco-Uribe, A., Hernandez, Q. L., Ochoa, J. and Yerena, G. E. (1998) *All that Glitters is Not Gold. Balancing Conservation and Development in Venezuela's Frontier For-ests*. World Resources Institute, Forest Frontier Initiative, Washington, DC.
35   Monela, G. C., Kayembe, G. C., Kaoneka, A. R. S. and Kowero, G. (1999) *Household Live-lihood Strategies in the Miombo Woodlands. Emerging Trends*. Sokoine University of Agriculture.
36   Myrdal, G. (1957) *Economic Theory and Underdeveloped Regions*. Gerald Duckworth, London.
37   Ndoye, O. and Kaimowitz, D. (2000) Macro-economics, markets and the humid forests of Cameroon, 1967–1997. *Journal of Modern African Studies*, vol 55, no 2, pp225–253.
38   Neumann, R. P. and Hirsch, E. (2000) *Commercialisation of Non-timber Forest Products: Review and Analysis of Research*. CIFOR and FAO, Bogor and Rome.
39   Ostrom, E. (1999) *Self-governance and Forest Resources*. CIFOR Occasional Paper no 20, Center for International Forestry Research, Bogor.
40   Pourtier, R. (1989a) *Le Gabon. Tome 1: Espace-histoire-société*. L. Harmattan, Paris.
41   Pourtier, R. (1989b) *Le Gabon. Tome 2: Etat et Développement*. L. Harmattan, Paris.

42   Ravallion, M. (1997) 'Good and bad growth: The human development reports'. *World Development*, vol 25, no 5, pp631–638.
43   Reardon, T. and Vosti, S. A. (1995) 'Links between rural poverty and the environment in developing countries: Asset categories and investment poverty'. *World Development*, vol 23, no 9.
44   Reddy, S. R. C. and Chakravarty, S. P. (1999) 'Forest dependence and income distribution in a subsistence economy: Evidence from India'. *World Development*, vol 27, no 1, pp1141–1149.
45   Redford, K. H. (1990) 'The ecologically noble savage'. *Orion Nature Quarterly*, vol 9, no 3.
46   Reed, D. (ed) (1992) *Structural Adjustment and the Environment*. WWF International, Westview Press, Boulder.
47   Ross, M. L. (2001) *Timber Booms and Institutional Breakdown in Southeast Asia*. Cambridge University Press, Cambridge.
48   Rudel, T. K. and Horowitz, B. (1993) *Tropical Deforestation: Small Farmers and Land Clearing in the Ecuadorian Amazon*. Columbia University Press, New York.
49   Rudel, T. K. (2001) 'Can the green revolution save the "green mansions"? The impact of increasing agricultural productivity on forest cover in the American south, 1935–1975', in Angelsen, A. and Kaimowitz, D. (eds) *Agricultural Technology and Tropical Deforestation*. CAB International, Wallingford.
50   Sayer, J. A. (1991) *Rainforest Buffer Zones: Guidelines for Protected Area Managers*. IUCN, Gland.
51   Sayer, J. A. (1995) *Science and International Nature Conservation*. Occasional Paper no 4, Center for International Forestry Research (CIFOR), Jakarta.
52   Schmidt, R., Berry, J. K. and Gordon, J. C. (eds) (1999) *Forests to Fight Poverty: Creating National Strategies*. Yale University Press, New Haven.
53   Shepherd, G., Arnold, M. and Bass, S. (1999) *Forests and Sustainable Livelihoods*. A contribution to the World Bank Forest Policy review process, September 1999. Internet document: wbln0018.world-bank.org/essd/forestpole.nsf/MainView ('Other stakeholder documents').
54   Sunderlin, W. D. and Rodriguez, J. A. (1996) *Cattle, Broadleaf Forests and the Agricultural Modernization Law of Honduras. The Case of Olancho*. CIFOR Occasional Paper no 7 (E), CIFOR, Bogor.
55   Sunderlin, W. D., Ndoye, O., Bikie, H., Laporte, N., Mertens, B. and Pokam, J. (2000) 'Economic crisis, small-scale agriculture and forest cover change in Southern Cameroon'. *Environmental Conservation*, vol 27, no 3, pp284–290.
56   Thomas, V., Dailami, M., Dhareshwar, A., Kaufmann, D., Kishor, N., Lopez, R. and Wang, Y. (2000) *The Quality of Growth*. Oxford University Press and World Bank, Oxford and New York.
57   Townson, I. M. (1994) *Forest Products and Household Incomes. A Review and Annotated Bibliography*. Oxford Forestry Institute, Department of Plant Sciences, University of Oxford, Oxford.
58   Warner, K. (2000) 'Forestry and sustainable livelihoods'. *Unasylva*, vol 51, no 202, pp3–12.
59   WCED (1987). *Our Common Future*. World Commission on Environment and Development. Oxford University Press, Oxford.
60   Wells, M. and Brandon, K. (1992) *People and Parks: Linking Protected Area Management with Local Communities*. IBRD/The World Bank, Washington, DC.
61   Westoby, J. (1987) *The Purpose of Forests. Follies of Development*. Basil Blackwell, Oxford.
62   Westoby, J. (1989) *Introduction to World Forestry*. Basil Blackwell, Oxford.
63   Whiteman, A. (2000). *Promoting Rural Development through Forestry Policy: Some Experiences from Developing Countries*. Presentation to the seminar: The Role of Forests and Forestry in Rural Development – Implications for Forest Policy, 5–7 July, Vienna.

64 World Bank (2001) *World Development Report 2000–2001: Attacking Poverty*. Oxford University Press, Oxford.

65 WRI (1994) *World Resources 1994–1995. A Guide to the Global Environment*. Oxford University Press, WRI, Oxford.

66 Wunder, S. (1997) *From Dutch Disease to Deforestation – A Macroeconomic Link? A Case Study From Ecuador, CDR WP 97.6*. Centre for Development Research, Copenhagen.

67 Wunder, S. (1999) *Value Determinants of Plant Extractivism in Brazil*. IPEA Discussion Paper no 682, IPEA, Rio de Janeiro, Brazil.

68 Wunder, S. (2000). *The Economics of Deforestation. The Example of Ecuador*. St Antony's Series, Macmillan & St. Martin's, London, June.

69 Wunder, S. (2000b) 'Ecotourism and economic incentives – An empirical approach'. *Ecological Economics*, vol 32, no 3, pp465–479.

70 Wunder, S. (2000c) *Big Island, Green Forests and Backpackers. Land-Use and Development Options on Ilha Grande, Rio De Janeiro State, Brazil*. CDR WP 00. 3 March, Centre for Development Research, Copenhagen, (copies from library@cdr.dk).

71 Young, C. E. F. (1995) *Public Policy and Deforestation in the Brazilian Amazon*. CREED/IIED, International Institute for Environment and Development, London.

72 Zimbabwe Trust, The Department of National Parks, CAMPFIRE Association (1994) *Zimbabwe: Tourism, People and Wildlife*. The Rural Extension Bulletin, no 5, August. University of Reading, Reading, UK.

6

Can Non-timber Forest Products Match Tropical Forest Conservation and Development Objectives?

J. E. Michael Arnold and M. Ruiz-Perez

Introduction

Historically, non-timber forest products[1] (NTFPs) were usually considered to be of little importance, a status reflected in their designation as 'minor' forest products. Much of their use was seen as being primarily of only local interest, and such commercial exploitation as took place was characterized as associated with lack of capital and technology, and often with exploitative use of labour (Homma, 1992). However, during the last 10–20 years there has emerged growing interest in attributes of NTFPs that appeared to be relevant to the growing focus on rural development and conservation of natural resources. This was articulated in three main propositions. One was that NTFPs contribute in important ways to the livelihoods and welfare of populations living in and adjacent to forests. Another was that exploitation of NTFPs is less ecologically destructive than timber harvesting and other forest uses, and could therefore provide a sounder base for sustainable forest management. The third was that increased commercial harvest of NTFPs should add to the perceived value of the tropical forest, thereby increasing the incentives to retain the forest resource.

Numerous authors stressed the apparent coincidence of conservation and development objectives that NTFPs appear to contribute to in these ways (see, for example, Myers, 1988; Nepstad and Schwartzman, 1992; Panayotou and Ashton, 1992; Plotkin and Famolare, 1992). Some valuation exercises suggested that the potential income from sustainable harvesting of NTFPs could be considerably higher than timber income, or income from agricultural or plantation uses of the forest sites (eg Peters et al, 1989; Balick and Mendelsohn, 1992). This resulted in the 'conservation by commercialization' hypothesis (see Evans, 1993) that has led to initiatives to expand and provide markets for NTFPs in order to tap an increasing share of this apparent store of sustainable harvestable wealth in tropical forests. It has also been argued that this potential could be considerably enhanced by drawing on indigenous knowledge and building on the sustainable systems of use

Note: Reprinted from *Ecological Economics*, vol 39, Arnold, J. E. M. and Ruiz-Perez, M., 'Can non-timber forest conservation and development objectives?' pp437–447, Copyright © (2001), with permission from Elsevier

that local people often seemed to have created (Posey, 1982; Prance, 1990; Stiles, 1994; Redford and Mansour, 1996). As a consequence, the heightened interest in NTFPs has been linked to the issue of empowering local people, and recognizing and legally securing their rights to manage their forest resources (see, for example, Dove, 1993).

In this chapter we review the evolution of the debate about these propositions and the lessons that appear to be emerging in practice, suggesting ways in which the original propositions might need to be revised.

Conservation

The ecological perspective

The maintenance of a forest-like structure associated with NTFP production is generally acknowledged as being positive, contributing to some of the classical forest environmental functions like carbon storage, nutrient cycling, erosion control and hydrological regulation (Myers, 1988; Gillis, 1992). Moreover, forests and home gardens managed for NTFP production can retain large amounts of plant and animal biodiversity (Michon and de Foresta, 1997), particularly when compared with alternative land uses (Boot, 1997), while providing an important source of income.

However, the propositions outlined above, and their interpretation, have raised concern that arguments about the relatively benign impact of harvesting for NTFPs have been overstated or misunderstood. Thus, the exploitation of forest resources has a differentiated effect, depending on the type of species and the parts being harvested. The extraction of bark can lead to the death of the individual, while the harvesting of fruits and flowers may have negative results in the whole population (Peters, 1994; Witkowski and Lamont, 1994). Some species are better able to sustain continuous offtake than others. In the case of plants, those exhibiting abundant and frequent regeneration and rapid growth will prevail (Cunningham and Mbenkum, 1993; Peters, 1994). Likewise, rodents, ungulates and other animals that have broad niches and rather prolific reproductive strategies are more able to stand heavy hunting (Bodmer et al, 1988; Fa et al, 1995).

There is a clear reduction in the composition and abundance of primary forest species and those of a more restricted habitat (Thiollay, 1996; García-Fernández et al, 2000). NTFP harvesting results in direct and indirect pressures on the forest, due to competition between humans and animals for some forest foods (Boot and Gullison, 1995). Though animals show different abilities to withstand pressure according to taxonomic groups, those that tend to be most heavily affected by hunting and other human activities include the most important predators and seed dispersers. Their depletion or removal can rapidly influence such forest characteristics as composition and structure of vegetation (Bodmer et al, 1988; Redford, 1992; Fitzgibbon et al, 1995). Finally, NTFP gathering also affects the genetic diversity of the population being exploited, especially when harvesting flowers or fruits that show differential traits resulting in different degrees of pressure (like larger fruits) (Peters, 1994).

Unless harvesting is controlled, some species will therefore become genetically impoverished or depleted much more rapidly than others. Over long periods, tropical forests can and do recover from even heavy use if allowed the time to do so without further disturbance. But this does not happen if there is repeated harvesting at short intervals relative to the forest's regeneration cycle (Poore et al, 1989), unless there is a monitoring and control system that provides a constant flow of information about the ecological response of species to varying degrees of exploitation (Peters, 1994). However, as was pointed out in the discussion about forest-derived agroforest systems, forests can be managed in ways that minimize the ecological impact of harvesting.

The impact of market forces

A number of researchers have been developing and testing models and hypotheses to assist in predicting how market forces are likely to have an influence on forest structure and use. Thus, Wilkie and Godoy (1996) argue that, with increased exposure to trade and markets, and per capita incomes rise, imported goods are substituted for some NTFPs and others are exploited primarily for sale. As alternative uses of labour become more attractive, use of the forest is increasingly concentrated on higher-value NTFPs. Thus, unless their use is controlled, or the species concerned are domesticated or replenished, their presence in the resource could diminish. In another influential model, based on Brazilian experience, Homma (1992) postulates that as commercial demand for a forest product emerges, output first expands; then, as quantities and quality from wild sources decline, prices will rise. Inelasticities of the supply of naturally occurring products then lead to development of domesticated sources and synthetic alternatives that replace the natural source.

Both of these models point to selective harvesting of those species that are more valued by the marketplace, and a consequent change in the composition of the remaining forest stock. In practice, these unidirectional evolutionary paths are not inevitable. Shifts in demand for forest products, for example, could reduce pressure on the resource or transfer it to another resource. Institutional measures to control the way in which the forest is used would also modify the impact of harvesting. For instance, forest management interventions, by increasing the productivity of the NTFP species, could prove to be an alternative to domestication, or could delay or modify the progression towards domestication. As Balée (1989) and Dufour (1990) have argued, the boundaries between wild and domesticated are not clear cut, giving ample room for a large variety of systems with good conservation potential. Some authors have proposed that we should think in terms of forest domestication rather than species domestication (Boot, 1997; Michon and de Foresta, 1997). Prance (1990) also argues that well-planned domestication integrated with extractive activities might help to curb the classical boom-and-bust cycles of extractive economies, contributing to their long-term maintenance. All these designs would allow retaining a good tree cover.

Nevertheless, it is clear that market demand is selective, and therefore works against the ecological objective of conserving the profile of biological diversity present in the untouched forest. Exposure to market pressures and opportunities is inescapably changing many subsistence-based use systems to market-oriented production systems, with clear losses of biodiversity (Rico-Gray et al, 1990; Lawrence, 1996; Bennett and Robinson, 2000). Moreover, as market prices seldom reflect the values of environmental and other 'external' costs and benefits, market demand may lead to short-term over-exploitation and even to local extinction of some plants and animals that provide highly desired products

(Vasquez and Gentry, 1989; Witkowski and Lamont, 1994; Fa et al, 1995; Hansis, 1998). This divergence between market and real economic and societal values casts doubt on the argument that the increased values attributable to tropical forests as a result of higher commercial demand for NTFPs necessarily encourage conservation of the resource.

Impacts of local uses

Some authors (González, 1992; Grenand and Grenand, 1996) point out that, though forest dwellers often appear to have evolved patterns of use that enable them to live in equilibrium with the forest, this does not mean that they are acting to protect nature in the sense understood today. Rather, it is because their system has a strong subsistence component, is based on the abundance and diversity of the resource and its ability to renew itself, and the human population density is relatively low.

Much harvesting of NTFPs is in forest systems that have in the past already been disturbed by human use to a greater or lesser degree. Most collecting and harvesting of NTFPs is by populations who combine this with some form of agriculture. It is therefore taking place not in pristine forest, but largely in secondary forests, bush fallow, farm bush or agroforest. This is partly explained by the proximity of these areas to the user communities and households, but also reflects the fact that in a number of respects such formations are more productive sources of desired species and products, and are more easily managed in a cycle of alternating cultivation and fallow (Posey, 1982; Davies and Richards, 1991).

In many situations, fallow land, farm bush and even the forest itself have in fact been found to be actively managed by local users to conserve or encourage particular species of value. The babaçú palm *(Orbygnia phalerata)* in north-east Brazil has long been integrated into local farmers' shifting cultivation systems (May et al, 1985), and farmers in the flood-plain forests of the Amazon area manage them to favour the economically more valuable species they contain (Anderson and Ioris, 1992). Damar, rattan and fruit gardens are examples of enriched forest management systems in Indonesia (Peluso and Padoch, 1996; Michon and de Foresta, 1997).

As the nature, rationale and consequences of managed local use have become better understood, it has been pointed out that much of what might be considered by ecologists and foresters to be degradation or depletion of a forest resource can be considered to be transformation and even improvement of the resource by those depending on it for inputs into their livelihood systems (Leach and Mearns, 1996). This has been accompanied by growing appreciation that associating conservation exclusively with such global values as biodiversity conservation has contributed to too narrow an assumption about linkages between human activity and forest change (Forsyth et al, 1998).

Development

NTFPs and rural household livelihoods

NTFPs are generally most extensively used to supplement diets and household income, notably during particular seasons in the year, and to help meet medicinal needs. NTFPs are also widely important as a subsistence and economic buffer in hard times. As is shown in Table 6.1, the importance of forest foods and incomes thus often lies more in its timing than in its magnitude as a share of total household inputs (Chambers and Leach, 1987; de Beer and McDermott, 1989; Falconer and Arnold, 1989; Scoones et al, 1992; FAO, 1995; Townson, 1995).

Use of some NTFPs is dwindling as people gain more access to purchased goods, as improved supplies of food crops have diminished the need to depend on forest foods, or as the opportunity cost of gathering foods, fuelwood and so on rather than purchasing them becomes higher. Supplies available for subsistence use can also fall as shortages emerge, and when their need for income forces the poor to sell products they would have otherwise used themselves (Falconer and Arnold, 1989; Ogle, 1996). Nevertheless subsistence use of NTFPs generally remains large and very important, as does their buffer role.

The role of income from forest products in household livelihood systems also changes, often rapidly, with changes in the demand for these products. Some forest products are goods that fall out of consumption patterns as incomes rise, for example those forest foods displaced by more convenient purchased foods. Others, such as mats, are vulnerable to competition from factory-made alternatives as improved transport infrastructure opens up rural areas to outside supplies (FAO, 1987). But demand for others, such as wooden furniture, rises with prosperity. Some products have large, diversified and stable markets; others face highly volatile markets, or demand that is seasonal and subject to sharp price fluctuations. While some products thus can provide a strong basis for livelihood systems, a number provide at best short-term opportunities, or generate only marginal returns to those engaged in their harvest and preparation.

Patterns of use differ among groups or households, and within households by gender and age. Forest foods and forest products income can be particularly important for poorer groups within the community (May et al, 1985; Siebert and Belsky, 1985; Fernandes and Menon, 1987; Jodha, 1990; Gunatilake et al, 1993; Cavendish, 2000). But the poor may not have access to the skills, technology or capital necessary to be able to benefit from the opportunities presented by growing markets for NTFPs. As a consequence, control over these opportunities, and over the resource, are often progressively captured by the wealthier and more powerful, and the households with the most labour, at the expense of the poorer within the community. Market forces can in this way create pressures on local collective systems of control over forest resources used as common property that can contribute to their breakdown, leading to uncontrolled and often destructive use of the resource (McElwee, 1994).

There is therefore a danger that poorly focused initiatives to increase commercialization of NTFPs could both disadvantage the very poor among local users, and encourage overuse of the forest resource. A great deal of the attention that has been given to NTFPs at the interface between conservation and development has been on ways of making trade in products for markets in developed countries more remunerative and stable to producers. However, these are trade flows that are very susceptible to changes in market requirements, to domination by intermediaries and to shifts to domesticated or synthetic sources of supply. Although the typical boom-and-bust sequence of responses to such short-term market opportunities may provide significant employment and income initially, in the longer term it can be very disruptive for rural economies, particularly where the trade

Table 6.1 *Forest outputs and rural livelihoods*

Livelihood inputs	Characteristics	Impacts of change
Subsistence Goods	Supplement or complement inputs of fuel, food, medicinal plant products etc from the farm system; often important in filling seasonal and other food gaps, particularly in hard times; forest foods enhance palatability of staple diets, and provide vitamins and proteins	Can become more important where farm output and/or non-farm income declines
Farm inputs	Forests provide starting point for rotational agriculture; on-farm trees provide shade, windbreaks and contour vegetation; trees/forests also provide low-cost soil nutrient recycling and mulch. Other inputs include arboreal fodder and forage, fibre baskets for storing agricultural products, wooden ploughs and other farm implements etc	Likely to decline in importance as incomes rise and supplies come increasingly from purchased inputs, or as increasing labour shortages/costs militate against gathering activities, or market demand diverts subsistence supplies to income-generating outlets. Trees can become increasingly important as a low-capital means of combating declining site productivity, and a low labour means of keeping land in productive use (eg home gardens)
Income	Many products characterized by easy access to the resource, and low-capital and skill entry thresholds; mainly low-return activities, producing for local markets, engaged in part-time by rural households, often to fill particular income gaps or needs; overwhelmingly very small, usually household-based enterprises (with heavy involvement of women, as entrepreneurs as well as employees) Some forest products provide the basis for more full-time and higher return activities; usually associated with higher skill and capital entry thresholds, and urban as well as rural markets	Increased capital availability, and access to purchased products, likely to lead to substitution by other materials (eg by pasture crops, fertilizer and plastic packaging) With increasing commercialization of rural use patterns some low-input low-return activities can grow; however, some produce 'inferior goods' and decline, others are displaced by factory-made alternatives, and others become unprofitable and are abandoned as labour costs rise

Table 6.1 *Forest outputs and rural livelihoods*

Livelihood inputs	Characteristics	Impacts of change
Reduced vulnerability	Some low-input gathering activities involve raw materials for industrial processes and external markets Can be important in diversifying the farm household economy – eg providing counter-seasonal sources of food, fodder and income Also important in providing a reserve that can be used for subsistence and income generation in times of hardship (crop failure, drought, shortage of wage employment etc), or to meet special needs (school fees, weddings etc)	Higher return activities serving growing demand are more likely to prosper, particularly those serving urban as well as rural markets; as this happens an increasing proportion of the processing and trading activity is likely to become centred in small rural centres and urban locations Gathered industrial raw materials tend to be displaced by domesticated supplies or synthetic substitutes The 'buffer' role of forests and trees can continue to be important well into the growth process Likely to decline in importance as government relief programmes become more effective, or new agricultural crops, or access to remittance incomes, make it less necessary to fall back on forest resources in times of need

Source: Based on Arnold (1998)

has encouraged people to move away from more diversified and less risky agriculture-based livelihoods (Browder, 1992; Homma, 1992).

Some of those commenting on cases where the adverse impacts of NTFP trades have been very pronounced have even argued that efforts to support development by promoting NTFP markets without securing the appropriate conditions (notably tenure and political rights) can be counterproductive (see, for example, Gray, 1990; Dove, 1993). As was noted above, trading NTFPs is likely to be appropriate only for those able to do so profitably. The existence in many poor and economically stagnant forest situations of huge numbers of people still engaged in low-return NTFP activities which have little prospect of other than short-term existence presents particular issues. Encouraging people to commit themselves to low-return commercial NTFP activities once higher-return or less arduous alternatives emerge could impede the emergence of better livelihood systems. It may be more fruitful to help people move into other more rewarding fields of endeavour rather than seeking to raise their productivity in their current line of work (Arnold et al, 1994).

Impacts of forest and environmental policies

Government policies often assert state control over the forest resource, or override local rights, thereby further undermining the authority and effectiveness of community-level institutions to control and manage forest use. Government policies can also constrain local efforts to realise more of the potential that NTFPs can contribute to household livelihoods. Because they give high priority to conservation objectives, many governments have set in place forest and environmental policies and regulations designed to limit rather than encourage production and sale of NTFPs (Dewees and Scherr, 1995). Restrictions placed on forest use in order to protect forests brought into community forestry schemes, and put them under sustainable forest management, can impose costs on local people which reduce their incentive to become involved. Allowable harvests may be reduced, and the structure of benefits changed as the composition of the forest changes under management. It is in fact difficult to find programmes that have not had at least a transitional adverse impact on those who have had to cut back or give up earlier gathering or grazing activities.

One widespread result of such changes in the policy and institutional situation has been ineffective local control of NTFP resources, and an environment in which household decision making and market forces fail to generate sustainable use of local forest resources. Moreover, it is often unclear which institutional models might be appropriate at present in situations marked by increasing conflict and lower commonality of purpose, and increasingly ineffective conflict resolution mechanisms that such policies and practices engender (Neumann, 1996). This obviously raises questions about the argument that increased local harvesting and trade of NTFPs necessarily increases effective commitment to conservation and sustainable management and use of natural resources (Jodha, 1990; Davis and Wali, 1993; Lynch and Talbott, 1995). In short, as information about the role of NTFPs in rural development has accumulated, it has become apparent that some forest products have economic characteristics that make them attractive to rural households – ease of access to the resource, low capital and skill thresholds to harvesting and processing, and outputs that help reduce households' exposure to risk. For those trapped in poverty NTFP activities can comprise an important part of their coping strategies. However, the high transactions costs associated with meeting market demand for many NTFPs mean that they are much less likely to be an attractive option for those emerging from poverty, and with alternative wealth-generating options available. Commercialization therefore does not necessarily provide opportunities for development for many of the rural poor in or adjacent to forested areas. Equally, commercialization of NTFPs could exacerbate rather than reduce the pressures that cause overuse of forest resources (Byron and Arnold, 1999; Cavendish, 2000).

Discussion

The discussion above suggests that the proposition that increased use of NTFPs is congruent with forest conservation needs to be qualified and elaborated. In practice, the different stakeholders with an interest in a forest and the NTFPs it can yield are unlikely to seek the same balance between developmental and conservation objectives. For instance, it is unlikely that the economic goals of local users will yield the same outcomes as the conservation goals of those concerned with preserving biodiversity (Wells et al, 1992).

It is important to recognize that divergence of interests between development and conservation does not necessarily mean that the different balances between the two that result are less or more 'sustainable' than the other. Rather it is the recognition that sustainability has a number of different dimensions. The objective of ecological sustainability is usually expressed in terms of maintaining forest cover and biodiversity. The goal of sustainable forest management has usually focused on maintaining a continuous flow of stated outputs, while retaining the productive capacity of the forest intact. Economists, on the other hand, tend to focus on the sustainability of economic benefits. As the benefits people seek to obtain from the forests change over time, pursuit of this objective is likely to entail changes to the resource base. Essentially, local management systems that alter the structure of the forest resource in favour of particular outputs can be seen to be giving priority to this economic objective.

Some have argued that a belief that there is a commonality of interest among different categories of users can arise from misunderstandings by local and environmental interest groups about each other. For instance, conservation NGOs failing to recognize that local communities give priority to tenurial and livelihood issues, and local communities mistakenly believing that conservation NGOs will provide assistance in meeting such needs (Stocks, 1996). It has also been suggested that conservation groups have on occasion sought to ally themselves with local development goals that are at variance with their interests as a way of 'buying time' until a better way is found of achieving conservation aims (Redford and Stearman, 1993). Similarly, forest dwellers may seek a common cause with conservationists where this can help them secure land titles and other guarantees (Mendes, 1992).

Another factor in shaping the initial proposition, and in explaining the strength of the support it received, can now be seen to be a measure of misunderstanding or misinterpretation of some of the data on which it was based. NTFP harvesting may frequently be less damaging than alternative land uses like cattle ranching or intensive logging, but it is not without impact. While it can help to preserve a tree cover that resembles a forest-like structure and performs several of its environmental functions, it does not maintain the same level of biodiversity and quality of species as a primary forest.

Likewise, in extrapolating from studies that arrived at high estimates of the potential value of offtake from particular forest situations, and arriving at conclusions about commercial revenues that might be generated, some of the features characterizing the situations to which the original point estimates referred have been overlooked or lost sight of (Simpson et al, 1996). The result has sometimes been to raise expectations beyond what can realistically be achieved. A recent study of experience with initiatives to encourage conservation-compatible types of forest production in Latin America concluded that, in practice, these provide only limited scope for enhancement of the incomes of those engaged in them, and so can have the effect of discouraging sustainable forest management. Thus, with the exception of some situations well endowed with commercially exploitable products, and well placed with respect to access to markets, harvesting and sale of NTFPs was found not to be financially rewarding (Phillips, 1993; Southgate, 1998).

In brief, it is now clear that strategies based on the assumption that developmental and conservation interests in NTFPs coincide can be unrealistic. It could be more effective to focus on understanding the areas in which they do coincide, and those in which they are in conflict, and in determining what balance between development and conservation is desirable and achievable. Different situations have different potentials and limitations, that call for different possible responses (Ruiz-Perez and Byron, 1999). NTFP gathering can contribute as a component of a wider conservation strategy that would encompass a spectrum from intensively transformed to little disturbed forests seeking for diversity both at species, ecosystem and landscape levels.

In doing so, it will be necessary to take account of the argument that the pursuit of conservation has been too much driven by Northern concepts and donor preoccupations, at the expense of those who depend on forests locally; and the argument that the conventional approach to the issue of the balance between conservation and development at this level has been based on flawed assumptions about how rural people and the 'environment' interrelate. It is argued that there is need for greater appreciation that the poor may experience their own environmental problems, which need to be addressed separately from environmental policies seeking to satisfy concerns about global values. To address these local concerns there is a need to move away from macroscale approaches and policies, to a more situation-specific focus, reflecting the protective mechanisms that local users themselves adopt, and the attributes of a resource that they value and seek to conserve (Forsyth et al, 1998). This could favour a shift from a predominantly protective orientation in forest management towards encouraging sustainable systems for production of livelihood benefits in as an 'environmentally friendly' a way as possible (Freese, 1997).

At the same time, we need to recognize the implications of the widely different roles that NTFPs play in the livelihoods of different categories of the poor who draw on forests. It may be necessary to plan separately for those among the very poor and disadvantaged who continue to rely on such NTFPs for survival, and for those engaged in NTFP activities that form part of the process of growth and development. In other words, it may often be necessary in designing and implementing policy and other institutional interventions to distinguish between those who can improve their livelihoods through NTFP activities, and those who have no other option but to continue to gather NTFPs in order to survive.

Acknowledgements

This paper builds on work on non-timber forest products carried out at the Center for International Forestry Research (CIFOR) over a number of years. The authors wish to express their appreciation to those of their colleagues, and other collaborators, who contributed to the development of the ideas expressed in this paper over that period. In particular, they wish to thank Lini Wollenberg, Neil Byron and Jenne de Beer for comments on an earlier draft paper on this topic.

Note

1   The expressions non-timber forest products, non-wood forest products, and minor forest products are frequently used interchangeably. The term non-timber forest products is used in this paper to denote any product other than timber dependent on a forest environment. It is restricted to tradable material products, and their processed derivatives, and does not include services derived from the forest such as carbon sequestration, nutrient cycling or amelioration of water flows.

References

1   Anderson, A. B. and Ioris, E. M. (1992) 'The logic of extraction: resource management and income generation by extractive producers in the Amazon', in Redford, K. H. and Padoch, C. (eds) *Conservation of Neotropical Forests: Working from Traditional Resource Use.* Columbia University Press, New York, pp179–199.
2   Arnold, J. E. M. (1998) 'Forestry and sustainable livelihoods', in Carney, D. (ed) *Sustainable Rural Livelihoods: What Contributions Can We Make?* Department for International Development, London, pp155–166.
3   Arnold, J. E. M., Liedholm, C., Mead, D. and Townson, I. M. (1994) *Structure and Growth of Small Enterprises Using Forest Products in Southern and Eastern Africa.* OFI Occasional Papers no 47, Oxford Forestry Institute, Oxford.
4   Balée, W. (1989) 'The culture of Amazonian forests', in Balée, W. and Posey, A. (eds) *Resource Management in Amazonia: Indigenous and Folk Strategies.* Advances in Economic Botany, vol 7, pp1–21.
5   Balick, M. J. and Mendelsohn, R. (1992) 'Assessing the economic value of traditional medicines from tropical rain forests'. *Conservation Biology,* vol 6, pp128–130.
6   Bennett, E. L. and Robinson, J. G. (2000) *Hunting of Wildlife in Tropical Forests. Implications for Biodiversity and Forest Peoples.* World Bank Biodiversity Series, no 76, World Bank, Washington, DC.

7   Bodmer, R. E., Fang, T. G. and Moya, I. L. (1988) 'Primates and ungulates: a comparison in susceptibility to hunting'. *Primate Conservation*, vol 2, pp79–83.

8   Boot, R. G. A. (1997) 'Extraction of non-timber forest products from tropical rain forests. Does diversity come at a price?' *Netherlands Journal of Agricultural Science*, vol 45, pp439–450.

9   Boot, R. G. A. and Gullison, R. E. (1995) 'Approaches to developing sustainable extraction systems for tropical forest products'. *Ecological Applications*, vol 5, no 4, pp896–903.

10   Browder, J. O. (1992) 'The limits of extractivism: tropical forest strategies beyond extractive reserves'. *BioScience*, vol 42, pp174–182.

11   Byron, N. and Arnold, J. E. M. (1999) 'What futures for the people of the tropical forests?' *World Development*, vol 27, no 5, pp789–805.

12   Cavendish, W. (2000) *Rural Livelihoods and Non-Timber Forest Products*. University of Oxford and Imperial College (unpublished).

13   Chambers, R. and Leach, M. (1987) *Trees to Meet Contingencies: Savings and Security for the Rural Poor*. Discussion Paper 228, Institute of Development Studies, University of Sussex, Brighton, UK.

14   Cunningham, A. B. and Mbenkum, F. T. (1993) *Sustainability of Harvesting Primus Africana bark in Cameroon*. People and Plants Working Paper no 2, May 1993. UNESCO, Paris, France.

15   Davies, A. G. and Richards, P. (1991) *Rain Forest in Mende Life: Resources and Subsistence Strategies in Rural Communities Around the Gola North Forest Reserve (Sierra Leone)*. A report to ESCOR, Overseas Development Administration, London, UK.

16   Davis, S. H. and Wali, A. J. (1993) *Indigenous Territories and Tropical Forest Management in Latin America*. Policy Research Working Paper Series no 1100, World Bank, Washington, DC

17   de Beer, J. H. and McDermott, M. J. (1989) *The Economic Value of Non-Timber Forest Products in Southeast Asia*. Netherlands Committee for IUCN, Amsterdam, The Netherlands.

18   Dewees, P. A. and Scherr, S. J. (1995) *Policies and Markets for Non-timber Tree Products*. Draft Working Paper, International Food Policy Research Institute, Washington, DC

19   Dove, M. R. (1993) 'A revisionist view of tropical deforestation and development'. *Environmental Conservation*, vol 20, pp17–24, 56.

20   Dufour, D. L. (1990) 'Use of tropical rainforest by native Amazonians'. *BioScience*, vol 40, pp652–659.

21   Evans, M. I. (1993) 'Conservation by commercialization', in Hladik, C. M., Hladik, A., Linares, O. F., Pagezy, H., Semple, A. and Hadley, M. (eds) *Tropical Forests, People and Food: Biocultural Interactions and Applications to Development*. MAB Series, vol 13, UNESCO, Paris and Parthenon Publishing Group, Carnforth, UK, pp815–822.

22   Fa, I. E., Juste, J., Pérez del Val, J. and Castroviejo, J. (1995) 'Impact of market hunting on mammal species in Equatorial Guinea'. *Conservation Biology*, vol 9, pp1107–1115.

23   Falconer, J. and Arnold, J. E. M. (1989) *Household Food Security and Forestry: An Analysis of Socioeconomic Issues*. Community Forestry Note 1, FAO, Rome, Italy.

24   FAO (Food and Agriculture Organization) (1987) *Small-scale Forest Based Processing Enterprises*. Forestry Paper 79, FAO, Rome, Italy.

25   FAO (Food and Agriculture Organization) (1995) *Report of the International Expert Consultation on Non-Wood Forest Products*. Non-Wood Forest Products, 3, FAO, Rome, Italy.

26   Fernandes, W. and Menon, G. (1987) *Tribal Women and Forest Economy*. Indian Social Institute, New Delhi, India.

27   Fitzgibbon, C. D., Mogaka, H. and Fanshawe, J. H. (1995) 'Subsistence hunting in Arabulo-Sokeke Forest, Kenya, and its effect on mammal population'. *Conservation Biology*, vol 9, pp1116–1126.

28 Forsyth, T., Leach, M. and Scoones, I. (1998) *Poverty and Environment: Priorities for Research and Policy, an Overview Study*, prepared for the UNDP and European Commission, Institute of Development Studies, Brighton (unpublished).
29 Freese, C. (eds) (1997) *Harvesting Wild Species: Implications for Biodiversity Conservation*. John Hopkins University Press, Baltimore, MD.
30 García-Fernández, C., Casado, M. A. and Ruiz-Perez, M. (2000) *Benzoin Gardens and Diversity in North Sumatra, Indonesia*. CIFOR, Bogor, Indonesia.
31 Gillis, M. (1992) 'Economic policies and tropical deforestation', in Nepstad, D. C. and Schwartzman, S. (eds) *Non-timber products from tropical forests: evaluation of a conservation and development strategy*. Advances in Economic Botany, vol 9, pp129–142.
32 González, N. (1992) 'We are not conservationists'. *Cultural Survival Quarterly*, vol 16, no 3, pp43–45.
33 Gray, A. (1990) 'Indigenous people and the marketing of the rainforest'. *The Ecologist*, vol 20, pp223–227.
34 Grenand, P. and Grenand, F. (1996) 'Living in abundance. The forest of the Wayampi (Amerindians from French Guiana)', in Ruiz-Perez, M. and Arnold, J. E. M. (eds) *Current Issues in Non-Timber Forest Products Research*. CIFOR-ODA, Bogor, Indonesia, pp77–196.
35 Gunatilake, H. M., Senaratne, A. H. and Abeygunawardena, P. V. (1993) 'Role of nontimber forest products in the economy of peripheral communities of Knuckles National Wilderness area in Sri Lanka: a farming system approach'. *Economic Botany*, vol 47, pp275–281.
36 Hansis, R. (1998) 'A political ecology of picking: non-timber forest products in the Pacific Northwest'. *Human Ecology*, vol 26, no 1, pp67–86.
37 Homma, A. K. O. (1992) 'The dynamics of extraction in Amazonia: A historical perspective', in Nepstad, D. C. and Schwartzman, S. (eds) *Non-timber Products from Tropical Forests Evaluation of a Conservation and Development Strategy*. Advances in Economic Botany, vol 9, pp23–32.
38 Jodha, N. S. (1990) 'Rural common property resources: Contributions and crisis', *Economic and Political Weekly*. *Quarterly Review of Agriculture*, vol 25, no 26, p65.
39 Lawrence, D. C. (1995) 'Trade-offs between rubber production and maintenance of diversity: the structure of rubber gardens in West Kalimantan, Indonesia'. *Agroforestry Systems*, vol 34, pp83–100.
40 Leach, M. and Mearns, R. (1996) 'Environmental change and policy: Challenging received wisdom in Africa', in Leach, M. and Mearns, R. (eds) *The Lie of the Land: Challenging Received Wisdom on the African Environment*. James Currey, Oxford, for The International African Institute, London, UK, pp1–33.
41 Lynch, O. J. and Talbott, K. (1995) *Balancing Acts: Community-Based Forest Management and Forest Law in Asia and the Pacific*. World Resources Institute, Washington, DC.
42 May, P. H., Anderson, A. B., Balick, M. J. and Unruh, J. (1985) 'Babaçu palm in the agroforestry systems in Brazil's mid-north region', *Agroforestry Systems*, vol 3, pp275–295.
43 McElwee, P. (1994) *Common Property and Commercialisation: Developing Appropriate Tools for Analysis*. MSc Dissertation, Oxford Forestry Institute, University of Oxford, UK (unpublished).
44 Mendes, F. (1992) 'Peasants speak: Chico Mendes—the defence of life'. *The Journal of Peasant Studies*, vol 20, no 1, pp160–176.
45 Michon, G. and de Foresta, H. (1997) 'Agroforests: pre-domestication of forest trees or true domestication of forest ecosystems?' *Netherlands Journal of Agricultural Science*, vol 45, pp451–462.
46 Myers, N. (1988) 'Tropical forests: much more than stocks of wood'. *Journal of Tropical Ecology*, vol 4, pp209–221.
47 Nepstad, D. C. and Schwartzman, S. (1992) 'Introduction: non-timber products from tropical forests: Evaluation of a conservation and development strategy', in Nepstad, D. C. and

Schwartzman, S. (eds) *Non-Timber Products from Tropical Forests: Evaluation of a Conservation and Development Strategy*. Advances in Economic Botany, vol 9, ppvii–xii.

48   Neumann, R. P. (1996) 'Forest products research in relation to conservation policies in Africa', in Ruiz-Perez, M. and Arnold, J. E. M. (eds) *Current Issues in Non-Timber Forest Products Research*. CIFOR-ODA, Bogor, Indonesia, p176.

49   Ogle, B. A. (1996) 'People's dependency on forests for food security. Some lessons learnt from a programme of case studies', in Ruiz-Perez, M. and Arnold, J. E. M. (eds) *Current Issues in Non-Timber Forest Products Research*. CIFOR-ODA, Bogor, Indonesia, pp219–241.

50   Panayotou, T. and Ashton, P. (1992) *Not by Timber Alone. Economics and Ecology for Sustaining Tropical Forests*. Island Press, Washington, DC.

51   Peluso, N. L. and Padoch, C. (1996) 'Changing resource rights in managed forests of West Kalimantan', in Peluso, N. L. and Padoch, C. (eds) *Borneo in Transition: People, Forests, Conservation, and Development*. Oxford University Press, Kuala Lumpur, Malaysia, pp121–136.

52   Peters, C. M. (1994) *Sustainable Harvest of Non-Timber Plant Resources in Tropical Moist Forest: An Ecological Primer*. Biodiversity Support Program, Washington, DC

53   Peters, C. M., Gentry, A. H. and Mendelsohn, R. O. (1989) 'Valuation of an Amazonian rainforest'. *Nature*, vol 339, pp655–656.

54   Phillips, O. (1993) 'The potential for harvesting fruits in tropical rainforests: New data from Amazonian Peru'. *Biodiversity and Conservation*, vol 2, pp18–38.

55   Plotkin, M. and Famolare, L. (1992) 'Preface', in Plotkin, M. and Famolare, L. (eds) *Sustainable Harvest and Marketing of Rain Forest Products*. Conservation International, Island Press, Washington, DC, ppxiii–xv.

56   Poore, D., Burgess, P., Palmer, J., Rietbergen, S. and Synnott, T. (1989) *No Timber without Trees: Sustainability in the Tropical Forest*. Earthscan, London, UK.

57   Posey, D. A. (1982) 'The keepers of the forest', *Garden*, vol 6, pp18–24.

58   Prance, G. T. (1990) 'Fruits of the rainforest'. *New Scientist*, vol 13, January, pp42–45.

59   Redford, K. H. (1992) 'The empty forest'. *BioScience*, vol 42, pp412–422.

60   Redford, K. H. and Stearman, A. M. (1993) 'Forest-dwelling native Amazonians and the conservation of biodiversity: interests in common or in collision?' *Conservation Biology*, vol 7, pp248–255.

61   Redford, K. H. and Mansour, J. A. (eds) (1996) *Traditional People and Biodiversity Conservation in Large Tropical Landscapes*. America Verde Publications for The Nature Conservancy, Arlington, Virginia.

62   Rico-Gray, V., Garcia-Franco, J. G., Chemas, A., Puch, A. and Sima, P. (1990) 'Species composition, similarity and structure of Mayan homegardens in Tixpeual and Tixcacaltuyub, Yucatán, México'. *Economic Botany*, vol 44, pp470–487.

63   Ruiz-Perez, M. and Byron, N. (1999) 'A methodology to analyze divergent case studies of non-timber forest products and their development potential'. *Forest Science*, vol 45, no 1, pp1–14.

64   Scoones, I., Melnyk, M. and Pretty, J. (1992) *The Hidden Harvest Wild Foods and Agricultural Systems: A Literature Review and Annotated Bibliography*. IIED, SIDA and WWF, London, UK and Gland, Switzerland.

65   Siebert, S. and Belsky, J. M. (1985) 'Forest product trade in a lowland Filipino village'. *Economic Botany*, vol 39, pp522–533.

66   Simpson, R. D., Sedjo, R. A. and Reid, J. W. (1996) 'Valuing biodiversity for use in pharmaceutical research'. *Journal of Political Economy*, vol 104, pp163–185.

67   Southgate, D. (1998) *Tropical Forest Conservation: An Economic Assessment of the Alternatives in Latin America*. Oxford University Press, New York.

68   Stiles, D. (1994) 'Tribals and trade: A strategy for cultural and ecological survival'. *Ambio*, vol 23, pp106–111.

69   Stocks, A. (1996) 'The Bosawas natural reserve and the Mayangna of Nicaragua', in Redford, K. H. and Mansour, J. A. (eds) *Traditional People and Biodiversity Conversation in Large Tropical Landscapes*. America Verde Publications for The Nature Conservancy, Arlington, Virginia, pp1–31.

70   Thiollay, J. M. (1996) 'Rain forest raptor communities in Sumatra: The conservation value of traditional agroforests', in *Raptors in Human Landscapes*. Academic Press, New York, pp245–262.

71   Townson, I. M. (1995) 'Forest products and household incomes: A review and annotated bibliography'. *Tropical Forestry Papers*, vol 31. CIFOR and OFI, Oxford, UK.

72   Vasquez, R. and Gentry, A. (1989) 'Use and misuse of forest harvested fruits in Iquitos area'. *Conservation Biology*, vol 3, pp350–361.

73   Wells, M., Brandon, K. and Hannah, L. (1992) *People and Parks: Linking Protected Area Management with Local Communities*. World Bank/WWF/USAID, World Bank, Washington, DC.

74   Wilkie, D. S. and Godoy, R. A. (1996) 'Trade, indigenous rain forest economies and biological diversity. Model predictions and directions for research', in Ruiz-Perez, M. and Arnold, J. E. M. (eds) *Current Issues in Non-Timber Forest Products Research*. CIFOR-ODA, Bogor, Indonesia, pp19–39.

75   Witkowski, E. T. F. and Lamont, B. (1994) 'Commercial picking of Banksia hookeriana in the wild reduces subsequent shoot, flower and seed production'. *Journal of Applied Ecology*, vol 31, pp508–520.

7

What Future for the Peoples of the Tropical Forests?

Neil Byron and Michael Arnold

Forest-Dependent People?

Over the past decade environmental and developmental concerns have converged, with increasing interest in both tropical forests as an important ecosystem, and in the well-being of people who live in or near them. The importance of forests and of non-industrial forest products to the quality of life and even survival of very large numbers of poor rural people in tropical developing countries now seems indisputable (Ruiz-Perez and Arnold, 1996).

Yet very great uncertainty persists about who these 'forest-dependent people' are, where they can be found, and what form their 'dependence' takes. Estimates of the numbers of people involved range from perhaps 250 million (Pimentel et al, 1997), to 500 million (Lynch and Talbott, 1995), to more than one billion (WCFSD, 1997). Much of the discrepancy can be explained by the ambiguity or complete lack of definitions.

Frequently the estimates of the 'numbers of people practising shifting cultivation' have been cited, although this is by no means synonymous with 'numbers of people generating much of their livelihood from forests'. (Estimates of the number of shifting cultivators suffer from equally confused terminology and absence of rigorous definitions. Some studies refer to the number of people who are totally dependent on forests for all their livelihood inputs, a definition that, though explicit, is difficult to apply. At the other extreme are estimates which include as 'forest-dependent' anyone who ever makes any opportunistic use of some product from the forest, or even trees outside forests. For example, Pimentel et al (1997) apparently include all people who collect use or sell any forest product, the people who harvest cultivated trees from farmlands, even workers on tea estates and banana plantations, as 'forest-dependent'. Lynch and Talbott (1995) estimate the number of people living in or near forest reserves at 500–600 million in Asia alone. 'Proximity to forests' is not, however, synonymous with 'forest dependence', although clearly the two are related.

Many urban households, particularly in Africa, rely on fuelwood that they purchase, and there seems to be no feasible substitute for the foreseeable future. Does this mean that they are 'dependent' on the forests from which their supplies come? Similarly many people rely on purchased, traditional forest-sourced medicines, for both cultural and price

Note: Reprinted from *World Development*, vol 27, Byron, N. and Arnold, M., 'What future for the peoples of the tropical forests?', pp789–805, Copyright © (1999), with permission from Elsevier

reasons. But it is equally questionable whether that necessarily means they have a relationship with the forest. Thinking about this type of 'commercial consumer/user dependency' illustrates what a diverse array of situations have been lumped together under this rubric, and that 'the line has never been drawn' about which groups are counted in the disparate estimates of forest-dependent peoples. Is it reasonable, or meaningful, to argue that both artisans and craftspeople who use forest products and subsistence hunter-gatherers living in forests are 'forest-dependent' – given that their linkages to the forest are clearly so different?

The existing literature is thus quite unsatisfactory for developing an understanding of the types of relationships that exist between tropical forests and the people who currently use them, benefit from them, and frequently protect and manage them. One purpose of this chapter is to elaborate the different contexts in which different types of people depend either on the forests per se, or specific products from them (eg medicines, fuel, fodder, craft woods). Even more importantly, we explore the way the people–forest–products relationships are changing over time.

Detailed case studies of specific contexts have been conducted and reported in the literature, for example an Amazonian tribe living in the forests; collectors and traders of nonfood forest products in West Africa, India, or South-East Asia. Yet many of these studies have only documented some aspects of the relationship between local forest users and the forests they use or depend upon. Even fewer have considered that relationship's future. From a policy perspective, even more important than such a 'snapshot' is to understand how these people–forest relationships might change over time, in particular economic, cultural and political contexts. While it is important to know whether there are 100 million or 1000 million people currently using forests and forest products, it is even more important to develop a deeper understanding of the relationships and their dynamics.

The condition of 'dependence' is defined as 'to rely for support' (Webster), or 'to be unable to do without' or 'need for success' (Oxford). Much use of forest outputs clearly does not carry this connotation. Many users have alternatives, and use forest outputs as a matter of choice, not necessity. Others, however, are dependent in the literal sense that their condition would worsen if they no longer had access to the forest outputs that form an integral part of their livelihood systems.

We argue that there are fundamentally different types of users, which can be summarized crudely as ranging from 'those who choose to generate much of their livelihood from forests because it is an attractive, viable option', across a spectrum to 'those for whom forest dependency is a livelihood of last resort – a symptom of their limited options and/or poverty – which they will abandon as soon as any plausible better option emerges'. In many cases, there is also a strong cultural or spiritual element in the people–forest relationship, while in others it is primarily an economic choice (broadly defining economic to include nonmonetary benefits).

Particularly in developing policies that devolve authority and responsibility for managing forests to local communities, it is crucial to assess whether the number of people willing and able to participate in forest management is likely to increase or decrease or, more realistically, why it might increase in some places while declining rapidly in others. It is extremely important, especially for assessing the impacts of policy reforms, to have a very clear idea of who and where are the affected groups, and how they might be affected. Which of them can, and most likely will, move away from their present levels of use of forests or forest products? Which of them need to continue to be able to draw on forests? Which of those are likely to do so more in the future? What are the implications for policy and management of these very different scenarios?

Another aspect that also has to be addressed is how much of the flows of 'forest' products actually comes from forests, as distinct from tree stocks and formations such as bush fallow and farm trees that occur outside the forest? Though our interest here is in people's linkages with forests, many users do not make this distinction, and we therefore need to recognize this continuum between forest and non-forest tree resources.

We therefore seek to deconstruct the term and the concept of 'forest-dependent people', to expose the great diversity of situations that have previously been lumped together, in order to better understand the current situations and, even more importantly, to recognize the divergent trends which appear to face different groups. We seek to clarify the debate by presenting a series of identifiable categories, illustrated by examples from specific contexts. But we will resist the temptation to undertake a global 'head-count', as we believe it is far more important to understand the nature and the significance of the relationships between local forest users/managers and their forests, and then how these relationships might change and adapt over time. Simply adding the numbers of 'forest-dependent people' is not very useful and almost certainly soon out of date, even if initially plausible, because of the rapid changes in these relationships which we observe in certain countries. The number of people concerned is not the only, or the best, indicator of the importance of this topic.

Moreover, the basis for making useful estimates of the numbers of people who draw on forest outputs for a specified proportion of their inputs is extremely weak. Global estimates advanced as numbers 'dependent' on forests, and of the extent of their dependency, are consequently likely to be very misleading. As most production trade and use of forest products for household consumption occurs in the informal sector, much of it outside the market economy, very little is recorded in production and trade statistics! There is a great deal of descriptive information, but this is generally concentrated in narrowly situation-specific case study accounts. Even these frequently encounter difficulties in capturing accurate information on household incomes, in providing information about subsistence uses, and in separating out that part of household inputs, labour allocation, incomes and costs that is attributable to forest products activities. As there is a great deal of variation in forest product use from location to location (and often within communities) it is also difficult to extrapolate from case study findings.

Nor do most censuses and surveys capture much information on household-level forest products activity and use. Extensive searches of such sources have identified only a small number of national or regional, informal sector small-enterprise surveys and some detailed household living standards surveys as yielding sufficient data to allow some measure of the magnitudes of forest products activities for large populations and areas. As discussed below in connection with the estimates for Africa that we develop using data from these sources, even these are limited by definitions that have been set by the objectives of the overall survey (ie they tend to collect information on only some forest products, and some relevant aspects of using households and enterprises).

Additional data, more focused on what is needed in the forest sector, could sometimes be obtained at low cost by 'piggy-backing' additional questions on to some of these national inquiries, and it would be useful if more attention could be given to doing so. More detailed studies in the forest sector to generate complementary information could be an additional step toward a better data base (see Townson (1995b) for such an exercise in Ghana). The primary focus for such work, however, should be on gathering a range of data that helps illuminate the nature and importance of forest products and activities in meeting household needs, and not just the numbers involved.

In the following section we summarize the diverse ways in which forests are extremely important to particular people in particular contexts and the nature and consequences of the changes in these relationships. We examine measures of importance that are far more informative (but more complex) than simple estimates of 'numbers of people generating all or part of their livelihood from forest activities' or 'living in/near forest reserves'. The third section establishes a 'typology' to identify the likely importance of forests to different categories of households. The fourth section considers possible alternative futures – how the different contexts described in the typology might develop in future. We suggest that the future of viable local institutional arrangements will depend greatly on how the different types of people–forest relationships respond to economic, social and political changes which either reinforce or erode people's incentives and capabilities to practise local forest management.

The Local Importance of Tropical Forests in Household Subsistence and Incomes[1]

For millions of people living in forest environments, the forest forms such a dominant part of their physical, material, economic and spiritual lives that its importance is not most appropriately described and assessed in terms of the individual products or services that the forest provides. The forest, as well as providing a wealth of material outputs of subsistence or commercial value, is the basis for livelihood systems based on hunting and gathering, or for rotational agriculture systems that depend on the ability of bush fallow to revive the productivity of the land. The forest thus constitutes an integral part of the habitat and of the social and cultural framework of those living within it. It should therefore be assessed and measured in toto, recognizing that physical, extractable non-timber forest products are only a limited subset of that whole.

A far larger proportion of users of forest outputs live in situations in which they are less intimately linked to forests. Forest products, though important, form a lesser part of the household livelihood system. To understand their relationships to the forest we do need to examine and understand the nature and dynamics of their particular uses of forest products.

Subsistence uses of forest products

Forests and forest trees are the sources of a variety of foods that supplement and complement what is obtained from agriculture, of fuels with which to cook food, and of a wide range of medicines and other products that contribute to health and hygiene. Probably the majority of rural households in developing countries, and a large proportion of urban households, depend on plant and animal products of forests to meet some part of their nutritional, cooking and/or health needs.

Forest foods seldom provide the staple, bulk items that people eat. For the majority of rural people, forest foods add variety to diets, improve palatability, and provide essential vitamins, minerals, protein and calories. The quantities consumed may not be great in comparison to the main food staples, but they often form an essential part of people's diets.

Forest foods are most extensively used to help meet dietary shortfalls during particular seasons in the year. Many agricultural communities suffer from seasonal food shortages, which commonly occur at the time of year when stored food supplies have dwindled and harvest of new crops is only just beginning. Forest and farm tree products are also valued during the peak agricultural labour period, when less time is available for cooking and people consume more snack foods.

Forests are especially important as a source of foods during emergency periods such as floods, famines, droughts and wars. Often these food resources differ from resources exploited in other periods. In famine periods, roots, tubers, rhizomes and nuts are most sought after. They are characteristically energy rich, but often require lengthy processing.

Supplies of wood fuels influence nutrition through their impact on the availability of cooked food. If there is less fuel (or time) for cooking, consumption of uncooked and reheated food may increase. This may cause a serious rise in disease incidence as few uncooked foods can be properly digested, and cooking is necessary to remove parasites. A decrease in the number of meals provided may have a particularly damaging effect on child nutrition, as children may be unable to consume enough of often an overstarchy staple food in one meal (Cecelski, 1987; Falconer, 1989).

Medicinal usage of forest products tends to overlap with that of forest foods: indeed particular items added to foods serve both to improve palatability and act as a health tonic or prophylactic. There are also often strong links between medicinal use and cultural values: for example, where illnesses are thought to be due to spiritual causes, or plants have acquired symbolic importance as treatments.

Forests are also a major source of construction materials such as poles and fibres for baskets and other storage structures. Forests and farm trees also provide fodder and mulch and additional inputs into farm systems, such as soil nutrient cycling and protection.

Income from forest products activities

Very large numbers of households also generate some of their income from selling forest products. Most such commercial forest products activities are conducted part-time by farm households which cannot raise enough to be food self-sufficient year round. The information available suggests that for most users, the importance of forest products income is usually more in the way it fills gaps and complements other income, than in its absolute magnitude or share of overall household income.

Income-earning activities based on marketable forest products may be seasonal or year-round, or may be occasional when supplementary cash income is needed. There are several dimensions to the seasonally of forest-based income-generating activities. Some are governed by seasonality induced cash requirements, such as the need for income to buy food during the 'hunger period' between harvests, or to purchase farm inputs. Many activities are seasonal largely because the crop or material can only be gathered at certain times of year. The fluctuation in timing of some forest products activities is dictated by the seasonality of other activities, such as the need for baskets at harvest time, and the surge in demand for many items as agricultural incomes peak. Activities may also be linked to changes in availability of labour and consequently decline in agricultural and planting seasons, or are phased to take advantage of slack periods. Often these pressures work in conjunction one with another, within sequences in which one activity's output becomes another's input, for example generating income from forest products in order to purchase seeds, hire labour for cultivation or generate working capital for trading activities (eg Leach and Fairhead, 1994). Forest products can also provide a source of 'windfall' income: a good crop providing a valuable injection of cash, enabling people to clear their debts or accumulate some capital.

Forest products activities also can provide an important supplemental source of income that people can fall back on in times of crop failure or shortfall, or in order to cope with some other form of emergency. Forests are therefore often very important as an economic buffer and safety net for poor households.

Where access to forests has been relatively unrestricted, forest foods and income from forest products are often particularly important for poorer groups within the community. Though it is often the wealthier in a community, with more resources to devote to forest product gathering and production, who are the heaviest users (Madge, 1990; Cavendish, 1996; Ogle, 1996), the poor usually derive a greater share of their overall needs from forest products and activities (Siebert and Belsky, 1985; Fernandes et al, 1988; Hecht et al, 1988; Jodha, 1990; Gunatilake et al, 1993). The characteristics of easy access to the resource and low entry thresholds enable many women to generate income from forest products activities. Such activities are often an important source of the income that women need to meet the costs of feeding and clothing the family, and they can be more dependent on such income than men (eg Hopkins et al, 1994).

In short, if we are to arrive at meaningful estimates of the importance of forests and forest products to people in their vicinity, we need to focus on measures that reflect the diversity of situations that exists, and the fact that for most this importance is best expressed in qualitative rather than quantitative terms. To do otherwise could prove to be very misleading. In the one region in which household survey results over large areas and populations do give us some basis for estimating numbers of people involved, Africa south of the Sahara, it appears that roughly 15 million people, about 4 per cent of the rural population, obtain some of their income from forest products.[2] (The conditions of extreme poverty that underlie much of forest products employment in Africa are not so widespread in Asia or Latin America. Thus it seems unlikely that these regions would exhibit such high densities of forest products involvement, except in more remote parts of the Indian subcontinent, Indochina and inland China.)

Only a small proportion of this number is living in or adjacent to forests; most are in more sparsely wooded parts of the region, and much of their supply of 'forest' products comes from outside forests. A study in a forested area in Sierra Leone, for instance, found that only 14 per cent of all hunted or collected foodstuffs and 32 per cent of medicinal plants were derived from the forest itself (Davies and Richards, 1991). To focus solely on numbers of people obtaining income directly from forests would therefore lead us to conclude incorrectly that a rather small share of the rural population uses forest products.

To narrow our concern down just to those who obtain the greater part of their inputs from forest products, as is done in some estimates of 'forest dependency', would also significantly understate the extent and nature of the links between people and supplies of forest products. In a survey of the forest zone in southern Ghana, for instance, where 10 per cent of the population were generating some income from forest products activities, only a minority of those involved reported that it was a major source – but more than 70 per cent stated that it was important in helping them meet particular needs, either because of its timing, or in absolute terms (Townson, 1995b).

The need to focus on the manner and importance of forest product inputs into users' livelihoods becomes even more evident when we look at how the relationships between people and forests are changing.

Patterns of change in use of forest products and access to forests

Changes in subsistence use

In some situations subsistence use of forest products appears to be dwindling, as people rely to a greater extent on food purchasing, as famine relief programmes become more effective, or as improved supplies of food crops have diminished the need to depend on forest foods. In Vanuatu, for instance, the introduction of the sweet potato, which can be planted at any time and produce an edible crop within three months, and manioc, which can be left unharvested for up to two years, has made the traditional emergency foods of wild taro, arrowroot, wild yams and sago virtually obsolete (Olsson, 1991).

Other changes that reduce the role that forest food plays in household nutrition may reflect penetration of rural markets by new food products, changing tastes, or decreased availability. But the latter can reflect changes in the availability or allocation of a household's supply of labour rather than physical shortage of the product. As the pressures on women's time increase they may no longer have as much time for gathering forest foods. As the value of labour rises with increasing wealth, the opportunity cost of continuing to spend time gathering foods, rather than purchasing them, becomes increasingly unattractive.

A decline in consumption of forest food can also reflect reduced knowledge about its use. As children spend more time in school than in the fields and the bush, the opportunity to learn about which plants can be consumed, and which cannot, is reduced. Settlement of previously nomadic populations in a fixed location is another widespread change that distances people from the food sources they used to be familiar with, constraining people's use of these foods even when they are still available and important for dietary balance (eg Melnyk, 1993).

A frequent cause of reduced subsistence use is likely to be shortages. These may be physical shortages due to overuse, shortages due to increasing restrictions on access to supplies, or economic shortages due to rising costs and/or growing competition for supplies. The needs of the poor for income from forest products activities can result in the diversion of supplies from own consumption to the market. A recent village study in Vietnam, for instance, found that forest vegetables, bamboo shoots and mushrooms that were collected and *eaten* by wealthier households had to be sold by poorer households in order to be able to buy rice (Nguyen Thi Yen et al, 1994).

Some changes in subsistence use therefore reflect choice; part of the process of evolution to a different livelihood level in which forest inputs have a lesser role. Some are responses to pressures that make it less possible for the household to maintain the same level of use. It is clear, however, that, in general, subsistence use continues to be very large, even where people are becoming increasingly integrated into the market economy. Moreover, the buffer role of the forest – as a resource that people can draw upon for emergency supplies of food, fodder and sources of income during periods of agricultural shortfalls or unemployment – continues to be very important for many people (Falconer, 1994; Ogle, 1996).

Patterns of change in income-generating activities

In some situations households are becoming more reliant on income from tree products activities, while in others they are moving away from involvement. At the same time, some kinds of forest products activities are expanding while others are stagnating or declining. If we are to understand where access to forest outputs is likely to be important in the future, the capacity to identify and understand these patterns of differential change will be necessary.

Market factors

The level of output in some activities is changing because of the nature of the markets into which the product is sold. Though some products have large, diversified and stable markets, others face highly volatile markets, or demand that is seasonal and subject to sharp price fluctuations. Production of some of the 'extractive' products for industrial markets, such as babaçu oilseed from the Amazon, can prove susceptible to major changes in market requirements, and to shifts to domestic or synthetic sources of supply (May et al, 1985; Afsah, 1992; Browder, 1992; Homma, 1996).

Domestic markets for forest products may provide more stable avenues for development. The large component of forest products activities in the rural sector reflects the size of rural markets for these products, and their dispersion across large areas with a relatively poor transport infrastructure, so that they are more effectively supplied locally (FAO, 1987). Market transactions in forest products grow as use of products that were not previously sold in rural areas, such as fuelwood and forest fruits, become increasingly commercialized. Most growth, however, is usually associated with expansion of urban demand. This tends to be based on a number of staple products that formed part of rural use patterns, and which continue to be consumed as people move to the towns, particularly when poverty constrains their access to more expensive formal sector products. A 1992 national household survey in Ghana found that even in the city of Accra more than three-quarters of all households purchased one or more of five forest products[3] (with as high or higher proportions of purchasers recorded in the rest of the country) (Townson, 1995b).

Some forest products used domestically, however, are 'inferior goods' that fall out of consumption patterns as incomes rise – some forest foods being displaced by more convenient purchased foods for example. Others are vulnerable to competition from factory-made alternatives as improved transport infrastructure opens up rural areas to outside supplies.

But demand for other ('normal' or 'luxury') goods rises with prosperity. The market prospects for products will also differ according to whether they are being sold into emergent, expanding, mature or declining markets. Thus, in the countries surveyed in eastern and southern Africa, employment in woodworking, a relatively high input and output activity serving expanding urban as well as rural demand, was growing ten times as fast as employment in grass, cane and bamboo activities (mats, baskets etc), which are low input and output processes tied to agricultural demand and subject to competition from alternative products (Arnold et al, 1994).

Production factors

The evolution of some activities is conditioned by the fact that features of their production or distribution process enable or prevent the component enterprises increasing in size, or adding extra value by diversifying into additional stages of the process, or organizing the process more efficiently. For example, the fact that most woodworking activities involve processes that can be readily upgraded and scaled up when faced with increased demand and competition, while mat and basket-making usually cannot, seems to contribute to the differential rates of growth between the two noted in the previous section. Other reasons for growth or decline are to be found within the individual enterprise. The opportunities to generate income from expanding forest products activities may require managerial or particular technical skills, or access to capital or credit, and will therefore be available only to some. Despite the importance of such activities to the poor, they may be less likely than wealthy neighbours or outsiders to be able to exploit the opportunities that new or upgraded markets present. This is of particular importance given the concentration of poor households in forest output activities.

Another powerful factor is the availability, and relative attractiveness, of alternative ways of earning income. Many forest products activities are time consuming, often tedious and arduous, and generate very low returns. They are consequently likely to be abandoned once more rewarding and congenial alternatives become available, or as increasing pressures on household labour resources make such low-value, labour-intensive activities no longer competitive. Others are likely to be attractive only temporarily, for example, wood fuel production and sales practised by immigrants or young men in the process of clearing land in order to create their own farms.

Access to forests

Nearly everywhere users of forest products are faced with a decline in the size or quality of the resource from which they obtain their supplies. This reflects clearance of land for agriculture and pasture as growing populations put pressure on the land base, timber harvesting that damages or destroys other components of the forest, and measures that alienate the resource to the state or result in de facto privatization by the wealthier and more powerful of the users.

In recent times the reduction in availability of forest resources has nearly everywhere accelerated. Increasing pressures on what is left have frequently led to progressive degradation. In a seminal study of dryland plain areas in several Indian states, Jodha (1990) found that the area of common pool land available to villagers as a source of forest products had declined by margins ranging from 31 per cent to 55 per cent in the 30 years after 1951. Concurrently, traditional methods of access control, usufruct allocation and conflict resolution had widely become ineffective or had disappeared, undermined by political, economic and social changes within the village and nation. Increasing population pressure and in-migration of outsiders, greater commercialization of the products of the resource, and technological changes that encourage alternative uses of the land were among the factors underlying these trends.

As access to forest resources declines, many users are progressively restricted in their choice to resources available in bush fallow and farm bush on lands over which they have some measure of individual control – and to resources they can create by growing trees on their farms. Historically, the place of trees on farms has been shaped primarily by growing pressures on limited amounts of arable land. But, as farm households have increasingly to depend on income earned from employment off-farm, labour rather than land is widely becoming the main resource limitation determining farmer options. Because the growing of trees requires lower inputs of labour to establish and maintain than most other crops, such shifts in the ratio of labour to land can encourage greater reliance on tree crops, in a number of different circumstances (Arnold and Dewees, 1995).

In many places, the focus of supply of some 'forest' products to rural households is steadily shifting from the forest to the farm. This shift, however, is only possible for those who have access to land, and sufficient resources to work the land. Reduced access to forest sources of the products they need is consequently often reducing the ability of small farmers, landless households and others among the poor to continue to participate in forest output activities. In addition, farm trees can provide only some of the outputs that people previously obtained from forests; they are therefore not a full substitute for the latter.

Reassessing 'forest dependency'

To summarize, a very large share of rural populations, and many urban households, in developing countries still include one or more forest or tree foods in their diets, cook their food using wood fuels, rely on traditional medicines from plant and animal products, and use wood and fibre for construction. Though the share of such products in overall use may often be declining, and supplies are increasingly coming from managed tree stocks rather than natural forests, the numbers of users may be expanding. Tens of millions of people also generate some of their income from forest products; and the importance of this clearly is growing.

Some of these changes reflect shifts in consumption patterns and habits, and the emergence of more productive or attractive options, as economies grow and incomes rise. Others are changes enforced by competition, rising costs and declining availability. With such diversity of situations, it is difficult and could be misleading to try and draw conclusions of general application. This is all the more so because of the wide variation in needs and use of forest outputs between richer and poorer within a community, and within a household between men and women, and even between age groups.

Nevertheless, some general patterns can be discerned. In situations where population is growing faster than per capita incomes, forest products activities emerge largely to absorb people unable to obtain income, or sufficient income, from agriculture or wage employment. This situation is likely to be characterized by labour-intensive, low-return, typically household-based, activities such as collecting and mat-making.

In situations where per capita incomes are rising, growth is more likely to be demand-driven, and low-return, labour-intensive activities tend to give way to more productive and remunerative activities such as vending, trading and activities to meet growing and diversifying rural and urban demands. At that stage, production and selling of forest products increasingly shifts from a part-time activity by very large numbers of people to more specialized year-round operations by a smaller share of the population (Haggblade and Liedholm, 1991; Liedholm and Mead, 1993).

These patterns can be modified in a number of ways, for example, where worsening urban poverty temporarily increases demand for low-cost forest products which normally would have been displaced in urban markets (Vosti et al, 1997). Over time, however, we can expect some forest products to become increasingly important, while others will fall out of use, and cause some related activities to become redundant and decline. In particular, those that generate only marginal returns to those engaged in their harvest and sale are unlikely to survive as costs rise and competition intensifies, or will persist only as long as the participants have no better option.

This shift is complicated by the fact that forests and forest products activities often play an important buffer role during the process of growth and change; a source of products and income that people can fall back on temporarily if need be. It therefore cannot be assumed, as people's use of forest products diminishes, that this necessarily removes the 'safety net' role of the forest.

These complex relationships cannot be assessed by unidimensional criteria, and cannot be merely described as 'forest dependency' without a serious loss of important information. In the following section we attempt a typology of these relationships in which people use and rely on the forests, and the reciprocal side of the relationship – what each group puts back into the forests, in terms of their protection and management. Then we can consider the forces for change in people's use of the forests, and consider the implications of such changes for the question of: who will manage the forests, for what purposes? In preparing for the devolution of decision making and responsibility for forest stewardship to local forest-based communities, it will be essential to assess, for each specific context, the continuing interest, willingness and capacity to protect and manage forests, as the available set of socioeconomic alternatives expands with development.

Types of People–Forest Relationships

In this section we propose a typology of different kinds of users, identify for each type the principal aspects of the relationship to forests and forest outputs, the importance of the forest outputs in the livelihood system in question, and the likely impacts of change. There are several dimensions of such a typology (Table 7.1), reflecting different aspects of their role and importance:

- Participation in forest output activities: reflecting the frequency or timing of use of forest products. And the extent to which household labour is allocated to these activities.
- Role of forest products in household livelihood systems: their importance as a share of household inputs, and in meeting household livelihood strategy objectives.
- Impact of reduced access to forests: does the forest serve as an economic and ecological buffer for its users, or are there alternatives, such as trees outside the forests or non-forest tree sources, of needed inputs and income?

• Likely future importance of forest outputs: do users face a growing or declining demand for forest outputs, or the potential for expanded or decreased involvement in production and trade in forest products?

These factors underlie the assessment set out below of a number of broad types of people–forest situations. The first covers populations living within a forest environment; principally hunter-gatherers and those practising long-rotation shifting cultivation. The second encompasses populations in a predominantly agricultural landscape, and practising food systems in which they continue to rely on access to adjacent areas of forest and woodland to supplement what they can produce on-farm. The third includes people and households who draw on forests indirectly, for example traders, further processors, employees, and are not necessarily living in or near forests.

There are no clear-cut boundaries either between or within these component parts of the whole. For instance, at the edges of forests shifting cultivation grades gradually into rotational agriculture. Similarly, the agriculture-plus-forest-input category ranges all the way from rotational bush fallow subsistence agriculture to predominantly commercial agriculture systems. Equally importantly, there is often a wide range of different levels and patterns of people–forest relationship even within a single situation. Nevertheless, such a framework does help to capture some of the salient features of the variation in people–forest relationships.

Populations living within forests

Hunters and gatherers

Forests provide the main sources of food, and are usually of very great cultural importance. Any change in the extent and quality of the forest, or in access to traditional forest areas, is likely to be very disruptive for traditional use and activity patterns. Some populations (eg in parts of the Amazon basin; see Grenand and Grenand, 1996) have managed to retain predominantly subsistence and self-reliant ways of life. Most, however, are increasingly affected by exposure to market forces. Where this is the case they tend to be highly dependent on middlemen for access to outside markets for sale of their products, and for supplies of outside goods (Peluso, 1986; Afsah, 1992; Browder, 1992). Despite romantic notions of 'affluent subsistence' life is typically tough and short, with generally low returns per effort expended.

The direction and impacts of change are likely to vary widely. When exposed to external pressures and opportunities (eg from logging companies), some indigenous peoples are assimilated, others cling to traditional ways, while others (such as Dayaks in Malaysian and Indonesian Borneo) seek to find a middle road – to adapt to and absorb what suits them from modern industry and society, while retaining traditional cultural and social values, by alternating between the forest village and the industrial workplace.

Shifting cultivators

The forest forms the starting point for crop agriculture in areas of cleared forest, alternating with long periods of forest fallow. Such shifting cultivation is supplemented by gathering and hunting inputs from the forest, and from the bush fallow. The cultural importance of the forest is typically very strong.

Table 7.1 *Some criteria for assessing the importance of degree of reliance on forest outputs (with illustrative examples)*

Criterion	Indicator	Example
Participation in forest output activity [labour allocation]	*Year round*	–full-time activity (eg carpenter, trader, employee) or continuous part-time component of household activities
	Periodic	–to fill seasonal gaps or to exploit seasonal availability
	Temporary	– new farmers selling wood fuel while establishing farms
	Occasional	–a buffer in hard times, meeting one-off costs (eg marriages) or simply opportunistic
Role in livelihood systems	*Central/ fundamental*	–forest-dwelling hunter-gatherer and subsistence (true shifting cultivation) populations
	Major/important	–substantial share of household inputs; important supplementary role (seasonal income, dietary inputs); basis for livelihood enhancement (eg more profitable activity)
	Minor but significant	–improves palatability of diets; opportunistic/windfall source of inputs/income
	Risk limitation	–subsistence and economic buffer in hard times; 'safety net/ last resort' source of income; diversifies household input base
	Declining	–items falling out of household consumption patterns; unprofitable activities being abandoned as better alternatives become available
Impact of reduced access to forests	*Critical*	–threatens the existence of a community in its present form, eg forest dwellers
	Severe	–causes serious worsening of livelihood situation, at least temporarily, eg lack of reserve of forest foods, income, in time of extended drought or other calamity or collapse of full-time/major activity based on forest raw material
	Modest (transitional)	–can switch to source outside forest (eg bush fallow) or can switch to non-forest activity/product
	Minimal/ none	–no longer competitive or better alternatives now available
Likely future importance of forest outputs	*Could increase/ stay as important as at present*	–no better options emerge (eg in low-income, low-skill, stagnant economy) or depletion of non-forest raw materials increases 'dependence' on forest products (eg as bush fallow becomes depleted)
	Changes in structure: use concentrated on fewer products	–growth in use of selected products (providing opportunities for some but not others), or domesticated and non-forest resources become more important

Table 7.1 *Some criteria for assessing the importance of degree of reliance on forest outputs (with illustrative examples)*

Criterion	Indicator	Example
	Declining, but still important as buffer	–in earlier still unstable stages of evolution to a higher income, more dynamic economy as emergency relief/avoidance measures still not adequate
	Not important	–users phasing out forest product activities, due to no remaining buffer role; declining demand for many forest products; forest sources no longer competitive and replaced by domesticated sources or substitutes; or better livelihood alternatives

Any decline in the forest area available for the system, typically because of alienation by the state, encroachment and/or increasing population pressure, is likely to lead to shortened fallow periods, declining productivity, progressive degradation of remaining forest, and increasing reliance on fallow and farm as the source of 'forest' products. Exposure to the needs and opportunities to earn income from outside wage employment can also diminish the availability of *labour* needed to sustain the system, leading to a reduced fallow rotation, and causing the abandonment of low-value, labour-intensive forest product gathering activities.

Shifting cultivation systems, as with hunting-and-gathering, consequently seldom provide a basis for livelihood improvement and they are often difficult to sustain, even at present levels. Both are likely to decline as expanding rural infrastructure improves the prospects for more intensive agriculture and wage employment. Increasing market opportunities and pressures may, however, enable people to stabilize their situation by shifting to agroforest systems that provide more productive use of the forest area remaining available to them (eg rubber, fruit and dipterocarp agroforests in Indonesia; Michon and de Foresta, 1995).

Farming communities drawing upon the forest

Off-farm forest/woodland is drawn upon for inputs that cannot be produced on-farm, or that can be more efficiently supplied from off-farm resources. This reliance is likely to increase when crop yields have been poor, and other sources of income are not available. Increasing exposure to market forces, the access to new markets and new technology that this brings, and the growing internal differentiation within communities that usually follows, can have sharply different impacts on people–forest relationships for different groups within a rural community.

Wealthier farmers and landowners

These may be substantial users of forest outputs (eg fodder). With greater access to labour and capital, they are also usually in a better position to exploit new or expanded market opportunities for forest products as these arise. There are numerous recorded instances of economic or political elites within rural communities restricting the access of poorer members to forest resources which were previously available communally, as the market value of the latter becomes more attractive (eg Fernandes et al, 1988). Those with sufficient land, and other income, are also better able to create their own resource of planted trees. The wealthier are also more likely to want to exploit opportunities that new markets and agricultural technologies provide to change the use of land under forest or woodland (eg Jodha, 1990).

Poor farmers and landless families

These are likely to continue to rely on access to nearby forests or woodland. Indeed, their reliance will tend to increase as smaller farm size or declining farm productivity reduces farmers' food self-sufficiency. As this happens, their dependence on forest products as a source of income may increase to the point at which some have to sell products that previously they collected for their own use. The poor and landless are also the people for whom the role of the forest as a buffer on which they can draw in times of hardship is particularly important. At the same time, increasing pressures to seek wage employment, in order to meet income needs, are likely to mean that they have less time for labour-intensive forest products activities, and so become less able to exploit forest products opportunities.

Increasing conflicts consequently often arise between those able and wanting to privatize communal land, resources or products, and those who continue to need to have access to such resources. This has contributed centrally to the decline of historical collective management systems, and of the resources controlled in this way, and to the difficulties encountered in creating viable contemporary systems (Arnold and Stewart, 1991; Shepherd, 1992; Davis and Wali Alaka, 1993). In short, communities made up of component groups with widely different needs, interests and power find it very difficult to agree on how to manage a local forest resource collectively.

A fundamental feature of many situations is thus that it is the poor who would benefit most from being able to continue to draw on forest products, but they are often faced with a diminishing resource and a declining capacity to exploit it. This contributes to the factors – increasing cost and opportunity cost, and declining markets and competitiveness – that often result in forest products activities playing a reduced role in their livelihood systems.

Livelihoods based on commercial forest products activities

Artisans, traders and small entrepreneurs

Much production, processing and sale of forest products occurs as a part-time activity within farming households. Activities such as mat and basket-making, and wood fuel vending, which have low skill and capital thresholds of entry, figure prominently among non-farm commercial activities in which the rural poor engage in the absence of other employment opportunities (Fisseha, 1987; Liedholm and Mead, 1993).

For those products for which rural and urban markets grow, processing tends to shift from artisans working part-time from the home to a more full-time workshop basis. Both production and trading also tend to become located in larger rural settlements and towns. With this increasing specialization and concentration, some products previously traded by households are progressively taken over by outside traders and entrepreneurs.

A growing share of the numbers involved in forest products activities is thus outside the forest – for instance, in areas surveyed in southern and eastern Africa, 14 per cent of all those processing and selling forest products were located in urban areas (Arnold et al, 1994). As these are mainly full-time activities, their association with forest products is strong, but at least one stage removed from the forest. For instance, most of those in carpentry and woodworking, even in rural areas, are likely to acquire their wood raw materials through the timber trade rather than from the forest. While they are likely to be affected by decline in supplies, they are likely to have the resources to be able to shift to other sources, other products or other locations.[4]

Employees in forest industries

By far the greater share of those who obtain employment in forest industry activities are in the small enterprises discussed above, rather than in the formal sector.[5] But these are typically very small, averaging less than two persons per enterprise wherever information on their size exists (Fisseha, 1987; Townson, 1995a). The bulk of those engaged are the entrepreneurs and members of their families. Where larger, modern-sector, forest industries have become established, they can provide a wage employment alternative to local people previously cut off from such options, and enable them to reduce their reliance on more arduous and less rewarding forest products and shifting cultivation activities. Many forest industry jobs, however, tend to go to outsiders because of the skills required; in addition, employment in logging and primary processing in a particular area tends to be relatively short term – they can be classical 'boom-and-bust' activities. The employment and income they provide for a while needs to be set against the disruption that they can cause to existing livelihood systems in forest areas.

What are the Development Options for Forest People?

Thus far we have recognized that there are hundreds of millions of people drawing on products of forests, or trees outside forests, to different levels of intensity, with different degrees of freedom to choose. Even where the quantity of forest products used is small availability is infrequent, lack of access to forests can cause severe hardships in emergency conditions. This may be interpreted as a reflection of the poverty and marginalization of many of the people involved – who have few buffers or safety nets but the forest.

We have also noted that, though access to forest products is so widely important in enabling people to survive in a situation of poverty, this set of activities may be less important in helping people escape from poverty. Many are arduous, labour intensive, and less rewarding than alternatives that become available as forest regions are opened up. Moreover, because of their poverty, the poor are often not in a position to take advantage of those growth opportunities that some forest products activities do provide.

Another key point that has emerged is that 'forest' products are increasingly coming from sources other than forest – bush fallow, farm trees and other tree stocks outside forests as normally defined. Though the purpose of this chapter has been to examine just the extent of people's dependence on forests, it is clear that we can only understand the role of forests in meeting people's needs if we recognize this continuum. On occasion, management may most appropriately take the form of transforming forest into some type of 'agroforest', or creating resources with some of the attributes of forests within agricultural systems – home gardens, compound farms, farm woodlots etc.

The essential question for any government or development agency anxious to help those called 'forest-dependent people' is: what would one do? First, we would argue strongly that there is no general panacea or formula, but rather that a detailed assessment needs to be prepared, by (or at least with) those people concerned. This assessment would cover the complete range of the relationships between the people and the forests that they use and/or manage, the current limitations to their livelihoods, and the potentials and desire for change. A number of broad, overlapping types of situation can be identified:

1   Forests continue to be central to livelihood systems: local people are or should be the principal stakeholders in these forest areas; meeting their needs is likely to be the principal objective of forest management, and this should be reflected in control and tenure arrangements which are centred on them (Peluso and Padoch, 1996).
2   Forest products play an important supplementary and safety net role: users need security of access to the resources from which they obtain these products, but are often not the only users in that forest area; forest management and control is likely to be best based on resource-sharing arrangements among several stakeholder groups.
3   Forest products play an important role but are more effectively supplied from non-forest sources: management of forests may need to be geared toward agroforest structures; policies that constrain growing of trees outside forests need to be revised.
4   Participants need help in exploiting opportunities to increase the benefits they can obtain from forest products activities: constraints in the way of smallholder access to markets need to be removed (Dewees and Scherr, 1996); improved access to credit, skills, marketing services etc, may be required (Liedholm and Mead, 1993).
5   Participants need help in moving out of forest products activities of declining importance and limited potential: help is needed to provide them with new options, which are quite likely to be outside forestry.

This wide range of variation in the situation of people who draw upon forests and forest products clearly needs to be matched by a selective approach to intervention and support. For the first category, secure local tenure and user group control over management constitute necessary and sufficient conditions for the continuing wellbeing of both the people and the forest. In contrast, for the second category a solution that reflects multiple stakeholders' shared control of the forests is more likely to be appropriate. For many in the third category tenure and control inside the forest are likely to be less important than what happens outside the forest. The future of those in the fifth category may not be in the forests at all.

As situations are moving between these categories, often quite rapidly, it is also very important that we do not put in place, or encourage, institutional arrangements which, while they may be relevant at present, are likely to be inconsistent with changes that are taking place or are likely to take place. For instance, does devolution of responsibility for forest management to local communities make sense in those circumstances where the role of forest products is likely to decline? What would happen to control and management of forests if the former users' interest in forest products, and their time available for forest management and protection, declines after responsibility has been effectively transferred to them from the state?

Similar considerations arise over the more technical aspects of interventions to support people to optimize benefits from the forests they draw upon. In designing programmes to provide support to small-scale forest products activities, it is important to recognize that there are different potential target groups with different needs and opportunities. Those in the process of starting up a commercial activity face different problems and constraints than those seeking to expand. New entrants driven by supply-side forces, as people search for activities where they can sustain themselves, face different issues than those who are responding to market opportunities. Among those enterprises that are growing, those seeking to expand from a one-person beginning have different needs for assistance from those that aspire to graduate to larger scales of operation.

Particular issues arise in trying to help the very large numbers of people engaged in low-return forest and tree products activities that can offer no more than marginal, unsustainable livelihoods. Support to such activities once higher return or less arduous alternatives emerge could impede the development of better livelihood systems for the participants. That being the case, it may be more fruitful to help people move into other more rewarding fields of endeavour rather than seeking to raise their productivity in current activities. Care needs to be taken in such a case to ensure that not only current income levels but also future growth prospects are indeed better in alternative products to which people are being encouraged to move (Liedholm and Mead, 1993; Arnold et al, 1994).

Conclusions

Huge numbers of people draw upon forest products, or similar products from tree cover outside forests, to meet part of their subsistence and income needs. Better estimates of the numbers of people in each of the categories we have described would help clarify options for development initiatives and policy reform, but would be costly and difficult to obtain. But, the importance of this people–forest relationship is not best measured or understood through attempts to estimate the numbers of people who 'depend' on forest outputs for a specified share of their livelihood inputs. Even where data could be generated on which one could base sound estimates of such magnitudes, such a focus would fail to recognize that the importance of many of these product and income flows lies in their timing and quality in terms of the livelihood strategy of the household in question, not in their magnitude.

In seeking to establish more useful ways of assessing the significance of people's use of, and reliance on, the outputs of forests and trees, a number of other fundamental factors need to be taken into account. One is the need to distinguish between those uses which do reflect actual dependency on the forest, in the sense that the users would be left seriously worse off in their absence, and those uses which reflect choice, and the presence of adequate alternatives. Much of what gets included in most estimates of 'forest-dependent people' is actually choice. Failure to recognize and clarify this can make it more difficult to focus on the problems of the main group of people who are dependent – namely the large numbers who rely on forests for some of their subsistence inputs, and who are involved in low-input, low-output commercial forest products activities because they simply do not have alternative ways of generating income.

A second factor that has major implications is the impact of change on the different categories of user. It is becoming increasingly clear that, as we would expect from related economic theory (Homma, 1996; Wilkie and Godoy, 1996), increasing integration into market systems, and the resulting opportunities to meet some local needs by purchasing from outside rather than by local production, and to generate income by selling other products previously consumed locally, can materially and rapidly alter patterns of forest product use. Activities based on low-value, labour-intensive forest products and processes will usually decline, while those based on higher-valued products in demand in the markets should increase. Information about the patterns of use that we are observing in the present, that are usually dominated by low-value products, is therefore likely to be of only limited usefulness in identifying what demands people will make on the forest in the future. Clearly this has major implications for the kind of institutional, support and management interventions that might be needed and appropriate.

Another important aspect that bears on how we adapt forest management to future local needs is the shift in the focus of production of many 'forest' products to bush fallow, trees on farms and other categories of tree cover and tree stocks outside the forest. Managing natural resources to supply 'forest' products may in some situations need to focus more on these 'agroforest' systems than on forests. This argument needs to be qualified, however, to take account of the continuing role of the forest as a buffer. Some of those who are moving away from stagnant minimal levels of livelihood, as better alternatives emerge and their incomes rise, will continue to need the forest as a buffer to fall back on temporarily in times of hardship. An important challenge is likely to be learning how to manage forests both for growth, and also as a safety net.

A final factor, following on from the others, is that of correctly matching institutional arrangements to these changing patterns of demand, use and supply. In what ways will a sharp decline in local uses of forest products, and trends towards individualization rather than collective local control of tree resources, affect the present thrust for governments to 'hand back' forest management to local people, or at the very least to give them a decisive voice in management decisions and practices? Will 'models' for local participation in decision making that rest on local collective institutions still make sense in these situations? There can be no uniform answers to such a question, but the variety of situations that is evolving will surely require a considerable degree of choice and flexibility in developing appropriate institutional arrangements.

Acknowledgements

The authors gratefully acknowledge the valuable comments and suggestions of Professor Bruce Campbell at the University of Zimbabwe, Harare, Dr R. J. Fisher at RECOFTC, Bangkok, and Dr Manuel Ruiz-Perez and Dr Lini Wollenberg at CIFOR.

Notes

1   This section draws on studies which have synthesized large amounts of material from case studies from ethnobotany, anthropology, geography and other disciplines on forest products output and use in particular situations (eg de Beer and McDermott, 1989; Falconer and Arnold, 1989; Falconer, 1990; FAO, 1995; Townson, 1995a).

2   This estimate is based on the findings from comparable recent surveys by the GEMINI project of small enterprises in Botswana, Kenya, Lesotho, Malawi, Swaziland and Zimbabwe (Liedholm and Mead, 1993; Arnold et al, 1994). These disclosed that, on average, 2.3 per cent of rural populations and 0.8 per cent of urban populations were obtaining income from small-scale forest products activities. Given that about half the population is in the economically active age range of 15–64, this implies an involvement by nearly 5 per cent of those who were economically active. The information on the size and composition of the workforce engaged in these enterprises suggests that therefore at least 10 per cent of rural households had some income from these forest products activities. When allowance is made for what was not covered by the surveys (which recorded only those who sold at least half of their production, and included only those products classified in the ISIC code as forest products – mainly woodwork, wood fuels and cane products) it would appear that about 3.5 per cent of the rural population (rather than the 2.3 per cent recorded), and 15 per cent of rural households, could have been gaining some income from forest products activities. If this ratio were to be applied to the total rural population in Africa south of the Sahara of about 360 million people, it implies that about 12.5 million may generate some income from forest products (and about another million in urban areas). Given that the density of involvement in the high forest zone across West and Central Africa is considerably higher than in the region surveyed, this in turn is likely to be an under- rather than an overestimate. This suggests that the total number earning some income from forest products in the region could be in the order of 15 million persons.

3   In order of importance by value: wood fuels, palm wine and spirit distilled from palm wine, bushmeat, snails, furniture and shea butter (Townson, 1995b).

4   The African surveys showed that, although there were high rates of closure among small forest product enterprises, nearly one-half of the entrepreneurs involved started new enterprises – some in forest products, most in non-forest products activities (Arnold et al, 1994).

5   For instance, the estimate from a national small enterprise survey of 237,000 persons employed in small woodworking, carving, wood fuel, and cane and grass product enterprises in Zimbabwe in 1991 compares with a reported 16,000 employed in forestry and forest industries in the country in that year (Arnold et al, 1994).

References

1   Afsah, S. (1992) *Extractive reserves: Economic-environmental issues and marketing strategies for non-timber forest products.* Draft. ENVAP, The World Bank, Washington, DC.

2   Arnold, J. E. M. and Dewees, P. A. (eds) (1995) *Tree Management in Farmer Strategies: Responses to Agricultural Intensification*. Oxford University Press, Oxford.

3   Arnold, J. E. M. and Stewart, W. C. (1991) *Common Property Resource Management in India*. Tropical Forestry Paper no 24, Oxford Forestry Institute, Oxford.

4   Arnold, J. E. M., Liedholm, C., Mead, D. and Townson, I. M. (1994) *Structure and Growth of Small Enterprises using Forest Products in Southern and Eastern Africa*. OFI Occasional Papers no 47, Oxford Forestry Institute, Oxford and GEMINI Working Paper no 48, Growth and Equity through Microenterprise Investments and Institutions (GEMINI) Project, Bethesda, Maryland.

5   Browder, J. O. (1992) 'The limits of extractivism: Tropical forest strategies beyond extractive reserves'. *Bioscience*, vol 42, pp174–182.

6   Cavendish, W. P. (1996) *Environmental Resources and Rural Household Welfare*. Mimeo, Centre for the Study of African Economies, University of Oxford.

7   Cecelski. E. (1987) 'Energy and rural women's work: Crisis, response and policy alternatives'. *International Labour Review*, vol 12, pp41–64.

8   Davies, A. G. and Richards, P. (1991) *Rain Forest in Mende Life: Resources and Subsistence Strategies in Rural Communities around the Gala North Forest Reserve (Sierra Leone)*. A report to ESCOR, Overseas Development Administration, London.

9   Davis, S. H. and Wali Alaka (1993) *Indigenous Territories and Tropical Forest Management in Latin America*. Policy Research Working Paper Series no 1 100. World Bank, Washington, DC.

10   de Beer, J. and McDermott, M. (1989) *The Economic Value of Non-timber Forest Products in Southeast Asia*. The Netherlands Committee for IUCN, Amsterdam.

11   Dewees, P. A. and Scherr, S. J. (1996) *Policies and Markets for Non-timber Tree Products*. EPTD Discussion Paper 16. International Food Policy Research Institute. Washington, DC.

12   Falconer. J. (1989) *Forestry and Nutrition: A Reference Manual*. Forestry Department, FAO, Rome.

13   Falconer. J. (1990) *The Major Significance of 'Minor' Forest Products: The Local Use and Value of Forests in the West African Humid Forest Zone*. Community Forestry Note 6, FAO, Rome.

14   Falconer, J. (1994) *Non-timber Forest Products in Southern Ghana: Main Report*. Republic of Ghana Forestry Department and Overseas Development Administration, Natural Resources Institute, Chatham.

15   Falconer, J. and Arnold. J. E. M. (1989) *Household Food Security and Forestry: An Analysis of Socioeconomic Issues*. Community Forestry Note 1, FAO, Rome.

16   FAO (1987) *Small-scale Forest Based Processing Enterprises*. Forestry Paper 79, FAO, Rome.

17   FAO (1995) *Non-wood Forest Products in Nutrition*. Paper prepared for the FAO/GOI Expert Consultation on Non-Wood Forest Products, Yogyakarta, Indonesia, 17–27 January.

18   Fernandes, W., Menon, G. and Viegas, P. (1988) *Forests, Environment and Tribal Economy: Deforestation. Impoverishment and Marginalisatiion in Orissa*. Indian Social Institute, New Delhi.

19   Fisseha. Y. (1987) *Small-scale Forest Based Processing Enterprises*. Forestry Paper 79, FAO, Rome.

20   Grenand, P. and Grenand, F. (1994) 'Living in abundance. The forest of the Wayampi (Amerindians from French Guiana)', in Ruiz-Perez, M. and Arnold, J. E. M. (eds) *Current Issues in Non-Timber Forest Products Research*. CIFOR-ODA, Bogor, Indonesia, pp177–196.

21   Gunatilake, H. M., Senaratine, D. M. A. H. and Abeygunawardena, P. (1993) 'Role of non-timber forest products in the economy of peripheral communities of Knuckles National Wilderness area of Sri Lanka'. *Economic Botany*, vol 47, pp275–281.

22   Haggblade, S. and Liedholm, C. (1991) *Agriculture, Rural Labor Markets, and Evolution of the Rural Nonfarm Economy*. Working Paper no 19, Growth and Equity through Microenterprise Investments and Institutions (GEMINI) Project, Bethesda, MD.

23   Hecht, S. B., Anderson, A. and May, P. (1988) 'The subsidy from nature: Shifting cultivation, successional palm forests and rural development'. *Human Organization*, vol 47, pp 25–35.

24   Homma, A. K. O. (1996) 'Modernisation and technological dualism in the extractive economy in Amazonia', in Ruiz-Perez, M. and Arnold, J. E. M. (eds) *Current Issues in Non-Timber Forest Products Research*. CIFOR-ODA, Bogor, Indonesia, pp59–81.
25   Hopkins, J. C., Scherr, S. J. and Gruhn, P. (1994) *Food Security and the Commons: Evidence from Niger*. Draft report to USAID Niger, IFPRI, Washington, DC.
26   Jodha, N. S. (1990) 'Rural common property resources: Contributions and crisis'. *Economic and Political Weekly. Quarterly Review of Agriculture*, vol 25, no 26.
27   Leach, M. and Fairhead, J. (1994) *The Forest Islands of Kissidougou: Social Dynamics of Environmental Change in West Africa's Forest-savanna Mosaic*. Report to ESCOR, Overseas Development Administration, London.
28   Liedholm, C. and Mead, D. C. (1993) *The Structure and Growth of Microenterprises in Southern and Eastern Africa*. Working Paper no 36, Growth and Equity through Microenterprise Investments and Institutions (GEMINI) Project, Bethesda, MD.
29   Lynch, O. J. and Talbott, K. (1995) *Balancing Acts: Community-Based Forest Management and National Law in Asia and the Pacific*. World Resources Institute, Washington, DC.
30   Madge, C. (1990) *Money, Medicine and Masquerades: Women, Collecting and Rural Development in the Gambia*. PhD thesis, School of Geography, Faculty of Arts of the University of Birmingham, Birmingham.
31   May, P. H., Anderson, A. B., Balick, M. J. and Unruh, J. (1985) 'Babaçu palm in the agroforestry systems in Brazil's mid-north region'. *Agroforestry Systems*, vol 3, pp275–295.
32   Melnyk, M. (1993) *The Effects of Sedenterization on Agriculture and Forest Resources in Southern Venezuela*. Network Paper 16b, Rural Development Forestry Network, Overseas Development Institute, London.
33   Michon, G. and de Foresta, H. (1995) 'The Indonesian agro-forest model: Forest resource management and biodiversity conservation', in Halladay, P. and Gilmour, D. A. (eds) *Conserving Biodiversity Outside Protected Areas: The Role of Traditional Agro-ecosystems*, pp90–106. The IUCN Forest Conservation Programme, IUCN, Gland, Switzerland.
34   Nguyen Thi Yen, Nguyen Quang Duc, Vu Manh Thien, Dang Due Phuong and Ogle, B. A. (1994) *Dependency on Forest and Tree Products for Food Security. A Pilot Study in Yen Huong Commune*. Ham Yen District, Tuyen Quang Province, North Vietnam. Working Paper 250, Swedish University of Agricultural Sciences, Uppsala.
35   Ogle, B. (1996) 'People's dependency on forests for food security. Some lessons learnt from a programme of case studies', in Ruiz-Perez, M. and Arnold, J. E. M. (eds) *Current Issues in Non-Timber Forest Products Research*. CIFOR-ODA, Bogor, Indonesia, pp219–241.
36   Olsson, G. (1991) *Forests and Forest Product Use in Vanuatu and Tonga*. Working Paper 2 – RAS/86/036, South Pacific Forestry Development Programme, FAO/UNDP, Port Vila, Vanuatu.
37   Peluso, N. L. (1986) *Rattan Industries in East Kalimantan, Indonesia*. Draft Paper commissioned by FAO, FAO, Rome.
38   Peluso, N. L. and Padoch, C. (1996) 'Changing resource rights in managed forests of West Kalimantan', in Padoch, C. and Peluso, N. L. (eds) *Borneo in Transition: People, Forests, Conservation, and Development*. Oxford University Press, Singapore, pp121–136.
39   Pimentel, D., McNair, M., Buck, L., Pimentel, M. and Kamil, J. (1997) 'The value of forests to world food security'. *Human Ecology*, vol 25, pp91–120.
40   Ruiz-Perez, M. and Arnold, J. E. M. (eds) (1996) *Current Issues in Non-timber Forest Products Research*. CIFOR-ODA, Bogor, Indonesia.
41   Shepherd, G. (1992) *Managing Africa's Tropical Dry Forests: A Review of Indigenous Methods*. Agricultural Occasional Paper 14, Overseas Development Institute, London.
42   Siebert, S. and Belsky, J. M. (1985) 'Forest product trade in a lowland filipino village'. *Economic Botany*, vol 39, pp522–533.

43 Townson, I. M. (1995a) *Forest Products and Household Incomes: A Review and Annotated Bibliography*. Tropical Forestry Papers 31, CIFOR and OFI, Oxford.

44 Townson, I. M. (1995b) *Patterns of Non-timber Forest Products Enterprise Activity in the Forest Zone of Southern Ghana*. Draft report to the ODA Forestry Research Programme, London.

45 Vosti, S. A., Witcover, J., Gordon, A. and Fereday, N. (1997) *Domestic Market Potential for Tree Products from Farms and Rural Communities: An Executive Summary*. Final report to UK Department of International Development, International Food Policy Research Institute (IFPRI), Washington, DC.

46 WCFSD (1997) *Our Future Our Forests*. Final Report Draft, World Commission on Forests and Sustainable Development, Geneva.

47 Wilkie, D. S. and Godoy, R. A. (1996) 'Trade, indigenous rain forest economies and biological diversity. Model predictions and directions for research', Ruiz-Perez, M. and Arnold, J. E. M. (eds) *Current Issues in Non-timber Forest Products Research*. CIFOR-ODA, Bogor, Indonesia, pp83–102.

Part 3

Threats and Opportunities

A number of forest issues have captured public attention in recent years. In some cases they have become inflated in international debates and even in the media, and have been portrayed as looming tragedies. This section reviews some of these issues and presents a series of chapters that set the problems in their true perspectives. The intention is not to pretend that these are not serious problems, but rather to demonstrate the complexity of the underlying issues and to show that solutions may not be as simple as one might imagine.

Chapter 8 by Kaimowitz reviews the question of forest law enforcement. Illegality in the forest sector has been prominent on the international forest agenda in the past few years – the extent of the problem is truly horrific. But the knee-jerk reaction of negotiators and the public is that we need to clamp down on all abuses of forest law. The normal way of doing this is by strengthening the law enforcement capacity of forest departments. The problem is that one must ask why these laws are so frequently violated – perhaps because many of them lack legitimacy in the eyes of significant sections of the population. If laws favour the rich and penalize the poor, or if laws are put into place by corrupt dictatorships, then it is small wonder that many people will try to bypass them. Kaimowitz argues that laws must be just and that it is often the rural poor who are the victims of both bad laws and the abuse of laws by influential elites. What is needed is a thorough revision of many aspects of forest governance.

A second issue that has received a lot of international attention in recent years is that of the poaching of forest wildlife. A bushmeat crisis has been described, especially in Africa, where the sight of large numbers of dead animals being transported to market on logging trucks has aroused public indignation – at least in the Western media. In Chapter 9 Brown and Williams show that bushmeat is a significant source of animal protein in many forest areas and they argue that the best way forward is to seek ways of sustainably managing this resource. Rural people have relied on wildlife for their meat supplies for thousands of years and it is unlikely that they are about to stop now. In addition, the empirical evidence for the imminent extinction of bushmeat species is still weak. It appears that once the populations of wild animals drop below a certain threshold, people switch to other quarry or to completely different activities.

Arnold and Persson, in Chapter 10, give a balanced account of the so-called 'fuelwood crisis'. Prophets of doom have been saying for a couple of decades that poor people in the dry tropics were about to run out of wood for cooking and heating. Indeed, there are major problems in some areas but as usual the bigger picture is more complicated.

When trees begin to become a scarce resource then measures to husband them begin to emerge spontaneously. A huge amount of fuelwood no longer comes from forests – it comes from trees on farms and along roadsides. In addition people use by-products of agricultural activities as a source of fuel. The other interesting lesson from the fuelwood crisis is that the efforts of governments to regulate or provide improved access to fuelwood were rarely successful. Solutions that come from the people themselves generally worked better.

Forest fires have been prominent in the media in recent years. Some of the worst fire incidents ever recorded have occurred in the past decade. Fires in forests are increasing and climate change and general disruption of forest systems are making the situation worse. In Chapter 11, Goldammer shows that here again the situation is more complicated than it is often portrayed and that the solutions are rarely those that the popular media and political decision makers tend to favour. What is needed is management of the flammable materials that create the conditions for fires – we need more investment in fire management and less in fire fighting. We also need much more information on where and why fires occur, and a more pragmatic attitude to fires in general. Some fires have always been part of the basic ecology of many forest systems and should continue to be so.

8

Forest Law Enforcement and Rural Livelihoods

David Kaimowitz

Introduction

International concern about illegal forestry activities has grown markedly. Asian, African and European governments have held high-level regional conferences on Forest Law Enforcement and Governance (FLEG). Indonesia has signed path-breaking Memoranda of Understanding on illegal logging with the United Kingdom, China and Norway. The Convention on Biological Diversity, the United Nations Forum on Forests, the International Tropical Timber Organization, and the G8 have all issued forceful statements, and incorporated the issue in their work plans. The European Commission has committed itself to formulating a European FLEG Action Plan. Japan and Indonesia have initiated an Asian Forest Partnership, with a major focus on illegal logging. Global Witness, the Environmental Investigation Agency, Transparency International, Greenpeace, Global Forest Watch and Friends of the Earth have raised public awareness about the problem.

There are good reasons for concern. Illegal forestry activities deprive governments of billions of dollars in tax revenues. They also cause environmental damage and threaten forests which many people depend on. Forest-related corruption and widespread violation of forestry laws undermine the rule of law, discourage legitimate investment and give the wealthy and powerful unfair advantages, due to their contacts and ability to pay large bribes. Money generated from illegal forestry activities has even been used to finance armed conflict.

Nonetheless, greater enforcement of forestry and conservation laws also has the potential to negatively affect rural livelihoods. That is because:

- Existing legislation often prohibits forestry activities such as small-scale timber production, fuelwood collection and hunting that millions of poor rural households depend on.
- Most small farmers, indigenous people and local communities are ill equipped to do the paperwork required to engage in forestry activities legally or to obtain the technical assistance needed to prepare management plans.
- Millions of rural households live on lands that governments have classified as state-owned forestland or protected areas, and existing laws often consider them encroachers even though their families may have lived there for generations.

Note: Reprinted from *International Forestry Review*, vol 5, Kaimowitz, D., 'Forest law enforcement and rural livelihoods', pp199–210, Copyright © (2003), with permission from the Commonwealth Forestry Association

- Forestry and wildlife departments generally enforce forestry and protected area legislation more vigorously and with less respect for due process and human rights when poor people are involved.
- In some countries, forestry and wildlife officials engage in illegal activities that harm the poor. Measures that empower these officials and give them more resources could make it easier for them to act with impunity.

The magnitude of these risks varies greatly from country to country. Some countries have little interest or capacity to enforce their forestry and conservation laws and the increased international attention to forest law enforcement will probably not change that. Others focus their regulatory efforts almost exclusively on curtailing abuses by large logging companies, which is less likely to have major negative effects on rural livelihoods. But there are many countries where existing efforts to enforce forestry and conservation laws already have significant negative impacts on rural livelihoods. In those cases greater law enforcement efforts might make the problem worse.

This chapter addresses complex and difficult problems. In many cases attempts to solve one set of problems will create others. Policies that work well in one location may have unanticipated or disastrous consequences in others. Clearly there are situations where the positive benefits from enforcing forestry and conservation laws outweigh the negative impact this may have on livelihoods, so governments and communities sometimes need to take measures that restrict the options of poor rural households. Similarly, it would be unwise to be naive about how easy it is to get communities themselves to effectively regulate the use of forests. Still, there are good reasons to question many of the existing and proposed efforts to regulate forests, and to take steps to ensure that regulations do not simply justify wealthy and powerful groups gaining a monopoly on access to forest resources, rather than protecting the resource.

Even though the chapter illustrates its arguments with examples, the author has no desire to single out or criticize specific countries or individuals. The examples used happen to be ones the author had information on. Some are out of date. In other cases the author cannot fully verify the accuracy of the information taken from published sources. Thus, it is important that the reader focus on the broad issues the chapter raises and not on the examples as such. The examples have only been included to demonstrate that the arguments have an empirical basis and are not simply conjecture.

Forestry and Rural Livelihoods

According to the UK's Department for International Development (DFID, 2001), 'A livelihood comprises the capabilities, assets and activities required for a means of living.' It is considered that someone has a better livelihood if he or she:

- has a higher income;
- receives more government services;
- has their physical security respected;
- has better health;

- has adequate food;
- is less vulnerable to changes in markets or their environment;
- relies on natural resources that are managed sustainably;
- can participate in political processes; and
- can maintain their cultural heritage and self-esteem.

These criteria link up with the discussion about forest law enforcement and rural livelihoods in the following ways:

- *Income:* in the context of forestry activities, higher incomes for low-income rural households can come from small-scale forest-based activities or wage labour.
- *Government services:* tax revenues from forestry can finance government services for the rural poor.
- *Physical security:* respecting forest users' physical security implies not physically mistreating them or imprisoning them without adequate due process.
- *Food, health and vulnerability:* poor rural households with access to wild meat, vegetables, fruits and medicinal plants and animals are likely to have better food security and health. This is especially crucial in situations where families have already exhausted the food from their last harvest and in periods of economic crisis, war or crop failure.
- *Sustainable natural resource management:* if people use forest resources sustainably rural families should be able to maintain the benefits forests provide over time.
- *Participation and cultural heritage:* governments and other groups can enfranchise low-income forest-dependent people politically, protect their cultural heritage and legal rights, and encourage their self-esteem by providing institutional mechanisms for participation in decision making and respecting their rights, cultures and opinions.

DFID's livelihood approach also postulates that people with more natural, physical, financial, human and social capital generally have better livelihoods. In the context of this paper, natural and social capital merit the greatest attention. One might expect poor rural households to live better if they have secure access to forest resources and if they have effective and efficient social mechanisms to regulate forest use, manage their forests and distribute the benefits.[1]

Given the above, this chapter assumes forest law enforcement policies favour the forests' contributions to rural livelihoods if they:[2]

1 increase the amount of forest products poor rural households can sell and the prices they receive;
2 increase wage labour in forestry and the salaries forestry workers earn;
3 increase tax revenues from forestry companies;
4 decrease the number of poor rural people physically mistreated, forced to pay bribes or inappropriately arrested or fined;[3]
5 increase poor rural households' access to forest resources and make it more secure, including access by women;
6 help maintain the long-term supply of forest products and services poor households use;
7 promote poor people's participation in decision making and collective action; and
8 respect poor households' rights, cultures and traditions.

Negative Impacts of Illegal Forestry Activities on Rural Livelihoods

Illegal forestry activities often negatively affect rural livelihoods. Indeed, that is the main reason development agencies whose primary mission is poverty alleviation have become increasingly concerned about the problem.

Lost income from forest products

Situations of widespread corruption and disrespect for the rule of law typically favour groups that have sufficient resources to pay bribes, develop informal links with government officials, and hire armed guards (World Bank, 1997). This puts households that engage in small-scale forestry activities at a clear disadvantage. Often they can only operate if they agree to sell their products to wealthier 'patrons' who protect them from forestry officials and provide credit.[4,5] These patrons' assistance comes at a high price. They pay producers much less than they would receive if they could borrow from formal lending agencies and sell their products legally to whomever they wanted to.

Lost job opportunities

Illegal logging may generate employment in the short term but in the longer term it can contribute to the depletion of timber resources and the subsequent collapse of forest industries. This has already happened in several West African and South-East Asian nations, and could well happen in others such as Cambodia and Indonesia.

Less government revenue

Every year developing country governments lose billions of dollars in revenues due to illegal tax evasion in the forestry sector and unauthorized timber harvesting in publicly owned forests.[6] This leaves governments less money to spend on services such as health, education, roads, electricity and agricultural extension. Lack of transparency in government budgets in countries with widespread corruption makes it less likely that whatever funds governments do receive will go to services for the poor. Weak rule of law and corruption also limit long-term economic growth, which further reduces tax revenues (Thomas et al, 2000).

Threats to physical security

When local people complain about illegal forestry activities the implicated parties often respond with threats or even violence. In addition, corrupt government officials sometimes take action against local people to protect their interests or those of illegal loggers and poachers.

Loss of access to forest resources

In heavily corrupted systems the only way to get access to forest resources may be through bribes and connections. People who are unwilling or unable to use those mechanisms cannot access forest resources or risk fine and arrest by accessing them illegally. Without transparency in decision making and a functioning system of legal due process, poor rural households have little recourse when government officials or private companies and individuals illegally deny them access to forest resources.

Forest loss and degradation

Illegal forest clearing, poaching and failure to respect timber-harvesting regulations can deplete the natural resources poor rural households rely on such as wild fruits and vegetables, bushmeat, medicinal plants, fuelwood and timber. Illegal forestry activities can also negatively affect environmental services important to poor rural households such as the provision of clear water, pest and disease control, pollination and regulation of the climate, stream flow and groundwater levels.

Box 8.1 *Illegal logging and social capital in Sumatra, Indonesia*

Communities in northern Aceh in Sumatra traditionally regulated their forests without any need for outside intervention. Customary authorities kept people from logging near rivers and in the upper watersheds and charged fees to outside loggers, which they used for local development.

In the early 1990s, collusion between local officials and entrepreneurs involved in illegal forestry operations undermined that system. The local police, army, forestry and district officials allowed anyone who wanted to log, transport wood or operate a sawmill to do so as long as they paid monthly informal payments and gave small sums of money to police and army officials who checked the permits of logging trucks at posts along the road. Those that paid bribes received permits and protection from harassment in return. Anyone trying to operate legally found it too expensive and time-consuming, and faced problems with local officials. Local tax collectors pocketed most forestry taxes, passing only a small portion to the official government coffers.

Groups engaged in illegal logging paid off village heads or gave them a share of the business. This caused conflicts within the villages. Some villagers permitted illegal logging because they realized it was futile to try to stop it since it had the backing of the local authorities. Others ended up working for the illegal loggers and pressuring their village heads to let the activity continue. This greatly undermined traditional village forest management.

McCarthy, 2000

Political disenfranchisement and loss of social capital

Corruption and lack of respect for the rule of law subverts the democratic process. Elected public officials lose influence or fail to represent the interests of those who elected them, while small elite groups can use bribery and private business associations with government officials to influence policies in their favour (Contreras-Hermosilla, 2002). When individuals or groups within a community engage in illegal forestry activities or support others involved in such activities, that may create discord and undermine pre-existing mechanisms for regulating the use of forest resources. Formal community forestry initiatives have difficulty competing with groups that operate illegally, since the latter can sell their products cheaper because they don't pay taxes, prepare management plans or devote resources to paperwork. Ironically, formal community forestry groups often find it more difficult to get government permits than illegal operators do because they are less able or willing to pay bribes to obtain them.

Negative Livelihood Impacts of Forest Law Enforcement

Even though illegal forestry activities can be bad for rural livelihoods, so can enforcing the existing forestry laws, and doing it more effectively may make the problem even worse. This applies particularly to situations where legislation and/or law enforcement practices discriminate against poor rural households.

Most small-scale commercial forestry activities in developing countries are illegal or have unclear status under existing laws. Those involved generally do not have permits or formal management plans and do not pay taxes, and they often work without permission in forests claimed by governments or large landholders.

A large but unknown number of people engage in informal forestry activities. Poschen (1997) calculated that fuelwood and charcoal activities employed something in the order of the equivalent of 13.3 million people full-time in the early 1990s; the vast majority of them outside the legal framework. No one knows how many villagers practise small-scale informal timber harvesting, but it is probably in the millions.

Box 8.2 *The informal sawnwood sector in Cameroon*

Small-scale informal wood processing with chainsaws and small mobile sawmills provides some 60 per cent of the sawnwood used in Cameroon's two largest cities, Douala and Yaoundé, as well as small quantities for other markets. Most of the wood comes from farmers' fields and community forests, not the official permanent forest estate. This activity causes little environmental damage and directly generates an estimated 3000 permanent jobs. Nevertheless, in 1999 the Ministry of Environment and Forestry made it illegal. This apparently did not reduce the activity's scale, but greatly increased the bribes demanded by police, soldiers, forestry officials and others.

Plouvier et al, 2002

The high transactions costs associated with operating legally are a major factor that typically confines small-scale commercial forestry to the informal sector. Existing laws and regulations require extensive paperwork, payments and visits to government offices. Professional foresters must sign certain papers and the offices that process those papers are typically far away. When low-income people go there the officials they need to talk to may be away or unwilling to receive them. It frequently takes a long time to get any response, and officials may send papers back several times for corrections.

Enforcing existing forestry laws sometimes reduces small-scale producers' incomes by discouraging them from engaging in forestry activities or forcing them to sell their products illegally for lower prices. It also increases their costs associated with avoiding detection and paying fines and bribes.

Box 8.3 *Red tape for small-scale producers in Brazil and Nicaragua*

The municipality of Paragominas in the eastern Brazilian Amazon has long been one of the country's main logging and wood processing areas. Nevertheless, the local forest service (IBAMA) office is not authorized to approve management plans or most licences and forestry permits. Only the office in the state capital, Belem, several hours away, can do that.

Big logging companies usually hire representatives in Belem to handle their relations with IBAMA. These representatives visit IBAMA regularly and know the rules and officials well.

Small-scale loggers find it difficult and expensive to visit Belem. Even if they get there they have a hard time seeing officials and often discover they lack the proper documents or have filled them out incorrectly. As a result, they may have to stay in hotels for days or travel back and forth several times. Few manage to get through the system.

Similarly, for a small farmer in Nicaragua to sell a single tree he or she has planted requires several administrative steps and permits. But as one observer notes, 'Your typical farmer does not know anything about administrative procedures, has no telephone, lives far away from the urban centre where the government offices are, and cannot get mail delivered to their house. To do all the paperwork requires a huge effort and takes several days of work'.

Brazil: Interviews by author with IBAMA officials in Paragominas, February 2001. Nicaragua: Nitlapán 2002, cited in Contreras, 2003

Lost job opportunities

Restricting the activities of larger commercial forestry and agricultural operations may negatively affect rural livelihoods to the extent it limits employment opportunities for low-income rural people. This is most apparent in the case of logging bans where formal sector timber production and the associated jobs disappear completely. (See below.)

Threats to physical security

In some countries forestry officials and police inappropriately arrest low-income people for violating forestry and protected area legislation, forcibly expel them from their houses and fields, hit them, rape them or even kill them. Unless governments take measures to prevent this, attempts to encourage forest law enforcement could easily worsen this problem.

Loss of access to forest resources

Many governments essentially tolerate poor families living in forestlands and protected areas claimed by the government, but there are also many cases where families have been evicted from such areas, often forcibly.[7] If governments were to strictly apply existing forestry and conservation laws restricting poor rural households' access to forest resources, that could have dramatically negative impacts on them.

Forest loss and degradation

In some instances enforcing or attempting to enforce existing forestry legislation may encourage forest destruction. This may occur because government law enforcement efforts undermine the existing community-based mechanisms for regulating forest use or because the law promotes tenure regimes and forestry practices that negatively affect forest conditions. This may be the case, for example, where the law favours large-scale industrial logging and the conversion of forests to agro-industrial plantations, but discourages small-scale low impact forestry activities.

Box 8.4 *The negative influence of forestry laws on community forests in Cameroon and Honduras*

In theory, Cameroon's 1994 Forestry Law permits communities to manage community forests. However, communities must submit a management plan before they start any activity. That is practically impossible for communities that lack the support of foreign donors and extremely difficult even for those which enjoy such support. In contrast, industrial loggers with 'ventes de coupes' are allowed to log areas of up to 2400ha without a management plan and companies can log for up to three years in forest concessions of up to 200,000 ha before they must submit a management plan.

Existing forestry laws and the way they are applied have also undermined community forests in Honduras. In 1977, the Honduran forestry department (COHDEFOR) helped to found the 'Honduran Regional Agroforestry Cooperative of Colon and Atlantida' (COATLAHL). The cooperative was supposed to assist 500 small foresters organized into ten groups to do the paperwork required by the law and to market their timber. In recent years, COATLAHL has run into serious problems. Many members have found it more profitable to operate illegally. Forestry officials have tried to force the cooperative to pay bribes in order to transport their timber. Illegal loggers and wealthy coffee growers have shot at cooperative members to keep them out of certain forests. Instead of trying to help the cooperative overcome its problems, COHDEFOR has created additional administrative requirements the cooperative must meet. This along with some internal problems within COATLAHL has led most members to abandon the cooperative. At present, only 106 remain and some of them are no longer active. Meanwhile, illegal logging is thriving.

Cameroon: WRM, 2001. Honduras: Castillo, 2002

Table 8.1 *Threats to rural livelihoods from illegal forestry activities and from forest law enforcement*

	Illegal forestry activities	Forest law enforcement
Forest product income	Small foresters earn less because they pay bribes and depend on patrons	Small foresters earn less because governments stop their 'illegal' activities or greater law enforcement leads them to have to pay higher bribes or depend more on patrons
Wages from forestry	Over-harvesting makes the forestry sector collapse	Government actions reduce logging
Government revenues	Tax evasion and illegal logging in public forests lowers revenue	Reduced logging due to law enforcement lowers revenue
Physical security	Illegal loggers and corrupt officials threaten and attack villagers	Officials inappropriately threaten, attack, arrest or expel villagers and destroy their crops and houses
Access to forest resources	Wealthy groups and officials illegally deny access to forests and due process	Wealthy groups and officials 'legally' deny access to forests and due process
Long-term supply of forest goods and services that poor households use	Damage to forests by illegal logging reduces forest product supply and disrupts environmental services	Damage to forests due to loss of social capital reduces forest product supply and disrupts environmental services
Collective action and participation	Illegal logging undermines local forest management institutions. Bribes and influence peddling replace democratic process	Laws that fail to recognize local forest management institutions undermine them and police action substitutes for dialogue
Respect for cultures and tradition	Illegal logging undermines traditional institutions	Laws that don't allow traditional practices or respect local institutions undermine peoples' traditional cultures
Economic growth	Widespread failure to respect rule of law reduces investment and growth	Limiting forestry activities reduces (short-term?) economic growth

Political disenfranchisement and loss of social capital

Considering the residency or livelihood activities of large numbers of rural people as illegal essentially 'criminalizes' those people and makes it easier to deny them their political and legal rights and the opportunity to participate in decisions related to natural resource management. As noted above, trying to enforce laws that fail to recognize and build upon pre-existing 'informal' mechanisms to collectively regulate the use of forest resources may also undercut those mechanisms and make it more likely that forest resources will become essentially open access.

Lack of respect for local culture and traditions

Many forestry and conservation laws fail to recognize indigenous and nomadic peoples' rights over the territories they have historically occupied and to take into account their traditional farming, hunting, fishing, grazing and gathering practices. That makes it harder for many local people to maintain their traditional diets, health practices and ways of life. One common example of this, particularly in Asia, are laws that prohibit swidden cultivation (also known as shifting cultivation or slash-and-burn cultivation). Swidden cultivation forms an integral part of the traditional practices of many peoples, and in many cases is the main livelihood option that people have available to them.

Does Legal Mean Sustainable?

Discussions about forest law enforcement sometimes practically equate sustainable forest management with complying with forestry laws, but the two differ markedly. A large portion of forestry legislation focuses on administrative requirements, fees, taxes and property rights, rather than on how forests are really managed. Some regulations actually encourage unsustainable management and some people that violate forestry laws manage forests sustainably.

Successfully enforcing laws that prohibit forest clearing, logging, hunting and collecting vegetable products usually, although not always, directly helps to protect the forest resources involved, at least in the short run.[8] The situation is less straightforward when it comes to laws and regulations that specify annual allowable cuts, harvesting rotations and minimum harvesting diameters. Enforcing these regulations usually will not suffice to sustain commercial timber production and environmental services over the long term, and may even make things worse.[9] Many existing prescriptions for tropical forest management have a surprisingly weak scientific basis (Fredericksen, 1998; Putz et al, 2000; Sist et al, 2001). Frequently they fail to take into account the regeneration requirements of commercial timber species and the role of animals in seed dispersal, pollination and pest and disease control (Sheil and Van Heist, 2000). Allowable cuts usually reflect political, economic and administrative concerns as much as the biological capacity of a forest to sustain timber production. Most legally sanctioned approaches to designing forest management plans assume forest ecosystems are in a steady state, rather than being path dependent outcomes of episodic disturbances. In principle, foresters are supposed to adapt management plans to the dynamic of each forest, but most foresters in developing countries lack the training and information required to do that, and forestry officials often will not accept the plans when they do.

In theory, having a formal management plan, getting it approved, implementing it and tracking the timber harvested in accordance to what it prescribes form one coherent system. In reality, there is often little connection between what the plan says, having the required permits, and what happens in the forest. Having the paperwork in order, per se, says little about how a forest is managed, especially where forestry officials rarely visit the forest and/or sign the papers in return for bribes.

In summary, there is little doubt that enforcing some forest laws could encourage sustainable forest management. Nevertheless, the relation is less clear and direct than most people think. While effectively enforcing some forestry laws and regulations may have a positive effect, enforcing others may make things worse. In many instances enforcing the laws is unlikely to affect how forests are managed at all. A great deal of forestry laws and regulations that discriminate against small-scale farmers and foresters and local communities have no scientific basis for doing so. Nonetheless, proponents of such regulations typically justify such inequitable rules on environmental grounds.

The Effects on Rural Livelihoods of Enforcing Different Forestry and Conservation Laws and Regulations

Logging bans and moratoriums

Logging bans are simple. Once one bans all logging in a region or a country the authorities can safely assume any logging that continues must be illegal. Nonetheless, such bans have had only mixed success at reducing environmental destruction (Boyer, 2000; FAO, 2001). Few countries have the political will and capacity to stop all logging in the designated areas, in part because of political pressure from the people the bans affect and the local governments that represent them.

Box 8.5 *The social impacts of China's logging ban*

In 1998, the Chinese Government banned logging of natural forests in the upper reaches of the Yangtze River and the middle and upper reaches of the Yellow River in an attempt to slow down environmental degradation. No reliable figures exist about how many forestry workers lost their jobs, but analysts initially estimated the ban would affect 1.1 million forestry workers. The government provided workers that lost their jobs with a lump sum severance pay equal to three times the average local wage. Those that cannot find new jobs receive unemployment benefits for up to three years. Some workers have found new jobs at higher wages with government assistance, but those with few skills and little experience have found it hard to find work. Small farmers are no longer able to harvest timber and fuelwood. Local and provincial tax revenues in the regions affected have declined and many social services such as education and health care, which were previously subsidized by state-owned forestry enterprises, have deteriorated since the enterprises stopped operation. As of 2001, the Chinese government was trying to design new measures to address these problems.

Youxian, 2001

Where logging bans have been implemented in places where many people depend on forestry activities for their livelihoods, great hardship has resulted. China represents the clearest case, even though its government has made major efforts to compensate those most affected. Similar problems have arisen in several South-East Asian countries, although there the governments have generally implemented the bans less effectively and the commercially valuable timber was already largely exhausted when the government imposed the bans.

Partial logging bans sometimes deny access to timber to small-scale loggers and loggers that lack political connections, while giving access to others. In fact, some partial logging bans end up becoming little more than an excuse for ensuring that only those favoured by key individuals within the government can have access to the resource.

Strict enforcement of all timber harvesting laws and regulations

In many countries attempting to strictly enforce all existing forestry laws and regulations affecting timber harvesting would be tantamount to imposing a logging ban. The laws and regulations are so demanding that loggers would find it practically impossible to comply with them and still earn a profit, if indeed they could comply at all. There are simply too many requirements, they are too difficult and costly to meet, and some even contradict each other. Without bribes to avoid inspections and speed up the paperwork, the approval of plans and permits would slow down significantly. That implies that truly rigorous forest law enforcement would put practically everyone engaged in forestry out of business, both in the formal and informal sectors.

As with logging bans, strictly enforcing all the existing forestry laws would have decidedly mixed effects on rural livelihoods. Under the unlikely assumption that governments were able to achieve this, there would be a lot less forest loss and degradation, but forestry workers and people engaged in small-scale forest-based activities would lose jobs and income. There would be very little forestry tax revenue. To get everyone to strictly obey all the laws and regulations might require repression. That could threaten households' physical security and undermine traditional mechanisms of forest management. Communities and poor rural households would lose access to forests where they currently live and work without legal recognition.

Strict protection of conservation areas

Creating strictly protected conservation areas is similar in many aspects to establishing a logging ban that applies only to one particular area. Completely prohibiting activities makes it easier to detect when a law has been violated. In this case the prohibitions may include clearing forest for agriculture, hunting, fishing, cattle grazing and harvesting forest plants, as well as logging.[10] Focusing law enforcement efforts on protected areas has the advantage of allowing officials to concentrate on a limited number of compact geographic locations and permits them to devote their attention to laws that link directly to what happens in the forest, rather than to administrative requirements.

However, greater enforcement of existing restrictions associated with protected areas could easily deprive many poor rural households of their incomes, access to forest resources and ability to maintain their traditional customs and lifestyles, and would be likely to lead to large numbers of arrests and human rights violations.

Outcome-oriented approaches to using forestry laws to improve commercial logging

As noted above, many forestry regulations focus on aspects that have little direct relation to how people manage forests and what happens as a result (Bennett, 1998). Recently, however, some international groups have emphasized enforcing those laws that most influence forest management and tax revenue. Rather than concentrating on whether logging companies meet all the multiple administrative requirements, they emphasize whether companies:

- have management plans based on serious forest inventories and only harvest logs specified in those plans;
- follow government restrictions concerning annual allowable cuts, minimum diameters, rotation periods and conservation areas;
- monitor and track each log from when it is harvested until it reaches its final destination; and
- pay all the mandated taxes and fees.

This approach has been designed largely for industrial logging companies, particularly those involved in export markets. There is still little practical experience with its implementation and limited data on its cost and effectiveness.

A priori it is difficult to predict how such an approach would affect rural livelihoods. That would depend largely on:

- how effectively the initiative was implemented and with what degree of fairness and transparency;
- whether the initiative substantially improved how forests were managed or simply made sure that management was well documented;
- how the initiative affected the sector's profitability and harvest levels;
- the characteristics of the informal forestry sector and how law enforcement efforts affected it; and
- the extent to which the new system re-enforced and legitimized maintaining control over forest resources in the hands of large-scale logging companies, rather than local communities, indigenous people and small-scale foresters and farmers.

Cracking down on informal timber and fuelwood harvesting

As noted previously, millions of poor rural households engage in fuelwood, charcoal and timber activities that are officially illegal or of uncertain legality, even though many countries allow families to harvest small amounts of forest products for their own consumption. Under normal circumstances, most countries make little effort to regulate these informal forestry activities. Local officials may sporadically make their presence felt, particularly when looking for bribes, but otherwise they turn a blind eye.[11]

However, to the extent that forestry officials *do* occasionally enforce some laws this usually harms rural livelihoods. Local people have to pay bribes, sell their products for lower prices and face problems of intimidation and threats to their physical security. Formal government structures that contradict traditional mechanisms regulating forestry activities undermine the latter.

Differences Across Countries and Contexts

Forest law enforcement efforts will affect rural livelihoods differently depending on the context. Key variables that influence these outcomes are the characteristics of the forestry:

- *sector* (eg size and characteristics of the forest itself, forest tenure, type of producers involved, product composition, market orientation and types of links between harvesters, processors, traders, lenders and investors);
- *legislation* (eg who it assigns property rights to, how large a technical and administrative burden it presents, and to what extent it restricts small-scale forestry activities); and
- *institutions* responsible for law enforcement (eg their territorial presence, technical capacity, level of decentralization and degree of transparency, corruption and respect for human rights).

Places rich in commercially valuable timber with large processing facilities attract regulators' interest since they are potential sources of tax revenues and informal payments. In such contexts, forestry legislation and institutions often help certain groups grab the resources at the expense of others (Ross, 2001). Governments usually allocate large forestry concessions to private companies with strong political connections with little regard for rural livelihoods. This applies particularly to Central Africa and East Asia. Forestry laws in these contexts typically legitimize the more powerful and wealthy groups' monopoly over forest resources. Since governments already enforce these laws sufficiently to ensure that poor rural households have only limited access to commercial forest resources, greater law enforcement in the more commercially valuable areas would only really pose a threat to the rural poor if the government went after informal sector activities that are relatively marginal in terms of commercial timber production.

Indonesia represents a special case as a country with a large forestry sector where the national authorities and large forestry conglomerates have recently lost much of their control over forest resources. That has opened up many new opportunities for regional and local elites and in some cases local communities and poorer households, who operate in violation of what the national authorities consider the law. To a certain extent the call for the 'restoration of law and order' in such circumstances represents a call for restoring the monopoly over forest resources by national public and private elites – and has uncertain impacts on rural livelihoods.

In China, India, Nepal and several other Asian countries governments have traditionally given substantial attention to regulating forest use. These nations have large, powerful and deeply entrenched state bureaucracies with strong historical traditions and limited transparency and accountability. The countries are forest poor but large numbers of people rely heavily on forest resources. The potential risk to rural livelihoods from increased forest law enforcement may be greatest in such contexts since people depend heavily on forests and at times the governments have demonstrated both the will and capacity to take measures that limit access of poor households and ethnic minorities to those forests.

In most other countries, efforts to regulate forests have been more limited and sporadic. Forestry departments are heavily under-staffed and have little political power or influence. Forestry law enforcement efforts are unlikely to be effective, but by the same token are also less likely to have major negative impacts on rural livelihoods – although they may still make some producers' lives more difficult and increase the costs of engaging in small-scale forestry activities.

Policy Options

Forestry law reform

One key element of ensuring that enforcing laws and regulations relating to forests does not harm rural livelihoods is to reform the laws and regulations so they discriminate less against low-income households, ethnic minorities and women. Key elements include:

- Establishing simple and low cost mechanisms to formally recognize the rights of local communities and smallholders over forest resources they already manage, and allocate additional resources to them. This must include, among other things, appropriate mechanisms for resolving competing claims.
- Reducing the number of administrative and technical requirements, simplifying them and allowing decisions about them to be made at the local level.
- Exempting small-scale activities from various technical and administrative requirements, including fees and taxes. (Additional measures may have to accompany this to ensure that wealthier actors do not abuse these exemptions.)
- Improving financial sector and money laundering laws and regulations to encourage banks to conduct full due diligence before lending to large companies that may be involved in illegal forestry activities.
- Establishing clear and accessible legal mechanisms to allow people to seek redress for government decisions and actions that may have harmed them illegally.
- Empowering local community organizations to monitor compliance of forestry laws with support from government authorities.
- Guaranteeing full public availability and transparency of government information related to forest regulation.
- Ending prohibitions on swidden cultivation and on processing timber with chainsaws and permitting rural households to engage in activities that form part of their cultural heritage.
- Formally recognizing and implementing international laws, treaties and agreements that support the rights of indigenous peoples, ethnic minorities and women.

Institutional reform

Reforming laws and regulations will have limited impact unless one also reforms the institutions charged with implementing them.[12] These institutions have to become more efficient and outcome oriented, less corrupt, more transparent and accountable, and more responsive to the needs of smallholders and local communities.

Experience shows that it is not easy to achieve this. Many government officials depend economically on formal and informal payments associated with the existing regulatory regimes. They may be reluctant to give up the authority associated with their discretionary ability to enforce or not enforce existing legislation, and the status associated with their supposed scientific understanding of how to manage forests. Working for wealthy and powerful forestry companies and farmers provides greater status and benefits than working for small farmers and foresters and indigenous people. Forestry officials have been trained and socialized under existing paradigms and many aspects of their institutional cultures reinforce them.

Donor support for agencies and officials that take into account livelihood concerns can encourage reform in government forestry agencies. So can the appointment of reform-minded officials, the implementation of training programmes, sanctioning of corrupt officials, and the recruitment of younger and more idealistic forestry officials. As much as possible, it is important to try to make government officials feel part of the reform process and that they benefit from it, rather than feeling threatened by it.

Besides working with the government agencies, it is also important to strengthen civil society organizations that independently monitor government agencies and forestry companies, provide legal and technical assistance to communities and smallholders, and promote multistakeholder dialogues and informal mechanisms for resolving conflicts. These organizations include NGOs, the mass media, professional associations and grassroots organizations. In addition to providing services directly, these organizations can encourage government agencies to be more accountable and transparent.

Focusing on the biggest violators

One obvious suggestion for reducing the potential harm to rural livelihoods from forest law enforcement would be to concentrate enforcement efforts on the largest violators – especially those that provide limited employment. In some, but certainly not all, contexts these are also the groups responsible for the greatest amounts of forest destruction and most of the tax evasion.

Enforcing laws that favour rural livelihoods

Some forest-related laws specifically favour poor rural households and ethnic minorities so those groups should benefit from their enforcement. For example, over the last few decades many governments in Latin America recognized indigenous peoples' rights over large territories, but indigenous people often find it difficult to protect those territories from encroachment by loggers, miners and farmers. Greater efforts to protect the indigenous peoples' rights could improve their situation and help guarantee their continued access to the forest products they depend on. The same applies to the legal rights of people living in extractive reserves in Brazil, indigenous peoples and community-based forestry organizations in the Philippines, and other similar groups.

One problem that rural communities and small farmers frequently suffer in many countries is the failure of logging companies to fulfil their promises to construct roads, fund social services and scholarships, pay fees and provide other benefits in return for permission to log their forests. Establishment and enforcement of legal contracts could potentially go a long way towards solving this problem.

Community-based law enforcement

In many countries rural communities play an increasing role in monitoring and reporting forestry law violations to government officials, confronting law violators themselves and regulating forest use among community members. In some cases they collaborate closely with government officials; in other cases the two conflict.

Examples of the potential for local communities to act effectively in law enforcement come from Honduras, Mexico and other countries where local communities have organized to expel outside logging companies they accuse of illegal forestry activities without the support of government departments, often with success. Generally this has involved communities that depended on the forests and were not involved in the logging operations themselves.

In principle, organizing communities to defend their own interests in relation to forest law enforcement should be an important element of any strategy to make sure that law enforcement efforts do not negatively affect rural livelihoods. Nonetheless, it would be important to synthesize the lessons from existing experiences before drawing any definitive conclusion. Communities are not homogeneous entities, and some community law enforcement efforts may negatively affect the livelihoods of poorer and weaker groups.

Sequencing

Clearly something needs to be done about illegal forestry activities and the weak rule of law in forested regions. Just as clearly many existing forestry and conservation laws discriminate against small-scale farmers and foresters and indigenous people, and enforcing those laws more effectively would only make the problem worse.

In principal, the logical thing to do would be to reform the laws and the institutions that implement them and then have the reformed institutions enforce the new laws. But illegal forestry activities are causing major damage now and reforming the forestry legislation and the institutions that implement it could easily take years, or fail completely. That raises a serious sequencing dilemma. Is it better to wait until the laws and institutions are improved before pressing for greater law enforcement, or would it be better to push existing institutions to enforce the present laws now, even though that could have a negative impact on rural livelihoods?

There is no easy answer. One probably has to work on both simultaneously, but it is important not to lose sight of the fact that enforcing many existing laws can have negative consequences. Forestry agencies and civil society organisations must work hard to focus on those law enforcement activities that have the greatest potential for improving forest management and tax revenues with the least negative impact on livelihoods.

Adaptive management and learning

There is still much to learn about how efforts to regulate forest use affect rural livelihoods and what can be done to get people to manage their forests more sustainably without making life harder for groups whose lives are already difficult enough. We need much more information about how forestry and conservation laws and regulations are currently enforced and what the impacts have been, as well as to learn from interesting experiences and examples of best practices. At present it is practically impossible to answer questions like: how do most forestry officials and park guards spend their time? How many people do they fine or arrest? How common and large are the bribes people pay? How likely is it that forestry violations are detected and prosecuted and result in punishment? How common are human rights abuses linked to forest law enforcement? How do these dynamics affect small-scale forestry producers' costs and incomes? How many people depend on forestry activities that are currently illegal and in what ways? How much deforestation and forest degradation that negatively affects rural livelihoods results from illegal forestry activities? Until there are more answers to such questions it will be hard to design appropriate forest law enforcement strategies that take into account the implications for rural livelihoods.

Most of the existing information about illegal forestry activities is anecdotal or speculative. While it has been extremely useful for increasing public awareness about the problem and for stimulating action in particular cases, it is less useful for coming up with appropriate policy responses.

To answer the more systematic questions about the links between forest law enforcement and rural livelihoods, more formal research will be required. But it will also require well-organized multistakeholder study tours, improved data collection in forest law enforcement agencies and workshops and visits where people learn from each other's experiences, among other things.

Conclusion

Governments and communities must regulate the management and use of forests to ensure that their useful functions are maintained over time, benefits are shared equitably, conflicts are resolved in a fair and transparent manner and sufficient tax revenues are obtained to pay for necessary public expenses. The widespread violation of existing forest laws and regulations has major negative impacts on forests, livelihoods, public revenues and the rule of law. Something must be done about that.

The problem is that many existing forests and conservation laws themselves have unacceptable negative impacts on poor people, ethnic minorities and women; and in many places they are enforced in a fashion that is discriminatory and abusive. Ways must be found to address the problems associated with illegal forestry activities that at

Table 8.2 *Options to address threats to rural livelihoods from illegal forestry activities and from forest law enforcement*

	Illegal forestry activities	Forest law enforcement	Options
Forest product income	Small foresters earn less because they pay bribes and depend on patrons	Small foresters earn less because governments stop their 'illegal' activities or greater law enforcement leads them to have to pay higher bribes or depend more on patrons	Reduce and simplify forestry regulations. Exempt small-scale activities from some regulations. Focus regulatory efforts where problems are greatest. Greater transparency in regulation
Wages from forestry	Over-harvesting makes the forestry sector collapse	Government actions reduce logging	Regulators give preference to labour-intensive activities
Government revenues	Tax evasion and illegal logging in public forests lowers revenue	Reduced logging due to law enforcement lowers revenue	Progressive and transparent tax collection
Physical security	Illegal loggers and corrupt officials threaten and attack villagers	Officials inappropriately threaten, attack, arrest or expel villagers and destroy their crops and houses	Strengthen human rights institutions, grass-roots organizations and independent judiciary and oversight. Provide legal assistance and promote legal literacy
Access to forest resources	Wealthy groups and officials 'illegally' deny access to forests and due process	Wealthy groups and officials 'legally' deny access to forests and due process	Tenure policies that increase community and smallholder access to forests. Recognize indigenous territories and increase efforts to protect them from encroachment. Multistakeholder dialogues to resolve conflicts and increase access to forests for the poorest groups
Long-term supply of forest goods and services that poor households use	Damage to forests by illegal logging reduces forest product supply and disrupts environmental services	Damage to forests due to loss of social capital reduces forest product supply and disrupts environmental services. Forest management institutions undermine them and police action substitutes for dialogue	Focus regulatory efforts on maintaining forest resources of value to poor families. Support local efforts to protect forest resources legally, politically and financially

Table 8.2 *Options to address threats to rural livelihoods from illegal forestry activities and from forest law enforcement*

	Illegal forestry activities	Forest law enforcement	Options
Collective action and participation	Illegal logging undermines local forest management institutions. Bribes and influence peddling replaces democratic process	Laws that fail to recognize local forest management institutions, undermine them and police action substitutes for dialogue	Recognize and support community efforts to protect forests. Multistakeholder dialogue and informal mechanisms to resolve conflicts. Compensate communities for environmental services
Respect for cultures and tradition	Illegal logging undermines traditional institutions	Laws that don't allow traditional practices or respect local institutions undermine peoples' traditional cultures	Avoid regulations that unduly restrict peoples' traditional activities. Implement international agreements concerning indigenous peoples
Economic growth	Widespread failure to respect rule of law reduces investment and growth	Limiting forestry activities reduces (short-term?) economic growth	Make regulatory systems more transparent, democratic and equitable. Promote small-scale forestry activities and partnerships between companies and communities

least do not aggravate the negative impacts of existing regulatory efforts on the rural poor. That will not be easy, but it will certainly be completely impossible unless the challenge is recognized from the outset. If this chapter contributes to that recognition, it will have served its purpose.

Acknowledgements

The author gratefully acknowledges the contributions of Greg Clough, Marcus Colchester, Filippo del Gatto, Luca Tacconi and Adrian Wells to this chapter and the generous financial support from the Department for International Development (DFID) of the United Kingdom. The paper does not necessarily reflect the official positions of either CIFOR or DFID. The author accepts sole responsibility for all errors.

Notes

1   The concept of social capital is justifiably controversial. As used in the context of this chapter the concept is synonymous with the local institutional capacity to regulate and manage forest resources effectively, efficiently and equitably.

2   One could also argue that policies favouring economic growth contribute to rural livelihoods. This chapter occasionally refers to the impact of policies on growth.

3   For the purposes of this chapter, an arrest or fine is considered 'inappropriate' if it (1) does not contribute to sustainable forest management, (2) does not follow due process, or (3) results from laws or law enforcement practices that discriminate in favour of wealthier or more powerful groups.

4   Private banks, NGOs and government credit agencies generally will not lend money to independent small-scale foresters, in part because of concerns about the legality of such activities.

5   Obidzinski (2003) documents in great detail how such patronage networks operate in East Kalimantan in Indonesia and how they affect the distribution of benefits.

6   Contreras-Hermosilla (2002) estimates the total annual loss from such illegal activities as being at least US$10 billion dollars each year. That figure includes developed and transition countries as well as developing countries.

7   One particularly complex and difficult example of this at present is that of India. The country's Supreme Court has ordered that by 31 May 2003 government officials should evict all the families encroaching upon the country's reserve forestland. The author was unable to locate any reliable estimate of the number of people this might affect; however, it is clearly in the hundreds of thousands and might be even more (Sharma, 2003).

8   Nonetheless, there are situations where failure to log, hunt or harvest plants can lead to ecological imbalances, fire hazards or other problems. One should also remember forests are not static. With or without further human disturbance they change over time.

9   For example, regulations that encourage companies to avoid large canopy gaps may impede regeneration of major commercial timber species such as mahogany. There are also situations where it is less profitable to manage forests sustainably than to manage them unsustainably because of the high cost of preparing the required management plans and similar documents. For example, Davies (1998) has shown that the main reason it is not commercially viable for landowners in northern Costa Rica to promote the natural regeneration of secondary forest for timber production on abandoned pastures is because of the high costs of the associated paperwork the law requires.

10   Although not all categories of protected areas prohibit all these activities.

11   This should be considered a hypothesis to be verified. There are practically no data on the extent or consequences of regulation of the informal forestry sector.

12   This may be only partially true when it comes to legal reforms that limit the institutions' functions and authority.

References

1　Bennett, E. I. (2002) 'Is there a link between wild meat and food security?' *Conservation Biology*, vol 16, no 3, pp590–592.
2　Boyer, C. R. (2000) *Conservation by Fiat: Mexican Forests and the Politics of Logging Bans, 1926–1979*. manuscript.
3　Contreras-Hermosilla, A. (2003) *Barreras a la Legalidad en los Sectores Forestales de Honduras y Nicaragua*. manuscript.
4　Department For International Development (DFID) (2001) *DFID Sustainable Livelihoods Guidance Sheets*, October, London.
5　FAO (1998) *Asia–Pacific Forestry Towards 2010*. Report of the Asia–Pacific Forestry Sector Outlook Study, FAO, Rome.
6　FAO (2001) *Forestry Outlook Study for Africa: A Regional Overview of Opportunities and Challenges Towards 2020*. FAO, Rome.
7　Fredericksen, T. S. (1998) 'Limitations of low-intensity selective and selection logging for sustainable tropical forestry', *Commonwealth Forestry Review*, vol 77, pp262–266.
8　Poschen, P. (1997) *Forests and Employment, Much More than Meets the Eye*, World Forestry Congress, Antalya, Turkey, 13–27 October, 4, 20. FAO, Rome. www.fao.org/montes/foda/wforcong/PUBLI/V4/T20E/1–5.HTM#TOP
9　Putz, F. E., Redford, K. H., Robinson, J. G., Fimbel, R. and Blate, G. M. (2000) 'Biodiversity conservation in the context of tropical forest management'. *Conservation Biology*, vol 15, pp7–20.
10　Ross, M. (2001) *Timber Booms and Institutional Breakdown in South East Asia*. Cambridge University Press, Cambridge.
11　Sheil, D. and Van Heist, M. (2000) 'Ecology for tropical forest management'. *International Forestry Review*, vol 2, pp261–270.
12　Sist, P., Fimbel, R., Nasi, R., Sheil, D. and Chevallier. M. H. (2001) *Towards Sustainable Management of Mixed Dipterocarp Forests: Moving beyond Minimum Diameter Cutting Limits*. Draft.
13　Thomas, V. et al (2000) *The Quality of Growth*. Oxford University Press, Oxford.
14　World Bank (1997) *Helping Countries to Combat Corruption, The Role of the World Bank*. The World Bank, Washington, DC.

9

The Case for Bushmeat as a Component of Development Policy: Issues and Challenges

D. Brown and A. Williams

Introduction

This chapter is concerned with an often-neglected class of NTFPs with important liveli-hoods dimensions, namely animals hunted for consumptive use linked to local liveli-hoods. For simplicity's sake, the colloquial African term 'bushmeat' is used to describe the class, despite its geographical restrictions and (for some) alleged colonial connota-tions. Non-consumptive uses are not considered here.

The chapter focuses on the question: 'Why should bushmeat be of concern to inter-national development policy?' It is suggested that the answer is twofold. On the one hand, the safety-net functions of bushmeat and similar forest products are crucial to the livelihoods of the poor in the humid tropics, particularly in areas with little immediate prospect of transformation out of poverty. And on the other, progress in the manage-ment of internationally high profile and emotive resources such as wildlife raises impor-tant issues in governance and may leverage broader governance benefits. The chapter argues the case for the centrality of a livelihoods perspective in the development of pub-lic policy on forest governance.

The chapter is in three sections. The first provides a reminder of the scale of the bushmeat trade, and the conservation and welfare issues which this raises. The second examines the evidence concerning the role played by bushmeat in the livelihoods of the poor. And the third considers the potential which exists for bushmeat to figure as a component of improved governance.

The Present Realities of the Bushmeat Trade

Bushmeat and international aid policy

The tensions which exist between conservation and pro-poor development are well-illus-trated in the title of a recent paper on the theme of hunting for wild meat in the humid

Note: Reprinted from *International Forestry Review*, vol 5, Brown, D. and Williams, A., 'The case for bushmeat as a component of development policy: Issues and challenges', pp148–155, Copy-right © (2003), with permission from the Commonwealth Forestry Association

tropics. Written by two eminent conservationists, it is entitled 'Will alleviating poverty solve the bushmeat crisis?' (Robinson and Bennett, 2002). From a livelihoods perspective, such a proposal would seem something of an inversion of priorities. While the point is taken that economic development provides no guarantee of sound environmental stewardship (indeed, the evidence is often in the opposite direction), it would seem more appropriate to ask what roles bushmeat and other forest goods might play in poverty alleviation. The authors' response to their rhetorical question is not encouraging. They write:

> The only way out of this crisis will be offered by long-term, integrated efforts to provide alternative sources of protein and income for the rural poor, curtail the commercial trade in wildlife, secure wildlife populations in protected areas, educate hunters and buyers. . . (2002).

The people whose behaviour must change, they say, are 'the millions of people at the margins of the cash economy in Asia, Africa and Latin America ... whose lives are intertwined with natural areas'. There is 'no "silver bullet" for the twin goals of conserving wildlife across the humid tropics and preventing the people whose lives now depend on wildlife from being driven further against the wall' (ibid). Coming at a time when most international aid agencies are seeking to limit their fields of activity, and to concentrate on those investments which offer the greatest chance of delivering poverty alleviation with maximum efficiency (see SoS, 2002), such a pessimistic scenario is challenging to those who have sought to deploy international aid resources in order to reconcile global interests in tropical forests with national and local realities. It raises important questions about the trajectory of conservation policy, and the focus of policy change. What benefits can be offered to justify the engagement of development assistance with the bushmeat issue, and what are the implications at the level of policy?

A success story?

On the surface, the scale, vigour and penetration of the bushmeat trade might be viewed in a highly positive light, as one of the great success stories of autonomous food production in the developing world, and a testimony to the resilience and self-sufficiency of its populations.

The volume of consumption

Bushmeat is one of those minor forest products that have been shown to have major significance for rural communities, particularly in the humid and sub-humid tropics. The levels of offtake vary by ecological zone, country and continent but in general levels of offtake are highest in the humid forests of West-Central Africa, and lower (though still significant) in Asia and South America. A number of factors have been held to account for these differences. Intercontinental variations can be explained partly in terms of the productivity of the forest ecosystems, the Central African forests being considerably more productive than those of South America (Fa and Peres, 2001). The relatively high ratio of sea coast to land area in the Asian case – and hence, increased potential for penetration by sea fisheries – may account for the historically lower contribution of mammalian meat to the diet on this continent, and the higher contribution elsewhere (Robinson and Bennett, 2000). The higher offtake levels from humid forest than savanna ecosystems

Table 9.1 *Relative importance of game meat in Africa (1994)*

Ecological region	Population (millions)	Game meat production Total (m tonnes)	Game meat production Average/person (kg/person/year)	All meat production Total (m tonnes)	All meat production Average/person (kg/person/year)
Savannah	344	405,421	1.2	4,857,133	15.2
Savannah/ Forest	163	533,763	3.3	1,571,732	9.7
Forest	54	287,225	5.3	418,527	7.8
Islands	16	3,846	0.2	378,029	22.7
Total	577	1,230,255	2.1	7,225,422	12.5

Source: Chardonnet et al, 2002

may seem paradoxical, given the much greater productivity of the latter. However, this may be explicable in terms of the high potential of savanna ecosystems to support domesticated fauna, and the associated cultural preference for farmed meat in such areas (Chardonnet et al, 1995). Other factors such as abundance and accessibility also influence levels of human dependence. For example, the relative importance of fish over bushmeat in human diets in the Amazon basin can be explained not only in terms of the high productivity of the Amazonian river systems, but also by the fact that the mammalian biomass there is not only low in volume but, being predominantly arboreal, is also inaccessible (Fa and Peres, 2001).

Estimates for the size of the harvest in the core production areas vary greatly, but by any standards the offtake is economically and ecologically of major significance (see Bennett and Robinson, 2000). Wildlife is estimated to play a significant and direct part in the livelihoods of up to 150 million people, much of this from consumptive use (ibid). One recent estimate puts the continent-wide production for Africa at over a million tonnes per annum (Elliott, 2002), though this could be a major underestimate. The current harvest in Central Africa alone, for example, may be in excess of two million tonnes annually (Fa et al, 2003); the subregional human population is 33 million, which would imply in excess of 60 kg/person/year if all of it is locally consumed. Table 9.1 provides one Africa-wide perspective on consumption – one of the more modest of recent estimates – categorized by ecosystem, and relative to other animal sources of dietary protein.

The offtake on other continents is often conspicuously lower, and less evenly distributed. However, high levels of dependence occur in South America, particularly among the indigenous populations. For example, annual consumption of wild animals by local peoples in Amazonas State, Brazil, has been estimated as 2,800,000 mammals, 531,000 birds and about 500,000 reptiles (Robinson and Redford, 1991).

South-East Asia presents a comparable picture. The trade in animal products from Laos to China is valued at upwards of US$12 million per year. In Sarawak, the economic value of wild meat consumed by rural populations, expressed in domestic meat equivalents, has been estimated at US$75 million per year; meat to the value of over US$5 million is sold on the open market (Bennett and Robinson, 2000).[1]

A positive scenario?

The virtues of bushmeat as a livelihoods activity include low barriers to entry and high social inclusion. The poor may benefit less than the rich but – it can be argued – this is always the case, and is a feature of inequality not of bushmeat. Unregulated and decentralized in trade, a fair proportion of the value of the product is retained by the primary producer (the hunter) – much more so indeed than has historically been the case with other forest products such as timber and beverage crops. Unlike domestic animal husbandry, the labour inputs that it requires are discontinuous and easily reconciled with the agricultural cycle. Bushmeat is the product of a system of farm/forest management that collectively offers high returns from a range of activities. For the risk-averse small farmers to whom labour is the major constraint, all this has much to commend it.

The trade is likewise low risk and flexible, with minimal capital costs, and thus particularly attractive to the poor. Markets are often indiscriminating and species-wide. Extractive technology is generally low level and accessible. Gender aspects are also remarkably positive. In most situations, it is men who do the hunting, but women who take charge of all the downstream processing and commerce, to the point of sale in the scores of 'chop bars' and restaurants which are a familiar feature of the urban scene in the South. Bushmeat has excellent storage qualities, in a manner compatible with the storage of agricultural produce. It is easily transportable and with a high value/weight ratio, bushmeat fits in with the realities of rural life in the tropics in other respects, particularly for the poor.

Arguably, the starting point in any analysis of the bushmeat trade should be these positive benefits, and any attempt to improve its management should take the preservation of them as its fundamental parameter.

In reality, however, the international profile of bushmeat is almost entirely negative, and its livelihood benefits are largely discounted both nationally and internationally. Bushmeat hardly figures at all, except repressively, in public policy in the range states (ie producer countries). The fact that the positive values of such a major commodity are unacknowledged by most policy makers, failing to appear in national economic statistics or to be subject to budgetary allocations by the state, reveals much about the political economy of natural resource exploitation in the tropics, as well as about the historical evolution of tropical governance. Bushmeat and other products of the hunt tend to feature among those goods conceded by range state governments, as part of a tacit agreement which separates 'traditional' products for domestic consumption and the generation of lower-level public sector rents from 'modern', industrial commodities which enter into the circuits of national wealth generation and political patronage, and over which the population at large has no established right of voice. One of the questions that needs to be posed when evaluating new initiatives relating to bushmeat is whether they challenge or support this marginalization within policy discourse. We will return to this issue below.

A bushmeat crisis?

One factor (though not necessarily the most important one) in accounting for the peripheral nature of bushmeat in policy terms is the issue of resource degradation. A trade in primary products on this scale and of a fundamentally extractive type raises important issues of future sustainability. It is beyond the scope of this chapter to evaluate in detail the evidence for and against the view that this has now reached 'crisis proportions'. Suffice to say that while the conservation interests, which are driving much of the international policy agenda, may be prone to exaggerate the problem, it would be surprising if there were not a looming crisis, given what is known of the potential of forest ecosystems to support the animal populations at stake, and the lack of management systems in place.

Total carrying capacity of tropical forests varies according to the context, but Robinson estimates that the maximum biomass of larger mammals in evergreen forests is unlikely to exceed $3000kg/km^2$ (in open mosaic forests carrying capacity may be five times as much). Assuming a fairly generous diet of animal protein, this suggests that human carrying capacity for people totally dependent on bushmeat for protein would be unlikely to exceed one person per km^2 (approx. $150kg/km^2/year$, 65 per cent of it edible meat) (Robinson and Bennett, 2000).

If these figures apply then existing rates of offtake must be severely threatening the stock. Much of conservation opinion is in line with this view. A recent study by Fa and Wilkie, (2002) suggests that 60 per cent of mammalian species are harvested unsustainably. Offtake levels in the Congo Basin are estimated to be almost 50 per cent higher than production and possibly more than four times sustainable levels (ibid). Fa (2003) and his colleagues predict that bushmeat protein supplies will drop by 81 per cent in all the central African countries in less than 50 years, and that only three countries would be able to maintain a protein supply above the recommended daily requirement of 52g/person/day.

Even if these estimates prove excessive, the outlook must surely still be very bleak. Among other things, the present situation is usually characterized by a complete lack of management by the state, as well as strong disincentives for local populations to attempt to regulate the harvest. The chances are, therefore, that there is an impending bushmeat crisis of significant proportions.

What does not necessarily follow, however, is the view that the way to manage this very large and lucrative, if unsustainable, trade is to attempt to ban it altogether. The arguments against such an approach are both welfare-related (in terms of livelihood and economic benefits) and practical (the low likelihood of success). Yet this is the strategy that has been most strongly advocated by the conservation lobby (eg Oates, 1998).

Bushmeat and Livelihoods

Donor interest in bushmeat would undoubtedly be increased if it could be shown that the trade could contribute to poverty eradication on a substantial scale. This is an under-researched theme in the literature. However, there are good reasons to doubt it as a general proposition. A distinction needs to be drawn between those livelihood benefits for the poor that relate primarily to survival strategies and safety-net functions, and those that have potential to contribute to growth and transformation in the rural economy. The evidence suggests that the long-term prospects for bushmeat relate to values of the former type, and that it is unlikely to figure strongly in any process of capital accumulation. As with many other NTFPs, important livelihood benefits relate to timing and compatibility with the types of multiple enterprise that are the bedrock of the peasant economy, independently of the overall volume of the trade (Arnold and Ruiz-Perez, 2001).

A range of factors, including nature of the resource, the character and volumes of its trade and the availability of alternatives, are of relevance here. Paradoxically, it may be partly the virtues of bushmeat as a livelihoods asset (low thresholds of entry, leading to broad participation, but also tight margins) that mean it is unlikely to figure strongly in rural transformation. There would also seem to be few opportunities for value to be added in processing, through technical sophistication or increased investments of labour (in this respect, bushmeat may differ from, say, artisanal woodworking). This is particularly the case where the trade is treated as de facto illegal, and thus has to be pursued largely underground. The international stigma of bushmeat also limits the potential for export-oriented processing and value added (though see below). Viewed from the perspective of volume, bushmeat also offers an unencouraging prospect. Even if the projections of sustainable offtake are overcautious, they are often so far below existing offtake levels as to make it most unlikely that sufficient capital could be generated from the sector to sustain long-term economic change.

On the other hand, urban demand in the range and neighbouring states is high and possibly growing, and it can feature as a luxury item in the trade, even where substitutes exist at a competitive price (Fa and Wilkie, 2002). There is historical evidence that bushmeat can play a secondary role in supporting economic change. Asibey (cited in Cowlishaw et al, 2003; but compare Oates, 1998), for example, notes the part it played in underwriting the development of the Ghana cocoa industry and opening up the forest frontier. Thus, even a decapitalizing stock can have a role in economic growth and structural change of long-term benefit to the poor.

Nevertheless, even discounting these uncertainties, it seems most unlikely that the main future justification for donor involvement with bushmeat is going to come from a positive assessment of its poverty eradication potential on any major scale, sustainable or not. The primary interest must rather lie with the livelihoods dimension, both in its own right – as an aspect of food security to vulnerable human populations, both rural and urban – and also as a means of underpinning and reinforcing the drive to improved public governance. The major challenge is to address the public governance dimensions without surrendering the livelihoods focus.

Bushmeat and the poor?

Until recently, livelihoods considerations have tended to figure only peripherally in wildlife research in the tropics. However, a number of recent research studies have addressed the livelihoods aspects directly, and linked them to issues of policy. These do not necessarily support the view that dependence is greatest among those living in the most extreme poverty, as had been previously proposed (see Scoones et al, 1992). However, the situation is complex, and defies simple poverty/non-poverty labelling.

A case in point would be the study undertaken by De Merode in the Zande area of Eastern Democratic Republic of Congo (De Merode et al, 2003). While significant variations in patterns of consumption and sale were found, correlated with relative wealth, all the families in the study could be classed as 'living in extreme poverty' by the standard international test (income of less than US$1 per day). Thus the variations were only relative.

De Merode's study sought to address three questions:

- whether wild foods (including bushmeat) were valuable to households, in terms of both consumption and sales;
- whether the value varied according to the season;
- whether the value was greater to the poorer or less poor.

In summary, it was found that while wild foods in general formed a significant proportion of household production, most was sold on the market and not consumed. This was particularly the case with bushmeat and fish, where more than 90 per cent of production was sold. Consumption levels varied by household, with both seasonality and wealth effects.

Consumption of wild foods increases significantly during the hungry season (particularly bushmeat, where consumption rose on average by 75 per cent). Bushmeat and fish consumption were fairly even across all wealth ranks, except the poor (who consumed very little of their own production, though they made up for this through bushmeat gifts). Bushmeat sales were influenced by the wealth rank of the household, with the richer households more likely to be involved in market sales. This was unrelated to questions of land access and tenure (all families had equal theoretical access to the production zones, and – unlike with fishing – there were no restrictions on activity related to non-membership of a craft guild). However, it was strongly correlated with access to capital (eg ownership of shotguns and nets) and to the wealth required to generate a surplus over consumption needs. Interestingly, both fish and bushmeat exhibited the characteristics of 'superior goods' (ie luxury items consumption of which increased exponentially with increasing wealth). By contrast, wild plants were 'inferior goods', in that increasing wealth implied decreasing household consumption.

A particularly interesting finding of the study was that, for families living in extreme poverty, market sales of bushmeat were more important than household consumption. This challenges the view put forward in certain conservation quarters that the way to reconcile the interests of the poor with those of conservation, in a 'win-win' scenario, is to prohibit market sales, but turn a blind eye to subsistence use.

For a view of the pro-poor dimensions of the bushmeat commodity chain, there is some useful evidence from the joint Zoological Society of London/University College London concerning mature bushmeat markets in Ghana (Cowlishaw et al, 2003a, 2003b; Mendelson et al, 2003). In a series of papers on the bushmeat trade in Takoradi, SE Ghana, Cowlishaw, Mendelson et al, discuss the relative incomes and influence of five primary actors – commercial hunters and farmer hunters (always men) and wholesalers, market traders and 'chop bar' operators (always women). In terms of numbers, this is a medium level enterprise, with about 1000 persons involved in the trade in the catchment area of the study, most of them non-professional farmer/hunters (75 per cent of the total).[2] While the chop bar owners handle the largest volume of produce, profitability is highest, in percentage terms, among the hunters. Hunters are found to capture 74 per cent of the final sales price in Takoradi chop bars. Taking into account the fact that most hunters are also farmers, bushmeat provides an important supplement to rural incomes in this group.

An interesting proposition of this work is that the sustainability of the current market profile is the result of the operation of an extinction filter. That is to say, vulnerable species (such as the slower-reproducing and larger, more vulnerable ungulates) have already been wiped out in the more accessible areas, so that only more robust and fast-reproducing species (chiefly rodents and smaller, farm-bush resident ungulates) now figure in the trade. The possibility then opens up of sustaining the industry using robust stocks of non-vulnerable species in the highly productive farm-bush, and using forest protection more selectively to preserve the vulnerable species in isolated forest areas. It would clearly be unwarranted to extrapolate from these findings or to draw conclusions as to the lack of connection between biological diversity and human welfare, but this does at least offer a positive scenario, allying conservation with consumption and not opposing it. The best strategy for achieving a win-win scenario might be to compromise early with the industry's needs, rather than – as is so often the case – to assume that the only way to save the weak and endangered species is to rigidly oppose the trade.

Conservation strategies and local livelihoods

Conservation strategies have, at least until recently, tended to discount these livelihoods dimensions. Typically, the dominant legal framework is one which supports the preservation of nature, with a strong emphasis on interdiction and non-consumptive uses, and restricting, not positively managing, public access to the resource (see Brown, 1998). The principles that are typical of forest laws, and which have long characterized (at least in theory) exploitation of timber – regulated access, realistic licensing, financing arrangements to minimize investor risk, and rules to discourage damage to the resource while permitting consumptive use – are rarely in evidence. Though some international conservation agencies have attempted a more positive approach (WWF-Cameroon's Conservation Programme being a promising example, work by the Wildlife Conservation Society with the timber concession of the company *Congolaise Industrielle de Bois* (CIB) in Congo-Brazzaville another), successes have been few and far between.

Much greater efforts have been expended, in fact, on compensatory approaches to justify interdiction of the trade. Alternative income-generating strategies have been much favoured (ranging from captive rearing of game species, to promotion of domesticated livestock, to handicraft development). These alternatives are often vulnerable to criticism precisely because of their lack of cognizance of livelihood interests. They are unlikely, used in isolation, to reduce hunting pressure unless they are compatible with the resource users' capital base, offer superior benefits to them, and successfully compete for their labour time. There could well come a time when such options do prove attractive to investors in the range states, but the evidence is that the time is still far off for the areas of greatest conservation concern.

The Approach from Governance

The second area of possible justification for donor interest in the bushmeat trade is the potential that it can contribute to improved governance. This final section of the paper considers some of the possibilities and risks which accompany the attempt to address the issue of forest governance as it pertains to bushmeat, through developing management models and other means.

The challenges are undoubtedly immense. On the one hand, most of the source areas are subject to major failures of governance which affect the utilization of all forms of natural resources, not just wildlife. Approaches that require the pre-existence of effective systems of governance are thus unlikely ever to take root. On the other hand, there is little by way of successful experience of wildlife management in any comparable situation on which to build. The management models which have proven most successful (such as 'CAMPFIRE' in southern Africa (Metcalfe, 1994)) are not well adapted to the circumstances which pertain in the bushmeat heartlands, with very restricted tourist markets, an imbalance between national and international tourist trade, frequently poor security and an almost total absence of the necessary access and infrastructure.

Governance considerations feature in the debate principally in two ways: first, as adjuncts to other initiatives that impinge on the issue of bushmeat, and its role in livelihoods; and second, in the way that they inform strategies to improve on bushmeat management. Forest certification provides an example of the former, attempts to institute new management models for wildlife harvesting an example of the latter. We shall consider each in turn.

Forest certification

To date the debate in relation to forest certification has centred not on the bushmeat chain of custody itself, but on the role which certification might play in mitigating the negative impacts of the timber industry on the wildlife resource. While the former approach would have virtues, agreed channels for a well-managed and sustainable trade are such a distant prospect as to make this of conjectural interest at the present time. By contrast, certification of tropical timbers is an active issue and commends itself more readily as a means of influencing the bushmeat supply. Being a non-state mechanism, it largely avoids the paradox of requiring the pre-existence of the public governance systems that it is intended to promote.

Wildlife and hunting already figure among certification principles and criteria, or can be construed as such, in relation to such issues as forest use rights, indigenous peoples' rights and environmental impacts (for example FSC principles 2, 5 and 6). In addition, a number of proposals are currently on the table for the more explicit integration of wildlife management standards into forest certification. The most prominent is the Ape Alliance's 'Code of Conduct to minimize the impact of hunting in logging concessions' (1998). The more recent proposal by Woodmark on behalf of Fauna and Flora International proceeds on essentially similar lines (Dickson, pers. comm.).

These proposals need to be assessed not only for their relevance and effectiveness as industry instruments, but for the assumptions which they make about forest livelihoods, and their likely impacts upon them. In a context where industrial interests are dominant, the interests of other resource users have to be fought for on whatever terms are available. The disincentive for range state governments to advance progressive legislation is particularly evident. Having often inherited the beneficial legacy of a legislative framework which denies almost all tenurial rights to traditional users, these are rarely of a mind to surrender their power to endeavours which would, at best, represent only a secondary level of economic activity, as well as one for which the revenues would be much more difficult to capture. This is particularly the case where the precedents created would have political resonance. Thus, whatever the intentions, there will be strong pressures to channel governance changes towards the needs of the industry, downplaying its other dimensions. This limits their potential to act as a catalyst for more general reform.

Work in this field is inevitably constrained by existing legal frameworks, which are inappropriate and unworkable as these colonial inheritances often are. The Ape Alliance Code is designed only to curb the industry, and makes no concessions to the interests of other forest users. It presumes and requires the sovereignty of the industry operator over the total land area, not merely the timber resource. It also accepts the validity of the existing legal frameworks and official notions of legality and illegality as a basis for the control of the wildlife resource, regardless of their public legitimacy, and encourages the development of an enclave economy (albeit a more disciplined one).

Where traditional ownership claims are accepted as valid (as in the FSC principles), these depend heavily on the notion of indigenous peoples' rights, as well as optimistic assumptions as to the clarity of their tenurial claims. Particularly in the context of West-Central Africa, where there is no history of alien conquest, and no segregation of 'indigenous' and 'settler' economies, such notions are fundamentally problematic. The situation here is more likely to be one in which traditional user rights are usurped by legislative principles that grant the state close to monopoly powers in managing resources ostensibly on the public's behalf. In the case of Cameroon, for example, the recently imposed national land-use system or zonage disregards historical patterns of land usage in favour of a division of the forest estate into two zones: the permanent estate, reserved for forest production and conservation, where the rural majority have no rights; and the much smaller non-permanent estate, conceded for immediate or eventual conversion to other forms of usage (Brown, 1999). It allows no recognition that there may be entirely legitimate land claims outside of this imposed structure of rights.

In summary, while there are strong grounds to support a better integration of different forms of land use, including hunting, through non-state mechanisms such as certification, present models are conceived within governance frameworks, which are not well adapted to the contexts typical of the bushmeat trade, and unlikely to prove 'pro-poor'. Among the major challenges which present themselves are the need for a more constructive engagement between the industrial and local economies, and recognition that there are legitimate forms of land use that are not based on close proximate residence.[3]

An attainable goal? – effective management of the bushmeat resource

The experience with certification warns of the difficulty of recognizing livelihood imperatives in contexts dominated by powerful industrial and conservation interests. An alternative – and in principle, more promising – route to governance reform is through the active management of the bushmeat resource.

Despite an unpromising institutional context,[4] there are indications that progress can be made, provided two conditions are met. The first is the revision of the legal frameworks to create legitimate channels for the bushmeat trade. The present situation – of presumed illegality at all levels – is neither conducive to the development of participatory management models nor to broader governance reform. So long as no legal channels exist, then any attempts to tighten up on management are likely only to drive the trade further underground (Dickson, 2003). This creates increased incentives for rent-seeking behaviour by officials, and encourages a further deterioration in standards of public governance. It also diminishes the potential for development of the commodity chain through export-oriented processing.

The second condition is that changes in the legal context should favour community participation in management not just for this one resource but more broadly. In this way, the social capital created for any one enterprise could become available to the others – a classic joint production issue, and hence a means of lowering transaction costs where they might otherwise be prohibitive. At the same time, broadening the range of legitimate interests in the forest estate is also likely to have beneficial effects on the quality of forest governance. In the case of Cameroon, for example, the allocation of forest exploitation rights to local communities may well lead both to better management of timber and plant NTFPs, and improved control over hunting and bushmeat. Increasing the voice of this constituency over key livelihood resources could then have knock-on effects on governance, both through the enlargement of the public interest in the future condition of forest resources, and linked effects on decentralized local government. Local awareness of the extent of the benefits, including revenues, which may be derived from forests is a necessary precondition for accountability and transparency in the use of public goods (see Brown et al, 2002).

Conclusion

To resolve the bushmeat crisis, effective management of the bushmeat resource, according to principles specific to the sector and giving major priority to livelihood concerns, must be of paramount importance. The case for international assistance to support the development of a well-regulated bushmeat industry must be based in the first instance on a recognition of its important livelihood benefits, and in the second, on its potential to contribute positively to the growth of good governance of the broader forest resource. An essential prerequisite for the latter must be to bring the bushmeat trade into the open and clearly identify the possibilities for legal and legitimate trade.

What has been lacking to date is an understanding of the centrality of social interests to conservation goals. As others have noted, sustainability is not, at the end of the day, an issue of purely biological concern (Hutton and Dickson, 2002). To argue that social and livelihoods issues are more pressing is merely to acknowledge that the decisions regarding what resources to retain and what to consume will ultimately be made not by conservationists but by those whose lives are directly affected by their day-to-day contact with the wildlife resource.

Notes

1   This is only an estimate. One of its assumptions is that one quarter of the population of Sarawak depends significantly on wild meat.

2   The totals for each category are: farmer/hunters (746); professional hunters (50); wholesalers (14); market traders (16); chop bar owners (143). (Cowlishaw et al, 2003a).

3   Hunting zones in tropical forests are not necessarily different from, say, North Atlantic fisheries in this regard.

4   Arnold and Ruiz-Perez (2002) point to the distinction between benign and active forest management; the past existence of the former did not necessarily imply the latter. In many situations, only the most sedentary animals show much evidence of active management by local users; in Ghana, this class is restricted to invertebrates such as snails and crabs [Falconer, pers. comm.].

References

1   Ape Alliance (1998) *The African Bushmeat Trade – A Recipe for Extinction*. Fauna and Flora International, Cambridge, 74p.

2   Arnold, J. E. M. and Ruiz-Perez, M. (2001) 'Can non-timber forest products match tropical forest conservation and development objectives?' *Ecological Economics*, vol 39, pp437–447.

3   Bennett, E. and Robinson, J. (2000) 'Hunting for the Snark', Chapter 1 of Robinson, J and Bennett, E (eds) *Hunting for Sustainability in Tropical Forests*. Columbia University Press, New York, pp1–9.

4   Bowen-Jones, E., Brown, D. and Robinson, E. (2001) *Assessment of the Solution-Orientated Research Needed to Promote a More Sustainable Bushmeat Trade in Central and West Africa*, research report to DEFRA, Bristol.

5   Brown, D. (1998) 'Participatory biodiversity conservation: rethinking the strategy in the low tourist potential areas of tropical Africa'. *Natural Resource Perspectives*, no 33, ODI.

6   Brown, D. (1999) *Principles and Practice of Forest Co-management: Evidence from West-Central Africa*. EU Tropical Forestry Papers no 2, ODI, London.

7   Brown, D., Malla, Y., Schreckenberg, K. and Springate-Baginski, O. (2002) *From Supervising 'Subjects' to Supporting 'Citizens': Recent Developments in Community Forestry in Asia and Africa*. ODI Natural Resource Perspectives 75, February.

8   Chardonnet, P., Fritz, H., Zorzi, N. and Feron, E. (1995) *Current Importance of Traditional Hunting and Major Contrasts in Wild Meat Consumption in Sub-Saharan Africa. Integrating People and Wildlife for a Sustainable Future, Proceedings of the First International Wildlife Management Conference* (eds Bissonette, J. A. and Krausman, P. R.). The Wildlife Society, Bethesda, Maryland, pp304–307.

9   Chardonnet P., Des Clers, B., Fischer, J., Gerhold, R., Jori, F. and Lamarque, F. (2002) 'The value of wildlife'. *Review sci. techn. Off. int. Epiz.*, vol 21, no 1, pp15–51.

10   Cowlishaw, G., Mendelson, S. and Rowcliffe, M. (2003a) *Market Structure and Trade in a Bushmeat Commodity Chain*, draft, Zoological Society of London.

11   Cowlishaw, G., Mendelson, S. and Rowcliffe, M. (2003b) *Evidence for Post-Depletion Sustainability in a Mature Bushmeat Market*, draft, Zoological Society of London.

12   De Merode, E., Homewood, K. and Cowlishaw, C. (2003) *The Value of Bushmeat and other Wild Foods to Rural Households Living in Extreme Poverty in the Eastern Democratic Republic of Congo*, draft, University College London.

13   Dickson, B. (2003) 'What is the goal of regulating wildlife trade? Is regulation a good way to achieve this goal?' in Oldfield, S (ed) *The Trade in Wildlife: Regulation for Conservation*. Earthscan, London, pp2–31.

14   Elliott, I. (2002) *Wildlife and Poverty Study*. DFID Rural Livelihoods Department, London.

15 Fa, J. E. and Peres, C. A. (2001) 'Game vertebrate extraction in African and Neotropical Forests: an intercontinental comparison', in Reynolds, J. D., Mace, G. E., Redford, K. H. and Robinson, J. G. (eds) *Conservation of Exploited Species*. Cambridge University Press, Cambridge.

16 Fa, J. E. and Wilkie, D. (2002) *Reducing Demand by Developing Economically-acceptable Alternatives to Wildlife Meat*, paper presented at the 16th Annual Meeting of the Society for Conservation Biology, Canterbury, July 2002.

17 Fa, J. E., Currie, D. and Meeuwig, J. (2003) 'Bushmeat and food security in the Congo Basin: Linkages between wildlife and people's future'. *Environmental Conservation*, vol 30, no 1, pp71–78.

18 Fa, J. E., Peres, C. A. and Meeuwig, J. (2002) 'Bushmeat exploitation in tropical forests: An intercontinental comparison'. *Conservation Biology*, vol 16, no 1, February, pp232–237.

19 Fairhead, J. and Leach, M. (1998) *Reframing Deforestation: Global Analysis and Local Realities: Studies in West Africa*. Routledge, London.

20 Forest Stewardship Council (2000) *FSC Principles and Criteria*. FSC Secretariat, Mexico (www.fscoax.org/ principal.html).

21 Hutton, J. and Dickson, B. (2002). 'Conservation out of exploitation: a silk purse out of a sow's ear?' in Reynolds, J. D., Mace, G. E., Redford, K. H. and Robinson, J. G. (eds) *Conservation of Exploited Species*. Cambridge University Press.

22 Mendelson, S., Cowlishaw, G. and Rowcliffe, M. (2003) *Anatomy of a Bushmeat Commodity Chain: Actors on the Urban Market Stage of Takoradi, Ghana*, draft, Zoological Society of London, London.

23 Metcalfe, S. (1994) 'The Zimbabwe Communal Assets Management Programme for Indigenous Resources (CAMPFIRE)', in Western, D and Wright, R. M. (eds) *Natural Connections: Perspectives in Community-based Conservation*. Island Press, Washington, pp161–191.

24 Oates, I. E. (1998) *Myth and Reality in the Rainforest*. Chicago University Press.

25 Robinson, J. and Bennett, E. (eds) (2000) *Hunting for Sustainability in Tropical Forests*. Cornell University Press.

26 Robinson, J. and Bennett, E. (2000) 'Carrying capacity limits to sustainable hunting in tropical forests', Chapter 2 of Robinson and Bennett (eds), pp13–30.

27 Robinson, J. and Bennett, E. (2002) 'Will alleviating poverty solve the bushmeat crisis?' *Oryx*, vol 36, no 4, p332.

28 Robinson, J. and Redford, K. (1991) *Neotropical Wildlife Use and Conservation*. Chicago University Press.

29 Scoones, I., Melnyk, M. and Pretty, J. (1992) *The Hidden Harvest: Wild Foods and Agricultural Systems: A Literature Review and Annotated Bibliography*. IIED, SIDA and WWF, London, UK and Gland, Switzerland.

30 Secretary Of State For International Development, UK (2002) *Speech by the Rt. Hon Clare Short MP to the UK Bushmeat Campaign Conference 28 May*. Zoological Society of London.

10

Reassessing the Fuelwood Situation in Developing Countries[1]

M. Arnold and R. Persson

Introduction

Understanding the fuelwood situation has always been hampered by lack of reliable information. Only a very small fraction of fuelwood production is recorded. The greater part of consumption is by poor households and so is also seldom reported. Assessment of the actual magnitude of fuelwood use, and the impacts on forests and rural livelihoods, has consequently been difficult to determine, and has been the subject of considerable debate.

In the mid-1970s, estimates that huge and growing numbers of people in developing countries depended on fuelwood as their principal domestic fuel led to predictions of potentially devastating depletion of forest resources. The perceived widening shortages of fuelwood were also expected to have serious negative socioeconomic consequences for the rural poor. Recommendations for widespread, rapid action to avert or reverse 'fuelwood gaps' called for large-scale plantations to be established near urban and other concentrated sources of market demand, with community and individual plantings to meet more localized rural needs. Interventions to bring this into effect rapidly emerged in both donor and national forestry programmes, attracting large funding flows.

By the mid-1980s much of this initial assessment had been quite radically revised. It was argued that the rate of growth of consumption had been overestimated, and that as much of fuelwood supplies came from outside forests the impact that fuelwood use had on forests had also been overstated. In addition, it was argued that the scope for intervening to encourage more tree planting for fuelwood use was more limited than had been assumed, because there were often lower cost alternatives (Dewees, 1989).

In response, fuelwood-oriented forestry programmes were widely scaled back during the 1990s, and attention to the fuelwood situation has been reduced. Nevertheless, a considerable amount of relevant new information has continued to be assembled, in energy as well as forestry studies, which provides a basis for examining the situation once again. In this chapter we look at what light recent information sheds on household usage of fuelwood and charcoal in developing countries, the supply systems associated with this, and the impacts on forests and livelihoods. We do not cover major industrial uses of these

Note: Reprinted from *International Forestry Review*, vol 5, Arnold, M. and Persson, R., 'Reassessing the fuelwood situation in developing countries', pp379–383, Copyright © (2003), with permission from the Commonwealth Forestry Association

fuels, or forms of energy generated from wood that are emerging as environmentally clean alternatives to fossil fuels in some applications, in particular in developed countries.

Woodfuel Consumption Trends[2]

Over the past few years, FAO has been carrying out a major revision of its published fuelwood and charcoal data and has developed more rigorous and realistic analytical and projection models (Whiteman et al, 2002). The number of countries reporting national fuelwood and charcoal production remains limited, and the accuracy of much of the reported data is still poor. However, use of data from more detailed studies in particular situations has enabled more realistic estimates to be developed for non-reporting countries, and the generation of analyses of trends in consumption that introduce a wider range of explanatory variables.[3]

One general result emerging from this work was that income consistently turned out to be an important influence on the level of woodfuel use. Although there are great variations between countries, consumption of both fuelwood and charcoal usually decrease with an increase in income. In addition, urbanization typically decreases fuelwood use and increases charcoal consumption, and per capita fuelwood consumption increases as the proportion of land under forest cover increases.

Further information has become available from analysis of data from surveys of about 25,000 households in 46 cities in 12 developing countries, carried out by the UNDP/World Bank Energy Sector Management Assistance Programme (ESMAP). As shown in Figure 10.1, demonstrating the relationship between household income and energy use, there is a general transition by urban users from heavy use of fuelwood to more convenient fuels as incomes rise. Charcoal is often the main 'transition' fuel to which they shift first (Barnes et al, 2002).

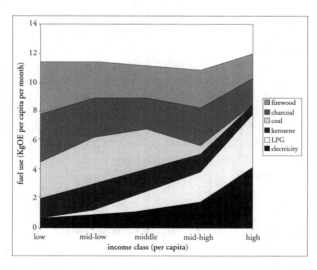

Figure 10.1 *Relationship between household income and energy use*

Table 10.1 *FAO projections of woodfuel consumption to 2030 in the main developing regions*

	1970	1980	1980	2000	2010	2020	2030
Fuelwood (million m³)							
South Asia	234.5	286.6	336.4	359.9	372.5	361.5	338.6
South-East Asia	294.6	263.1	221.7	178.0	139.1	107.5	81.3
East Asia	293.4	311.4	282.5	224.3	186.3	155.4	127.1
Africa	261.1	305.1	364.6	440.0	485.7	526.0	544.8
South America	88.6	92.0	96.4	100.2	107.1	114.9	122.0
Charcoal (million tonnes)							
South Asia	1.3	1.6	1.9	2.1	2.2	2.4	2.5
South-East Asia	0.8	1.2	1.4	1.6	1.9	2.1	2.3
East Asia	2.1	2.3	2.3	2 2	2.1	2.0	1.8
Africa	8.1	11.0	16.1	23.0	30.2	38.4	46.1
South America	7.2	9.0	12.1	14.4	16.7	18.6	20.0

FAO projections using the revised data and models show that the impact of these and other factors is reflected in markedly different trends in different regions (Table 10.1). In Asia, which accounts for nearly half of the world's woodfuel consumption, aggregate consumption of fuelwood is declining. This reflects a significant decline in China and much of East and South-East Asia since the 1980s, and the fact that consumption in South Asia appears to be at or close to its peak. In Africa, where fuelwood use per capita is on average considerably higher than in Asia, the rate of growth in consumption is declining but is still quite high in many countries. In South America, where fuelwood is a less important fuel, its overall consumption appears to have been rising only slowly.

Charcoal consumption is often growing faster than fuelwood consumption. Though small relative to fuelwood in most of Asia, charcoal use is becoming a much larger part of the woodfuels total in Africa and South America. In Africa, the aggregate of consumption of fuelwood and wood for charcoal is growing at a rate close to that of population growth.

Though use of woodfuels is thus generally not growing at the rates assumed in the past, the quantities used, and the numbers using them, are still huge. Moreover, in the main consuming regions these magnitudes will often continue to be very large. The International Energy Agency recently estimated (IEA, 2002) that the number of people using fuelwood and other biomass fuel in Africa will rise by more than 40 per cent between 2000 and 2030 to about 700 million, and that in the latter year there will still be about 1,700 million users in Asia.[4]

Patterns of Supply and Impacts on Forests

Over the past two decades quite a few countries have developed more up-to-date and complete estimates of national fuelwood supply and demand 'balances', often within broader energy projection exercises. The more comprehensive exercises attempt to assess output from tree resources both within and outside forests. They also estimate the contribution and potential of other biomass fuels, such as crop residues and animal dung, which can be used as fuelwood substitutes.

The results for most of the countries studied show that wood and related biomass fuel resources exist in sufficient abundance to provide more than adequate physical coverage of woodfuel needs. Trees outside forests appear to supply a large share of overall woodfuel output, highlighting the importance of non-forest resources. For instance, the FAO/Netherlands Regional Wood Energy Development Programme in Asia found for 16 countries studied that total potential woodfuel supplies exceeded woodfuel demand in 1994, and are likely to continue to do so in all but two in 2010 (RWEDP, 1997). In five of the countries more than 75 per cent of fuelwood production came from trees and other wood sources outside forests, and in another two more than 50 per cent.

Where fuelwood is being sourced from forests, studies have shown that it is usually from land being cleared for farming – areas close enough to urban markets to supply fuelwood often being areas under pressure from clearance for agriculture to supply food to the same markets. In general, recent information thus tends to support the view that demand for fuel is seldom the primary source of depletion or removal of forest cover on a large scale.

Recent reviews of findings from studies on the causes of deforestation also support this. For example, an assessment of economic models of tropical deforestation, while indicating the existence of multiple rather than single causes and noting that evidence regarding fuelwood is weak, points to it being an occasional cause, mainly in parts of Africa (Kaimowitz and Angelsen, 1998). An analysis based on a wide range of case studies in tropical countries also found fuelwood harvesting to be important mainly in some situations in Africa where deforestation is associated with wood extraction (Geist and Lambin, 2002).

These situations tend to be where there is growing urban demand for charcoal. In parts of Africa where improvements in rural infrastructure and transport have enabled production of charcoal to be located in more distant dry-land forest and woodland areas, its harvest can materially alter the structure and productivity of the woodstocks being drawn upon. Studies in areas supplying a number of cities in southern and eastern Africa have shown that harvesting for charcoal can lead to downgrading of woodland to bush, and bush to scrub, over very large areas. However, re-growth on all but the areas put under continuous cultivation ensured a considerable measure of regeneration of wood resources. Moreover, harvesting was found to be within sustainable limits over much of the area (SEI, 2002). Thus, even where wood for charcoal is the main forest product it may have only a limited impact on forest or woodland loss. However, by depleting preferred species, favouring coppicing species etc, it can adversely affect the composition and biological productivity of the resource.

Resources planted to provide fuelwood supplies are still limited in extent. The continued ability in many situations to expand production into existing woodstocks as improvements in infrastructure and transport make them economically accessible, appears to have contributed to the real price of woodfuels remaining little changed in urban markets over lengthy periods of time (SEI, 2002). Where this is the case, the usual economic signals of shortage that price rises provide are not present, which helps explain why depletion of physical woodstocks seldom triggers investment in resource renewal.

However, in some areas low in natural wood resources, such as the province of Cebu in the Philippines and parts of southern Ethiopia, much of fuelwood production is now based on farm-grown trees. In addition, in parts of dryland India the tree shrub *Prosopis juliflora,* originally introduced as protective groundcover, has spread rapidly and now forms a major new source of fuelwood on communal and private lands.

Impacts on Subsistence Users

Most use is still of a rural subsistence nature. Fuelwood is still the main domestic fuel in rural households in most developing countries, and gathered supplies still constitute the households' main source of fuelwood. Recent household surveys over large areas in India found that wood accounted for 56 per cent of their energy use (Chopra and Dasgupta, 2000), and that about 55 per cent of household needs for fuelwood were collected free (ESMAP, 2002).

The observational evidence that shortages of fuelwood for subsistence users are becoming more pronounced, particularly for the landless and those with little land, is considerable. In both Africa and Asia formal and informal privatization of land and wood resources previously available for use by fuelwood gatherers as common property is on a wide scale reducing rural households to what they can produce on their own land, or purchase or steal. Such tenurial changes can encourage those with land to produce more, but leave the landless and those with very little land worse off.

The poorest may also be disadvantaged by shifts to bring remaining common pool resources under local control and sustainable management. Fuelwood tends not to be of high priority in programmes such as Joint Forest Management. Harvesting of fuelwood is usually restricted in the process of bringing a local resource under management, and women's needs for fuelwood commonly have lower priority than those of men for forest products for sale when setting longer-term management goals.

Women seldom list fuelwood shortage as being high among their concerns, but it is still likely to involve a cost to them. Households generally respond to fuelwood shortages by purchasing more of their supplies, or increasing the time spent on fuelwood collection. Some households also move down to burning straw, dung and other less favoured fuels, while wealthier households shift to alternative fuels. Measures to economize fuelwood use are also adopted, for example using foods that take less time or fuel to cook.

While interventions to encourage the adoption of more fuel-efficient stoves have had some impact in urban areas, success has been limited in rural areas. Some evidence suggests that where stoves are seen as saving money (in towns) they are popular, but where they are merely saving time or biomass (in rural areas) men are not prepared to spend money purchasing them. Recently attention to improved stoves has shifted from increasing efficiency of woodfuel use to reducing damage to health from airborne particulates and noxious fumes associated with the burning of wood and charcoal (IEA, 2002). However, the evidence suggesting this linkage has been questioned. Thus there is uncertainty about the extent and nature of the impacts of improved stoves, and interventions in this area have often been reduced or discontinued.

Programmes to support farm growing of trees for fuelwood have also been scaled back, as evidence emerged that the spontaneous responses to fuelwood shortages that households adopt involve lower costs, and are more efficient than farm forestry interventions in addressing the constraints they face. However, there is growing evidence that tree management by farmers for other purposes is on the rise in many situations, and that some of the resulting output is going towards increasing household fuelwood supplies. This suggests that access to a broader menu of low-cost, multi-purpose woody species and husbandry options to choose from might assist farmers to increase their supply of fuelwood as a co- or bi-product of their strategies for incorporating and managing on-farm trees and shrubs.

Woodfuels as Sources of Household Income

The sale and trading of woodfuels provide an income for huge numbers of people. With ease of access both to the resource and markets, very large numbers of the landless and very poor gather and sell wood for fuel, and large numbers of farmers harvest and sell it as well. Much of woodfuel retailing is small scale and accessible to the urban poor too. Overall, it is a major source of income for the poor and can be one of the main sources from forest product activities.

For some people engaged in woodfuel production, selling or trading, such activities represent their principal source of income. This was found to be the case, for instance, for about 125,000 people producing or selling charcoal for use in the city of Dar es Salaam, Tanzania, in the 1990s (SEI, 2002). For others, fuelwood or charcoal provides a supplemental, transitional or seasonal source of income, or serves as a 'safety net' in times of hardship.

Though urban demand usually is much larger, rural demand for purchased woodfuel is also growing. For instance, in Ghana in 1991–1992, 27 per cent of all fuelwood purchased by households and 13 per cent of charcoal was bought by households in rural areas (Townson, 1995).

As was noted earlier, urban demand for fuelwood is now declining in some areas, notably in Asia. In Hyderabad, India, a threefold increase in population over the 13 years to 1994 was accompanied by no growth in overall fuelwood consumption, and a 60 per cent fall in household consumption of fuelwood (Barnes et al, 2002). Changes in market demand of this magnitude clearly must have substantial impacts on those engaged in the supply chain; an aspect of woodfuel trading that perhaps should receive more attention than at present.

Though a source of some income for so many people, producing and trading woodfuel is seldom very remunerative. Low prices and the high levels of competition that ease of entry into the activity usually create often mean that woodfuel selling generates little surplus for those engaged in it. This keeps most of those engaged in it poor, and discourages investment in more efficient production (or in sustainable management or renewal of the resource). Fees and other government charges, and regulations governing the sale and trading of woodfuels, can also impose significant costs and constraints on who can participate. In addition, weak tenurial rights over the resource can mean that poor rural producers and traders are progressively excluded from access to the resource and markets as the trade grows.

In principle, the transition to participatory local management or co-management of forest and woodland resources should help alleviate some of these constraints. In practice, the results have been variable. The low priority relative to other outputs that often is associated with fuelwood and fuelwood producers within some such programmes has been noted earlier. In addition, where civil society institutions are weak, the benefits of woodfuel trading may be captured by outsiders or an elite, rather than by the more needy members of the community.

In an ambitious set of programmes supported by the World Bank and other donors in parts of West Africa where wood for charcoal is the principal forest output, communities are granted formal control over natural woodland and given exclusive rights to the sale of all woodfuel produced. In return, they enter into an agreement to manage the woodland sustainably. Funding comes from differential taxation in urban markets of woodfuel sales from these 'fuelwood markets' and from uncontrolled sources. However, management plan and control costs are high and the differential taxing system is in practice difficult to administer. Progress to date has also been hampered by the scope for corrupt manipulation and difficulties in preventing competition from cheaper, uncontrolled sources (Bertrand, 2002).

Overall, such experiences tend to emphasize the importance of strong tenure and governance as a prerequisite for progress in this area. Some have also argued that there could be difficulties in absorbing the higher transaction costs of more effective management measures as long as the value of the woodfuels traded remains so low (SEI, 2002). However, higher woodfuel prices could have negative implications both for poor users, and for the comparative advantage that poor producers presently possess.

Conclusions

In general, the new information outlined in this chapter supports arguments developed in the late 1980s that there is not a 'fuelwood crisis' of such a magnitude, and with such potentially dire consequences in terms of forest depletion, as to require major interventions to maintain or augment supplies. More accurate and better defined data, and more realistic analytical and projection models, show that demand is not growing at the rates estimated earlier. Increasing urbanization and rising incomes are reflected in a slowing down of the rate of increase in fuelwood use, and in some areas consumption is now in decline. Supplies are in practice being drawn from a much wider base than just forests, and users have access to a range of responses that enable them to adjust to changes in the availability of fuelwood without necessarily needing to invest in additional wood resources. Charcoal use, on the other hand, is usually continuing to grow and is becoming a much larger part of the woodfuels total in some regions, and where this is happening it often does warrant particular attention.

However, while the arguments that special programmes and initiatives to address woodfuel demand were neither needed nor appropriate in most circumstances have been widely acted upon, the accompanying argument that there was a need to incorporate woodfuels more fully into the forestry mainstream appears not to have been acted on to the same extent. It has been suggested that this may have been in part because the arguments for moving away from the earlier fuelwood focus in forestry have been misinterpreted as meaning that woodfuel use is rapidly diminishing. It may also reflect arguments in the energy sector that reliance on such sources of energy acts as a constraint to livelihood enhancement and broader economic improvement, and that the focus should be on helping users to move from woodfuels to more 'efficient' fuels (IEA, 2002). Another factor may be that fuelwood has in a sense fallen between timber and non-timber in recent approaches to forestry and development.

In practice, huge numbers of people continue to rely on woodfuels as a source of energy or income, and will continue to do so. Similarly, woodfuels remain one of the largest outputs of the forest sector. The task implicit in the 1980s arguments is to see this as an integral part of forest management, rather than being in need of responses developed separately from the rest of forestry.

The issues that the previous discussion has identified are in fact ones that should logically be of concern within one or more of the main thrusts that participatory and livelihood-oriented forestry strategies are taking: moving to effective and equitable local management of common pool wood resources; supporting farmer management of on-farm tree resources; and generation of incomes through production and trading of non-timber forest products. However, despite the growing focus on giving forestry a stronger livelihood orientation, woodfuels seldom appear to have received attention commensurate with their importance in this connection.

In short, woodfuels may be less of a concern to the security of the forest estate than was previously feared. However, they are a larger component of the contribution that forestry can make to poverty alleviation than appears to be currently reflected in most forestry and agroforestry policies and programmes.

Notes

1 This chapter is based on a study by M. Arnold, G. Kohlin, R. Persson and G. Shepherd, 'Fuelwood Revisited: What Has Changed in the Last Decade?' published as an Occasional Paper no 39 by CIFOR 2003

2 The term 'woodfuel' is used here to cover both fuelwood and charcoal.

3 The new figures use the non-modelled FAO data between 1970 and 1998 as a starting point. Altogether, 1056 records of fuelwood data and 370 records of charcoal data were collated. An extensive search of a wide variety of sources was also undertaken to unearth and incorporate as many additional records relating to actual usage as possible. This information was used in developing new analytical models to revise estimates of per capita consumption. For some countries where sufficient total national consumption data were available, models of consumption at this level were used. For others, estimates of non-household use were added to modelled household consumption figures, to arrive at estimates of total calculate prospective national consumption figures. The household consumption models for each country included 'dummy variables' related to either 'national' or 'regional'

consumption, according to the data available (Broadhead, Bahdon and Whiteman, 2001).

4 The IEA study estimated that, in 2000, 80–90 per cent of the biomass used as fuel in Africa was woodfuel, whereas in Asia more than half was agricultural residues and dung (IEA 2002).

References

1 Barnes, D. F., Krutilla, K. and Hyde, W. (2002) *The Urban Energy Transition: Energy, Poverty and the Environment in the Developing World*. World Bank, January 2002 (Draft).

2 Bertrand, A. (2002) *A New Perspective on Sustainable Woodland Management and Woodfuel Energy*. Paper prepared for the AFTEG/AFTRS Joint Seminar on Household Energy and Woodland Management, World Bank, April 2002.

3 Broadhead, J., Bahdon, J. and Whiteman, A. (2001) 'Woodfuel consumption modelling and results'. Annex 2 in *Past trends and future prospects for the utilisation of wood for energy*, Working Paper no: GFPOS/WP/05, Global Forest Products Outlook Study, FAO, Rome.

4 Chopra, K. and Dasgupta, P. (2000) *Common Property Resources and Common Property Regimes in India: A Country Report*. Institute of Economic Growth, New Delhi (Draft).

5 Dewees, P. A. (1989) 'The woodfuel crisis reconsidered: Observations on the dynamics of abundance and scarcity'. *World Development*, vol 17, no 8, pp1159–1172.

6 ESMAP (2002) *Energy Strategies for Rural India: Evidence from Six States*. Joint UNDP/World Bank Energy Sector Management Assistance Programme, World Bank, August 2002.

7 Geist, H. J. and Lambin, E. L. (2002) 'Proximate causes and underlying driving forces of tropical deforestation'. *BioScience*, vol 52, no 2, pp143–150.

8 International Energy Agency (IEA) (2002) 'Energy and poverty', Chapter 13 in *World Energy Outlook 2002*, OECD, Paris.

9 Kaimowitz, D. and Angelsen, A. (1998) *Economic Models of Tropical Deforestation: A Review*. CIFOR.

10 RWEDP (1997) *Regional Study on Wood Energy Today and Tomorrow in Asia*. Field Document no 50, Regional Wood Energy Development Programme in Asia, FAO, Bangkok.

11 SEI (2002) *Charcoal Potential in Southern Africa*. CHAPOSA: Final Report. INCO-DEV, Stockholm Environment Institute, Stockholm.

12 Townson, I. M. (1995) *Patterns of Non-timber Forest Products Enterprise Activity in the Forest Zone of Southern Ghana: Main Report*. Report to the ODA Forestry Research Programme, Oxford.

13 Whiteman, A., Broadhead, J. and Bahdon, J. (2002) 'The revision of woodfuel estimates in FAOSTAT'. *Unasylva*, vol 53, no 4, pp41–45.

11

Forest Fires: A Global Overview

Johann G. Goldammer

Introduction

Fire is a prominent disturbance factor in most vegetation zones throughout the world. In many ecosystems fire is a natural, essential and ecologically significant force, organizing physical and biological attributes, shaping landscape diversity, and influencing energy flows and biogeochemical cycles, particularly the global carbon cycle. In some ecosystems, however, fire is an uncommon or even unnatural process that severely damages vegetation and can lead to long-term degradation. Such ecosystems, particularly in the tropics, are becoming increasingly vulnerable to fire due to growing population, economic and land-use pressures. Some recent well-known examples include widespread wildfire occurrence in evergreen rainforests in the Amazon Basin of Brazil, Indonesia and Central America. Moreover, the use of fire as a land-management tool is deeply embedded in the culture and traditions of many societies, particularly in the developing world. Given the rapidly changing social, economic and environmental conditions occurring in developing countries, marked changes in fire regimes can be expected, with uncertain local, regional and global consequences. Even in regions where fire is natural (eg the boreal zone), more frequent severe weather conditions have created recurrent major fire problems in recent years. The incidence of extreme wildfire events is also increasing elsewhere the world, with adverse impacts on economies, livelihoods and human health and safety that are comparable to those associated with other natural disasters such as earthquakes, floods, droughts and volcanic eruptions. Despite the prominence of these events, current estimates of the extent and impact of vegetation fires globally are far from complete. Several hundred million hectares of forest and other vegetation types are estimated to burn annually throughout the world. These fires consume several billion tons of dry matter and release emission compounds that affect the composition and functioning of the global atmosphere and human health. However, the vast majority of these fires are not monitored or documented. Informed policy and decision making clearly requires timely quantification of fire activity and its impacts nationally, regionally and globally. Such information is currently largely unavailable.

The primary concerns of policy makers focus on questions regarding the regional and global impacts of excessive and uncontrolled burning, broad-scale trends over time, and the options for instituting protocols that will lead to greater control. Other key

Note: Reprinted from *An Overview of Vegetation Fires Globally*, Goldammer, J. G. Adapted from a report from the Working Group on Wildland Fire, the United Nations International Strategy for Disaster Reduction, Copyright © (2004), with permission from the United Nations

questions involve determining under what circumstances fires poses a sufficiently seri-ous problem to require action, what factors govern the incidence and impacts of fires in such cases, and what might be the relative costs and benefits of different options for reducing adverse impacts?

The Current Extent and Impact of Wildland Fire Globally

Global wildland fire assessments

The most recent inventory of global fire activity was carried out under the framework of For-est Resources Assessment 2000, conducted by the Food and Agriculture Organization of the United Nations (FAO) for the 1990–2000 period. This report, the *Global Forest Fire Assess-ment 1990–2000* (FAO, 2001), prepared in cooperation with the Global Fire Monitor-ing Center (GFMC) is the most complete to date, including for the first time information on fires in non-forest vegetation types (eg savannah fires and agricultural burning).

However, the wildland fire statistics collected are not complete for all countries: in several cases the accuracy of the estimates is unknown and the format is not consistent. Further, statistical datasets providing number of fires and area burned do not begin to meet the level of information required to assess the environmental and economic conse-quences of wildland fires. For example, current formats for fire statistics collection do not include parameters that would permit conclusions on economic damage or impacts of emis-sions on the atmosphere or human health. The pathways of vegetation succession following fire are complex and include the cumulative impacts of anthropogenic and environmental stresses. It is not possible at this time to conclude from existing statistical data whether long-term changes can be expected in terms of site degradation and reduction of carry-ing capacity of fire-affected sites. Thus, a new system for fire data collection that would meet the requirements of a growing number of different users is urgently needed.

The recently published Global Burned Area Product 2000 derived from a space-borne sensing system (SPOT Vegetation) is the first important step towards obtaining prototype baseline data on the extent of global wildland fires for the year 2000 (JRC, 2002). The accuracy of the satellite-derived data set has also yet to be determined. However, initial analysis indicates that approximately 351 million hectares globally were affected by fire in the year 2000. For individual countries the number of fires and area burned differ consid-erably from the data provided by national agencies. This is likely to be a result of the fact that statistical datasets of national agencies in many countries primarily include fire incidences in managed forests, with only a few countries providing fire statistics that cover non-forest ecosystems. Furthermore most countries do not have in place appropriate means to sur-vey wildland fire occurrence impacts. On the other hand the area burned, as derived from the spaceborne instrument, also fails to provide information on the environmental, eco-nomic or humanitarian impacts of fire. An appropriate interpretation of satellite-gener-ated fire information requires additional information layers, particularly on ecosystem vulnerability and recovery potential, that are not yet available at a global level.

Major conclusions of the 1990–2000 Global Forest Fire Assessment can be summarized as follows:

- Drought years in the 1990s caused widespread burning in tropical rainforests with significant impacts on natural resources, public health, transportation, navigation and air quality.
- While many countries, and regions have a well-developed system for documenting, reporting and evaluating wildfire statistics in a systematic manner, but lack information on the effects of these fires, many other countries do not yet have such a system, largely because of more pressing social issues. Satellite systems have been used effectively to map active fires and burned areas, especially in remote areas where other damage assessment capabilities are not available.
- Even those countries supporting highly financed fire management organizations are not exempt from the ravages of wildfires in drought years. When wildland fuels have accumulated to high levels, no amount of firefighting resources can make much of a difference until the weather moderates (as observed in the United States in the 2000 fire season).
- Uncontrolled use of fire for forest conversion and for agricultural and pastoral purposes continues to cause a serious loss of forest resources, especially in tropical areas.
- Some countries are beginning to realize that inter-sectoral coordination of land-use policies and practices is an essential element in reducing wildfire losses. There were numerous examples in the 1990s of unprecedented levels of inter-sectoral and international cooperation in helping to lessen the impact of wildfires on people, property and natural resources.
- Examples exist where sustainable land-use practices and the participation of local communities in integrated forest fire management systems are being employed to reduce resource losses from wildfires.
- In some countries, volunteer rural fire brigades are successful in responding quickly and efficiently to wildfires within their range; and residents are taking more responsibility to ensure that homes will survive wildfires.
- Although prescribed burning is being used in many countries to reduce wildfire hazards and achieve resource benefits, other countries have prohibitions against the use of prescribed fire.
- Fire ecology principles and fire regime classification systems are being used effectively as an integral part of resource management and fire management planning.
- Fire research scientists have been conducting cooperative research projects on a global scale to improve understanding of fire behaviour, fire effects, fire emissions, climate change and public health.
- Institutions like the Global Fire Monitoring Center have been instrumental in bringing the world's fire situation to the attention of a global audience via the Internet.

In reviewing the global fire situation, it is apparent that continued emphases on the emergency response side of the wildfire problem will not address current needs and will only result in a continuation of the current trend towards large and damaging fires. The solution to the emergency response dilemma is to couple emergency preparedness and response programmes with more sustainable land-use policies and practices. Only when sustainable land use practices and emergency preparedness measures complement each other will long-term natural resource benefits for society be realized.

Global observation and monitoring of wildland fires

From the changing role of fire in the different vegetation zones described above it can be concluded that fire management is becoming increasingly important with respect to global issues of resource management, disaster reduction and global change. With this increasing importance comes the need for a concerted effort to put in place the international global observation and monitoring systems needed to give early warning and identify disastrous fire events, inform policy making and to support sustainable resource management and global change research (Justice, 2001). The observation systems will need to include both ground-based and space-based monitoring components. Advances in information technology now make it easier to collect and share data necessary for emergency response and environmental management. Current satellite assets are underutilized for operational monitoring and fire monitoring falls largely in the research domain. Increasing attention needs to be given to data availability, data continuity, data access and how the data are being used to provide useful information.

There is no standard in-situ measurement/reporting system and national reporting is too variable and inadequate to provide a regional or global assessment. It is also often hard to relate the satellite and in-situ data reporting. Reliable information is needed to inform policy and decision making, and management policies should be developed based in part on a scientific understanding of their likely impacts. Meetings are needed for exchange of information on monitoring methods, use of appropriate technology, policy and management options and solutions. A continued and informed evaluation of existing monitoring systems, and a clear articulation of monitoring requirements and operational prototyping of improved methods, is also required.

The Global Observation of Forest Cover/Global Observation of Land Dynamics (GOFC/GOLD) programme, a part of the Global Terrestrial Observing System (GTOS), is designed to provide such a forum. The GTOS, which is sponsored by the International Global Observing System Partners, has its Secretariat at the FAO in Rome. GOFC/GOLD has its Secretariat in the Canadian Forest Service. The GOFC/GOLD programme is an international coordination mechanism to enhance the use of earth-observation information for policy, natural resource management and research. It is intended to link data producers to data users, to identify gaps and overlaps in observational programmes, and recommend solutions. The programme will provide validated information products, promote common standards and methods for data generation and product validation and stimulate advances in the management and distribution of large volume datasets. Overall it is intended to advance our ability to obtain and use environmental information on fires and secure long-term observation and monitoring systems.

GOFC/GOLD has three implementation teams: fire, land cover and biophysical characterization. The principal role of GOFC/GOLD is to act as a coordinating mechanism for national and regional activities (Justice et al, 2003). To achieve its goals GOFC/GOLD has developed a number of regional networks of fire data providers, data brokers and data users. These networks of resource managers and scientists provide the key to sustained capability for improving the observing systems and ensuring that the data are being used effectively. GOFC/GOLD regional networks are being implemented through a series of regional workshops. These regional network workshops are used to engage the user community to address regional concerns and issues, provide a strong voice for regional needs and foster lateral transfer of technology and methods within and between regions. Networks are currently being developed in Central and Southern Africa, South-East Asia, Russia, the Far East and Central and South America.

The GOLD-Fire programme has a number of stated goals:

- To increase user awareness by providing an improved understanding of the utility of satellite fire products for resource management and policy within the United Nations and at regional, national and local levels.
- To encourage the development and testing of standard methods for fire danger rating suited to different ecosystems and to enhance current early warning systems for fires.
- To establish an operational network of fire validation sites and protocols, providing accuracy assessment for operational products and a test bed for new or enhanced products, leading to standard products of known accuracy.
- To enhance fire product use and access, for example, by developing operational multi-source fire and Geographic Information System (GIS) data and making these available over the Internet.
- To develop an operational global geostationary fire network providing observations of active fires in near real time.
- To establish operational polar orbiters with fire monitoring capability. Providing:

    - operational moderate-resolution long-term global fire products to meet user requirements and distributed ground stations providing enhanced regional products. These products should include fire danger, fuel moisture content, active fire, burned area and fire emissions; and
    - operational high-resolution data acquisition allowing fire monitoring and post-fire assessments.

- To create emissions product suites, developed and implemented providing annual and near real-time emissions estimates with available input data.

It is particularly important to improve the quality, scope and utility of GOFC-Fire inputs to the various user communities through:

- gaining a better understanding of the range of users of fire data, their needs for information, how they might use such information if it was available, and with what other data sets such information might be linked;
- increasing the awareness of users with respect to the potential utility of satellite products for global change research, fire policy, planning and management; and
- developing enhanced products based on ongoing interaction with representatives of the various user communities.

Vegetation fire emissions: atmosphere, climate and human health

Research efforts under the Biomass Burning Experiment (BIBEX) of the International Geosphere-Biosphere Programme (IGBP), International Global Atmospheric Chemistry (IGAC) Project, and a large number of other projects in the 1990s were successful in the sampling and quantification of fire emissions, and the determination of emission factors for various fuel types. However, global and regional emission estimates are still problematic, mostly because of uncertainties regarding amounts burned. Most recent estimates indicate that the amount of vegetative biomass burned annually is in the magnitude of 9.2 billion tonnes (Andreae and Merlet, 2001; Andreae, 2002).

Vegetation fires produce a range of emissions that influence the composition and functioning of the atmosphere. The fate of carbon contained in fire-emitted carbon dioxide (CO_2) and other radiatively active trace gases is climatically relevant only when there is no regrowth of vegetation – deforestation or degradation of sites, for example. Alternatively, NO_x, CO, CH_4 and other hydrocarbons are ingredients of smog chemistry, contribute to tropospheric ozone formation and act as 'greenhouse gases', while halogenated hydrocarbons (eg CH_3Br) have considerable impact on stratospheric ozone chemistry and contribute to ozone depletion.

Fire-emitted aerosols influence climate directly and indirectly. Direct effects include (a) backscattering of sunlight into space, resulting in increased albedo and a cooling effect, and (b) absorption of sunlight that leads to cooling of the Earth's surface and atmospheric warming. As a consequence convection and cloudiness are reduced as well as evaporation from ocean and downwind rainfall. The key parameter in these effects is the black carbon content of the aerosol and its mixing state.

Indirect effects of pyrogenic aerosols are associated with cloud formation. Fire-emitted aerosols lead to an increase of Cloud Condensation Nuclei (CCN) that are functioning as seeds for droplet formation. Given a limited amount of cloud water content this results in an increase of the number of small droplets. As a first consequence clouds become whiter, reflect more sunlight, thus leading to a cooling effect. As a second indirect effect of 'overseeding' the overabundance of CCN coupled with limited amount of cloud water will reduce the formation of droplets that are big enough (radius ~14µm) to produce rain; consequently rainfall is suppressed. In conclusion it can be stated that:

- the fire and atmospheric science community has made considerable progress in determining emission factors from vegetation fires;
- global and regional emission estimates are still problematic, mostly because of uncertainties regarding amounts of area and vegetative matter burned;
- fire is a significant driver of climate change (as well as a human health risk);
- fire, climate and human actions are highly interactive;
- some of these interactions may be very costly both economically and ecologically.

Smoke from burning of vegetative matter contains a large and diverse number of chemicals, many of which have been associated with adverse health impacts (Goh et al, 1999). Nearly 200 distinct organic compounds have been identified in wood smoke aerosol, including volatile organic compounds and polycyclic aromatic hydrocarbons. Available data indicate high concentrations of inhalable particulate matter in the smoke of vegetation fires. Since particulate matter produced by incomplete combustion of biomass are mainly less than one micrometers in aerodynamic diameter, both PM_{10} and $PM_{2.5}$ (particles smaller than 2.5μm in aerodynamic diameter) concentrations increase during air pollution episodes caused by vegetation fires. Carbon monoxide and free radicals may well play a decisive role in health effects of people who live and/or work close to the fires.

Inhalable and thoracic-suspended particles move further down into the lower respiratory airways, and can remain there for a longer period of time, leaving deposits. The potential for health impacts in an exposed population depends on individual factors such as age and the pre-existence of respiratory and cardiovascular diseases and infections, and on particle size. Gaseous compounds adsorbed or absorbed by particles can play a role in long-term health effects (cancer) but short-term health effects are essentially determined through particle size. Quantitative assessment of health impacts of air pollution associated with vegetation fires in developing countries is often limited by the availability of baseline morbidity and mortality information. Air pollutant data are of relatively higher availability and quality but sometimes even these data are not available or reliable.

Vegetation fire smoke sometimes overlies urban air pollution, and exposure levels are intermediate between ambient air pollution and indoor air pollution from domestic cooking and heating. Because the effects of fire events are nation- and region-wide, a 'natural' disaster can evolve into a more complex emergency, both through population movement and through its effects on the economy and security of the affected countries. The fire and smog episodes in South-East Asia during the El Niño of 1997–1998, in the Far East of the Russian Federation in 1998, and again in Moscow Region in 2002, are striking examples.

The World Health Organization (WHO), in collaboration with the United Nations Environment Programme (UNEP) and the World Meteorological Organization (WMO), has issued comprehensive guidelines for governments and responsible authorities on actions to be taken when their population is exposed to smoke from fires (Schwela et al, 1999). The guidelines give insights into acute and chronic health effects of air pollution due to biomass burning, advice on effective public communications and mitigation measures, and guidance for assessing the health impacts of vegetation fires. They also provide measures on how to reduce the burden of mortality and preventable disability suffered particularly by the poor, and on the development and implementation of an early air pollution warning system.

Changing Fire Regimes

The following section provides a review on changing fire regimes in some of the most important global vegetation zones: the northern boreal forest, temperate forests, tropical rainforest and tropical/subtropical savannah regions.

The fire situation in boreal forests

The global boreal forest zone, covering approximately 12 million km², stretches in two broad transcontinental bands across Eurasia and North America, with two thirds in

Russia and Scandinavia and the remainder in Canada and Alaska. These extensive tracts of coniferous forest have adapted to and become dependent upon periodic fire for their physiognomy and sustainable existence. These forests provide a vital natural and economic resource for northern circumpolar countries boreal forests are estimated to contain ~37 per cent of the world's terrestrial carbon. These forests have become increasingly accessible to human activities – including natural resource exploitation and recreation – over the past century, with the export value of forest products from global boreal forests accounting for 47 per cent of the world forest product trade.

The largest boreal forest fires are extremely high-intensity events, with very fast spread rates and high levels of fuel consumption, particularly in the deep organic forest floor layer. High-intensity levels are often sustained over long burning periods, creating towering convection columns that can reach the upper troposphere and lower stratosphere, making long-range smoke transport common. In addition, the area burned annually by boreal fires is highly episodic, often varying by an order of magnitude between years.

For many reasons boreal forests and boreal fires have increased in significance in a wide range of global change science issues in recent years. Climate change is foremost among these issues, and the impacts of climate change are expected to be most significant at northern latitudes. Forest fires can be expected to increase sharply in both incidence and severity if climate change projections for the boreal zone prove accurate. These changes will act as a catalyst for a wide range of processes controlling boreal forest carbon storage, causing shifting vegetation, altering the age class structure towards younger stands, and resulting in a direct loss of terrestrial carbon to the atmosphere.

Over the past two decades forest fires in boreal North America (Canada and Alaska) have burned an average of 3 million hectares annually. While sophisticated and well-funded fire management programmes in North America suppress the vast majority of fires whilst they are still small, the 3 per cent of the fires that grow larger than 200 hectares in size account for 97 per cent of the total area burned. There is little likelihood of improving suppression effectiveness since the law of diminishing returns applies. The 3 per cent of fires that grow larger do so because they occur under extreme fire danger conditions and/or in such numbers that suppression resources are overwhelmed, and applying more funding would have no effect. In addition, Canada and Alaska have vast northern areas which are largely unpopulated and with no merchantable timber, and where fires are monitored but suppression is unwarranted and not practised. The current high levels of fire activity across the North American boreal zone, and the restricted suppression effectiveness, together with recognition of the need for natural fire, mean that it is all but certain that climate change will greatly exacerbate the situation. The only option will be adaptation to increasing fire regimes (eg Kasischke and Stocks, 2000).

The fire problem in Eurasia's boreal forests – primarily in Russia – is similar in some ways to the North American situation, but there are significant differences (Goldammer and Furyaev, 1996; Goldammer and Stocks, 2000; Shvidenko and Goldammer, 2001). The Russian boreal zone is close to twice as large as its North American counterpart, stretching across 11 time zones, with a wide diversity of forest types, growing conditions, structure and productivity, and anthropogenic impacts that define different types of fires and their impacts. For example, surface fires are much more common in the Russian boreal zone than in North America where stand-replacing crown fires dominate the area burned, and many Russian tree species have adapted to low-intensity fires. While Canada has a strong continental climate in the boreal region of west-central Canada, where most large fire activity occurs, Eurasia has much stronger continentality over a much larger land base. Low precipitation and/or frequent droughts characterize the climate of the major land area of the Russian Federation during the fire season, and more extreme fire danger conditions pertain over a much larger area. In addition, large areas of the Russian boreal zone are not protected or monitored, so fire activity is not recorded for these regions.

Official Russian fire statistics for the past five decades typically report annual areas burned between 0.5 and 1.5 million hectares, with very little interannual variability. In fact, as Russian fire managers agree, these numbers are a gross underestimation of the actual extent of boreal fire in Russia. There are two reasons for this: a lack of monitoring/documentation of fires occurring in vast unprotected regions of Siberia, and the fact that the Russian reporting structure emphasized under-reporting of actual area burned to reward the fire suppression organization. In recent years, with the advent of international satellite coverage of Russian fires in collaboration with Russian fire scientists, more realistic area burned estimates are being generated. For example, during the 2002 fire season satellite imagery revealed that about 12 million hectares of forest and non-forest land had been affected by fire in Russia. However official sources reported only 1.2 million hectares of forests and 0.5 million hectares of non-forest land burned in the protected area of 690 million hectares (Goldammer, 2003; Davidenko and Eritsov, 2003; Sukhinin et al, 2003). During the early summer of 2003 remote sensing data indicate a total vegetated area affected by fire in Russia exceeding 22 million hectares (GFMC, 2003). Based on recent remote sensing data, it appears that the annual area burned in Russia can vary between 2 and 15 million hectares/year.

In recent years there has been an increase in the occurrence of wildfires under extreme drought conditions (eg the Trans-Baikal Region in 1987 and the Far East in 1998), and the severity of these fires will greatly disturb natural recovery cycles. There is growing concern that fires on permafrost sites will lead to the degradation or disappearance of eastern Siberian larch forests. Russia has close to 65 per cent of the global boreal forest, an economically and ecologically important area that represents the largest undeveloped forested area of the globe. With the vast quantities of carbon stored in these forests, Russian forests play a critical role in the global climate system and global carbon cycling. Projected increases in the severity of the continental climate in Russia suggest longer fire seasons and much higher levels of fire danger. The inevitable result will be an increase in the number, size and severity of boreal fires, with huge impacts on the global carbon cycle and the Russian economy.

At the same time, recent political and economic changes in Russia have led to an extreme reduction in their fire suppression capability, with the State Forest Service showing a huge debt, and the outlook for future funding increases bleak. As a result, *Avialesookhrana*, the aerial fire protection division of the Forest Service, has been forced to reduce aerial fire detection and suppression levels, with operational aircraft flying hours and firefighter numbers now less than 50 per cent of 1980s levels. Consequently, the occurrence of larger fires is increasing, and the areas being burned annually are unprecedented in recent memory.

Increasing fire risks in the boreal forests of Russia are a major threat to the global carbon budget, and this requires significant national and international attention (Kajii et al, 2003). Management and protection of these vital resources should not be given solely to the private sector or delegated to regional levels. The establishment and strengthening of a central institution to protect forests must be a priority supported by Russia and the international community.

Recent Intergovernmental Panel on Climate Change (IPCC) reports have emphasized the fact that climate change is a current reality, and that significant impacts can be expected, particularly at northern latitudes, for many decades ahead. Model projections of future climate, at both broad and regional scales, are consistent in this regard. An increase in boreal forest fire numbers and severity, as a result of a warming climate with increased convective activity, is expected to be an early and significant consequence of climate change. Increased lightning and lightning fire occurrence is expected under a warming climate. Fire seasons are expected to be longer, with an increase in the severity and extent of the extreme fire danger conditions that drive major forest fire events. Increased forest fire activity and severity will result in shorter fire return intervals, a shift in forest age class distribution towards younger stands, and a resultant decrease in terrestrial carbon storage in the boreal zone. Increased fire activity will also likely produce a positive feedback to climate change, and will drive vegetation shifting at northern latitudes. The boreal zone is estimated to contain 35–40 per cent of global terrestrial carbon, and any increase in the frequency and severity of boreal fires will release carbon to the atmosphere at a faster rate than it can be re-sequestered. This would have global implications, and must be considered in post-Kyoto climate change negotiations.

Increased protection of boreal forests from fire is not a valid option at this time. Fire management agencies are currently operating a maximum efficiency, controlling unwanted fires quite effectively. There is a law of diminishing effects at work here though, as increasing efficiency would require huge increases in infrastructure and resources. While it is physically and economically impossible to further reduce the area burned by boreal fires, it is also not ecologically desirable, as fire plays a major and vital role in boreal ecosystem structure and maintenance. Given these facts, it would appear that, if the climate changes as expected over the next century, northern forest managers will have to constantly adapt to increasing fire activity. The likely result would be a change in protection policies to protect more valuable resources, while permitting more natural fire at a landscape scale (Stocks et al, 2001; Stocks, 2001).

The fire situation in temperate forests

In recent years fires have been increasing in number and severity across the temperate zone, with significant fire events becoming more common in the United States, the Mediterranean Basin and Mongolia for example.

Temperate North America: the United States

Over most of the past century the United States has made a huge investment in wildland fire management, developing a sophisticated fire suppression capability. As the 20th century progressed, the United States became increasingly effective in excluding fire from much of the landscape. Despite numerous human-caused and lightning fires the area burned was greatly reduced from the early 1990s. However, the price for successful fire exclusion has proven to be twofold. The most obvious price was the huge cost of developing and maintaining fire management organizations that were increasingly requiring higher budgets to keep fire losses at an acceptable level. The second cost, hidden for decades, is now apparent, as the policy emphasis on fire exclusion has led to the build-up of unnatural accumulations of fuels within fire-dependent ecosystems, with the result that recent fires burned with greater intensity and have proven much more difficult to control. Despite extensive cooperation from other countries, and huge budget expenditures, intense droughts in 2000 and 2002 contributed to widespread wildfires in the western United States that burned between 2.5 and 3 million hectares in each year. Losses from these fires now include substantial destruction of homes, as the trend towards living in fire-prone environments grows, and the wildland urban interface is expanding.

The Mediterranean Basin and the Balkans

Within the Mediterranean Basin fire is the most important natural threat to forests and wooded areas. Mediterranean countries have a relatively long dry season, lasting between one and three months on the French and Italian coasts in the north of the Mediterranean, and more than seven months on the Libyan and Egyptian coasts to the south. Currently, approximately 50,000 fires burn throughout the Mediterranean Basin, and burn an annual average of more than 600,000 hectares; both statistics are at a level twice that of the 1970s. In countries where data is continuous since the 1950s, fire occurrence and area burned levels have shown large increases since the 1970s in Spain, Italy and Greece. Human-caused fires dominate in the Mediterranean Basin, with only 1–5 per cent of fires caused by lightning. Arson fires are also quite prevalent.

Paradoxically, the fundamental cause of the increasing vulnerability of vegetation of the Southern European countries bordering the Mediterranean Basin is linked to increased standards of living among the local populations (Alexandrian et al, 2000). Far-reaching social and economic changes in Western Europe have led to a transfer of population from the countryside to the cities, a considerable deceleration of demographic growth, an abandonment of arable lands, and a disinterest in the forest resource as a source of energy. This has resulted in the expansion of wooded areas, erosion of the financial value of wooded lands, a loss of inhabitants with a sense of responsibility for the forest, and a resultant increase in the amount of available fuel.

The demographic, socioeconomic and political changes in many countries of Southeast Europe and the neighbouring nations of the Balkans have resulted in an increase of wildfire occurrence, destabilization of fire management capabilities and increased vulnerability of ecosystems and human populations. The main reasons for this development include the transition from centrally planned to market economies, national to regional conflicts, creation of new nations involving political tensions and war, land-use changes, and regional climate change towards an increase in the frequency of extreme droughts. New solutions are required to address the increasing fire threat on the Balkans. Regional cooperation in the Mediterranean region and the Balkans must address the underlying causes of changing fire regimes and a more economic transnational use of fire management resources (Goldammer, 2001b).

Temperate steppes to boreal forests: Mongolia

With an area of 1,565,000km² and a population of 2.3 million, Mongolia is one of the least populated countries in the world, yet significant fire problems exist there. With an extremely continental climate, poor soil fertility and a lack of surface water, wildfires in Mongolia have become a major factor in determining the spatial and temporal dynamics of forest ecosystems. Of a total of 17 million hectares of forests, 4 million hectares are estimated to be disturbed, primarily by fire (95 per cent) and logging (5 per cent). Forests are declining in Mongolia as continual degradation by wildfire turns former forests into steppe vegetation. The highest fire hazard occurs in the submontane coniferous forests of eastern Mongolia, where highly flammable fuels, long droughts and economic activity are most common. In recent years fire activity in Mongolia has increased significantly, due to economic activity on lands once highly controlled or restricted. In Mongolia, only 50–60 forest fires and 80–100 steppe fires occur annually on average. The major underlying cause of increasing wildfires is the fact that urban people are now seeking a livelihood in forests due to the collapse of the industrial sector. During the 1996–1998 period, when winter and spring seasons were particularly dry, Mongolia experienced large-scale forest and steppe fires that burned over 26.3 million hectares of forest and steppe vegetation, including pasture lands, causing significant losses of life, property and infrastructure. The loss of forest land in Mongolia is increasing, with severe economic consequences, and a growing realization that a precious ecological resource which contains virtually all rivers, protects soils and rangelands, and provides essential wildlife habitat is at risk.

Australia: New vulnerabilities

Australia's fire problems are currently a growing focus of fire managers and policy makers in the Australasian region. The continent is facing a dilemma: On the one hand Australia's wildlands have evolved with fire and thus are extremely well adapted to fire. In the late 1990s more than 345,000 wildfires burned an average of circa 50 million hectares of different vegetation types every year. In many of Australia's wildlands frequent fires of moderate intensity and severity are important to maintain properties and functioning of ecosystems and to reduce the accumulation of highly flammable fuels. Thus, it is generally accepted that fire protection (fire exclusion) in Australia's wildlands will lead to fuel accumulation and, inevitably, to uncontrollable wildfires of extreme behaviour, intensity and severity. These ecosystems need to be burned by natural sources or by prescribed fire.

On the other hand there is a trend towards the building of homes and infrastructure in the wildlands around Australian cities. This exurban settlement trend has created new vulnerabilities and conflicts concerning the use of fire as a management tool. The extended wildfires occurring in the south-east of Australia (New South Wales, ACT, Victoria) in 2002 and 2003 had limited impacts on native vegetation but had a significant impact on values at risk at the wildland–residential interface as well as on plantation forests.

Tropical rainforests

Fires in tropical evergreen forests, until recently, were considered either impossible or inconsequential. In recent decades, due to population growth and economic necessity, rainforest conversion to non-sustainable rangeland and agricultural systems has proliferated throughout the tropics (Mueller-Dombois and Goldammer, 1990; UNEP, 2002). The slash-and-burn practices involve cutting rainforests to harvest valuable timber and burning the remaining biomass repeatedly to permanently convert landscapes into grasslands that flourish for a while due to ash fertilization, but eventually are abandoned as non-sustainable, or to convert rainforest to valuable plantations. Beyond this intentional deforestation there is a further, more recent problem, as wildfires are growing in frequency and severity across the tropics. Fires now continually erode fragmented rainforest edges and have become an ecological disturbance leading to degradation of vast regions of standing forest, with huge ecological, environmental and economic consequences. It is clear that, in tropical rainforest environments, selective logging leads to an increased susceptibility of forests to fire, and that the problem is most severe in recently logged forests. Small clearings associated with selective logging permit rapid desiccation of vegetation, and soils increase this susceptibility to fire. Droughts triggered by the El-Niño-Southern Oscillation (ENSO) have exacerbated this problem, and were largely responsible for significant wildfire disasters in tropical rainforests during the 1980s and 1990s.

Tropical rainforests cover ~45 per cent of Latin America and the Caribbean. Between 1980 and 1990, when the first reliable estimates were made, the region lost close to 61 million hectares of forest, 6 per cent of the total forested area. During the 1990–1995 period a further 30 million hectares were lost. The highest rates of deforestation occurred in Central America (2.1 per cent per year) but Bolivia, Ecuador, Paraguay and Venezuela also had deforestation rates above 1 per cent per year, while Brazil lost 15 million hectares of forest between 1988 and 1997.

Landscape fragmentation and land cover change associated with this massive deforestation combine to expose more forest to the risk of wildfire, and fires are increasing in severity and frequency, resulting in widespread forest degradation. This change in tropical fire regimes will likely result in the replacement of rainforests with less diverse and more fire-tolerant vegetation types. Although quantitative area-burned estimates are sporadic at best, it is estimated that the 1997–1998 El Niño-driven wildfires burned more than 20 million hectares in Latin America and South-East Asia. The widespread tropical rainforest wildfires of 1998 have changed the landscape of Latin America's tropical evergreen forests by damaging vast forested areas adjacent to fire-maintained ecosystems. Fires will likely become more severe in the near future, a fact not yet appreciated by resident populations, fire managers and policy makers. The Latin American and Indonesian problems have much in common, and indicate the problems in tropical rainforests worldwide.

Smoke pollution from tropical rainforest fires, as is the case in many other fire regions of the world, greatly affects the health of humans regionally, with countless short- and long-term respiratory and cardiovascular problems resulting from the lingering smoke and smog episodes associated with massive wildfires.

The environmental impacts from tropical forest fires range from local to global. Local/regional impacts include soil degradation, with increased risks of flooding and drought, along with a reduced abundance of wildlife and plants, and an increased risk of recurrent fires. Global impacts include the release of large amounts of greenhouse gases, a net loss of carbon to the atmosphere, and meteorological effects including reduced precipitation and increased lightning. A loss of biodiversity and extinction of species is also a major concern.

The economic costs of tropical forest fires are unknown, largely due to a lack of data, but also attributable to the complications of cause and effect: negative political implications definitely discourage full disclosure. These can include medical costs, transportation disruption, and timber and erosion losses. They can also, in the post-Kyoto era, include lost carbon costs which, in the case of the 1998 fires in Latin America, can be crudely estimated at US$10–15 billion.

The driving force behind the devastating Indonesian fires of 1982–1983 and 1997–1998 was droughts associated with ENSO. This combined with the exposure of rainforest areas to drought as a result of selective logging. It has been estimated that the overall land area affected by the 1982–1983 fires, in Borneo alone, was in excess of 5 million hectares. In non-ENSO years fires cover 15,000–25,000 hectares. The 1997–1998 fire episode exceeded the size and impact of the 1982–1983 fires. During 1997 large fires occurred in Sumatra, West and Central Kalimantan and Irian Jaya/Papua. In 1998 the greatest fire activity occurred in East Kalimantan. In total, the 1997–1998 fires covered an estimated area in excess of 9.5 million hectares, with 6.5 million hectares burning in Kalimantan alone. These widespread fires, all caused by humans involved in land speculation and large-scale forest conversion, caused dense haze across South-East Asia for an extended period. Severe respiratory health problems resulted, along with widespread transport disruption, and overall costs were estimated at US$9.3 billion. Carbon losses were particularly severe due to high levels of fuel consumption, particularly in peatlands.

Southern African savannahs/grasslands

Fire is a widespread seasonal phenomenon in Africa (van Wilgen et al, 1997). South of the equator, approximately 168 million hectares burn annually, nearly 17 per cent of a total land base of 1014 million hectares, accounting for 37 per cent of the dry matter burned globally. Savannah burning accounts for 50 per cent of this total, with the remainder caused by the burning of fuelwood, agricultural residues and slash from land clearing. Fires are started both by lightning and humans, but the relative share of fires caused by human intervention is rapidly increasing. Pastoralists use fire to stimulate grass growth for livestock, while subsistence agriculturalists use fire to remove unwanted biomass while clearing agricultural lands, and to eliminate unused agricultural residues after harvest. In addition, fires fuelled by wood, charcoal or agricultural residues are the main source of domestic energy for cooking and heating.

In most African ecosystems fire is a natural and beneficial disturbance of vegetation structure and composition, and in nutrient recycling and distribution. Nevertheless, substantial unwarranted and uncontrolled burning does occur across Africa, and effective actions to limit this are necessary to protect life, property and fire-sensitive natural resources, and to reduce the current burden of emissions on the atmosphere with subsequent adverse effects on the global climate system and human health. Major problems arise at the interface between fire savannas, residential areas, agricultural systems and those forests which are not adapted to fire. Although estimates of the total economic damage of African fires are not available, fires crossing borders from a fire-adapted to a fire-sensitive environment are increasingly destroying ecologically and economically important resources. Fire is also contributing to widespread deforestation in many southern African countries.

Most southern African countries have regulations governing the use and control of fire, although these are seldom enforced because of difficulties in punishing those responsible. Some forestry and wildlife management agencies within the region have the basic infrastructure to detect, prevent and suppress fires, but this capability is rapidly breaking down and becoming obsolete. Traditional controls on burning in customary lands are now largely ineffective. Fire control is also greatly complicated by the fact that fires in Africa occur as hundreds of thousands of widely dispersed small events. With continuing population growth and a lack of economic development and alternative employment opportunities to subsistence agriculture, human pressure on the land is increasing, and widespread land transformation is occurring. Outside densely settled farming areas, the clearance of woodlands for timber, fuelwood and charcoal production is resulting in increased grass production, which in turn encourages intense dry season fires that suppress tree regeneration and increase tree mortality. In short, the trend is toward more fires.

Budgetary constraints on governments have basically eliminated their capacity to regulate from the centre, so there is a trend towards decentralization. However, the shortage of resources forcing decentralization means there is little capacity for governments to support local resource management initiatives. The result is little or no effective management and this problem is compounded by excessive sectoralism in many governments, leading to uncoordinated policy development, conflicting policies and a duplication of effort and resources. As a result of these failures, community-based natural resource management is now being increasingly widely implemented in Africa, with the recognition that local management is the appropriate scale at which to address the widespread fire problems in Africa. The major challenge is to create an enabling rather than a regulatory framework for effective fire management in Africa, but this is not currently in place. Community-based natural resource management programmes, with provisions for fire management through proper infrastructure development, must be encouraged. More effective planning could also be achieved through the use of currently available remotely sensed satellite products.

These needs must also be considered within the context of a myriad of problems facing governments and communities in Africa, including exploding populations and health (eg the AIDS epidemic). While unwarranted and uncontrolled burning may greatly affect at the local scale, it may not yet be sufficiently important to warrant the concern of policy makers, and that perception must be challenged as a first step towards more deliberate, controlled and responsible use of fire in Africa.

Global peatlands

The world's peatlands play a significant role in biodiversity patterns, socioeconomic development and livelihood, water storage and supply, flood control and climate regulation. Peatlands are fragile ecosystems, vulnerable to fire and impacted heavily by uncontrolled drainage (Parish et al, 2002). The increased conversion of forests on tropical peatlands to agricultural cultivations and the presence of large areas of drained peatlands in the boreal zone of is a growing concern for fire managers. Peat soils can reach a depth of 10 metres or more and thus contain burnable volumes in excess of 100,000m³ of biomass per hectare.

Recent fire episodes in the tropical and boreal peat-swamp biomes have revealed that the combination of extreme droughts, especially in conjunction with the El Niño phenomenon or extended summer droughts in boreal Eurasia, lead to severe and often irreversible damage to wetland ecosystems. This results in transfer of terrestrial carbon to the atmosphere and to severe smoke pollution affecting human health and security.

Projected regional climate changes, coupled with continuing trends of land-use changes, indicate that the vulnerability of peatlands will increase and the loss of peatlands by excessive, uncontrolled fire will accelerate.

However, there are options to proactively protect and maintain peatlands. To avoid major fire disasters in peatlands, there is a need to educate land managers in these areas about best fire management practices on peat soils.

Peat is a renewable natural resource, which will regenerate if the burned area is rehabilitated and properly protected, even if the re-accumulation of peat is a fairly slow process.

Conclusions: Policy and Wildland Fire Science

The challenge of developing informed policy that recognizes both the beneficial and traditional roles of fire, while reducing the incidence and extent of uncontrolled burning and its adverse impacts, clearly has major technical, social, economic and political elements. In developing countries better forest and land management techniques are required to minimize the risk of uncontrolled fires, and appropriate management strategies for preventing and controlling fires must be implemented if measurable progress is to be achieved. In addition, enhanced early warning systems for assessing fire hazard and estimating risk are necessary, along with the improvements in regional capacity and infrastructure to use satellite data. This must be coupled with technologies and programmes that permit rapid detection and response to fires.

A better understanding by both policy makers and the general population of the ecological, environmental, sociocultural, land-use and public health issues surrounding vegetation fires is essential. The potential for greater international and regional cooperation in sharing information and resources to promote more effective fire management also needs to be explored. The recent efforts of many UN programmes and organizations are a positive step in this direction, but much remains to be accomplished. In the spirit and fulfilment of the 1997 Kyoto Protocol to United Nations Framework Convention on Climate Change, the 2002 World Summit for Sustainable Development (WSSD) and the UN International Strategy for Disaster Reduction (ISDR), there is an obvious need for more reliable data on fire occurrence and impacts. Remote sensing must and should play a major role in meeting this requirement. In addition to the obvious need for improved spaceborne fire-observation systems and more effective operational systems capable of using information from remote sensing and other spaceborne technologies, the remote sensing community needs to focus its efforts more on the production of useful and meaningful products.

Finally, it must be underscored that the traditional approach in dealing with wildland fires exclusively under the traditional forestry schemes must be replaced in future by an intersectoral and interdisciplinary approach. The devastating effects of many wildfires are an expression of demographic growth, land use and land-use changes, the sociocultural implications of globalization and climate variability. Thus, integrated strategies and programmes must be developed to address the fire problem at its roots, while at the same time creating an enabling environment and developing appropriate tools for policy and decision makers to proactively respond to fire.

What are the implications of these conclusions on fire science? The first major interdisciplinary and international research programmes were undertaken back in the early 1990s. They included intercontinental fire-atmosphere research campaigns such as the Southern Tropical Atlantic Regional Experiment (STARE). The Southern Africa Fire-Atmosphere Research Initiative (SAFARI) in the early 1990s (JGR, 1996) paved the way for the development of models for a comprehensive science of the biosphere. At the beginning of the third millennium it is recognized that progress has been achieved in clarifying the fundamental mechanisms of fire in the global environment. This includes the reconstruction of the prehistoric and historic role of fire in the genesis of planet Earth and in the co-evolution of the human race and nature.

However, at this stage we have to examine the utility of the knowledge that has been generated by a dedicated science community. We have to ask this at a time when it is becoming obvious that fire plays a major role in the degradation of the global environment. It follows from the statement of Pyne (2001):

> Fire has the capacity to make or break sustainable environments. Today some places suffer from too much fire, some from too little or the wrong kind, but everywhere fire disasters appear to be increasing in both severity and damages.

that we must ask whether wildland fire is becoming a major threat at the global level. Does wildland fire at a global scale contribute to an increase of exposure and vulnerability of ecosystems to secondary/associated degradation and even catastrophes?

The regional analyses provided in this chapter reveal that environmental destabilization by fire is obviously accelerating. This trend goes along with an increasing vulnerability of human populations. Conversely, humans are not only affected by fire but are the main causal agent of destructive fires, through both accidental, unwanted wildfires, and the use of fire as a tool for conversion of vegetation and reshaping whole landscapes.

This trend, however, is not inevitable. There are opportunities to do something about global fire because – unlike the majority of the geological and hydro-meteorological hazards – wildland fires represent a natural hazard which is primarily human-made, can be predicted, controlled and, in many cases, prevented.

Here is the key for the way forward. Wildland fire science has to decide its future direction by answering a number of basic questions: What is the future role of fundamental fire science, and the added value of additional investments? What can be done to close the gap between the wealth of knowledge, methods and technologies for sustainable fire management and the inability of humans to exercise control?

From the perspective of the author the added value of continuing fundamental fire science is marginal. Instead, instruments and agreed procedures need to be identified to bring existing technologies to application. Costs and impacts of fire have to be quantified systematically to illustrate the significance of wildland fire management for sustainable development.

Fire science must also assist in understanding which institutional arrangement would work best for fire management in the many new nations that have been created over the past dozen years. For instance fire management capacity needs to be rebuilt in the nations that emerged from the collapse of the Soviet Union and Yugoslavia, countries that emerged from the democratization of former dictatorships, or that resulted from dramatic regime changes. The questions to be addressed include:

- What kind of fire policies and fire institutions should such nations adopt?
- What research programmes are suitable?
- What kind of training yields the biggest results?
- What kind of fire management systems are appropriate for what contexts?
- What kind of international aid programmes achieve the best outcomes?
- How should such countries reform in a way that advances the safety of their rural populations and the sustainability of their land and resources?

So far no such study – no such field of inquiry, the political ecology of fire – exists. Yet there are ample examples available from history, especially Europe's colonial era, and many experiments over the past 40 years. There is the record of policy and institutional reforms for the major fire nations such as the United States, Russia, Canada and Australia. There are scores of FAO-sponsored projects. What is needed is a systematic collection and analysis of these experiences and data. This is something that can be achieved with a modest investment of scholarship and money.

Similarly, a compelling need exists to understand better the impact of industrialization that involves the burning of fossil biomass. Both developed and undeveloped countries are struggling to understand the consequences of fossil fuel use for fire management of this transformation. How, precisely, does burning fossil biomass change the patterns of fire on the land, for good or ill? We understand something about the relationships and cumulative effects between biomass burning and fossil-fuel burning in the atmosphere; we do not understand the mechanics of their competition on the Earth's vegetated surfaces. Modern transportation systems can open forests to markets, and lead to extreme fires. Equally, chemical fertilizers, pesticides and mechanized ploughs can remove fire from agricultural fields. The replacement of biofuels for cooking and heating in some regions by fossil fuels has led to a vast accumulation of hazardous fuels in wildlands. In other regions the availability of fossil or solar energy has eased the pressure of vegetation depletion. Yet both the introduction of fire and its removal have ecological consequences. These are linked problems for which there are no models or theory.

Governments sponsor most of the current fire research. This is because those governments have responsibility for large tracts of public land. These landscapes matter because their fires can (and do) threaten communities, because the mismanagement of fire can undermine the ecological health of the protected biota, and because they influence carbon cycling and global warming. But most of the world's fires reside in the developing world and are embedded within agricultural systems or systems subject to rapid logging for export or conversion to plantations. These are the scenes of many of the worst fires and most damaging fire and smoke episodes. Traditional research into fire fundamentals has scant value in such conditions, which are the result of social and political factors. Yet these are circumstances in which even a small amount of research could produce large and immediate dividends.

This implies that the scientific focus has to be shifted. The fire domain has been governed for a long time by interdisciplinary natural sciences research. Engineering research has contributed to a high level of development in the industrial countries. What is needed in future is a research focus at the interface between the human dimension of fire and the changing global environment. The new fire science in the third millennium must be application-oriented and understood by policy makers, a science that bridges institutions, politics and ecology. Continued research in fire fundamentals, fascinating as it is, cannot address these matters.

The establishment of a dedicated Working Group on Wildland Fire of the United Nations reflected this recommendation. Between 2001 and 2003 the Working Group operated in support of the Inter-Agency Task Force for Disaster Reduction of the ISDR. It brings together an international consortium of UN agencies and programmes, representatives from natural sciences, humanities, fire management agencies and non-government organizations (ISDR, 2001). The terms of reference of this group was, among others, to advise policy makers at national and international levels on the reduction of the negative effects of fire on the environment, in support of sustainable management of the Earth system. The activities include a major global networking activity – the Global Wildland Fire Network – facilitated through the Global Fire Monitoring Center (GFMC, 2002) and supported by the science community. With the end of its lifetime the Working Group transited to the UN-ISDR Wildland Fire Advisory Group in 2004. Its main function is to serve as a liaison between the Global Wildland Fire Network and the UN system.

The contribution of global wildland fire science to the way forward must lead towards the formulation of national and international public policies that will be harmonized with the objectives of international conventions, protocols and other agreements. The Convention on Biological Diversity (CBD), the Convention to Combat Desertification (UNCCD), United Nations Framework Convention on Climate Change (UNFCCC) and the UN Forum of Forests (UNFF) are all concerned. The wildland fire community must also search for efficient, internationally agreed upon solutions to respond to wildland fire disasters through international cooperative efforts.

References

1 Alexandrian, D., Esnault, F. and Calabri, G. (2000) 'Forest fires in the Mediterranean area'. *UNASYLVA*, vol 59, no 197. www.fao.org/docrep/x1880e/x1880e07.htm

2 Andreae, M. O. (2002) *Assessment of Global Emissions from Vegetation Fires*. Input paper for the Brochure for Policy Makers, UN International Strategy for Disaster Reduction (ISDR) Inter-Agency Task Force for Disaster Reduction, Working Group 4 on Wildland Fire.

3 Andreae, M. O. and Merlet, P. (2001) 'Global estimate of emissions from wildland fires and other biomass burning'. *Global Biogeochem Cycles*, vol 15, pp955–966.

4 Davidenko, E. P. and Eritsov, A. (2003) 'The fire season 2002 in Russia: Report of the Aerial Forest Fire Service Avialesookhrana'. *Int. Forest Fire News*, no 28, pp15–17.

5 Food and Agriculture Organization of the United Nations (FAO) (2001) FRA Global Forest Fire Assessment 1990–2000. Forest Resources Assessment Programme, Working Paper 55, FAO, Rome, pp 189–191. www.fao.org.0/forestry/fo/fra/docs/Wp55_eng.pdf

6 Global Fire Monitoring Center (GFMC) (2002) www.fire.uni-freiburg.de

7 Global Fire Monitoring Center (GFMC) (2003) Webpage on current and archived significant global fire events and fire season summaries: www.fire.uni-freiburg.de/current/globalfire.htm

8 Goh, K. T., Schwela, D. H., Goldammer, J. G. and Simpson, O. (1999) *Health Guidelines for Vegetation Fire Events: Background Papers*, published on behalf of UNEP, WHO and WMO. Institute of Environmental Epidemiology, Ministry of the Environment, Singapore.

9   Goldammer, J. G. (1993) *Feuer in Waldökosystemen der Tropen und Subtropen*. Birkhäuser-Verlag, Basel-Boston.

10   Goldammer, J. G. (2001a) *Africa region fire assessment*, in FRA Global Forest Fire Assessment 1990–2000, Forest Resources Assessment Programme, Working Paper 55, pp 30–37, FAO, Rome.

11   Goldammer, J. G. (2001b) 'Towards international cooperation in managing forest fire disasters in the Mediterranean region', in Brauch, H. G., El-Sayed Selim, M., Liotta, P. H., Chourou, B. and Rogers, P. (eds) *Security and the environment in the Mediterranean; conceptualising security and environmental conflicts*. Springer-Verlag, Berlin-Heidelberg-New York, March 2003, in press. Prepublication: www.fire.uni-freiburg.de/GlobalNetworks/Mediterrania/Mediterrania_0.html

12   Goldammer, J. G. (2003) 'The wildland fire season 2002 in the Russian Federation'. An assessment by the Global Fire Monitoring Center (GFMC). *Int. Forest Fire News*, no 28, pp2–14.

13   Goldammer, J. G. and Crutzen, P. J. (1993) 'Fire in the environment: Scientific rationale and summary results of the Dahlem Workshop', in Crutzen, P. J. and Goldammer, J. G. (eds) *Fire in the environment: The ecological, atmospheric, and climatic importance of vegetation fires*. Dahlem Workshop Reports. Environmental Sciences Research Report 13. John Wiley, Chichester, pp1–14.

14   Goldammer, J. G. and Furyaev, V. V. (eds) (1996) *Fire in Ecosystems of Boreal Eurasia*. Kluwer Academic Publishers, Dordrecht.

15   Goldammer, J. G. and Stocks, B. J. (2000) 'Eurasian perspective of fire: Dimension, management, policies, and scientific requirements', in Kasischke, E. S. and Stocks, B. J. (eds) *Fire, Climate Change, and Carbon Cycling in the Boreal Forest*. Ecological Studies 138, Springer-Verlag, Berlin-Heidelberg-New York, pp49–65.

16   International Forest Fire News (IFFN): www.fire.uni-freiburg.de/gfmc/iffn/iffn.htm

17   ISDR Working Group on Wildland Fire (2002) UN International Strategy for Disaster Reduction (ISDR) Inter-Agency Task Force for Disaster Reduction, Working Group 4 on Wildland Fire, Report of the Second Meeting, Geneva, 3–4 December 2001. www.unisdr.org/unisdr/WGroup4.htm

18   Joint Research Center of the European Commission (JRC) (2002) Global Burnt Area 2000 (GBA2000) dataset: www.gvm.jrc.it/fire/gba2000/index.htm

19   Journal of Geophysical Research (JGR) Special Issue (1996) 'Southern Tropical Atlantic Regional Experiment (STARE): TRACE-A and SAFARI'. *Journal of Geophysical Research*, vol 101, no D19, pp23, 519–524, 330.

20   Justice, C. O. (2001) Global Observation Land Dynamics (GOLD-Fire): An International Coordination Mechanism for Global Observation and Monitoring of Fires, in ISDR Working Group on Wildland Fire 2002. UN International Strategy for Disaster Reduction (ISDR) Inter-Agency Task Force for Disaster Reduction, Working Group 4 on Wildland Fire. Report of the Second Meeting, Geneva, 3–4 December 2001. Source: www.unisdr.org/unisdr/WGroup4.htm

21   Justice, C. O., Smith, R., Gill, M. and Csiszar, I. (2003) 'Satellite-based fire monitoring: current capabilities and future directions'. *International Journal of Wildland Fire* (in press).

22   Kajii, Y, Kato, S., Streets, D. G., Tsai, N. Y., Shvidenko, A., Nilsson, S., McCallum, I., Minko, N. P., Abushenko, N., Altyntsev, D. and Khodzer, T (2003) 'Boreal forest fires in Siberia in 1998: Estimation of area burned and emission of pollutants by advanced very high resolution radiometer data'. *Journal of Geophysical Research*, vol 107 (in press).

23   Kasischke, E. S. and Stocks, B. J. (eds) (2000) 'Fire, climate change, and carbon cycling in the boreal forest'. *Ecological Studies*, vol 138, Springer-Verlag, Berlin-Heidelberg-New York.

24   Mueller-Dombois, D. and Goldammer, J. G. (1990) 'Fire in tropical ecosystem and global environmental change', in Goldammer, J. G. (ed) *Fire in the tropical biota. Ecosystem processes*

and global challenges. Ecological Studies, Springer-Verlag, Berlin-Heidelberg-New York, vol 84, pp1–10.

25 Parish, F., Padmanabhan, E., Chee Leong D. L. and Chiew, T. H. (2002) *Prevention and Control of Fire in Peatlands*. Proceedings of a Workshop held at the Forestry Training Unit, Kepong, Kuala Lumpur, 19–21 March 2002, Global Environment Centre & Forestry Department Peninsular Malaysia.

26 Pyne, S. (2001) *Challenges for Policy Makers, in ISDR Working Group on Wildland Fire 2002*. UN International Strategy for Disaster Reduction (ISDR) Inter-Agency Task Force for Disaster Reduction, Working Group 4 on Wildland Fire. Report of the Second Meeting, Geneva, 3–4 December 2001. Source: www.unisdr.org/unisdr/WGroup4.htm

27 Schwela, D. H., Goldammer, J. G., Morawska, L. H. and Simpson, O. (1999) *Guideline Document*. Published on behalf of UNEP, WHO and WMO. Institute of Environmental Epidemiology, Ministry of the Environment, Singapore. Double Six Press, Singapore. Also available at: www.who.int/peh/air/vegetation_fires.htm

28 Shvidenko. A. and Goldammer, J. G. (2001) 'Fire situation in Russia'. *International Forest Fire News*, no 24, pp41–59.

29 Steffen, W. L. and Shvidenko, A. Z. (eds) (1996) 'The IGBP Northern Eurasia study: prospectus for integrated global change research'. *The International Geosphere-Biosphere Program: A Study of Global Change*. International Council of Scientific Unions (ICSU), IGBP, Stockholm.

30 Stocks, B. J. (2001) *Forest Fires, Climate Change and Carbon Storage in Boreal Forests*. ISDR Working Group on Wildland Fire 2002. UN International Strategy for Disaster Reduction (ISDR) Inter-Agency Task Force for Disaster Reduction, Working Group 4 on Wildland Fire. Report of the Second Meeting, Geneva, 3–4 December 2001. Source: www.unisdr.org/ unisdr/WGroup4.htm

31 Stocks, B. J., Wotton, B. M., Flannigan, M. D., Fosberg, M. A., Cahoon, D. R. and Goldammer, J. G. (2001). 'Boreal forest fire regimes and climate change', in Beniston, M. and Verstraete, M. M. (eds) *Remote Sensing and Climate Modeling: Synergies and Limitations. Advances in Global Change Research*. Kluwer Academic Publishers, Dordrecht and Boston, pp233–246.

32 Sukhinin, A. I., Ivanov, V. V., Ponomarev, E. I., Romasko, V. Y. and Miskiv, S. I. (2003) 'The 2002 fire season in the Asian part of the Russian Federation'. *International Forest Fire News*, no 28, pp18–28.

33 United Nations Environment Programme (UNEP) (2002) *Spreading Like Wildfire. Tropical Forest Fires in Latin America and the Caribbean. Prevention, Assessment and Early Warning*. UNEP Regional Office for Latin America and the Caribbean. See also: www.fire.uni-freiburg. de/GlobalNetworks/SouthAmerica/UNEP%20Report%20Latin%20America.pdf

34 van Wilgen, B., Andreae, M. O., Goldammer, J. G. and Lindesay, J. (eds) (1997) *Fire in Southern African Savannas. Ecological and Atmospheric Perspectives*. The University of Witwatersrand Press, Johannesburg, South Africa.

Part 4

The Challenge of Sustainable Management

The next group of chapters provides a critical account of some of the options that are commonly advocated for achieving the sustainable development of tropical forest resources. Together they demonstrate the complexity of the situation and the dangers of any 'one size fits all' solutions. Kaimowitz et al (Chapter 12) begin by reviewing the options for public policy changes and show that inappropriate policies have been the underlying cause of much forest loss. Their chapter shows that much is now known about the way many policies can have unintended and undesirable impacts on forests but that much still needs to be learned regarding how to fully incorporate forest issues in the policy process.

Sayer et al (Chapter 13) and Pearce et al (Chapter 14) show how changes in technologies can provide opportunities as well as challenges for sustainable forest management together with the fact that what is judged as sustainable in one place may well not be in another, and that what is considered to be unsustainable today might be considered sustainable in the future. These chapters reinforce the work of Kaimowitz et al in showing that the extra-sectoral policy environment has a major impact on sustainability. If the full costs of environmental externalities are applied to some forest management but not to competing sources of raw materials – whether forest-based or not – then clearly the economic sustainability of the more regulated management system will be jeopardized.

Zuidema et al (Chapter 15) address the issue of protected areas. They argue that national parks and equivalent reserves must remain the principal strategy for conserving forest biodiversity but that in densely populated poor countries it may be a mistake to expand protected areas too much. Protected areas compete for land with agriculture and they rarely provide immediate development benefits. Getting a smaller, better-targeted set of protected areas may be a smarter strategy in the long term than trying to protect as much as possible with risk of provoking a backlash later. What this chapter demonstrates is that considerable conservation gains can be had from even relatively small areas if they are well located and well managed.

In Chapter 16 Smith et al unravel some of the confusion and myths around the role of carbon sequestration in funding forest conservation. They outline the potential benefits but emphasize the difficult realities of adopting this path. There are many obstacles to securing the so-called 'co-benefits' claimed by advocates of carbon sequestration projects.

Forest sequestration projects that will yield real benefits to local people and contribute to biodiversity conservation will present numerous management challenges and may prove impossible to implement on a large scale.

Wilhusen et al, in the last chapter in this section, present an interesting and balanced perspective on the perennial debate about integrated conservation and development projects. They agree with some of the scepticism about the earlier attempts to solve conservation problems with local development activities. But they argue that this is not a justification for abandoning those approaches. Ultimately, if conservation programmes are to be sustainable then local development needs cannot be ignored. They concur with Zuidema et al that simply establishing more and more protected areas will not provide a long-term solution in the face of the pressing need to alleviate poverty.

The overall message from these six chapters is that sustainable management and conservation present varying sets of challenges. Responses have to be tailored to local conditions and they have to include interventions at all levels – from international and national policies through to actions at the level of the forest stand, protected area and local community.

Public Policies to Reduce Inappropriate Tropical Deforestation

David Kaimowitz, Neil Byron and William Sunderlin

Introduction

This chapter seeks to explain how, and to what extent, public policy measures might be used to reduce inappropriate tropical deforestation. In this introductory section we first define the terms we use (ie 'deforestation', 'agents' of deforestation, and 'inappropriate' deforestation) and then present our central argument and the structure of the chapter.

During the 10,000 years before 1950, forest cover declined from around 40 per cent to 30 per cent of the total land mass of the earth; it has taken only 50 additional years to reduce it by a similar amount again, to only 18 per cent (Bryant et al, 1997). Today, only 33 million km² of forest, approximately half of the original forest cover, remains (Sayer et al, 1992).[1]

Contemporary concern about deforestation focuses on tropical countries because that is where the majority of forest removal now occurs. Between 1980 and 1990, an estimated 137.3 million hectares of tropical forests were lost, about 7.2 per cent of the total that existed in 1980 (FAO, 1996, p55).[2]

Many authors assume tropical deforestation has increased in the last few decades (see for example WRI, 1996) but this has not been proven. Indeed, recent preliminary evidence from FAO suggests the deforestation rate between 1980 and 1990 was less than FAO's earlier estimates for that period and less than for 1970–1980. Any claims regarding the magnitude of global deforestation must be taken with caution, however, as there are serious data and definitional problems.

Regardless of the exact magnitude and location of tropical deforestation, it is clear deforestation often implies:

- the loss of livelihoods for forest-dependent people, many of whom may not wish to or be able to find other sources of employment;
- decreasing stocks of fuelwood and non-timber forest products (NTFPs) as well as of industrial timber;
- greater soil erosion and river siltation;
- substantial species and gene loss, in view of tropical forests' high level of endemic biodiversity;

- substantial emission of carbon dioxide which contributes to global warming; and possibly
- other types of local and regional climate change.

We follow FAO in defining deforestation as 'the sum of all . . . transitions from natural forest classes (continuous and fragmented) to all other classes' (FAO, 1996, p22). This definition focuses attention on the loss of natural forests and is framed in the context of specific definitions of forest and non-forest land-use classes. Analysis of transitions among these classes can shed light on the cause-and-effect relations underlying deforestation, and help determine optimal land-use choices. On this definition, the amount of deforestation in the period 1980–1990 was estimated at 8.2 per cent in Asia, 6.1 per cent in Latin America, and 4.8 per cent in Africa (FAO, 1996, p67).

Forest degradation is a crucial adjunct to deforestation. Forest degradation is a 'decrease of density or increase of disturbance in forest classes' (FAO, 1996, p21). Degradation implies a major loss of forest products and ecological services, even when there is little deforestation as such (ie outright conversion from forest to another land use). In what follows, we focus on deforestation, but degradation may, in the long run, be just as important.

The process of deforestation must be analysed at two levels: agents and causes. The agents of deforestation are the people who physically (or through decisions over their labour forces) convert forests to non-forest uses: small farmers; plantation and estate owners; forest concessionaires; infrastructure construction agencies; and so on. FAO (1996, ppxv–xvi) attributes most deforestation in Latin America to medium and large-scale operations (resettlement schemes, large-scale cattle ranching, hydroelectric dams etc) and notes that deforestation in that region is characterized by transitions from closed forest to non-forest land uses. In Africa, deforestation is largely related to the expansion of small-scale farming and rural population pressure (growing number of smallholders), associated with conversion from closed forest cover to short fallow farming. Deforestation in Asia is associated with both relatively large operations (as in Latin America) and rural population pressure (as in Africa), and involves conversion of closed forest to long and short fallow, plantations and non-forest uses. In the larger countries such as China, India and Indonesia, different types of processes dominate depending on the specific context of the region involved. There are no simple, universal single-cause explanations.

Different agents' relative contributions to deforestation are difficult to determine for two reasons. First, little reliable information exists on this subject. Second, even if it did, in many situations various agents operate in the same location, and cannot be analytically separated. Agents are often interdependent (eg cattle ranchers who supply smallholders with chainsaws) and different activities can take place sequentially in the same location (eg small farmers frequently occupy abandoned timber concessions and many farmers engage in salvage logging prior to land clearing for farming).

One reason for the intensity and confusion in the deforestation debate is that proponents of different views tend to emphasize the role of loggers, small farmers or large ranchers and portray these agents as being universally dominant, when in fact the different agents co-exist and closely interact in many countries, and their relative importance varies over time and between regions. Given such complex relationships, there are no clear guilty or innocent parties and no one should expect neat simple solutions.

The 'causes' of deforestation refer to the multiple factors that shape agents' actions and, in particular, their decision to deforest. These causes include: market forces (eg international price fluctuations of agro-export commodities), economic policies (eg currency devaluation), legal or regulatory measures (eg a change in land tenure laws), institutional factors (eg the decision to deploy more forest rangers to a particular area) and political decisions (eg a change in the way forest concessions are allocated), among others. This chapter discusses how certain causes can be manipulated to influence the behaviour of agents, so as to lessen the rate of inappropriate deforestation.

Two main reasons lead us to believe that some deforestation is inappropriate and that inappropriate deforestation is a significant problem. First, deforestation generally causes costs to society and long-term consequences which are rarely considered by individual deforesters. Second, the relative importance of these effects tends to grow over time if an increasing proportion of deforestation occurs in areas of only marginal value for agriculture but which sequester large amounts of carbon, have fragile soils or are high in biodiversity.

Any decision regarding which deforestation is appropriate or not is ultimately political in nature, and cannot be justified on purely technical grounds. Nevertheless, many forestry policy experts agree that forest clearing is more likely to be inappropriate when it involves the following:

- Areas which currently have little value for agriculture by virtue of their soil quality and/or the gradient of the land.
- Areas which have large amounts of biodiversity, particularly endemic biodiversity, which is not well represented in existing protected areas.[3]
- Areas with large numbers of forest-dependent people who show no inclination to abandon their existing livelihood strategies.
- Areas which, by virtue of their rich potential and comparative advantage for timber production, make this the most profitable land use, even after the area is logged for the first time.
- Fragile areas where the ecological cost of conversion resulting from 'downstream' effects outweighs any economic gain from non-forest land uses.
- Peri-urban areas where forests play a key role in conserving aquifers, providing fuel wood, and supporting recreational and tourism activities.

It is important to distinguish inappropriate from appropriate deforestation for three reasons. First, because it is often assumed (at least implicitly) that all tropical deforestation is inappropriate and this is not necessarily the case. Second, clarifying why deforestation is inappropriate can help specify the geographical areas (and sometimes socioeconomic groups) which should be the targets of policy designed to reduce inappropriate deforestation. Third, by clearly defining the areas appropriate for conversion, pressure may be eased on forests where conversion is inappropriate.

Table 12.1 offers a rudimentary conceptual framework for distinguishing inappropriate from appropriate deforestation. It takes into account three cross-cutting categories of valuation: biophysical, economic and political. No one category alone provides a sufficient basis for cogent decision making on whether a forest should be protected or cleared. Decision makers must take all of them into account simultaneously.

Table 12.1 *Conceptual framework for distinguishing 'appropriate' and 'inappropriate' deforestation*

Form of deforestation	Criteria influencing appropriateness		
	Biophysical	*Economic*	*Political*
Inappropriate	• high adverse local and/or downstream effects of conversion • high biodiversity • high biomass	• high utility to local people as forest • low and/or temporary agricultural (or other) potential	No potential stakeholders ultimately gain from the conversion ('lose-lose')*
Ambiguous	[intermediate status between the characteristics listed above and below]	[intermediate status between the characteristics listed above and below]	Some stakeholders win, others lose from the conversion ('win-lose')
Appropriate	• low adverse local and/or downstream effects of conversion • low biodiversity • low biomass	• low utility to local people as forest • high and lasting agricultural (or other) potential	All potential stake-holders gain from the conversion (Potential 'win-win')

* One might rightly ask: 'If the situation is "lose-lose", why would someone in this situation deforest?' Here we mean 'lose-lose' in two different possible senses. First, the agent(s) deforesting expected a gain but ended up moving (and therefore 'lost') because the soil could not support agriculture. Second, we envision situations where stakeholders appear to gain (obtain a source of livelihood through deforestation) yet lose in the relative sense (ie better livelihoods could have been obtained through means other than deforestation).

The categories in this table are idealized and do not necessarily correspond to the 'real world'. Because an ever larger share of remaining tropical forests are in remote and/or hilly areas, these forests will be less appropriate for conversion because of their relative unsuitability for agriculture or their high biodiversity value.[4] These same forests, however, will likely be targeted by a growing number of stakeholders (precisely because they are scarce), and the chance of 'win-lose' and 'lose-lose' outcomes will increase as well. Much remaining tropical forest cover falls into the 'ambiguous' zone described in Table 12.1[5] so identifying where deforestation is more or less acceptable is a tall challenge, precisely because it requires giving attention to three (not always mutually consistent) criteria.

This chapter's basic thesis is as follows:

Individuals and businesses deforest because they think it is their most profitable alternative. To get them not to deforest in situations where forest clearing is inappropriate, deforestation must be made less profitable or other alternatives (either based on retaining forests, or completely outside forest areas) must be made more profitable.

Deforestation can be made less profitable by:

• reducing the demand and/or prices for products produced from newly cleared land;
• increasing the unit costs and riskiness of activities associated with deforestation; or
• eliminating speculative gains in land markets.

Alternatives to deforestation can be made more profitable by:

- increasing the income stream to be obtained by maintaining forests;
- reducing the costs of maintaining forests; and
- increasing the opportunity costs of labour and capital which might otherwise be used in activities associated with deforestation.

The chapter's organization follows this typology of policy options for reducing inappropriate deforestation. All the policies analysed are evaluated on the basis of the following six criteria: effectiveness, ability to be targeted, equity, political viability, direct cost and indirect cost. In each section we provide a summary to evaluate the different policy reforms discussed.

We conclude there is no perfect or generalizable policy for reducing inappropriate deforestation. There are no 'first-best' options. Each national situation is different, much uncertainty remains about key cause and effect relations, and there are usually trade-offs between the different criteria for evaluating the policies. Some policies are effective, but difficult to target, costly, or not politically viable. Other policies are easier to target, but more costly or not politically viable, and so forth.

Making Deforestation Less Profitable

Policies to reduce prices and demand for tropical agricultural and forestry products

Demand for tropical products depends on the size and incomes of the consuming population, relative prices, quantitative trade restrictions and consumer preferences. Policies which affect any of these variables will affect demand. We first analyse the effect on forest clearing of policies which affect the demand for agricultural goods, then look at policies which influence the demand for forest products.

High prices and demand for tropical agricultural commodities can stimulate the clearing of forests to produce these commodities, as illustrated by the 'Hamburger Connection' in Central America or the Cassava fodder export schemes in Thailand. Therefore, under certain circumstances, policies which reduce the demand for tropical agricultural products could limit the conversion of forests to crop lands and pastures.

This type of approach has certain advantages. The policies involved are mostly national in scope, and governments can implement some of them easily and cheaply (at least with regards to their direct costs). Since the policies' direct effects are proportional to the quantity of products sold, larger producers bear most of whatever negative effects they may have on producers' incomes.

Nevertheless, these policies are blunt instruments, which are difficult to target, and thus likely to have perverse side effects. They rely entirely on the market to distinguish where and how supply should be reduced, and this is unlikely to efficiently meet environmental objectives. In some cases, lower prices may lead supply to fall more rapidly on lands which are economically marginal for agriculture, but this does not hold in other situations,[6] and land which is economically marginal for agriculture is not necessarily the most important land to maintain under forest cover to meet biodiversity, watershed protection or timber production objectives. Often, policies can be designed to promote more intensive agricultural production or concentrate production where it is most appropriate, rather than to reduce aggregate demand. In fact, depressing the demand for tropical agricultural products lowers the incentives for technological changes which might lead such products to be produced more intensively, on less land. When deforestation is driven largely by land speculation or production of subsistence or non-marketed goods, these policies will have little impact. Besides, it is difficult to win political support for policies which restrict the demand for or production of tropical products, particularly export products, and restrict the incomes of primary producers in tropical developing countries.

Where high agricultural prices result from direct or indirect government subsidies (as is often the case, for example, with sugar cane production) the argument for reducing demand by eliminating those subsidies is stronger. Guaranteed price support schemes have, in the past, increased the expected profitability of deforestation by both increasing the expected sale price of outputs and by reducing the commercial risks of such ventures. Market promotion and marketing boards have similar effects of reducing market risk and volatility and increasing long-term average returns to producers. Under these circumstances, eliminating subsidies or marketing assistance for products from newly cleared forestlands may be effective, low cost and socially progressive. But in some countries it may be politically unacceptable because of the perceived impacts on farmers' incomes. In particular, because most price subsidies accrue to large farmers, who produce most of the (subsidized) agricultural products, eliminating these subsidies is usually politically difficult.

Ceteris paribus, higher population implies more demand for agricultural products. Depending on the scope of the markets, local, national or international population levels may be relevant. However, income levels and distribution, consumer preferences, and other factors which affect relative prices generally influence demand even more, at least in the short or medium term. Population growth can also influence forest clearing through numerous other causal relations, which often have conflicting effects (Bilsborrow and Geores, 1994). After a considerable lag, policies which lead to lower population growth could potentially limit deforestation through lower demand for agricultural products, but the magnitude of that effect is unlikely to be large. Since population also affects many other important policy variables, governments are unlikely to design population policies principally with regards to their impact on deforestation.

Per capita income and economic growth greatly influence the demand for tropical products although this is only one way they affect deforestation. Several authors have speculated about an 'environmental Kuznets curve': at lower levels of per capita income, economic growth leads to greater deforestation due to increased demand for tropical products; but after income levels reach a certain threshold, deforestation declines because the economies become less dependent on agricultural and forest products, primary production becomes more intensive, new employment opportunities arise and demand for forest preservation grows (Capistrano, 1994).

The empirical evidence on this issue is weak, contradictory and based mostly on cross-sectional data and the few 'tiger economies of East Asia', which may not reflect how the majority of tropical countries are likely to evolve in the future. What evidence there is, however, suggests that given existing per capita income levels in most tropical countries, without a change in the pattern of development, economic growth is likely to lead to greater forest clearing in the medium term (Stern et al, 1996). This implies that one way to avoid a trade-off between economic growth and deforestation is for countries to promote patterns of economic development which are less dependent on primary agricultural and forestry commodities. The alternative is to develop a well-managed, ecologically and economically sustainable forests sector. In either case, it is essential to adopt other policies which limit inappropriate forest conversion.

Governments influence the relative prices of, and hence demand for, tropical products through various measures, including exchange rate policies, tariffs, non-tariff trade barriers, pricing of government utilities and price controls, among others. The trend in tropical countries in recent years has been towards trade liberalization and more free-market oriented policies. The European and US governments, on the other hand, continue to use tariff and non-tariff trade barriers to regulate imports of agricultural products to meet commercial, political and environmental objectives.

Real currency devaluations, designed to increase the relative price of tradable goods and services compared with non-tradables, generally increase real producer prices of tropical agricultural products, making it more profitable to convert forest land to agriculture (Capistrano, 1994; Wiebelt, 1994; Frickman et al, 1995). This problem is most likely to arise where export producers are important agents of deforestation, such as in certain countries of Central America (beef), South America (soybeans), and West Africa (cocoa) (Kaimowitz et al, 1996; Kaimowitz, 1996).

Exchange rate policies are broad and far-reaching, so concerns over inappropriate deforestation cannot be expected to be a major factor in their design, nor should they be. However, in situations where a large portion of anticipated export growth generated by devaluations comes from the expansion of agricultural production to areas currently under forests, the issue cannot be avoided. Policy makers should explicitly consider the trade-offs between the possible benefits of short-term increases in exports and the costs associated with forest conversion. Alternatively, they might try to devise simultaneous parallel measures to minimize the predictable adverse impacts.

A similar argument applies with respect to agricultural price policies focusing on production for domestic consumption, such as price controls, import restrictions, guaranteed minimum prices for food crops or public marketing of foodstuffs. In most cases, the issue of deforestation will (quite reasonably) receive little attention when designing these policies. There are some instances, however, where policy changes can be expected to shift relative prices in a way which will promote rapid agricultural expansion in forested areas. In Ghana, for example, Munasinghe and Cruz (1994) have shown that policies which improve agricultural prices promote extensive, rather than intensive, agricultural growth, the environmental costs of which may not justify the benefits. In such circumstances, policy makers should explicitly consider whether short-term agricultural growth is worth the long-term consequences, and/or implement 'damage control' measures alongside the necessary macroeconomic reforms.

The conversion of forests to pastures in Central America illustrates many of the previous points. Many authors have cited this as a case where strong demand for beef imports in the US in the 1970s and high international beef prices led cattle ranching to expand rapidly at the expense of forests (Myers, 1981). During the 1980s, however, Central American beef and dairy producer prices fell sharply as a result of increased protectionism and reduced per capita beef consumption in the US, domestic price controls, overvalued exchange rates, and international powdered milk donations. This, along with other factors, reduced the pace of forest conversion to pasture, but by no means stopped it. Ironically, when demand and prices declined, livestock production fell more rapidly in traditional cattle regions than along the agricultural frontier, because producers on the frontier had few alternatives to investing in cattle, could simultaneously speculate in land markets, and obtained high initial pasture yields in recently deforested areas (Kaimowitz, 1996). Thus, as with the cocoa example (Ruf and Siswoputranto, 1995) it cannot be generally assumed that the 'new' producers at the forest margin will be the first to become non-viable as demand contracts – although they are physically 'at the margins' they are economically less marginal than the producers in 'older' areas.

Similar arguments apply to the case of tropical timber, but there are also important differences. Since industrial logging in most tropical countries is selective and does not involve direct conversion of large forest areas to other uses, the effect on forest conversion of policies which affect the demand for forest products is much more indirect. Logging companies frequently facilitate forest conversion to other uses by building roads which assist farmers subsequently to enter and clear land or by providing investment capital to farmers or ranchers which can be used to clear land or purchase cattle.[7] But there are also other situations where they do not.[8] Moreover, policies which depress the demand for timber can be a disincentive for sustainable forest management, where that is a possibility, even though in the short term they limit some negative effects associated with current logging practices. The question therefore is not to reduce the aggregate demand for tropical timber, but to reduce demand for the tropical timber that is closely and visibly associated with inappropriate tropical deforestation and degradation, while, if possible, stimulating the demand for tropical timber proven not to be linked to deforestation. We return to this issue below.

As with agricultural products, larger populations, economic growth, exchange rate devaluation and more free-market pricing, policies generally increase both the domestic and export demand for timber, and hence logging. Large devaluations, in particular, have been associated with major increases in tropical timber exports from Bolivia, Central Africa and South-East Asia (Anderson et al, 1995; Tchoungi et al, 1995).

There has been much controversy over the use of log export taxes or bans to reduce logging of natural forests (among other objectives) (Barbier et al, 1994).[9] These policies are relatively easy to enforce (although there is always some slippage) and generally reduce the amount of logging in the short term, but at a high economic cost. Their medium- and long-term impacts are less certain, but many experts believe bans discourage efficient logging, wood processing and investment in sustainable forest management and encourage domestic timber consumption. This sometimes leads to even greater logging than would have occurred without the ban (as occurred in Indonesia).

Bans and taxation of tropical timber imports, currently being discussed in certain developed countries, are similar, with two notable differences. First, since not all countries which import timber are likely to adopt such a policy, its effect on the domestic prices of timber in exporting countries will probably be smaller than with an export ban (Buongiorno and Manurung, 1992). Second, the importing countries appear more prepared to distinguish timber from sustainably managed forests from that which is not. This could provide an incentive to invest in sustainable forest management which does not exist with traditional log export bans (see below). It could still prove difficult, however, to define sustainable forest management and monitor whether imported timber comes from forests which meet those standards.

Finally, we note the case of commercial firewood collection, which is sometimes directly responsible for forest clearing in drier peri-urban areas. Higher urban incomes generally lead to rapid substitution of fuelwood by other energy sources, and hence can actually reduce the pressure on forests (and vice versa). Attempts to promote the substitution of fuelwood by kerosene, butane or electricity by making these energy sources cheaper through subsidies, however, have had little success in reducing fuelwood consumption and been expensive for governments (Mercer and Soussan, 1992).

In summary, any policies designed to restrict forest conversion by reducing the demand for primary products are blunt instruments, which make them alternatives of last resort.

There may be circumstances, however, where they are the only realistic alternative for achieving rapid reductions in inappropriate deforestation. Policy makers must become more conscious of the potential impact of their macroeconomic, agricultural and trade policies on forest clearing, actively promote patterns of economic growth which rely less on the liquidation of natural capital, and take a more long-term view in assessing policy trade-offs.

Policies that increase the costs and risks of production associated with deforestation

Some policy measures can reduce inappropriate deforestation by increasing the costs and risks of production associated with deforestation. Often these policy measures involve eliminating government subsidies or other distortions generated by previous policies, and thus actually save governments money, rather than involving additional expenses. First we review policies related to agricultural production, such as agricultural input subsidies, colonization schemes, and research and extension; then we look at policies affecting logging in natural forests; finally, we examine the implications of road construction and maintenance and of zoning and protected areas, which affect both agriculture and logging.

Basically, any government actions that encourage agricultural intensification tend to slow deforestation at the forest-farmland boundary, while actions which encourage extensification of agriculture (bringing new land into production) are likely to promote deforestation. Intensification means increasing the amount of labour/capital/information applied per hectare of land. Whether the expansion of market demands and higher product prices lead to more or less pressure on the agriculture/forest frontier will depend on the farmers' choice of technology (that is, how much to intensify and how much to expand in area) in response to perceived opportunities. Assuming farmers respond rationally to the expected net returns from either means of increasing their output, any public policies that affect their input prices will affect their decisions. The effects of government subsidies for agricultural inputs such as fertilizer, water, seeds, pesticides, credit, fuel and transport on agricultural intensification versus extensification are far from uniform. Subsidized irrigation water supply generally facilitates intensification, because it is not feasible to supply irrigation water to remote areas at the forest frontier. Fuel and transport subsidies tend to favour extensification by making it more profitable to farm in remote areas. Livestock credit also tends to facilitate extensification, since tropical livestock production is typically an extensive activity. Fertilizers, seeds and agricultural credit subsidies can facilitate either intensification or extensification. It is generally accepted that the agricultural green revolution, whose success was at least partially due to the subsidized availability of the inputs required for intensification (credit, fertilizer, pesticides, transport) helped save considerable areas of forests in South-East Asia by making cultivation of unirrigated rice on forest hillsides much less profitable.

Governments of many countries have deliberately encouraged forest conversion through official resettlement and colonization programmes and some still do, for example the Trans-Amazonia Highway and the Rondônia colonization scheme in Brazil and Indonesia's Transmigration programme. Since so many governments have used these poli-cies in the past, it might seem at first glance that arguing for their removal would be hopeless. Moreover, many governments in Africa, Asia and Latin America have used 'national defence' arguments to justify their colonization schemes, and these arguments carry great political weight. Nevertheless, a political climate favouring the elimination of public subsidies and less government intervention tends to favour these reforms.

Removing credit and input subsidies and eliminating land colonization and resettle-ment schemes to reduce inappropriate deforestation would probably be moderately effec-tive, readily targetable, have no cost to the national treasury, and would actually result in substantial savings in budgetary outlays. How politically viable it would be is unclear. Subsidies or programmes which encourage extensive expansion of agriculture promote deforestation, while subsidies which favour intensification would tend not to encourage deforestation. Thus it would be appropriate to distinguish between agricultural subsi-dies, before concluding that their abolition would help slow deforestation. For example, removal of fertilizer subsidies in parts of sub-Saharan Africa may well accelerate defor-estation. It would be a mistake, therefore, to believe that eliminating subsidies alone is sufficient to solve the problem of inappropriate deforestation. Brazil and Central Amer-ica, for example, greatly reduced the tax incentives and subsidized credit for cattle ranch-ing in recent years and in both cases inappropriate deforestation declined, but remained high (Ozorio de Almeida and Campari, 1995; Kaimowitz, 1995).

Agricultural research and extension activities have rarely been explicitly aimed at promoting deforestation. Nevertheless sometimes recommended new technologies are adopted by farmers not only on their existing fields but also on new fields that were formerly forest (even protected forests and national parks). As Angelsen (1995) noted, if a new agricultural technology is worth adopting on existing farmland, it may also be viable on nearby lands which were previously left in forests. Sometimes they may make it profitable to farm in places where previously it was not. Improvements in soybean technology were an important factor in promoting the clearing of forest areas to expand soybean production in Brazil and Bolivia. New pasture varieties also made it profitable to clear forest for pasture in Latin America in places where otherwise it would not have been. One possible policy reform would be to ensure the content of agricultural exten-sion services was not biased to encourage deforestation. However, such measures are likely to be only moderately effective, they are difficult to target, and although having few cost or equity consequences, are unlikely to find a strong political constituency.

Frequently recommended policy reforms which reduce the profitability of industrial logging (and so presumably reduce the amount of inappropriate deforestation due to such logging) include increasing stumpage charges and other fees and reducing direct and indirect subsidies to timber processors. Timber stumpage charges and other timber fees are often extremely low. Governments frequently justify this as a way to attract new investment or to capture other social benefits like infrastructure development and employment creation. For example, the Bangladesh Forest Industries Development Corporation made only notional payments (a few cents per tonne) for areas, logs and bamboo for more than twenty years (FAO, 1993). The Hindustan Paper Corporation in Kerala (India) paid US$0.50 per ton of eucalyptus pulpwood compared to the Forest Department's direct production costs of almost US$25 per tonne (FAO, 1993). Naturally the industry operations expanded considerably, but when these stumpage charges were eventually increased to realistic levels, the company's activities contracted leaving excess capacity, unemployment and demands for more plantations and yet more logging of protected natural forests. Subsidizing log input prices is a very inefficient way of creating additional employment and discourages intensive recovery of finished products in mills.[10] The systems used for (non-competitive) allocation of logging concessions and licences has attracted almost as much criticism as the log-pricing issue, for the effects on mismanagement of forests and accelerated deforestation, as well as the revenues forgone by the State. The issuance of 500 timber concessions in Thailand by 1968, covering 50 per cent of the country, opened up vast new areas for encroachment and official agricultural development (FAO, 1993).

Substantial increases in the fees and charges set by governments for exploitation of natural forests might reduce logging in inappropriate areas and eliminate some of the least efficient operators from the industry, who may also be among the most destructive. But its effectiveness for reducing inappropriate deforestation could be mixed if it also became a deterrent to sustainable forest management, so it would need to be carefully targeted. There would be no direct costs, rather increases in government revenues – provided it did not lead to massive closures and reductions in volumes more than proportional to the price increases. While the equity implications could be very favourable, the political viability is, at best, uncertain.

As one form of indirect subsidy to local industry, countries have operated a two-price scheme, where logs for local processing are considerably cheaper than export logs. Accelerated industrialization has been accomplished in Indonesia and Malaysia through a suite of policy measures linked to trade and tariff policies, tax incentives, log export restrictions, 'cheap logs', worker training programmes and so on (de los Angeles and Idris, 1990). However, most independent studies suggest the short-term economic cost to Indonesia was extremely high (Lindsay, 1989; Douglas, 1995). Even Malaysia's successful and widely acclaimed policies for accelerated industrialization and 'downstream, value-added processing' were assessed by a World Bank team as economically inefficient and too ambitious. By making logging more profitable than it would otherwise be, the subsidies policy expanded the area logged annually. Further, if wood prices were set at artificially low levels, distorting wood-processing decisions and leading to excessive physical wastage (Douglas, 1995), then greater areas had to be logged to achieve the same final output of timber products.

Road construction plays a crucial role in deforestation, and has been the topic of much analysis and debate. It is often argued that new road constructions present the greatest threat to remaining forests by providing access to forested areas (Dudley et al, 1995; Bryant et al, 1997). Roads are often essential both to market farm outputs and deliver inputs to the farm, and can greatly increase the profitability of agricultural and wood production in frontier areas. Thus, one of the most common forms of assistance requested by farmers in many countries is the extension and upgrading of road-transport systems.

Sometimes, however, roads do not create opportunities for deforestation but rather the existence of those opportunities leads to the roads' construction. If prospective farmers perceive their best opportunity to be clearing some forest and converting the land to another land use they ask governments to construct roads to facilitate that. In this case roads facilitate deforestation, but they may not be its principal cause. Thus, in analysing the impact of public road construction on forests it is important to take into consideration why specific roads have been built. It some cases the prime impetus is to provide access to timber for logging, in others it is for farming, and in still others (such as the Trans-Amazonia or Trans-Borneo Highways) it is for military or 'national development' reasons.

The road-construction and transport policy reforms that might reduce deforestation by increasing the costs of deforesting activities do not necessarily involve less road-building but rather a change in the nature, type and location of roads built. In particular, existing road systems would have to be intensified, rather than expanding new roads into previously remote forested areas. Such reforms could be effective, targeted, budget neutral, and have low indirect costs. However since increased road intensity could lead to capital appreciation of existing farmlands. the effects may be regressive. In addition, because of the strong political demand for roads by landholders in remote areas and many governments' interests in promoting colonization in those areas, such reforms may not be politically palatable.

The topics of land-use zoning and the declaration and enforcement of protected areas warrant many volumes (eg Wells and Brandon, 1993). We confine ourselves to summary comments. The rate of deforestation can be substantially reduced where protected areas are effectively managed and have the support of surrounding populations. But where effective management and popular support do not exist, declaration of parks or national forests may actually accelerate deforestation. Many countries have proclaimed National Forests, Parks and Reserves but have been unwilling or unable effectively to enforce such claims. If governments were more effectively to enforce Reserve boundaries and were consistent in not recognizing 'squatters' within reserves, this could substantially increase the risks and reduce the expected profitability of deforestation. While such measures could be targeted, they may be expensive, ineffective and inequitable in developing countries with many poor landless people. Forest resource degradation was made worse by restricting peoples' rights to use land and neglecting their traditional uses and their capacities to preserve forests. A new approach which actively engages local people in decisions and management, and which ensures they are not disadvantaged by forest conservation or production activities, is currently being applied and evaluated worldwide.

It is not difficult to devise a suite of policies which would retard or discourage deforestation of protected forest areas by increasing the risks and consequences to those who deforest. However, the prevailing political tendency is to ignore deforestation and encourage the expansion of alternative land uses. Most such policy reforms would appear to be moderately effective, equitable, readily targeted and would reduce government expenditures. However, the reason such subsidies currently exist is that many governments apparently see expansion of agriculture and (eventually) higher incomes to farmers as desirable goals, irrespective of the deforestation consequences.

Policies which discourage forest clearing to establish property rights

Many people clear forests not only to reap the benefits of selling standing timber and/or producing crops and livestock on the cleared land, but also as a way to gain property rights over that land, or strengthen their existing rights (Mendelsohn, 1994; Angelsen 1995; Ozorio de Almeida and Campari, 1995). This allows them to earn an implicit rent from the land and perhaps later sell the acquired land at a substantial profit. In many Latin American countries, one condition for prospective landholders to obtain legal recognition of property rights over formerly public lands has been to demonstrate that they were 'productively' occupying those lands by clearing forest and planting pasture or crops. Deforested land has also been better protected from expropriation by agrarian reform agencies or invasion by squatters (Utting, 1992). In South-East Asia and West Africa, farmers frequently clear forests to plant perennial crops such as cocoa or rubber, partly to establish or improve their property rights over that land (Angelsen, 1995). Even when land is private, forests often belong to governments. This gives landholders an additional incentive to clear the forest to obtain broader property rights (Dorner and Thiesenhusen, 1992) or greater legal recognition of de facto claims.

In much of Africa, communal land tenure regimes and ethnically segmented land markets traditionally limited the relevance of land speculation as a factor in deforestation, although as noted above perennial crops were often planted to establish long-term usufruct rights. This situation has slowly begun to change, however, as integration into the global market economy and demographic pressures have increased, and governments have sought to establish private property regimes.

To discourage the clearing of forests in order to gain or improve property rights, governments can use measures which: break the link between forest clearing and tenure; tax non-forested (cleared) land more heavily than forested land; or reduce the opportunities for capital gains from land sales. This could potentially be achieved through a combination of land tenure, taxation, infrastructure, credit and macroeconomic policies.

Some Latin American governments have changed their land titling and agrarian reform policies in an attempt to decouple the link between forest clearing and increased tenure security. Formally, they no longer require public lands to be cleared for claimants to obtain or maintain property rights. They now deny the right to request land titles to farmers who have inappropriately deforested the land after the new policies came into effect (Sunderlin and Rodriguez, 1996). Land titling programmes for humid tropical areas have attempted to provide secure property rights to existing settlers, without them having to clear land for that purpose. Many Latin American countries require that people who receive property rights over previously public lands in tropical forest areas maintain a certain percentage of that land in forest. Brazil, for example, for many years required that 50 per cent of each rural property within Legal Amazonia remain forest, and now requires 60 per cent (Burford, 1991).

The impact of these policies has ranged from mildly effective to negative, with the main problems being weak implementation, forest clearing provoked by uncertainty surrounding the policy changes, and new migrants being attracted by the chance to obtain land titles. Governments find it somewhat expensive to implement these policies and hard to target them based on specific local conditions. Where governments are unwilling or unable to enforce legal property rights, land claimants keep clearing forests to ward off other potential claimants. While there are typically few strong objections to them at the national or international level, local elites usually oppose any attempts to restrict their ability to gain property rights over deforested lands. As a result, the policies may be less equitable in practice than they are in theory.

Throughout the tropical world, there is increasing interest in establishing or supporting existing common property regimes for land and forests to protect indigenous peoples and other forest-dependent people, encourage sustainable forest management and limit the scope of land markets. The literature on the topic rarely stresses this latter aspect, but it is potentially quite important.

Common property regimes can effectively limit deforestation related to land speculation in certain places. They are relatively cheap to establish legally where they already exist informally, and governments do not have to compensate other land claimants. However, they are difficult to create where informal common property regimes did not exist in the past (Richards, 1997). Policies which promote common property regimes are equitable in the sense of supporting groups which have traditionally been poor and marginalized, but often involve ensuring large amounts of resources to small numbers of individuals.

Theoretically, land taxes could be used to discourage forest clearing by taxing cleared lands at a higher rate than forested land. Land taxes and capital gains taxes could also potentially discourage land speculation by raising the cost of using landholding as a hedge against inflation or a source of capital gains, charging land owners for the costs of infrastructure improvement, and lowering the price of land (Strasma et al, 1987). Experience to date, however, has not been encouraging. Since few countries have used punitive land and capital gains taxes to discourage land speculation, it is hard to assess their practical effectiveness. Most governments have found it difficult effectively to administer such taxes, due to the large amounts of information required and the high potential for evasion, or have avoided them because of potential political opposition (Skinner, 1991).

Since the purchase of land is a common hedge against inflation, macroeconomic policies which reduce inflation may also limit the incentives for land speculation. This seems to be particularly relevant for certain Latin American countries, such as Brazil, where high inflation rates were previously common, but were not associated with general economic collapse. It is also true, however, that as long as infrastructure investments continue, regional markets grow, and public services gradually improve in agricultural frontier areas, real local land prices will rise, and that will encourage land speculation. There is ample evidence of similar processes at work in Indonesia, Thailand and the Philippines.

Certain government subsidies can increase land prices, and thus stimulate deforestation for speculative purposes where property rights can be secured or improved by clearing land (Binswanger, 1994). Road construction, in addition to facilitating access to forested areas, raises land values and offers opportunities for capital gains (Angelsen, 1995). Credit subsidies or tax incentives for agriculture or livestock which are preferentially available to landholders also increase the value of holding land. Government purchases of land for protected areas, refugee resettlement, infrastructure construction or other purposes often inflate local land markets. Removing these subsidies can sometimes greatly reduce inappropriate deforestation, especially where high subsidies existed in the past. Governments can actually save money by doing this and there are few indirect costs. Since most such subsidies presumably go mainly to wealthy families and corporations, except perhaps for road construction, eliminating them tends to favour equity. Important interest groups support these subsidies, but there are often others who oppose them.

In summary, eliminating policies which promote land speculation and strengthening either private property or common property regimes can sometimes be effective tools against inappropriate deforestation. Most other proposed policies which might limit forest clearing to gain secure tenure have still not been fully tested. Land taxation, in particular, merits greater consideration, given the notable gap between its theoretical potential and practical difficulties. Most policies discussed in this section are easy to target to specific regions, but much more difficult to adapt to the conditions of individual properties. Generally, the more effective (and less purely symbolic) these policies are likely to be, the less politically viable they are, although eliminating certain credit subsidies and tax incentives is apparently an exception.

Making Forest Retention More Attractive

Policies which increase the profitability of maintaining forests

Having considered many policy instruments which might reduce the profitability of removing tropical forests, we turn now to possible interventions that could increase the profitability of retaining them and managing them on a long-term sustainable basis. We consider three aspects in this regard: increasing timber revenues, increased revenues from non-timber forest products, and the possibility of payments for global environmental services.

To the extent that excessive logging (both in area and in degree of damage caused) contributes to deforestation, some recent efforts to reduce deforestation have focused on policies to encourage forest management and logging practices on concessions and state forests to be less destructive and more sustainable (eg Dudley et al, 1995; Bryant et al, 1997). Market forces could provide a significant incentive for more sustainable forest management, as the expected future values of remaining commercial forests increase greatly, and as accessible stocks of tropical forests are depleted. However, recent assessments (Byron and Ruiz-Perez, 1996; Sayer and Byron, 1996) of how forestry practices may change in response to changing markets for forest products concluded that 'frontier-logging' of relatively remote areas in the tropics may become less important in the future. Once the forests in more accessible areas (close to roads, rivers and ports) have been exploited, any cost advantage these forests might have had will be rapidly eroded. Remaining natural forests will become less able to compete with the outputs of the rapidly expanding plantation sector in the tropics and subtropics, except in speciality markets. In contrast to rising costs and declining quality of logs from natural forests, the volume and quality of plantation material will continue to improve and technological advances in plantation silviculture and wood processing will further lower unit production costs.

Thus although some market assessments have proposed that market forces will increase the potential timber revenues from retaining natural forests for high-value products, we see little basis for relying on that. This effect is likely to be too little, too late. Further, if timber concessionaires harbour doubts about their long-term security of tenure, they are likely to continue to behave in a short-term exploitative way; even if they agree that in the future, tropical timber values might soar, they doubt that they will be the beneficiaries of such price rises.

In the very short term, higher timber prices and roads may encourage unsustainable practices but in the long term may be necessary for sustainable practices to be financially viable. Short-term licences do not encourage long-term stewardship of forests, but even 25-year licences are less than many cutting cycles, destroying any motivation to nurture regrowth. 'Cleaning operations' every five to ten years can be observed. Even with very long-term licences, concessionaires in many tropical countries may behave as if their tenure is insecure, because of extreme uncertainty about political changes. In many countries, the logging rights are not transferable, reducing the incentive to manage the forests to maintain a high residual value of the forest enterprise 'as a going concern'. Transferability of licences might indirectly impose commercial penalties on licensees for not looking after the forest under their control and provide financial rewards (as higher transfer prices) for those who have not high-graded, who have protected advanced growth and installed good infrastructure. Systems like the 'Evergreen Licences' devised in British Columbia (Canada) with 'clawback' provisions – providing strong commercial incentives for long-term sustainable forest management – may warrant consideration and application in tropical developing countries.

In 1990 the International Tropical Timber Organization (an agency of UNCTAD) reached agreement between producer and consumer countries that by the year 2000, all tropical timber entering international trade would come from sustainably managed forests. Other initiatives (eg the Forest Stewardship Council) advocate policies which might make sustainable timber management more profitable (through independent inspection and certification of secure and longer-term concessions, accompanied by ecolabelling of products) to achieve premium prices or preferential market access for timber from these certified, well-managed forests. Since 1990 NGOs and, to a lesser extent, governments have urged consumers to purchase only certified or ecolabelled tropical timber. They wish to exclude or minimize sales of timber from poorly managed forests, while simultaneously obtaining premium prices for certified timber.

The certification approach has generated much debate: who certifies? on what basis? who controls the chain of custody between the forest and the final consumer to guarantee validity? how many consumers would pay what sort of premium for certified timber? and would this system be voluntary or compulsory (and if the latter, is it legal under the rules of the World Trade Organization?). In spite of the many unresolved issues, polices by both governments and NGOs to encourage and respect certification might give some positive commercial rewards for sustainable forest management. While generally politically acceptable, it remains unclear how effective (or even relevant) certification will be in reducing inappropriate deforestation on the much broader scale, where conversion to other land uses is the primary threat to forests.

To summarize, the policy interventions that might reduce deforestation by increasing expected long-term timber values rely on identification and development of niche markets for tropical timbers – small quantities of very high value wood, carefully removed from forests. Certification and ecolabelling are a special case of this which attempts to dissociate 'sustainable forest management' from exploitative logging causing deforestation. The feasibility, effectiveness and costs of this approach remain to be determined. If one of the major disincentives to sustainable forest management is insecurity of tenure or fear of expropriation, removal of such disincentives would appear an attractive, even necessary, policy option. But there is no evidence that this is a sufficient condition for slowing commercial deforestation.

Numerous international NGOs (eg Cultural Survival Enterprises and Conservation International) have argued that finding new products in the tropical forests, finding new markets for new or existing products, and/or more effective marketing of non-timber forest products (NTFPs) will make the tropical forests much too valuable to destroy. Both government institutions and NGOs have incorporated such efforts into their policies and programmatic agendas, ranging from the World Bank Pilot Project for Brazil (with a significant component aimed at supporting four Extractive Reserves in the Brazilian Amazon) to local NGO initiatives to promote new tree species of economic potential in the Highlands of Northern Sumatra. Numerous, very substantial policy and development initiatives are pursuing this proposition. Yet debate is raging about the extent to which such efforts to promote extraction, use and marketing of NTFPs represent a real option for slowing deforestation and improving the livelihoods of forest people. Dove (1993) has argued that incorporation of NTFPs into international markets will achieve neither socioeconomic nor environmental objectives, and may actually be counterproductive in both senses – at best it diverts attention from what he sees as the real causes of deforestation and degradation, and their political resolution.

The potential for these policy measures to slow deforestation depends on whether one accepts that forest people are destroying tropical forests because the stream of income from forests is too low (or because they simply do not understand the magnitude of these benefits). However Dove believes that forest destruction is caused by loggers, migrants and predatory outsiders – the forest people are the victims rather than the perpetrators of deforestation. Thus he argues that efforts to halt forest destruction should focus on the economic behaviour of the former, not the forest people. He asserts that increasing commercialization of NTFPs will foster predatory exploitation and destroy traditional management systems that might have helped protect forests.

Ruiz-Perez and Byron (1999) showed how crucial determinants of the outcomes from commercialization of NTFPs are the distribution of property rights and the ability (or otherwise) of local people to claim and enforce such rights. For a strong 'community' of NTFP users/collectors with organization and political empowerment, commercialization of NTFPs can enhance welfare and forest conservation. Conversely, the strength of any political elite who may seek to expropriate the NTFP values from the forests as soon as they realize that significant potential rents exist may lead to serious social and environmental exploitation.

Policy proposals and development actions aimed at slowing deforestation by increasing the stream of NTFP revenues to forest peoples could be effective, targeted, equitable, low cost and moderately politically acceptable to governments with inhabited tropical forests, but only given specific institutional and legal contexts.

It has also been proposed that mechanisms for payment by beneficiaries for the environmental services provided by tropical forests could provide a market-based stimulus for retaining forests and slowing deforestation. The proposed policy reform is to pay those who retain forests for the environmental services their actions provide to the wider public. Analogous to the upstream-downstream analyses for the beneficiaries compensating the providers of environmental externalities in a watershed or on a national scale, there have been numerous proposals for capturing the 'option and existence values' of tropical forests, as perceived by affluent residents of OECD countries, as a way of internalizing some of the global externalities associated with tropical deforestation. The UN Convention on Biological Diversity calls for further research on the 'equitable distribution of benefits and costs of forest conservation', which is seen as a prerequisite to devising suitable international or intranational compensatory schemes. This has to date focused on two main areas – biodiversity prospecting and carbon sequestration.

The agreement between Merck pharmaceuticals and the Biodiversity Institute (InBio) of Costa Rica for the commercial exploration and assessment of that country's biodiversity has attracted great publicity. It appears to be an example of using commercial, market-based transactions to make retaining tropical forests more valuable than destroying them. Despite the publicity about the potential of 'biodiversity prospecting' to reward forest managers and indigenous forest people for retaining forest biodiversity, it has recently been argued that the realistic potential of such payments is very small, apart from a very few very small niche markets (Simpson, 1997).

Similarly, the UN Framework Convention on Climate Change has authorized 'Activities Jointly Implemented' whereby carbon dioxide emitters (such as power utilities) in developed countries pay for carbon sequestration in developing countries, as offsets for their CO_2 emissions. To date these offsets have taken the form of payments for the establishment of new forests (such as the Dutch 'Forests Absorbing Carbon Emissions' (FACE) Foundation) or the payment to logging companies to adopt 'Reduced Impact Logging' techniques.[11] As yet, there is no legal obligation for power utilities (in the United States) to expend shareholders funds in Central America or Malaysia to offset their US emissions, nor are there tax credits at present.[12] Most studies have looked at the demand side based on the theoretical costs *if* companies had to pay carbon taxes or *if* they had to pay to reduce their emissions – which most do not, except in Norway and Denmark.

The FCCC has debated proposals for 'international tradable carbon emissions permits' which appear to be an efficient and low-transactions costs approach to determining who would emit and sequester carbon dioxide, and where. Yet countless practical difficulties remain. In the context of their efficacy in preventing tropical deforestation, it appears unlikely that payments (whether for carbon, biodiversity or watershed protection) would have much impact unless the payments actually went to those making the deforestation decisions on the ground, rather than to national governments.

In summary, policies to encourage the retention of tropical forests through capturing payments for the global environmental services provided remain of questionable effectiveness, targetability and cost. There are so many unknowns still that the whole area remains decidedly ambiguous.

Policies which increase the opportunity costs of labour and capital

We have argued that any policies that reduce the relative profitability of clearing forests, as compared to retaining them, will tend to slow deforestation rates. Since labour and capital are major factor inputs in forest conversion, but much less so in retention, increasing the opportunity costs of labour and capital could be expected to reduce deforestation rates. Such policies include general growth promoting policies, as well as specific employment promotion policies. In parts of East and South-East Asia (eg Korea and Malaysia) the opportunity costs of labour and capital have clearly increased and deforestation seems to have slowed there, and even been reversed. In other places where alternative returns to labour and capital remain low (eg Laos and Cambodia) deforestation rates remain high. This suggests possible connections between these costs and the extent of inappropriate deforestation.

Even within South-East and East Asian countries, there appears to be socioeconomic stratification and differentiation: remote, poor rural forest villages are being left out of the economic miracle. Most have experienced very low income growth, if any; people have little formal education, limited access to or control over valuable resources and poor transport and communications infrastructure, and struggle to maintain traditional living standards. One of the options for those remaining in the villages and forests is to clear forests for marginal agriculture, often temporary and unsustainable. The forest populations have been willing to clear land, to log illegally, to hunt/collect NTFPs to earn cash incomes. If the prices they receive are low, they increase the volumes harvested to earn the required incomes!

In cities and towns, high-tech industries, tourism and the service sector are booming as prosperous enclaves. This rapid economic growth in urban, non-land-based sectors is already having profound impacts on forests and deforestation rates. While rapid growth in agriculture, timber industries and plantation estates may contribute to deforestation, employment growth in sectors like tourism, textiles, manufacturing etc has increased labour opportunity costs to such an extent that it has reduced deforestation. However, where the deforesters have been left out of that growth, and no more attractive livelihoods have emerged, deforestation has continued. The evidence to date is circumstantial and anecdotal, but the policy implications could be that promotion of sustainable, non-rural employment options may reduce the types of deforestation caused by poor people trying to earn a meagre living from deforesting. Similarly, where deforestation is driven by the interests of large capital, the implication could be that with the emergence of more lucrative options (eg through the development of capital markets) these parties might lose interest in deforestation.

Given this diagnosis of who is clearing forests, where and why, and the observation that where more attractive alternative employment options have emerged, people (both large-scale and peasants) have voluntarily moved out of forest-degrading activities, a possible policy measure is suggested. Economic growth, with reasonable equity of access to new industries and resources may make certain types of deforestation in particular contexts sufficiently unremunerative that it slows voluntarily. The pursuit of equitable and sustainable economic development is a stated goal of most developing countries, so it would be fortuitous indeed if that process inherently encouraged forest conservation. Unfortunately there is little evidence that the solution is so simple. In terms of the criteria used here, policies to promote industrialization and employment generation in non-land-based sectors are already on the political agenda; while they might be effective in reducing deforestation, and equitable if they could be achieved, relatively few countries have achieved this. Indeed, promoting sustainable economic growth and employment creation has been more elusive than policies to slow deforestation directly but, in the long run, might be the eventual solution.

Conclusions

We conclude there is no perfect or generalizable policy for reducing inappropriate deforestation. There are no 'first-best' options. Each national situation is different, much uncertainty remains about key cause-and-effect relations and there are usually trade-offs between policies' effectiveness, ability to be targeted, political viability and direct and indirect costs. Most of the policy instruments discussed here are very blunt instruments in regards to stopping deforestation – governments will be forced to choose from a mix of measures, specifically crafted to local conditions. There is no magic 'silver bullet' policy reform that will slow all types of tropical deforestation. Most such panaceas which have been advocated have been revealed as merely possible partial solutions, only applicable under limited conditions.

From the discussion here, we argue that it is crucial to resolve the interrelations between deforestation, logging, road construction and conversion to alternative land uses. We suggest that logging will lead to increased forest conversion by an influx of migrants, if the following conditions apply simultaneously:

- roads are constructed that open up new areas of forests;
- the non-forest land use is much more profitable than retaining forests;
- very poor enforcement of forest boundaries by government agencies (eg Forest Service or National Parks Service) and an institutional or legal context in which people can expect that land which they occupy, claim or 'stake out', will eventually be recognized, even legalized by the government so creating an open-access 'frontier';
- a large pool of unemployed or landless people, or with very low incomes and prospects, who constitute potential migrants (whether as individuals moving spontaneously or sponsored by the State or private agro-industries). We hypothesize that the pace of colonization might be related to the difference between current incomes of potential migrants and the amount they expect to earn by colonizing forest.

All of the instances of which we are aware, where rapid forest clearance by 'squatters' has occurred, appear to be consistent with this hypothesis. Conversely, where these conditions do not apply (which admittedly is still rare in tropical developing countries), forests can be logged but subsequently remain under management and not be cleared or seriously degraded (eg in Peninsula Malaysia).

This assessment, if correct, suggests that the answer to the forest conversion issue is not necessarily to stop logging per se, or to stop logging in all new areas, or to ban all new road construction in forest areas, but rather to reform those policies and institutions that at present make forest colonization *seem* more attractive than the potential migrants' current activities. This might include the pull factors (to reduce the profitability of illegally clearing forests or of speculating in land that was supposed to be kept as forest), or the push factors (to increase the limited livelihood options outside of forests). The evidence from the rapid economic growth of Asian tiger economies is that as employment and income prospects outside the agriculture sector improve, fewer people want to undertake the dangerous, illegal, difficult and often unprofitable activities of temporary agriculture in forestlands. However, if the new land use is very profitable (eg growing cocoa, coffee, cinnamon, rubber or fruit trees, or even timber trees like Eucalypts, teak or Gmelina) and if the potential capital gains from 'capturing' some real estate from the government forests are high, it might be very difficult to slow the rate of forest conversion by such people. The case for intensification of agriculture as a primary instrument for reducing deforestation has been cogently summarized by Waggonner (1996).[13]

Thus the principal government policy reforms that could reduce inappropriate tropical deforestation are likely to be:

- Eliminate subsidies to agricultural and pastoral industries which encourage deforestation, and, if necessary, offset these by schemes targeted at intensification of existing agricultural areas. This also applies to road construction – that is intensification of existing rural road networks rather than expansion into hinterlands to open up new areas (unless they are specifically zoned for conversion from forest to agriculture).

- Eliminate existing legal and institutional incentives or requirements to clear forests as a basis for gaining recognized land tenure, and which reward 'squatting' or colonization of protected forest reserves.

- Reform forest industry concession and licences to provide incentives for long-term sustainable management, through increasing the duration and security of licences and making them tradable, while collecting more of the potential economic rents to remove the 'gold-rush' mentality which has characterized the industry for much of the past 40 years.

- Develop innovative institutional arrangements for devolving more decision making authority and responsibility to those whose livelihoods and quality of life are directly linked to the extent and quality of tropical forests. In association with this, both production and conservation activities (for the full range of consumptive and non-consumptive goods and services from forests) can be fostered, or at least not discouraged, from both government forests and private property outside of government forest reserves.

- Encourage voluntary market differentiation by consumers, that discriminates positively towards those products which have been sustainably produced from forests, and penalizes those products which have contributed to deforestation. This could

be applied not only to tropical timber, and to non-timber forest products like Brazil nuts, but also to other products such as chocolate, coffee and leather.
•   Facilitate the recognition and compensation for environmental services provided by forests – at global, national and local scales – through mechanisms directly linked to the amount of environmental services generated (whether carbon sequestration or biodiversity conservation or watershed protection) and ensure that the transfer payments actually are received by those making the decisions at the forest frontiers.

As the World Bank (1992) clearly showed, most of the major environmental problems in developing countries are not due to the pursuit of economic development, but rather are due to incorrect economic policies: poorly defined property rights; underpricing of resources; state allocations and subsidies; and neglect of non-marketed social benefits. Instead of trying to devise new policies to stop further resource and environmental deterioration while promoting real development, we should first try to eliminate those (legal, social, political and institutional) factors that cause or exacerbate the problems.

Notes

1   Durning (1993, p5), on the other hand, estimates that only one-third of the original forest area remains.
2   These figures are derived from the data in Table 4.2.1 in FAO (1996, p55). They are based on a definition of 'forest' which includes closed, open, long fallow and fragmented forest.
3   The official target of the International Union for the Conservation of Nature (IUCN) is for at least 10 per cent of each major forest ecological zone to be given Protected Area status. However, Iremonger et al (1997) concluded that of 138 regional/eco-regional zones worldwide, only in Insular South-East Asia was this target achieved.
4   This is not to imply that the areas of highest biodiversity value are necessarily remote and hilly sites undesirable for agriculture (or vice versa). Indeed, conflicts arise because the high biodiversity areas frequently are the remaining forest areas of high agricultural potential.
5   Note that the three categories are not discrete, but rather the endpoints and the midpoint of a gradient which ranges from 'clearly unsuitable' to 'extremely suitable'.
6   Ruf and Siswoputranto (1995) analysed the economics of growing cocoa on newly cleared forestlands, relative to existing farmlands. They found that newly cleared lands have a consistent price advantage of US$300–400 per hectare (the 'natural fertility subsidy'). This means that if world prices fall, it is the supply from growers on existing farmlands which contracts first – clearing new forest becomes the only viable option as prices fall.
7   For example, the highest rate of population increase in the Philippine uplands was in the municipalities with logging concessions (Cruz and Zosa-Feranil, 1988; cited in Garrity et al, 1993).
8   There are numerous examples, covering vast areas around the world, where forests have been logged, but are still reasonably productive and ecologically stable. Thus it

is overly simplistic, and incorrect, to equate commercial logging with deforestation. Industrial logging is most likely to lead to deforestation when associated with spontaneous small-scale settlement and conversion, or with large-scale agro-industry conversion.

9 In fact, log export bans are rarely intended as conservation measures, but as part of a package to encourage accelerated domestic processing, with employment generation and value-added, and hopefully higher net foreign exchange earnings.

10 Manasan (1989) observed that although South-East Asian governments justified their direct and indirect subsidies by the expectation that the expansion of forest industries would generate jobs, only the Philippines offered a direct financial incentive to employment, via tax rebates. Yet even this was ineffective because it was neutralized by other tax concessions which promoted capital-for-labour substitution. Overall, relatively few jobs have been created despite the substantial subsidies, and the impact on forests has been severe. (Lindsay, 1989; Douglas, 1995)

11 These can reduce the amount of damaged and decaying material left in forests, and increase the amount of future carbon sequestration by healthy, rapidly growing immature forests.

12 While spending US$1 million on offsets/tree-planting in tropical developing countries is cheaper than spending US$10 million on emission-control at home, there are no legal obligations to do either!

13 He concluded that if during the next 60–70 years the world farmer could reach the average yield of today's average US corn grower (8 tonnes of grain equivalent/hectare) then 10 billion people – twice the world's current population – would need only half the current area of cropland, while eating at today's average American calorie levels.

References

1 Angelsen, A. (1995) *The Emergence of Private Property Rights in Traditional Agriculture: Theories and a Study from Sumatra*, paper presented at the Fifth Common Property Conference: Reinventing the Commons, the International Association for the Study of Common Property, 24–28 May, Bodo, Norway.

2 Barbier, E. B., Burgess, J. C., Bishop, J. and Aylward, B. (1994) *The Economics of the Tropical Timber Trade*. Earthscan, London.

3 Bilsborrow, R. and Geores, M. (1994) 'Population, land-use, and the environment in developing countries: What can we learn from cross-national data?', in Pearce, D. and Brown, K. (eds) *The Causes of Tropical Deforestation. The Economic and Statistical Analysis of Factors Giving Rise to the Loss of Tropical Forests*. University College London Press, London, pp106–133.

4 Binswanger, H. (1994) 'Brazilian policies that encourage deforestation in the Amazon'. *World Development*, vol 19, no 7, pp821–829.

5 Buongiorno, J. and Manurung, T. (1992) 'Predicted effects of an import tax in the European Community on international trade in tropical timbers'. *Journal of World Forest Resource Management*, vol 6, pp 117–137.

6 Bryant, D., Nielsen, D. and Tingley, L. (1997) *The Last Frontier Forests: Ecosystems and Economies on the Edge*. World Resources Institute, Washington DC.

7   Byron, R. N. and Ruiz-Perez, M. (1996) 'What future for the tropical moist forests 25 years hence?' *Commonwealth Forestry Review*, vol 75, no 2, pp 124–129.

8   Capistrano, A. D. (1990) *Macro-economic Influences on Tropical Forest Depletion, A Cross Country Analysis.* PhD dissertation, University of Florida.

9   Constantino, L. F. and Kishor, N. M. (1995) *Stabilization, Structural Adjustment, and Bolivia's Forestry Exports.* Latin America Technical Department Environment Division Dissemination Note #13, The World Bank, Washington, DC.

10 de los Angeles, M. and Idris, R. (1990) *Investment Incentives of Southeast Asian Countries in the Wood-based Sector*, proceedings of the Malaysian Timber Industry Board Asia-Pacific Conference 1990, 24–26 September, Kuala Lumpur.

11 Dorner, P. and Thiesenhusen, W. C. (1992) *Land Tenure and Deforestation, Interactions and Environmental Implications.* Discussion Paper 34: UN Research Institute for Social Development, Geneva.

12 Douglas, J. J. (1995) *The Economics of Long Term Management of Indonesia's Natural Forest.* Mimeo, World Bank, Jakarta.

13 Dove, M. R. (1993) 'A revisionist view of tropical deforestation and development'. *Environmental Conservation*, no 20, pp1–17.

14 Dudley, N., Jeanrenaud, J. P. and Sullivan, F. (1995) *Bad Harvest: The Timber Trade and The Degradation of the World's Forests.* Earthscan, London.

15 Durning, A T. (1993) *Saving the Tropical Forests: What Will It Take?* Worldwatch Paper 117, Worldwatch Institute, Washington, DC.

16 FAO (1993) *Forestry Policies of Selected Countries in Asia and the Pacific.* FAO Forestry Paper 115, Food and Agriculture Organization of the UN, Rome.

17 FAO (1996) *Forest Resources Assessment 1990: Survey of Tropical Forest Cover and Study of Change Processes.* FAO Forestry Paper 130, Food and Agriculture Organization of the UN, Rome.

18 Frickman Young, C. E. and Bishop, J. (1995) *Adjustment Policies and the Environment: A Critical Review of the Literature.* CREED Working Paper Series no 1: International Institute for Environment and Development, London.

19 Garrity, D. P., Kummer, D. M. and Guiang, E. S. (1993) 'The Philippines', in *Sustainable Agriculture and the Environment in the Humid Tropics.* National Academy Press, Washington, DC, pp549–624.

20 Iremonger, S., Kapos, V., Rhind, J. and Luxmore, R. (1997) *A Global Overview of Forest Conservation.* Paper to World Forestry Congress, Turkey, October 1997.

21 Kaimowitz, D. (1996) *Livestock and Deforestation in Central America in the 1980s and 1990s: A Policy Perspective.* CIFOR Occasional Paper, no 12, Bogor.

22 Kaimowitz, D., Thiele, G. and Pacheco, P. (1996) *The Effects of Structural Adjustment on Deforestation and Forest Degradation in Lowland Bolivia*, mimeo.

23 Lindsay, H. R. (1989) 'Indonesian forestry policy: an economic analysis'. *Bulletin of Indonesian Economic Studies*, vol 25, no 1, pp99–115.

24 Manasan, R. G. (1989) *A Review of Investment Incentives in ASEAN Countries.* Working paper 88–27, Philippines Institute of Development Studies, PIDS, Manila.

25 Mendelsohn, R. (1994) 'Property rights and tropical deforestation'. *Oxford Economic Papers*, vol 46, pp750–756.

26 Mercer, D. E. and Soussan, J. (1992) 'Fuelwood problems and solutions', in Sharma, N. P. (ed) *Managing the World's Forests. Looking for Balance Between Conservation and Development.* Kendall/Hunt Publishing Company, Dubuque, Iowa, pp177–214.

27 Munasinghe, M. and Cruz, W. (1994) *Economy-wide Policies and the Environment.* World Bank Environment Paper 10, World Bank, Washington, DC.

28 Myers, N. (1981) 'The Hamburger connection: How central America's forests became North America's Hamburgers'. *Ambio*, vol 10, no1, pp3–8.

29 Ozorio de Almeida, A. L. and Campari. J. S. (1995) *Sustainable Settlement in the Brazilian Amazon.* Oxford University Press, New York.

30 Richards, M. (1997) 'What future for forest management based on common property resource institutions in Latin America?' *Development and Change,* vol 28, no 1, pp95–118.

31 Ruf, F. and Siswoputranto, P. S. (eds) (1995) *Cocoa Cycles: The Economics of Cocoa Supply.* Woodhead Publishing, Abington, UK.

32 Ruiz-Perez, M. and Byron, R. N. (1999) 'A methodology to analyse divergent case studies of NTFPs and their development potential'. *Forest Science,* vol 45, no 1, pp1–14.

33 Sayer, J. A., Harcourt, C. S. and Collins, N. M. (1992) *The Conservation Atlas of Tropical Forests – Africa.* Simon and Schuster, New York.

34 Sayer, J. A. and Byron, R. N. (1996) 'Technological advance and the conservation of forest resources', *Journal of World Ecology and Sustainable Development.* vol 3, no 3, pp43–53.

35 Simpson, R. D. (1997) 'Biodiversity prospecting: Shopping the wild is not the key to conservation'. *Resources,* no 126, pp12–15.

36 Skinner, J. (1991) 'If agricultural land taxation is so efficient, why is it so rarely used?' *The World Bank Economic Review,* vol 5, no 1, pp113–133.

37 Stern, D. I., Common, M. S. and Barbier, E. B. (1996) 'Economic growth and environmental degradation: The environmental kuznets curve and sustainable development'. *World Development,* vol 24, no 7, pp1141–1160.

38 Strasma, J. D., Alm, J., Shearer, E. and Waldstein, A. (1987) *Impact of Agricultural Land Revenue Systems on Agricultural Land Usage.* Associates in Rural Development, Burlington, Vt.

39 Sunderlin, W. and Rodriguez, J. (1996) *Cattle, Broadleaf Forests and the Agricultural Modernization Law of Honduras, The Case of Olancho.* CIFOR Occasional Paper, no 7, Bogor.

40 Tchoungui, R. et al (1995) *Structural Adjustment and Sustainable Development in Cameroon.* Working Paper 83, Overseas Development Institute, London.

41 Utting, P. (1992) *Trees, People and Power: Social Dimensions of Deforestation and Forest Protection in Central America.* United Nations Research Institute for Social Development, Geneva.

42 Waggoner, P. E. (1996) 'How much land can ten billion people spare for nature?' *Daedalus,* Summer 1996, pp73–93.

43 Wells, M. and Brandon, K. (1992) *People and Parks: Linking Protected Area Management with Local Communities.* World Bank, WWF and USAID, Washington, DC.

44 Wiebelt, M. (1994) *Protecting Brazil's Tropical Forest, A CGE Analysis of Macro-economic, Sectoral, and Regional Policies.* Kiel Working Paper no 638, The Kiel Institute of World Economics.

45 WRI (1996) *World Resources 1996–97.* World Resources Institute; UN Environmental Programme; UN Development Programme; World Bank. Oxford University Press, New York and Oxford.

13

Technologies for Sustainable Forest Management: Challenges for the 21st Century

J. A. Sayer, J. K. Vanclay and N. Byron

Introduction

Experience suggests that technological challenges will continue to confront foresters in the 21st century in their efforts to manage their forests sustainably, particularly if technology is interpreted in the broadest sense. In this chapter, we try to anticipate some of these issues, and to identify the research needed to provide the tools and techniques required for effective management in a changing environment. We anticipate change in several areas:

- production shifting from native forest to plantations;
- technological developments allowing more efficient processing, less waste and more recycling;
- end-use products becoming less dependent on the specific wood characteristics of raw materials;
- demand increasing, but fluctuating according to technologies in non-forest sectors;
- better information for decision makers, through the integration of remote sensing, GIS and other technologies into decision-support systems;
- more rapid shifts in loci of production and transformation as industries seek out areas of comparative advantage;
- more pragmatic and efficient options for conserving biodiversity.

We anticipate a significant shift in timber production from natural forests to plantations. This will be driven by several factors, including the reliability of supply, uniformity of raw material and competitive pricing. Technically, there is no reason why plantations cannot supply most of the world's wood requirements by early next century. Even if demand exceeds the most optimistic projections (ECE, 1996; Solberg et al, 1996), requirements may increasingly be satisfied from plantations, which offer both economic and environ-

Note: Reprinted from *Technologies for Sustainable Forest Management: Challenges for the 21st Century*, Sayer, J. A., Vanday, J. K. and Byron, N. CIFOR Occasional Paper, Copyright © (1997), with permission from CIFOR

mental advantages over natural forest production. Demand for the few speciality products that can be obtained only from natural forests may not increase greatly, and can probably be satisfied from ecologically sensitive logging operations in areas where forests are retained primarily for their environmental and amenity functions.

Technology promises many advances in processing, recycling and in other areas that may influence wood and paper consumption (eg office automation), but most of these will have relatively minor influences on the sustainability of forest management, other than their ability to drive demand. Technological developments in areas more closely allied to sustainable forest management (eg logging equipment) are also likely to advance relatively slowly and deliver modest gains.

Information systems are one area of technology that may have a big impact on forest management. Sustainable forest management relies on timely and accurate information, and the ability to obtain such information is likely to change significantly in the near future. We may expect major developments in remote sensing, in various forms of computer modelling and in the integration of these and other components into accessible decision-support systems.

Lastly, the globalization of economies will allow producers and manufacturers of forest products to shift their operations between countries and localities in response to changing economic and regulatory environments and to concentrate in areas where production, transport and labour costs are minimal.

Defining Sustainability: Criteria and Indicators

No one disputes that sustainability involves satisfying present needs without compromising future options, but it is not always obvious what this means in terms of forest management. It is not merely an issue of natural forests versus plantations, or clear felling versus selection logging systems, but involves more fundamental questions about the functions and services provided by forests, and about stakeholders, equity and expectations. Clarifying these issues, and establishing criteria and indicators (C&I) of sustainability, is a small but important step towards sustainable forestry.

The original drive to develop C&I came from the ecocertification lobby, but the potential application of C&I in the numerous national and international debates on forest sustainability is now widely accepted. This has led to a fertile debate internationally, and has provided sharper focus for efforts to develop techniques to reduce harvesting impacts in tropical forests. The C&I debate has led to a general recognition that forest management needs to become increasingly effective in adapting to local ecological and socioeconomic conditions. Governments and international agreements may impose limits within which forest management should take place, but decisions on product optimization and the degree to which forest systems can be modified from their natural state will have to be taken by local stakeholders. What is defined as sustainable forestry will vary greatly over space and time as society's needs and perceptions evolve. This has led CIFOR to focus its attentions not on producing a definitive list of C&I but rather on producing a 'tool box' which stakeholder groups can use to develop appropriate indicators once they have established their own management objectives and performance criteria (Prabhu et al, 1996).

C&I are tools which can be used to conceptualize, evaluate and implement sustainable use of all kinds of forests. In this context, CIFOR's research has assumed the broadest possible definition of sustainability. Thus sustainability has been taken to mean maintaining or enhancing the contribution of forests to human wellbeing, both of present and future generations, without compromising their ecosystem integrity, that is their resilience, function and biological diversity. This definition of sustainable forest management has allowed CIFOR researchers the flexibility to develop operational definitions in terms of C&I for selected forest sites, and in turn has enabled similarities and differences between sites to be examined.

An important objective of the C&I research is to disaggregate the sustainability goals into key components that can be measured either quantitatively or qualitatively on the ground. Taken together with the objectives of relevance and cost effectiveness, C&I are potentially powerful tools that can provide a practicable and relatively objective assessment of the sustainability of forest management. C&I are, however, not simply tools for the assessment of the status of the forest and the quality of forest management. Research is beginning to reveal the relative values of alternative management options in achieving sustainability. This information is helping to identify management options that will be most cost-effective in achieving sustainable wood production while reducing social and environmental externalities, and thus is enabling the debate on management options to proceed on a more objective basis.

Within CIFOR's research on criteria and indicators of sustainability at the forest management unit level, the approach to deriving appropriate indicators both for biodiversity and genetic resources has been to focus on those processes that maintain adequate levels of diversity in sustainably functioning landscapes. The use of 'pressure-state-response' indicators is widely accepted as being valuable in assessing sustainability. Indicators of pressure, for example, areas deforested/logged/burnt/grazed, are relatively simple to implement. Similarly, indicators of state, for example, current forest extent, existence of corridors, are also fairly straightforward. However, indicators of response are potentially the most valuable, as they indicate the likely future direction of the system. They are also the most difficult to develop, but CIFOR's emphasis on indicators related to processes that maintain bio/genetic diversity offers an opportunity to assess both state and response indicators simultaneously. Examples of proposed indicators arising out of CIFOR's work are given in Figure 13.1.

The existence of a C&I toolbox and of a transparent process for establishing and reviewing C&I at the national and management unit level are also fundamental to progress in achieving 'adaptive forest management'. Increasingly, foresters must adapt their management practices to satisfy people's changing expectations and needs for forest goods and services. This requires sensitive indicators of changes in forest attributes and the ability to predict responses of the forest to modified management regimes. Appropriate C&I might, for example, help measure the impact on biodiversity, amenity or water yields of a decision to change the magnitude, frequency or nature of harvests. The utility of C&I lies in their efficient handling of information, notably the explicit identification of goals and definition of performance thresholds, targets and processes based on the most appropriate management practices for a given area. This makes them key tools for adaptive management. A fundamental aspect of the C&I research and development process has been the recognition that they must be transparent in application and acceptable to stakeholders, so that they form a broad and effective platform for building consensus.

Genetic level:

> Spatial/temporal changes in levels of genetic variation
> Directional change in allele/genotype frequencies
> Capacity for migration among populations
> Changes in reproductive system

Population/species level:

> Temporal changes in community guild structures
> Temporal changes in selected (indicator) taxa
> Changes in population structure of key taxa
> Changes in nutrient cycling/decomposition
> Changes in water quality and quantity

Habitat level:

> Temporal changes in habitat diversity

Landscape level:

> Changes in area of each vegetation type
> Changes in landscape patterns (connectivity, dominance, edges)

Figure 13.1 *Key indicators of forest sustainability*

Overcoming Obstacles to Sustainability

The international debate on C&I has promoted rigorous thought on what is meant by environmental sustainability and has encouraged people to explicitly identify the various elements that collectively comprise sustainability. One positive outcome is that the broader debate on forestry issues has become less polarized. Many of the activist NGOs that were totally opposed to logging of natural forests just a few years ago are shifting to the middle ground and conceding that environmental and amenity values of forests can be maintained when timber is harvested. Similarly, most industries and forest services are recognizing that it is necessary and reasonable to modify forest harvesting practices to reconcile differences with a civil society preoccupied with environmental conservation. The important question is then, what obstacles remain, and how can they be overcome?

Competition for land

There is intense debate at present about the problem of finding enough agricultural land to meet world food needs in the 21st century (Brown, 1995; CGIAR, 1996; Waggoner, 1996). Experts differ widely on their views as to how much new agricultural land will be needed to feed the world and on the extent to which yield increases on existing farmland will provide for the world's future food needs. Some believe that biotechnology will allow for a new 'green revolution'. The overall assessment of FAO is nonetheless that very significant areas of forest will have to be converted to agriculture in coming decades (Figure 13.2). A further problem is that the best land that sustains the highest yields is limited, and has mostly been dedicated to agriculture for some time. The key question is whether further increases in agricultural output can be achieved through increased yields from existing lands or whether they will depend on expansion onto new, lower-potential lands. Whatever the outcome, the competition for good land will be intense and forestry will inevitably be pushed into poorer areas. Fortunately, forest plantations in the tropics and subtropics can succeed on land of intermediate productivity, much of which is unsuited to permanent agriculture. There is a great deal of such land available in the tropics. In particular, there is great potential for expansion of plantation forestry in the countries of

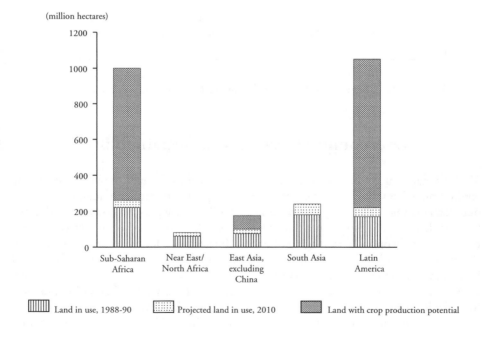

Source: Alexandratos (1995, pp162–163)

Figure 13.2 *Projected cropland expansion in developing countries by 2010*

the Guyana Shield and in the less densely populated areas of coastal Africa, from Gabon to Angola and Tanzania to Mozambique.

The observation that agriculture and plantation forestry may occupy different niches avoids two more fundamental questions: 'how much agriculture do we need?' and 'how much timber plantation do we need?' (eg Nilsson, 1996). Although important, these rather obvious questions may distract from the more elusive but compelling question, 'how much forest, natural or otherwise, do we need to maintain essential environmental services?' (eg Waggoner 1996). Currently, we have no answer to such questions, but we recognize them as important areas for research.

The increase in world demand for commodities derived from tree crops may greatly reinforce the competition for forest land. Very large areas in the humid tropics are currently being converted to oil palm plantations. Some analysts believe that this may lead to oversupply. But technologies for using palm oil to run internal combustion engines are improving and if this use becomes economically attractive the market for palm oil would be colossal. Any moves by governments to impose economic disincentives on the use of fossil fuels could make oil palm and many other renewable sources of energy and hydrocarbons much more attractive with consequent increases in demand for land upon which to cultivate them. In the face of such increased competition it would seem prudent to maximize the benefits that we can derive from forests and this must involve maximizing the range of products and services, although not necessarily on the same site.

Economic impediments

Technologies that were critical to achieving sustainability in the past may become less important under some scenarios for the world's forests in the future. In other papers (Byron and Ruiz-Perez, 1996; Sayer and Byron, 1996), we tried to anticipate how forestry practices may change in response to changing demands for forest products. We take a contrary view to that of the World Resources Institute (Bryant et al, 1997) that the major international forest issue is regulating frontier logging. We conclude that the frontier logging of relatively remote areas in the tropics, which has been a prominent feature of the timber industry in the late 20th century, may become less important in the future. We base this assessment on the fact that the technological requirement for large-diameter low-density hardwoods, common in many South-East Asian and African forests, will decline at the same time as the difficulties of exploiting the remaining stands of these forests increase. The technological problems which made the more highly diverse and higher wood-density timbers of Papua New Guinea and South America less attractive in the past have largely been solved. However, demand for the products of these forests will be limited by the cost of extraction. Once the forests in more accessible areas (close to roads, rivers and ports) have been exploited, any cost advantage that these forests might have had will be rapidly eroded. We expect them to become less able to compete with the outputs of the rapidly expanding plantation sector in the tropics and subtropics. In contrast to rising costs and declining quality of logs from natural forests, the volume and quality of plantation material will continue to improve while technological advances in plantation silviculture and wood processing continue to lower unit production costs. Some specialist products (eg durable timbers for marine applications) may circumvent this trend in the short term, but will in the longer term be displaced by

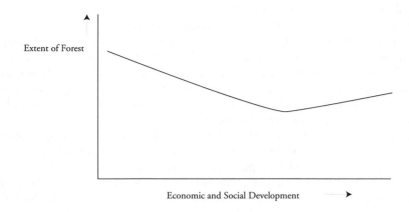

Figure 13.3 *Schematic representation of the inflexion in the forest area curve which occurs when thresholds of economic and social development are exceeded*

technologies such as plato®. Even if plato®, scrimber®, valwood® and other technologies fail commercially, the search continues for technologies to make high-value products out of cheaper and more readily available fibre.

At present about 15 per cent of the world's industrial wood production comes from about 25 million hectares of fast-growing plantations located in both tropical and warm temperate countries. High-yield forestry is a reality and the biological ability to shift most wood production to plantations exists and can be put into practice if prices of industrial wood rise high enough to justify it. In this context, the prime motivation for maintaining natural forests may be for amenity and environmental services in richer countries, and for non-timber forest products (NTFPs) and subsistence goods in poorer countries. However, it seems unlikely that logging of natural forest will disappear completely. Even in the most developed economies, the existence of forest industries, the cost of transporting timber products, and the desire to maintain employment in rural areas leads to continued logging of natural forests even when these are also valued highly for environmental services and amenity values. There will always be some areas of natural forests where the returns from logging are sufficiently high, and costs low enough, for them to be competitive with the plantation industry (Figure 13.3).

Maintaining the environmental and amenity values of natural forests will not be cheap, as significant investment and operating costs must be paid, either by the users directly, or by the public at large. In some cases, costs may be paid from timber receipts (on these or other forested lands, including plantations), but there is no inherent reason why timber revenues should pay for amenity uses. However, alternative sources of funding are often unavailable, and timber revenues represent the usual way to finance the maintenance of many other forest services.

Scenarios for natural forests

What will dictate the management of the remaining natural forests in the 21st century? In the past, forecasts of future timber needs have been an important consideration in determining the extent of national permanent forest estates. If timber is increasingly to come from plantations, one might expect that this would lead to a reappraisal of such policies, and to a reduction in the area of natural forest. However, few countries have responded in this way. Rather, they place renewed emphasis on the values of forest for amenity and environmental services at both the local and global levels. Thus several countries with little need to harvest timber from natural forests (eg New Zealand) strive to maintain extensive areas of forest in a near-natural condition. A consequence is a situation currently found in much of Europe where the mean annual increment of forests exceeds actual or predicted demand for timber (Solberg et al, 1996). There, forests are more valuable to society as aesthetic and amenity resources than as sources of industrial cellulose.

An interesting question is the extent to which similar trends will eventually emerge in the tropics. The answer to this question will depend on many factors originating outside the forest sector. Principal amongst these will be the rates of growth of both populations and economies, and the availability of alternative sources of employment for people who now derive their livelihoods from forests and forestry. It seems reasonable to assume that forest cover in countries whose economies are rapidly developing will in general follow a U- (or reverse-J) shaped curve, with forest areas declining until thresholds for industrial employment and the capacity to intensify agriculture and forestry are attained (Figure 13.3) (See for example Capistrano, 1990). The World Bank predicts that four of the most important tropical forest countries (Indonesia, Brazil, India and China) will join the world's largest economies early in the 21st century. Past trends suggest that attainment of this level of development may herald the bottom of the curve, and that decline in forest area may cease. However, one difference in these and many other tropical countries is that there may still be a large proportion of the population whose per capita income remains much lower than in those industrialized countries where forest areas are now expanding.

It is difficult to determine the point at which a reversal of the deforestation trend might occur. Such predictions are complicated by the fact that conditions within countries differ so greatly. However, it seems reasonable to anticipate that the tendency to deforest should cease when countries graduate into the 'middle-income category'. Important issues relate not only to where the minimum occurs, but also what its value is, what the recovery rate is, what the eventual equilibrium forest area may be, and how these points may be influenced by policy and technology. One of the critical questions for biodiversity is whether the forest area has to decline almost to zero before the response is triggered.

It seems possible that within a couple of decades, the 'tiger economies' of South-East Asia (and other countries with rapid economic growth) will attach a value to the biodiversity and watershed protection functions of some forests that will be higher than the potential revenues to be derived from these forests for timber production. This optimistic scenario would result in economics succeeding in achieving the objectives set out in many national forest policies where regulation has manifestly failed to do so. However, this approach examines only national trends, and it is much more difficult to assess global trends, as the timber trade may export the trend to deforest.

It is much more difficult to anticipate deforestation trends in Africa where economic growth is likely to be slower than in Asia and South America, and where most scenarios anticipate the persistence of a large population of poor subsistence farmers well into the 21st century.

Political and institutional constraints

Throughout history most forest products have been harvested for use locally. Although international trade in forest products is as old as history, it was only during the colonial period that tropical timber began to be traded in large volumes. The globalization of trade in the mid-1960s resulted in large quantities of forest products entering the international trade from remote, sparsely populated areas in the tropics. As populations in the South increase and economies further develop we may see a reversion to the situation where an increasing proportion of forest products are consumed closer to their point of production. This will almost certainly be reflected in an increase in the proportion of world trade in forest products being South–South as opposed to South–North. ITTO foresees an acceleration in the next few years of this trend towards increasing domestic log consumption in tropical countries (ITTO, 1996). If this is the case, and if the evolution of civil society in the expanding economies of the tropics parallels that in industrialized countries, we should see a situation where significant amounts of timber for local markets continue to be exploited from near-natural tropical forests.

How Technology Can Help

Increasing production

Wood productivity in natural forests ranges, in most cases, between 1 and 3m³/ha/yr. However, forest plantation productivity has increased spectacularly during the past few decades and continues to do so (Cossalter, 1996). Growth rates of 20m³/ha/yr are now routinely achieved with some tropical and subtropical fast-growing species while some industrial plantations of the tropics and subtropics exceed 30m³/ha/yr operationally. These gains have been realised in part through improved genotypes, but also through better silviculture and management of plantations. The implications for the land area required to service world timber demand are obvious (Figure 13.4).

Spectacular though they are, these technological achievements concerned a limited number of species grown on short rotations for the production of pulp-wood, chips, industrial charcoal and small-sized wood for other industrial uses. They represent a minute proportion of the tree species which can be planted in the tropics and subtropics. With few exceptions, timber species grown on medium and long rotations have not benefited from these technological advances. Long rotations have not appealed to commercial investors, and tree breeding and silvicultural research have focused largely on fast-growing species for industrial uses. Some reduction in rotation length can be achieved by intensive selection and breeding for rapid growth, and promising results in this regard have been obtained with species such as *Araucaria cunning-hamii*, *Araucaria hunsteinii*,

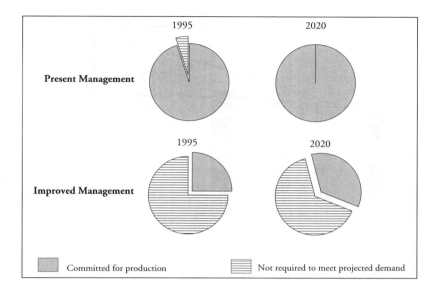

Note: Much of the world's forest is overutilized and undermanaged, so that average productivity (of industrial roundwood) is 1m³/ha/yr. This low productivity means that most of the world's forest need to be utilized to satisfy current demands for roundwood and fuelwood. A small increase in productivity in the more accessible and productive forests would free much forest from production obligations. The increase illustrated is equivalent to average Swedish levels (4.5m³) in temperate-boreal regions, and to existing well-managed standards (6m³) in the tropics.

Figure 13.4 *Scenarios for allocatio of forest lands*

Gmelina arborea, several Central American pines and *Paraserianthes falcataria.* However there are limits to the potential of tree breeding, and it seems unlikely that timber species could exceed 40cm dbh within 15 years. However, relatively modest increases in productivity will satisfy projected demands in the foreseeable future (Figure 13.5; Brooks et al, 1990).

Maintaining high yields from successive rotations of tree plantations while ensuring that the quality of the soil resource base and the resistance of the planting stock to pest and diseases is not declining is currently an important area of research for CIFOR.

Reducing impacts

There is a strong case to be made for further refinement of reduced-impact logging (RIL) technologies for application in the extensive areas of the tropics which we believe will be kept under forest primarily for environmental and amenity reasons but where timber production will continue to be a viable option. Most of the techniques embodied in RIL are not new, the innovation relates to the economics of using these technologies and to policies and incentives to promote their adoption. Increased use of these techniques should lead to a reduction in environmental impacts and greater productivity

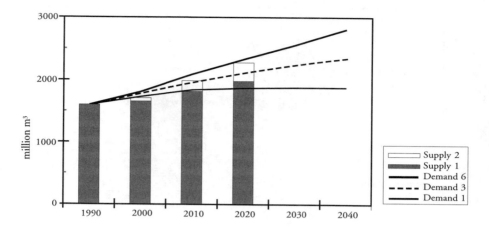

Figure 13.5 *Modest but realizable gains in productivity will satisfy projected global consumption in industrial roundwood (Data from Brooks et al, 1996)*

during future cycles. Thus, there is scope to promote the gains to be attained by reducing damage to trees and to the soil, by minimizing breakage and waste, and by reducing capital and operating costs of machinery.

Offering alternatives

Technologies outside the forest sector may also make a contribution towards sustainable forest management by reducing pressure for conversion of forestlands to other uses, by relieving demand for forest products, and by increasing options available to forest managers. Developments in the wood-processing industries may contribute to greater efficiency and less waste in both the factory and the forest. Technology may enable the utilization of a greater range of tree sizes and species, in turn creating new silvicultural opportunities for forest management. These may allow both commercial and environmental objectives to be realized with fewer compromises.

Reductions in per capita wood consumption need not imply a reduction in consumer satisfaction. Technology may reduce demand for wood without inconveniencing consumers by, for example:

- supplementing or replacing wood fibres (eg packaging) with agricultural by-products (eg straw);
- providing alternative energy sources (eg solar hot water, more efficient stoves);
- increasing re-use and recycling (more material, more cycles, greater recovery);
- improving office automation (better computer monitors, convenient 2-sided copying).

Our focus on technologies within the forestry sector does not diminish the importance of these and other initiatives in contributing to sustainable forest management by satisfying increasing consumer demands from smaller forest areas with less disturbance.

Providing information

Information is one of the keys to efficient forest management, and further progress towards sustainability may be achieved through better use of existing information technologies. Efficient management requires information that is relevant, timely, accurate and concise. In the past, forest managers have often relied on continuous forest inventory and similar monitoring systems to assess management impacts and sustainability. This is somewhat like driving a car with an opaque windscreen by monitoring the rear-vision mirror. Today, remote sensing enables early detection of many phenomena (fire, disease, logging etc) allowing glimpses from the side windows. However, forward-looking or adaptive management requires prediction systems (computer models as well as field trials) to give adequate information for efficient decision making. The C&I work now in progress will offer further guidance for efficient provision of key information. Conflict resolution is now recognized as an important part of forest planning, and decision-support systems should be able to evaluate alternatives to provide a factual support to the conflict resolution process. The technology to do this already exists, but needs to be implemented (Vanclay, 1993).

Some Strategic Questions

How much forest do we need for biodiversity conservation?

Market forces might be expected to be the main determinants of land allocation to agriculture and intensive forestry. However, experience over the past few decades has shown that markets are not efficient at achieving optimal allocation of land to maintain the environmental services of forests. The global environmental values of forests lie in their role in sequestering carbon and their role in maintaining biodiversity. Maximizing carbon sequestration requires maintaining or creating as much vegetation as possible, but it does not really matter what type of vegetation it is. Optimizing forests for biodiversity conservation presents more complex problems.

The nature conservation community is recognizing the lack of realism of its vision for biodiversity conservation which dated from the 1970s and 1980s. There is an increasing realization that it is unrealistic to aspire to extensive national parks in remote areas of developing countries from which all human use is excluded. Such parks are not viable in the current social and economic context of the countries concerned. At the same time, conservation biologists are demonstrating that the doomsday scenarios of biodiversity loss associated with macrolevel fragmentation of forests (eg the chapter by Myers in Meffe and Carroll, 1994) are not being realized in practice (Heywood and Stuart, 1992; Boyle and Sayer, 1995; Zuidema et al, 1997). Pulliam and Babbitt (1997) have shown that a remarkably small aggregate area of reserves would be adequate to conserve populations of most of North America's endangered species. Work currently underway at CIFOR suggests that this will also be true for tropical moist forests. The logical conclusion of this research is that the areas of old-growth forest that must be preserved to maintain most of the world's forest biodiversity is much lower than much popular conservation literature (eg Myers, 1989; Bryant et al, 1997) suggests. The challenge for

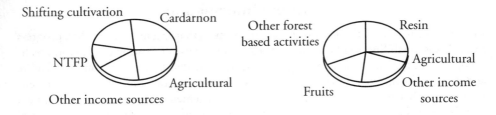

Figure 13.6 *Contribution of forest-based activities to total income*

biodiversity conservation is not to 'halt deforestation'; it is to secure a minimum set of strategically located old-growth reserves in representative areas with high diversity and high endemism.

Such a set of reserves is a good beginning, but is unlikely to maintain biodiversity unless the land surrounding these 'islands' also supports the conservation effort, while simultaneously meeting social and economic goals. Appropriate and effective policies and other incentives must be established to promote the role of biodiversity conservation outside protected areas, and this will be achieved only if adequate information is available. Not all areas within a landscape have equal conservation value, so technologies and methodologies to identify and manage critical parts of the landscape need to be developed. Similarly, since it is not possible to conserve all components of biodiversity, technologies to identify 'keystone' species or other components of special ecological value are needed. Finally, the need to 'manage' biodiversity means that information is required on the impacts of different human activities on components of biodiversity.

Recent reviews of the extent to which biodiversity can persist in logged forests (eg Sayer et al, 1996) further strengthens the case that most of the world's forest biodiversity can be retained without the draconian restrictions on all forest use that have often been the rallying cry of the conservation community (eg Myers, 1996). The greatest utility of forest systems will be achieved by an appropriate balance between relatively small areas of old-growth reserves and high-yielding plantations, and relatively large areas of other forest systems managed for multiple goods and services. The importance of these near-natural multiple-use areas is illustrated by studies by CIFOR and others that indicate that hundreds of millions of people depend on forest products for a significant component of their livelihoods. Many of these people are perceived by governments as being 'farmers' and the extent of their dependence on forests is often underestimated in taking decisions on land allocation (Figure 13.6). Many conservation organizations are now taking a serious interest in multiple-use conservation areas as a matrix containing a core set of representative biodiversity reserves. This is being increasingly seen as the best option in situations where the poverty of local populations renders people less interested in incurring the opportunity costs associated with the vast reserves sometimes advocated to preserve obscure species of animals and plants whose utility has yet to be demonstrated.

A number of conservation organizations are now embarking upon projects to try and achieve forest use in a way that is consistent with the maintenance of biodiversity values. One of the more difficult technological challenges facing us is to determine the extent to which environmentally motivated logging programmes are indeed consistent with retention of all or most biodiversity. In particular, we need to establish which elements of biodiversity will be lost even under very low-impact logging systems. We also need to have sensitive indicators to tell us when the disturbance caused by forestry activities is beginning to have negative impacts upon biodiversity and to signal clear warnings when we are approaching thresholds beyond which there may be inflections in the biodiversity retention curve.

Trees on non-forest land

It seems inevitable that, within the foreseeable future, the area of forest worldwide will continue to decrease, and that agricultural production systems (including forest plantations) will become more intensive. Despite these pressures, we want to maintain the essential ecosystem functions performed by trees and forests, and to minimize the disruption caused by the reduction and fragmentation of forest areas. There are several important questions relating to this realization:

- Is there a threshold at which ecosystem services and functions are disrupted, or is there a gradual diminution of services and functions?
- How can one tell when a system is approaching a threshold or critical level?
- Can planted trees on non-forest land help to maintain forest services and functions?
- What is the optimum configuration and spacing of trees and how does it depend on land use and distance to forest 'islands'?
- How should trees on non-forest land be managed to maximize ecosystem services and functions?
- How can this information be best communicated to policy makers, planners and the community at large, and how can they participate in the process?

What are the appropriate scales of management?

Optimizing forest outcomes in these complex situations will require that decisions taken about forests should be in a context which goes beyond the conventional management unit or stand. The forest management unit must be seen and managed in the context of the broader landscape. Product optimization is unlikely to be achieved by homogeneous treatments of extensive areas. Rather, it will require that different products are optimized in different parts of the landscape. This might lead us to a much finer balance in the allocation of land among its production, conservation and amenity functions. Some of the most exciting technologies which may emerge to address the needs of forest sustainability in the 21st century may be based upon models developed to predict outcomes of management interventions at a scale which is often referred to as the 'landscape scale'. This has been a subject of considerable research in the Pacific Northwest of the United States under the heading Forest Ecosystem Management. The Canadian Model Forest Program also seeks to reconcile the needs of multiple stakeholders in complex forest landscapes.

Conclusions

There is no 'silver bullet' or golden rule that can be applied universally to ensure sustainable forest management, in part because what is defined as sustainable will vary with time and space as society's expectations and aspirations change. However, a systematic approach to forest management offers an efficient way to progress towards sustainability. Such an approach should involve the usual planning cycle: formulation of objectives, preparation of a strategy, planning, implementing, monitoring and reappraisal. It also requires that managers understand their resource and their particular situation: this requires good resource assessment and decision-support systems. Social sustainability also requires that stakeholders participate in decisions, costs and benefits, and that effective procedures are used to resolve conflicts. Within an appropriate system, technical advances such as better machines and new implements may help to make a difference, but will not in themselves ensure sustainability. The important technologies for sustainable forestry are those relating to the broadest sense of the word; those that foster better communication between stakeholders and allow informed decisions spanning scales from the gene to the ecosystem. These technologies in themselves are not new; what is new is our realization of their importance, and our understanding of their effective deployment. This remains an important challenge for forest managers in their search for sustainability.

Acknowledgements

Tim Boyle, Christian Cossalter, Dennis Dykstra, Carol Pierce Colfer and Ravi Prabhu provided helpful comment on the draft manuscript.

References

1   Alexandratos, N. (ed) (1995) *World Agriculture: Towards 2010*. An FAO Study. FAO, Rome and John Wiley, Chichester, UK.
2   Boyle, T. J. B. and Sayer, J. A. (1995) 'Measuring, monitoring and conserving biodiversity in managed tropical forests'. *Commonwealth Forestry Review*, vol 74, pp20–25.
3   Brooks, D., Pajuoja, H., Peck, T. J., Solberg, B. and Wardle, P. A. (1996) 'Long-term trends and prospects in world supply and demand for wood', in Solberg, B. (ed) *Long-Term Trends and Prospects in World Supply and Demand for Wood and Implications for Sustainable Forest Management*. EFI Research Report 6, pp75–106.
4   Brown, L. R., (1995) *Who Will Feed China: Wake-up Call For A Small Planet*. Worldwatch Institute, Environmental Alert Series no 6, Washington, DC.
5   Bryant, D., Nielsen, D. and Tangley, L. (1997) *The Last Frontier Forests: Ecosystems and Economies on the Edge*. World Resources Institute, Washington, DC.
6   Byron, N. and Ruiz-Perez, M. (1996) 'What future for the tropical moist forest 25 years hence?' *Commonwealth Forestry Review*, vol 75, pp124–129.
7   Capistrano, A. D. (1990) *Macro-economic Influences on Tropical Forest Depletion, A Cross-Country Analysis*. PhD dissertation, University of Florida, Gainesville.

8   CGIAR (1996) *The CGIAR Research Agenda: Facing the Poverty Challenge.* Mid-term Meeting, 20–24 May 1996, Jakarta. Document SDR/TAC:IAR/96/6.1. CGIAR, Washington, DC.

9   Cossalter, C. (1996) 'Addressing constraints to the development of plantation forestry in the tropics: A role for tree improvement', in Dieters, M. J., Matheson, A. C., Nikles, D. G., Harwood, C. E. and Walker, S. M. (eds) *Tree Improvement for Sustainable Tropical Forestry.* Proc. QFRI-IUFRO Conf., Caloundra, Queensland, Australia. 27 October–1 November 1996, pp282–287.

10   ECE (1996) *Main Findings and Implications of the Study of European Timber Trends and Prospects: Into the 21st Century (ETTS V).* Economic Commission for Europe TIM/R.274 FO:EFC/96/10 15 July.

11   Heywood, V. H. and Stuart, S. N. (1992) 'Species extinction in tropical forests', in Whitmore, T. C. and Sayer, J. A. (eds) *Tropical Deforestation and Species Extinction.* Chapman and Hall, London, pp91–117.

12   ITTO (1996) *Annual Review and Assessment of the World Tropical Timber Situation 1995.* Document GI-7/95. International Tropical Timber Organization, Yokohama.

13   Meffe, G. K. and Carroll, C. R. (eds) (1994) *Principles of Conservation Biology.* Sinauer Associates, Sunderland, Massachussets.

14   Myers, N. (1989) *Deforestation Rates in Tropical Forests and Their Climatic Implications.* Friends of the Earth, London.

15   Myers, N. (1996) 'The world's forests: problems and potentials'. *Environmental Conservation*, vol 23, pp156–168.

16   Nilsson, S. (1996) *Do We Have Enough Forests?* IUFRO/IIASA Occasional Paper 5. Sopron.

17   Prabhu, R., Colfer, C. J. P., Venkateswarlu, P., Tan, L. C., Soekmadi, R. and Wollenberg, L. (1996) *Testing Criteria and Indicators for the Sustainable Management of Forests: Phase 1.* Final Report. CIFOR, Bogor.

18   Pulliam, H. R. and Babbitt, B. (1997) 'Science and the protection of endangered species'. *Science*, vol 275, pp499–550.

19   Sayer, J. A. and Byron, R. N. (1996) 'Technological advance and the conservation of resources'. *International Journal of Sustainable Development and World Ecology*, vol 3, pp43–53.

20   Sayer, J. A., Zuidema, P. A. and Rijks, M. H. (1995) 'Managing for biodiversity in humid tropical forests'. *Commonwealth Forestry Review*, vol 74, pp282–287.

21   Solberg, B., Brooks, D., Pajuoja, H., Peck, T. J., Wardle, P. A. (1996) *Long-term Trends and Prospects in World Supply and Demand for Wood and Implications for Sustainable Forest Management: A Synthesis.* EFI, Joensuu.

22   Vanclay, J. K. (1993) 'Inventory and yield prediction for natural forest management', in Thwaites, R. N. and Schaumberg, B. J. (eds) *Australasian Forestry and the Global Environment*, Proc. Institute of Foresters of Australia 15th Biennial Conf., Alexandra Headland, Qld, 19–24 Sept. 1993. IFAInc, Brisbane, pp163–169.

23   Waggoner, P. E. (1996) 'How much land can ten billion people spare for nature?' *Daedalus*, vol 125, no 3 (Summer 1996), pp73–93.

24   Zuidema, P., Sayer, J. A. and Dijkman, W. (1997) 'Forest fragmentation and biodiversity: The case for intermediate-sized conservation areas'. *Environmental Conservation*, vol 23, pp290–297.

14

Sustainable Forestry in the Tropics: Panacea or Folly?

David Pearce, Francis E. Putz, Jerome K. Vanclay

Introduction

Concern about the depletion of the world's forests has led to many international calls for radical efforts to reduce deforestation, including the United Nations Intergovernmental Forum on Forests of the UN Commission on Sustainable Development (1999), and the World Commission on Forests and Sustainable Development (1999). This concern reflects an appreciation of the ecological and economic functions of forests: as providers of timber and many non-timber products, as the habitat for much of the world's biological diversity, and as regulators of local, regional and global environments. These functions are at risk. Most forest clearance is in tropical areas where there are great demands for agricultural land. In temperate and boreal areas, the pressures from logging are more important. But in all areas, forestry itself has an important role to play, both as a partial cause of deforestation, and, if practised wisely, as a potential source of salvation for at least some of the world's forests. In terms of its causal role, forestry tends to open up primary forest areas, enabling colonists to move in, using roads forged by the timber companies. In some parts of the world, forests are converted not to agriculture but to biomass plantations of fast-growing trees or to other agro-industries based on tree-crop plantations such as palm oil and rubber. Here, the primary agent is not the peasant, but the richer elements of local and international society.

How, then, can the world's forests be used more wisely? It is this complex question that we seek to answer in this chapter. Some argue for outright protection, caricatured perhaps in the phrase 'fence and forget'. Others argue for 'sustainable forest management', and still others for systems of forest management that rely on acceptance of an initial period of exploitation of valuable species followed by outright protection. The issue, then, is the optimal use of forested land, which begs the question of what is meant by 'optimal'.[1] This is addressed shortly.

Forested land may be retained as forest or it may be converted to non-forest uses such as agriculture, grazing, urban expansion or industrial tree-crops. The first question, then, is under what circumstances is it better to convert forestland to non-forest uses, and when not?

Note: Reprinted from *Forest Ecology and Management*, vol 172, Pearce, D., Putz, F. E. and Vanclay, J. K., 'Sustainable forestry in the tropics: Panacea or folly?', pp229–247, Copyright © (2003), with permission from Elsevier

If it can be shown that forestland is best retained as forest – where 'best' needs to defined – the further issue arises of what kind of forestry is to be preferred. Here, the issue is clouded in terminological confusion because the words used in reference to forestry have come to mean different things to different people. But in order to focus the debate, we choose three archetypes familiar in the literature: conventional logging (CL), sustainable timber management (STM) and sustainable forest management (SFM). We adopt this terminology not because we think it is free from misinterpretation, but because the literature on the role of forestry in deforestation has adopted it, making it extremely difficult to elicit the lessons from that literature without using that language. We devote some time to explaining what we mean by the terms below and why, in an ideal world, we would prefer a different terminology. For the moment, we take CL to be more short-term in focus, less concerned with forest regeneration through management, and often lacking in government control. We take STM to be a forest management system that aims for sustained timber yields. We take SFM to be a system of forest management that aims for sustained yields of multiple products and services from the forest.

There have been recent challenges to the idea that conservation is best served through sustainable timber or forest management (Vincent, 1992; Kishor and Constantino, 1993; Howard et al, 1996; Rice et al, 1997; Bowles et al, 1998). One argument is that conservation can only be served by outright protection (Bowles et al, 1998), that is while SFM has the potential for protection, it is inferior to outright protection. Another view is that conservation might be better served by an initial period of well-managed logging followed by protection (Rice et al, 1998a, b; Cannon et al, 1998). Against this, it is argued that outright protection has a limited chance of being successful in face of the high costs of protection, pressures to use forests for profit, and human population growth (but see Bruner et al, 2001; Vanclay, 2001). It follows that in many places sustainable forest management offers the only chance of maintaining forests and biodiversity (Whitmore, 1999).

The Theme

In terms of the debate about the optimal use of forested lands, the existing literature tends to focus on the financial returns from STM, SFM and CL and on physical descriptions of the comparative ecological impact profiles of these forms of forest management. The focus on financial returns is justified insofar as actual forest use is determined by relative profits. The focus on ecological impact profiles is relevant for a full economic assessment.

An economic assessment makes several potentially major adjustments to a financial analysis: the existing financial costs and benefits are adjusted to 'shadow values' to reflect the true opportunity cost of the resources involved; and environmental and social costs and benefits ('externalities') at the national and global level are included.

1   The first modification adjusts financial costs and benefits to reflect shadow prices. A shadow price, say the price of labour or the exchange rate, differs from a financial price in that it reflects the true opportunity cost of the resources in question. For example, the ruling wage rate would be used in a financial analysis, but if the labour

employed would otherwise be unemployed, the shadow wage rate will tend to be closer to zero (since the wage in alternative employment is, effectively, zero). A shadow exchange rate is the rate that would prevail if trade were free and open, rather than, as is often the case, managed through trade quotas and tariffs. It is important to understand that this shift to shadow pricing alters the stakeholder perspective. Whereas financial costs and benefits are relevant to the logger or concessionaire, shadow-priced costs and benefits are relevant from the standpoint of the forest-owning nation.

2   The second modification adds in all environmental and social consequences that affect the wellbeing of anyone within the nation. Thus, if indigenous peoples are adversely affected by the forest development, their wellbeing must be counted in any economic study. Similarly, if logging contributes to soil erosion, flooding or loss of biodiversity, an economic analysis would attempt to take these into account. It is important to understand that ecological functions of forests have a parallel in economics – all ecological functions are economic functions.

3   The third modification constitutes a global analysis and would additionally include the gains and losses of people outside the country in which the forest is located. Thus, if individuals in another country experience a loss of wellbeing from knowing that deforestation, perhaps indirectly caused through logging, is taking place, that loss of wellbeing has also to be accounted for. This loss of wellbeing is relevant regardless of whether it emanates from a loss of any use value (eg ecotourism or carbon storage) or any loss of non-use value, that is wellbeing unassociated with any direct use of the forest.

It is not always appreciated that economic analysis is potentially quite different to financial analysis. An economic analysis might, for example, sanction an activity that is wholly unprofitable from a financial standpoint.

In this chapter we try to build up the overall picture as best we can, by beginning with financial analysis and extending it to full global economic assessment. It is important to understand that a global economic assessment is useful only insofar as it demonstrates the superiority of one form of forest land use over another, that is it shows, in an accounting sense, which land use is 'best'. Unless there are corresponding cash flows that capture those values, the exercise remains interesting but unlikely to change the way forests are treated. For example, SFM may turn out to be financially inferior to CL, but this does not mean that SFM is to be dismissed. An economic analysis that includes all social and environmental externalities can guide us to the relevant conclusion. Now suppose the economic analysis demonstrates that SFM is superior to CL, regardless of the contrary finding for the financial analysis. Since the financial costs and benefits 'drive' the land-use decision, SFM can only be introduced if forest land use is regulated in some way, or if forest land users are compensated for the difference between the profits under CL and the profits under SFM. In this chapter, we are concerned mainly with the demonstration phase. The broad issue of designing compensatory and 'capture' incentives is not addressed in detail here except indirectly by reference to the literature (eg Pearce, 1996). Capture mechanisms include carbon trading, watershed protection payments and forest certification.

The Terminology of Forest Management

As noted above, the terminology used in the debate over the appropriate use of forested land has become confusing. 'Logging' rightly refers to the process of harvesting timber from a forest, even though timber harvesting may appear more 'value-neutral'. Logging, however, can be a legitimate part of good or 'wise' forest management. In the same way, the literature now refers to 'conventional logging' as if it too characterizes undesirable treatment of forest. To a forester, however, conventional logging might characterize standard forest management practice as opposed to unconventional means of timber extraction, for example the use of helicopters. But some conventional logging is not practised wisely, so that it becomes possible to contrast poor management practice with, say, reduced impact logging (RIL). To a forester, RIL would simply be a feature of any good management system. In what follows we maintain the more popular image of conventional logging as meaning use of the forest for short-term timber supplies, aimed solely at short-term profits and without significant government control. Management plans may or may not exist for this type of timber harvesting, and, while the potential is there for switching to a more long-term sustained timber yield, it is more likely that forest degradation, forest loss and conversion to non-forest use will follow.

In terms of timber volumes, conventional timber harvesting may be sustainable or unsustainable. But the connotation of conventional logging is that it is often unsustainable, that is not focused on long-term timber supplies. Sustainable timber management (STM) therefore arises when a forest management plan is fully implemented for timber and focuses on long-term non-declining flows of timber.

Sustainable forest management embraces the view of the forest as yielding many different products and providing many different ecological services. Sustainable forest management will therefore produce an array of products and services that may or may not include timber. SFM therefore relates to the multiple use of the forest. To a forester, the term 'management' could relate to the management of resources, inventorying and yield calculation, and to silvicultural practice (eg climber cutting), so that, on some definitions, SFM is already embodied in good practice timber harvesting. Again, then, the terminology of SFM is not ideal but is retained here to convey the idea of multi-product uses and with a focus on the longer term. 'Protection' is also ambiguous. For ardent environmentalists, it almost certainly means the maintenance of the structure and composition of the forest without change caused by human intervention. For others, it risks being confused with 'conservation' which is the proper management of the forest for the sustained yield of some product(s), service(s) or some combination of products and services. Again, foresters would argue that they have always been in the business of conservation in this sense (as indicated by the traditional term of 'Conservator'). 'Protection' also conjures up the image of leaving a forest totally alone when, in practice, some management of fire and invasive exotic species is still likely to be required to conserve structure and composition.

The Meaning of 'Optimal' Forest Land Use

What it is 'optimal' depends on the viewpoint of the economic agent making the decision to convert forestland or to adopt a particular forestry regime. Hence we need to identify the stakeholders. From the standpoint of most logging companies a forest exists to *be* logged. In principle, a forest will not be logged if it is unprofitable to do so, although it is perfectly feasible that loggers may log land at a loss if subsidies prevail. If logging is profitable, the regime used will generally be that which maximizes profits, subject to any regulations on harvesting that may be in place. In addition to obligations under relevant legislation, logging companies may voluntarily attenuate maximum profits if they feel some obligation towards the environment. In some cases, it would appear that the most profitable regime is not employed. RIL might, for example, lower costs but not be used, perhaps because of ignorance, fear that investments in training may not be recouped, or other reasons that remain unclear (Putz et al, 2000).

Forest owners include national and regional governments, local communities, indigenous groups, individuals and companies not engaged in logging. They may have several motives relating to the forest, as a supply of products including timber, as an environmental resource, as a stewardship objective, and so on.

From the standpoint of forest dwellers, a forest exists to provide an array of ecological and economic functions ranging from timber, fuelwood and wildmeat, to protection against floods.

From the standpoint of poor agricultural colonists, forests exist for the land they provide for timber, crops and livestock, mainly the latter two. Such colonists may nonetheless have complex mixes of motives. For instance, Mourato (1999) showed that slash-and-burn cultivators in Peru exhibit a strong concern for the conservation values of their forests. Nor is the image of colonists being poor always correct; they may also be wealthy individuals or companies looking to exploit subsidies, to speculate on land values, or to anticipate conversion to lucrative plantations and agro-industries.

From the standpoint of the conservationist, the forest exists to provide ecological functions, amenity and the provision of wellbeing to forest peoples. Motivations vary and may range from a desire to make direct use of the forest (eg ecotourism) to a concern for the intrinsic rights of biodiversity to exist.

From the standpoint of national governments, forests may serve any of the above functions depending on the extent to which governments have the wellbeing of particular stakeholders at heart. They may prefer:

- logging to preservation because it provides employment and tax income;
- conversion because it may yield higher returns than timber production;
- conversion and colonization because of the need to establish political frontiers and accommodate migrants;
- conservation because of a concern for vulnerable indigenous groups, because of the potential income from sustainable uses of the forest, because there are financial inducements to conserve, or because they have forest protection as a general social objective.

From the standpoint of the world as a whole there may be a preference to log forests for the valuable timber they contain, or to conserve forests for their local and global ecological functions. In the latter case, there may be a preference to conserve forests because of their role in providing biodiversity and in storing carbon that would otherwise be released to the atmosphere, contributing to global warming. The relevant agents reflecting 'world' interests include some activities by bilateral and multinational aid agencies, various UN organizations not directly involved with aid, the Global Environment Facility, and international NGOs.

The viewpoints of the stakeholders necessarily conflict; otherwise, there would be no problem. Different uses of forested land are often incompatible. The following options are there for 'resolving' these conflicts of values.

1   To impose a given use on all stakeholders, regardless of the differences of viewpoint. This 'solution' is potentially unstable because one or more stakeholders will lose from the imposed land use. Hence, they have a continuing incentive to break the agreement by securing its subsequent rejection, or by 'illegally' using the forest for their own purposes.
2   To find an agreement which adopts a given use of the forestland and in which those who lose are compensated in some way for forgoing their use of the land. On this solution, all stakeholders are (ideally) better off with the agreed land use than they were without it.

The second solution suggests the meaning of 'optimal', it is a land use which is judged socially the most beneficial overall, but in which those who lose from the land use are compensated for their losses. This definition accords with elementary game theory (for a brief introduction see Perman et al, 1999).

In practice, actual compensation for losses is often not feasible. At the very least then, forestland should be allocated to those uses that maximize, as far as possible, the aggregate social value of the forestland. If gains and losses are measured in monetary terms, then this requirement is equivalent to a standard cost–benefit analysis approach and the compensation is potential rather than actual. Put another way, gainers have to be able to compensate losers and still have net gains to show (Pearce, 1986). In practice, while we may not be able to assign economic values to all functions, the cost–benefit approach is a reasonable way of organizing the framework for analysis.

A convenient language to describe stakeholders' interests is that of 'private' and 'social' gains and losses. 'Private' refers to the private interests of the stakeholder, this is what benefits him or her. 'Social' takes the wider, social perspective and the jurisdiction may be local, national, regional or global. In theory, governments or global agencies should take the social standpoint, but it is well known that this is not always the case. Both perspectives are relevant to determining 'optimal' forest land use because adopting a social perspective without acknowledging that some stakeholders' private interests may be compromised will, as noted above, be potentially unstable.

The Meaning of Sustainable Forest Management

Optimal land use is not necessarily the same thing as sustainable land use (Toman and Ashton, 1994; Pearce, 1999). The current debate about forestry is partly about sustainability, that is about making use of forested land in a sustainable fashion. In large part, converting forest land to agriculture – in tropical regions – is not sustainable because soils may not have the capacity to sustain agricultural activity indefinitely, although the case for perennials is far stronger in this respect. There are, however, many cases in which tropical forests have been converted and successful agriculture follows (see eg Schneider, 1995). In other cases, provided suitable fallow periods prevail, forested land can be converted for short-term agricultural use and then be cleared again and reused after the fallow period. This 'cycle' of agriculture and forest regeneration is also potentially sustainable.

As noted above, the debate mixes two different aspects of sustainability:

1 STM in which the focus is on a sustained yield of timber over long time periods;
2 SFM in which the focus is on the many products and services of the forest sustained over long periods of time.

While it is generally thought that STM is consistent with SFM, it is at least open to argument that STM may not maintain all the components of biodiversity, including ecosystem and landscape-level structures and processes (Putz et al, 2000). Thus, it is important to distinguish STM from SFM. In sustainable timber management, timber is extracted with regard to a continuous future supply of wood through investment in regeneration. STM also tends to be associated with minimization of damage to residual stands (see Vanclay, 1996a, b), possible investment in finding uses for currently non-merchantable species, and accelerated growth of merchantable species in managed stands.

Many writers have offered definitions of SFM (eg Dickinson et al, 1996; Reid and Rice, 1997; IFF, 1999), but perhaps the most complete definition comes from Bruenig (1996):

> ... management should aim at forest structures which keep the rainforest ecosystems as robust, elastic, versatile, adaptable, resistant, resilient and tolerant as possible; canopy openings should be kept within the limits of natural gap formation; stand and soil damage must be minimised; felling cycles must be sufficiently long and tree marking so designed that a selection forestry canopy structure and a self regulating stand table are maintained without, or with very little, silvicultural manipulation; production of timber should aim for high quality and versatility. . . The basic principle is to mimic nature as closely as possible to make profitable use of the natural ecosystem dynamics and adaptability, and reduce costs and risks. . .

The Context for the Analysis

The context for this chapter is one where the starting point is an existing forest. Thus, we do not discuss the optimal use of bare or degraded land. Additionally, although land uses that involve conversion of the forest are relevant to the analysis, they are incidental to the main focus which is on the appropriate form of forestry. It may or may not be the case that land conversion is socially or privately 'better' than a given forestry use. Finally, outright protection is also relevant to the analysis but is not the main focus. Like conversion, protection has to be part of the analysis because forest practices are capable of being a precursor to a protected area classification. Conventional logging, for example, is frequently a precursor to agricultural colonization, and hence to land conversion. One current argument is that protection might follow on from an initial period of logging.

The Private Interests of the Logging Companies

The empirical evidence

From the logging firm's point of view, the use of the forest will be dictated by the option providing the largest private financial rate of return. Empirical evidence relating to these rates of return is limited (but see Pearce et al, 2000). A particular problem concerns the fact that STM and SFM systems have rarely been in place long enough for an accurate picture on financial returns to be obtained (Dickinson et al, 1996). Furthermore, it is seldom to the logging firm's advantage to reveal the actual costs of logging operations. Double accounting is therefore commonplace when record books are accessible at all. Those few studies that compare STM/SFM and CL therefore tend to be based on financial model simulations. Although there is additional evidence on the rate of return to STM/SFM, authors often do not attempt a comparison with CL, contenting themselves with a demonstration that STM/SFM is profitable per se. A recent review of some 30 studies (Pearce et al, 2000) revealed that:

1   STM is potentially profitable at 'reasonable' discount rates of 5–10 per cent in real terms);
2   STM is almost systematically less profitable than 'liquidation' forestry and other forms of conventional logging.

These conclusions echo those already reached by other commentators, for example Bach and Gram (1996). Nonetheless, while this inequality of profitability explains the widespread preference of loggers for CL, it does not justify it. The reason for this, as indicated previously, is that the financial cost–benefit calculation of the logger is not the same as that for society generally and certainly not for the world as a whole.

Factors that could favour the financial profitability of STM

Advocates of STM have drawn attention to four main factors that might increase the financial return to STM relative to CL: discount rates, future price increases, incremental growth rates for timber volume and property rights.

The discount rate

One 'price' of potential importance for the SFM versus CL debate is the discount rate. For the financial perspective the relevant discount rate is that of the logger or concessionaire. For the national perspective, the relevant rate is the social discount rate. The two rates can be expected to differ, with the social rate being below the private rate (Pearce, 1986). Surprisingly little is known about discount rates in developing countries, but they tend to be so high that it is difficult to justify even the most conventional of development projects (Pearce et al, 2000). In the context of forestry, they are effectively fatal for any investment with a long-term focus.

While high personal discount rates appear to be the norm on the basis of the empirical evidence, it is important to stress that few studies exist that adopt rigorous methodologies for estimating those discount rates. Additionally, some poor communities do manage timber production on a non-exploitative basis, suggesting that communal discount rates may be markedly less than purely personal rates (see Pinedo-Vazquez and Rabelo, 1999).

Timber prices

If timber prices are expected to appreciate then there is some benefit to curtailing the harvest now in favour of the future (effectively, price increases can be thought of as a deduction from the discount rate). But future timber prices are unlikely to grow rapidly. Some of the high price increases simulated in the STM studies, for example Howard and Valerio (1996), are based on protected forest industries. World prices are a better guide. Moreover, world price ('border prices') would be the relevant magnitude in economic, as opposed to financial, studies. Rice et al (1998a, b) use 2 per cent per annum growth in real prices which may, however, be an exaggeration of future price increases. Work at Resources for the Future (Sohngen et al, 1998) suggests baseline price growth rates well under 1 per cent per annum for the next 100 years. Even with a high demand scenario, price increases barely exceed 1 per cent per annum over the next 60 years. These results are consistent with other estimates of long-term trends (eg see Brooks et al, 1996). Overall, it seems unlikely that future price increases will confer significant advantages on STM relative to CL. More generally, as long as timber is 'abundant', stumpage prices will be low, making STM financially vulnerable (Southgate, 1998).

Timber volume growth rates

Timber volume growth rates have an effect similar to real relative price increases. If growth rates are faster under STM then the difference can be regarded as the equivalent of a reduction in the discount rate. Rice et al (1998a, b) suggest 2 per cent per annum as an average for growth relevant to STM, that is 2 per cent per annum in volume of the stand. However, quotations of percentage growth rates are not very meaningful. First, growth depends on stand condition such that growth is an inverse 'U' shaped curve relative to stand condition (low at poor states and low again if there is high density and crowding, although the latter is rare in managed natural forests) – see Vanclay (1994a). Second, large trees may have small percentage growth rates but substantial incremental yields in terms of cubic metres of wood. Third, account has to be taken of damage in CL to residual trees (10–40cm dbh) that will form the next crop in polycyclic management operations. Surviving damaged trees grow slowly and will not contribute to the next commercial crop due to stem defects. Most of the growth benefit from RIL and STM derives from higher stocking and fewer weed-dominated areas, such as vine blankets. Overall, STM could easily result in volume increments of commercial species that are two to four times higher than after CL.

Property rights

It is widely argued that insecure or short-term property rights encourage CL, so that longer-term rights would encourage a switch to SFM, or at least STM. Rice et al (1998a, b) accept the argument in principle but argue that longer-term concessions would not alter the underlying financial costs and benefits, favouring CL. But tenure might also encourage the training of staff, or at least enable better choices to be made. Boscolo and Vincent (1998) simulate the effects of longer-term concessions on the timing of harvests in Malaysia and show that, in their model, it would make no difference. Generally, it does not pay to leave trees standing. Probably, the best way to accommodate the concession length issue is to regard longer concessions as an enabling device for STM, which, without additional incentives such as performance bonds, will nonetheless be unlikely to lead to STM.

Efficiency and best practice

One remaining issue concerns the extent to which the STM systems reflect 'best practice'. That is, what is being observed may not be the most efficient form of STM, making cost comparisons misleading. Various inefficiencies need to be addressed: (a) ratio of usable wood to cut wood; (b) ratio of cut wood to wood at the mill gate; and (c) ratio of mill output to mill input. Do (a) and (b) vary by type of management regime? How do loggers respond to efficiency improvements and why do not they adopt them automatically? Part of the problem seems to be that critics are damning STM as it has been practised. Defenders of STM and SFM are saying that past systems were poorly implemented, for example by excessive canopy opening, inappropriate log transport, inappropriate machinery, lack of training and planning etc. In other words, we need to know what would constitute an efficient system.

Efficient STM or SFM may be difficult to attain, since efficiency implies an agreed objective and the reality is that no such consensus exists. Objectives might, for example, embrace recreating the original stand, regenerating harvested species, conserving 'habitat' trees, minimizing gaps, and so on. The reality seems to be that SFM is itself an 'elastic' concept, making the criticism that it has not been practised when it should have been, difficult to evaluate.

Valorizing non-commercial timber species

It has been argued that 'valorizing' non-commercial timber stock will provide less incentive to use regimes that damage residual stands (Buschbacher, 1990). Rice et al (1998a, b) argue, on the other hand, that expanding the commercial range of species simply results in all species being exploited. The reality is complex and depends on prices for such species, supervision, and what the management regime is trying to achieve. In Queensland, the valorization of species took pressure off the best and most accessible areas, and good supervision of the operations meant that the additional species harvested did not result in degradation of the residual stand (Vanclay, 1996a, b). But it is hard to generalize on the valorization issue.

Concessionaires and confidence

Sustainable forest practices require confidence in the long-term future. Many of the world's most valuable forests are in areas where there is a potential for rapid political change and the insecurity that it engenders. The effect of lack of security will be to reduce confidence in the future and hence to favour short-run exploitation. The effect is the same as that of a high discount rate.

The National Perspective

The national perspective on land-use options differs from the financial analysis outlined above in several ways. First, logging may not be the 'best' use of the forestland, so that there may be broader options. Second, attention now has to be paid to the sequencing of land use. One possibility is that logging is followed by either protection or conversion (and to which we might add abandonment). Which one follows will determine the flow of costs and benefits to the nation. Third, financial gains and losses should no longer be as relevant as economic gains and losses, that is financial flows should be shadow priced. Fourth, all forest values other than timber values become relevant. Fifth, if global values exist and can be converted to resource flows of benefit to the nation, then they too become relevant. We address each modification in turn.

Widening the options to all forest land uses

In practice, there will be combinations of uses that should also be considered, for example agro-forestry, clearance for plantation forestry, oil palm and so on But the principle is the same; whereas a private logger might reasonably consider only the financial costs and benefits of logging options, the nation state should consider all options and evaluate their economic values.

Sequencing of land use

Surprisingly, little attention appears to be paid in the CL versus SFM debate to the sequencing issue. It is well known that logging opens up frontier land that may then be colonized by agriculturists. Sequencing may happen the other way round. The land may initially be cleared for agriculture and taking timber may be an ancillary operation aimed at helping recover the costs of conversion. Repetto (1990) is of the view that most deforestation arises from the initial action of logging which creates access to hitherto inaccessible forestland. In their review of econometric studies of deforestation Kaimowitz and Angelsen (1998) find that deforestation is higher when land is accessible, when timber and agricultural prices are high (encouraging logging and conversion), when rural wages are low, and when there are opportunities for long-distance trade. Of these factors, several – accessibility, timber prices and trade potential – all relate to logging. Southgate (1998) documents cases where most road construction in forested areas has come from loggers, encouraging conversion to crop-land and pasture. Low stumpage prices might contribute to conversion because even the modest rates of return that might be expected from agriculture compare favourably to forestry at low stumpage prices.

This picture contrasts with the one suggested by some analysts who argue that CL could be followed by outright protection (Rice et al, 1998a, b; Cannon et al, 1998). Their argument is that loggers are unlikely to return to the same area after the initial logging phase because:

1   they do not practise sustainable forestry, thus making future timber stands unlikely to be of commercial interest, and because
2   they have high discount rates, also making future yields unattractive.

The land is therefore potentially available for protection without the further threat of logging. The argument has some force, but there are several problems.

First, the picture of loggers entering an area only once is often not accurate. Loggers often return 5–10 years after the first harvest to harvest trees that have become commercially valuable because of changes in transport infrastructure, in milling methods and market potential. Re-entry loggers may also be different people to the first-time entrant: smaller operators with lower operating costs acting as agents for small mills working in formerly high-graded areas. The 'protect after logging' scenario thus has to relate to a context in which the threat of subsequent logging intervention remains.

Second, even if the threat of further logging is removed, the threat of colonization for non-timber purposes is not removed and, indeed, is, ex hypothesi, more likely. Colonization may be for subsistence agriculture but also for agro-industrial use such as oil palm. Additionally, the 'logging followed by protection' view is based on what could happen rather than on what actually has happened in the past. However, there is an argument that says protection is easier after CL because land prices fall once loggers withdraw. Land can then be bought cheaply for protection purposes.

Third, CL may degrade the forest to such an extent that what remains is not worth protecting. Substantial canopy opening increases susceptibility to fires and weed infestation. A spiral of degradation may become irreversible.

Conventional logging followed by protection has indeed occurred, for example in Queensland, in the Noel Kemp Mercado carbon offset project in Bolivia, and in parts of Africa. But how far these examples arise simply because there were funds available for subsequent protection and because there was low population pressure on available land is unclear. Indeed, it is hard to envisage many circumstances in which there will be limited pressures to convert the land. The choice is not between 'logging followed by protection' and SFM/STM, but between some form of continuous forestry and land conversion. The potential for logging to be followed by protection may therefore be smaller than some of the literature acknowledges, although Reid and Rice (1997) accept that STM will be best suited to areas where there are strong pressures to colonize the forest for conversion. And, of course, protection is costly and in no way avoids the need for continued management.

One other form of sequencing has strong arguments in its favour. Here, the aim would be to meet timber demand from plantations, leaving natural forests to be managed mostly for non-timber purposes. The sequence is then to afforest rapidly to establish plantations on degraded lands, accepting some loss of natural forest in the interim, then to protect the remaining natural forest whilst meeting demand for timber from plantations. Hunter (1998) describes this situation for New Zealand. The appeal of such a policy in New Zealand, and in the US where some environmental groups are calling for cessation of all logging on National Forests, carries the suggestion that this option may be best for countries with relatively high per capita incomes. A country like Malaysia, where there are already extensive lowland plantations, could adopt such a policy, but lower income countries are still likely to face formidable pressures for conversion of natural forest, so that finding profitable forest management systems is still important. The New Zealand and Malaysia examples raise the interesting issue of whether there might be an 'environmental Kuznets curve' (EKC) for forest protection. EKCs exist when growth in income per capita first results in environmental degradation and then, after some turning point, a reduction in degradation. Such curves have been found for air pollution. In the current case, the hypothesis would be that countries opt for more forest protection and more plantations as incomes pass a certain point.

Again, it needs to be recalled that protection is not costless. Not only are there continuing management costs, but there are capital costs of fencing and establishing management institutions. To these must be added the value of the protected land in uses forgone.

Shadow pricing private costs and benefits

The analyses of costs and benefits to loggers have typically all been in terms of financial rather than economic flows. Exceptions are the World Bank studies reported by Grut (1990).

Allowing for non-timber values

The recognition of non-timber values (NTVs) alters the focus of analysis from STM to SFM. In economic language, the relevant measure is now total economic value (TEV) from the different possible land uses. TEV comprises use and non-use values and both are capable of expression in monetary terms by estimating the relevant willingness to pay (WTP) for those functions (Pearce, 1993, 1996). The basic argument is that, even if STM is 'worse' than CL in financial terms, if the WTP for the incremental non-timber benefits of SFM exceeds the financial deficit, SFM will be preferred from a national perspective. More formally,

$$\text{SFM} > \text{CL}, \quad \text{if } \text{WTP}_{\text{NTV}} > \Pi_{\text{CL}} - \Pi_{\text{STM}} \tag{1}$$

where Π is profit.

Contrary to the contention of Bawa and Seidler (1998) that there is no experience of timber regimes that integrate NTPs into the management system, Romero (1999) found that RIL had no effect on the available biomass of epiphytic bryophytes that are harvested and sold by local people in Costa Rica. Similarly, Salick (1995) found that non-timber forest products (NTFPs) and natural forest management for timber were compatible in Nicaragua. In small-scale natural forests, integration of NTFPs with timber is more the rule than the exception (see Pinedo-Vazquez and Rabelo, 1999). However, Putz et al (2001) note that management invariably involves favouring some wildlife species over others.

The evidence on environmental impacts logging regimes

The presumption in inequality (1) above is that environmental benefits under STM/SFM exceed those under CL. This has been challenged by Rice et al (1997, 1998a, b). They argue that the physical effects of CL on the forest were relatively mild for the case they studied in lowland Bolivia. However, that case relates to extremely low-intensity mahogany harvesting. and it would be hard to envisage that it would also hold for the much more intensive harvesting characteristic of the eastern Amazon or the dipterocarp forest of South-East Asia. Rice et al (1998a, b) and Reid and Rice (1997) argue that STM/SFM can be just as destructive of the total forest as CL, a view supported by Bawa and Seidler (1998). Uncontrolled logging, it is argued, may be comparatively benign, especially on flat lands that are logged when soils are dry and where there is a low density of commercially exploitable species. However, under less favourable conditions, CL may be very destructive.

STM can also be destructive, especially in the short term, if it involves major canopy clearance in an effort to encourage regeneration of light-demanding species, but much depends here on the management system in place. The Malayan Uniform System (Manokaran, 1998; Ashton and Peters, 1999) removed most of the canopy, but successfully regenerated the basal area of primary forest. Manokaran (1998) described the effects of selective logging during the 1950s at Pasoh, Malaysia, and observed that by the mid-1990s, the regenerated forest was well stocked with commercially valuable dipterocarp species, unlike the less successful selective management system practised in the hill dipterocarp forests. In contrast, the selective logging practised in Queensland caused relatively little disturbance and was successful in providing a viable harvest and maintaining the forest (Vanclay, 1994b, 1996b).

Biodiversity

Part of the problem with the discussion on environmental impacts concerns the characterization of the environmental objectives of forest management. Whereas Rice et al (1997) talk primarily in terms of the maintenance of biodiversity, others would argue that what matters is the maintenance of ecosystem functions and ecological processes. Some of the concern about avoiding management-induced changes in tropical forest composition is based on the concept of 'climax communities', and the idea that tropical forests are unchanging and lacking in resilience. Given the long history of substantial human impacts on tropical forests and the large areas of tropical forest currently under some sort of silvicultural management by people, the incompatibility of management for timber with biodiversity maintenance seems largely unfounded. For example, many researchers have reported that stands with mahogany almost certainly suffered severe natural disturbances in the past, or regenerated in agricultural clearings that were abandoned centuries ago (eg Lamb, 1966). Despite this evidence for ecological resilience, much more research is needed on how to mitigate the deleterious impacts of forest management operations. For example, how should untreated reserves be distributed within managed forests, and is it preferable, from a biodiversity maintenance perspective, to concentrate or disperse logging operations?

Unfortunately, the issue is far from clear cut, even at the level of biodiversity (Putz et al, 2001). First, all forest management is likely to reduce biodiversity relative to pre-intervention conditions, or at least to change species composition. The very term 'management' means that something is being done to the forest that would not have happened without intervention. Second, the issue then becomes one of comparing the biodiversity profile – that is the nature and extent of diversity – under the different management regimes, say SFM and CL. Thus. Rice et al (1998b) make a biodiversity conservation argument in favour of introducing protection after logging has taken place. They suggest that even CL followed by protection is superior to STM/SFM because the former halts the process of forest domestication. But this is a double-edged argument, for CL could just as easily result in the loss of species dependent on large canopy openings. Certainly, CL stands are especially prone to weed infestations due to excessive damage and lack of pre- or post-logging treatments to discourage weeds and encourage potential crop trees. Even though it does not constitute STM or SFM, RIL would be a substantial step in the right direction. Thus, pre-felling vine cutting can substantially reduce post-logging incidence of serious vine infestations, and also reduce logging damage where vines tie together tree crowns. RIL thus constitutes a major step forward.

Additionally, there has been a tendency to generalize from single localities and case studies. The work of Rice et al relies heavily on observations in the dry forests of northern Bolivia where, if sustainable timber exploitation for the species currently harvested is to be practised, substantial canopy manipulation is required in order to provide the conditions for the regeneration of light-demanding species, especially *Swietenia macrophylla*. The structure and composition of the forest would thus have to change in a substantial way to avoid loss of the canopy species currently most valuable. Even for these forests, it is unclear that species loss due to management need be significant. These forests have survived major disturbances in the past, and proper zoning of the forest should ensure biodiversity is retained.

Not all commercially valuable timber tree species require substantial canopy disturbance for regeneration. In South-East Asian dipterocarp forests, for example, minimizing damage to residual stands is important to protect the abundant advanced regeneration of commercial species present before harvesting and to reduce the likelihood of vine infestation once the canopy is opened. The valuable canopy dominants in the forests of much of the Guyana Shield area in northern South America are also negatively affected by the substantial canopy openings required to regenerate mahogany in northern Bolivia.

Thus, it is not possible to say that STM systems necessarily result in less biodiversity than CL systems. Without careful management, they may do so. If one of the aims is to conserve biodiversity, then management systems should be capable of achieving that aim. In doing so, it may well be the case that the financial returns to STM fall since there will often be a trade-off between biodiversity objectives and maximum financial return. Given that biodiversity conservation figures prominently in Forest Stewardship Council certification, any price or marketing gains from certification will also reduce the profit differential between CL and SFM.

An interesting study by Stephens (1999) in South Australian *Eucalyptus regnans* forests shows that conservation strategies for Leadbeater's possum are best effected through forest reserves, with a close second-best approach being modified harvesting regimes. 'Best' and 'better' in this context are measured by the survival probability of Leadbeater's possum (the 'effectiveness' of the strategy) and the forgone timber values (the cost). The modified harvesting regimes involved the retention of small habitat patches within harvested areas, suggesting, again, that careful logging is consistent with biodiversity protection.

Carbon

Forest-based carbon sequestration, both by conserving carbon already stored in forests and by sequestering additional carbon by stimulating tree growth, has become an important focus for foresters because of the role that forest carbon release plays in accelerating global warming. This role was given official recognition in the 1997 Kyoto Protocol to the 1992 Framework Convention on Climate Change. Countries can benefit in terms of achieving their emission reduction targets by engaging in 'carbon trades' whereby they offset some of their own emission targets by reducing net emissions in another country. The issue here is how different management regimes affect carbon storage.

Silvicultural practices affect carbon balances. Thus, if vines are cut and left to decompose, carbon is released in the short-term from the decomposition process, but can be offset later by the faster growth of the trees freed of vines. SFM, practised properly, involves minimum site preparation and extended rotations, whereas, clear-cutting results in loss of necromass and soil organic matter which may well not be offset by subsequent sequestration through faster tree growth. In general, then, silvicultural practice benefits the carbon balance. Exceptions might include savannah woodlands (for example the pine forests of Central America) that are invaded by shade-tolerant and less fire-resistant trees. Restoring the open stands will result in carbon releases that may not be compensated for by increasing sequestration in the remaining trees. Dixon (1997) conducted a survey of experience with silvicultural practices in 40 countries, showing carbon sequestration benefits ranging from 5 t of carbon/ha per annum on average at high latitudes to 401 t of carbon/ha per annum at low latitudes. If these practices were applied to the 600 million ha of land suitable for forest management in the nations surveyed, conservation of carbon would be of the order of 100–300 mt of carbon per annum over a 50-year period. The economic efficiency of such measures is unlikely to compare favourably to the costs of reducing greenhouse gas emissions through energy-related schemes from plantations, but Dixon suggests that the practices in question could sequester carbon at some US$3/t of carbon, lower than the US$30/t of carbon often quoted.

In comparative simulations of carbon sequestration in Malaysian forests logged by trained crews following RIL guidelines and with CNV methods, Pinard and Putz (1996), see also Boscolo and Vincent, 1998) showed that the use of RIL techniques conserved carbon in the harvested stands and resulted in substantially greater rates of post-harvesting sequestration due to higher stocking of potential crop trees and fewer problems with vines and other weeds.

If forests were managed according to their mixed timber and carbon value, as opposed to their timber value alone, significant changes would need to occur in management practices. Standing volumes and rotation ages would need to be increased, and there would need to be substantial increases in investment in silriculture. Leakage issues would loom large under such a scenario, given the large and growing international demand for paper and other wood fibre products. With long rotations, natural forest succession would occur which, in some cases, would reduce the attractiveness of forests for early successional species of plants and animals, as well as for recreation, so some of the 'carbon gains' would be offset by other factors. Despite these misgivings, the clear implication is that the attachment of economic values to carbon via trading and/or meeting domestic emission reduction targets could substantially favour better managed forests. However, a more sceptical view is offered by Smith et al (1999b).

Numerous 'carbon offset' projects exist either in actual or simulated form. Details of the various deals and their costs per tonne of carbon reduced can be found in Pearce et al (1998). Few of the deals relate to forest management. The offset projects can be analysed to elicit the average costs per tonne of carbon equivalent reduced. However, since the deals have not generally been developed on a cost-efficient basis (selecting the cheapest options first), such an analysis is not particularly helpful. Estimates of the prices at which carbon will trade under the 'flexibility mechanisms' in the Kyoto Protocol do, however, suggest that around US$10/t of carbon is likely to be a mean price, provided substantial trading takes place and it is likely that carbon may trade at between US$5 and US$15/t of carbon. The importance of these figures can readily be seen. For forest conservation as a whole, compared to conversion, forests may secure a carbon 'credit' of US$750–2250/ha on this basis (assuming 50–150 t of carbon is emitted by conversion). For the comparison between SFM and CL, of course, the gains would be far more modest. If the figures of Dixon (1997) are adopted, incremental gains over CL may range from US$5–41/t of carbon, or US$25–615/ha.

However, great care needs to be taken in multiplying carbon storage or sequestration estimates by unit money values for traded carbon. The procedure is correct if the price of traded carbon is an equilibrium price, that is one that equates supply and demand for traded carbon. That price will be sensitive to the number of deals done. If there are vast 'offers' of carbon from countries seeking to capitalize on the carbon value of sustainable forestry, it may have the effect of forcing the price of carbon down, thus reducing the economic returns from carbon conservation. These 'system-wide' effects have been stressed in a number of studies, for example Sohngen et al (1998).

WTP for certified timber

There are two approaches to securing an economic measure of the value of non-timber forest values. The first rests on what people are willing to pay for timber certified as coming from 'sustainable' forests, the idea being that 'sustainable' timber embodies all NTVs in some premium over the world price for timber. The second attempts directly to estimate the economic value of the various forest functions independently.

If consumers of wood products are willing to pay a premium to guarantee that the forests supplying the products are sustainably managed, then that WTP can be thought of as an approximation of the WTP in the equation above. Similarly, if forest companies are willing to adopt sustainable practices in order to secure some marketing gain from certification, then the costs of certification provide a lower bound of the additional value of certification.

Certification schemes exist to guarantee the sustainability of various forests, akin to 'eco-labelling' of various products. Certification costs are around US$0.2–1.7/ha for developing countries (Crossley and Points, 1998) and US$0.23/ha for assessment and US$0.2–0.07/ha per annum for licensing and auditing in the US (Mater et al, 1999). Accordingly, any WTP above this level of cost represents the 'net premium' for SFM. Evidence of WTP for certified timber by premium consumers is equivocal, and there is no clear evidence that consumers are willing to pay more (Barbier et al, 1994; Forsyth et al, 1999).

Crossley and Points (1998) suggest that certified products are securing premia of 5–15 per cent in some cases, but that the real benefits of certification for industry lie in securing greater market share and longer-term contracts. There is some evidence that companies gaining certification secure higher company value, that is the value of certification shows up in share prices on the stock exchanges.

If we take the range 5–15 per cent as a likely measure of premium, the argument in Gullison (1995) and Rice et al (1997) is that this is far from sufficient to compensate for the additional profitability of CL over SFM. But their argument is suspect, as a typical ratio of financial profits for CL relative to STM would be, say 1.5. For STM to become competitive it is not necessary for prices to rise by 50 per cent. A hypothetical numerical example shows that the price premium need only be a fraction of the difference in profits between CL and STM. If costs of CL are 75 and those for STM are 100, but both face the same market price of 150, then the ratio is 75/50 =1.5. The net price premium that will make profits equal is given by

$$p^* = \left(\frac{C_{ST} - C_{CL}}{P} \right) \tag{2}$$

where P is the common log price. In the numerical example, $p^* = 17$ per cent (25/150), which is considerably different to 50 per cent. It is true that this premium is gross of certification costs, so that the true price premium required for parity between profits in the two regimes is higher than 17 per cent.

WTP for non-timber products and services

A more direct approach is to seek the WTP that people express for the particular non-timber products and services from forests. We review what is known about these WTPs, but note that they need to be applied to the differential flow of environmental products and services from SFM as opposed to forests that are just logged. The total values would be relevant if all such services were lost, as they might be from forest conversion. The differential values are relevant for the CL/SFM comparison. The importance of this distinction is that we know a reasonable amount about the economic value of forest services, but, apparently, little about the differential flows of those services according to different forest management regimes.

Environmental economists have made great progress in eliciting economic values for forest products and services. Recent surveys include Godoy et al (1993), Pearce and

Table 14.1 *Indicative values of forestland for non-timber uses*

Category of use	US$/ha per annum
Non-timber extractive values	50
Non-extractive	
Recreation	5–10
Ecological	30
Carbon	600–4400
Non-use	2–27

Moran (1994), Lampietti and Dixon (1995), Southgate (1996), Chomitz and Kumari (1996), Pearce (1998) and Pearce et al (1999). There are of course substantial difficulties in reaching general conclusions from WTP studies, primarily because appropriate guidelines for carrying out such studies, such as those set out in Godoy et al (1993) and Godoy and Lubowski (1992) have not been followed. The result has been a mixture of legitimate and illegitimate valuation procedures. The types of mistake made have included generalization from studies of a small area of forest to wider areas, with little regard for (a) the fact that the area in question will not be typical of the whole forest area simply because of variations in distance to market, and (b) ignoring the fact that, in a hypothetical world where the whole forest was exploited for non-timber products, the prices, and hence the profitability, of non-timber production would fall. Another methodological issue is the extent to which values are based on maximum sustainable yield or on actual harvests, which are often very much less, that is, the values that emerge that is are sensitive to what is assumed about the management regime in place. Godoy et al (1993) also point out that some studies value the stock and some the flow, the former being an interesting measure for wealth accounting but of little value when comparing competing land-use values. Studies also vary as to whether they report gross revenues or revenues net of labour and other costs. Finally, little account has been taken in many of the studies of the extent to which the relevant non-timber activity is itself sustainable, so that what is being compared may well be two non-sustainable land-use options.

It is clearly hazardous to try to find some kind of consensus from the estimates of Pearce et al (2000). Table 14.1 reports indicative annual values, but care needs to be taken in generalizing any of these numbers (Southgate et al, 1996). Whichever way the analysis is done, the major role of carbon values is revealed. Should, for some reason, global warming not remain a serious issue of concern, then tropical forests might be found to have measured environmental value of around US$100/ha, far from enough to justify outright protection on economic grounds.

One of the few studies that attempts to place an economic value on the differential flows of goods and services from CL and SFM is Kumari (1995, 1996) for the peat swamp forests of North Selangor in Malaysia. The analysis relates to the differential benefits of moving from an existing unsustainable timber management system, based on Malaysian Stateland forest practice, to sustained forest management. The sustainable systems are markedly better than the unsustainable system, showing a 13 per cent improvement on CL. Although there is a decline in timber revenues, non-timber (mainly agro-hydrological and rattan) and global benefits (carbon and conservation of endangered species) increase more than enough to offset the losses. It is of course open to question whether this analysis is typical for forests generally. Importantly, the global benefits will not accrue to forest owners or concessionaires without institutional change which compensates them for storing carbon in the biomass. Existing carbon trades, and those expected under the Kyoto Protocol, thus become extremely important. In essence, Kumari's analysis provides the 'demonstration' phase, but not the capture phase of the analysis, as she herself notes. Nonetheless, Kumari's approach, which is essentially a traditional incremental cost–benefit analysis, is the correct one and is likely to be the only one that can capture all the relevant changes in the multiple outputs of different forest management regimes.

Smith et al (1999a) (see also Mourato, 1999) conducted a contingent valuation study of slash-and-burn farmers in the Ucayali region of the Peruvian Amazon. They sought the farmers' willingness to accept (WTA) compensation simultaneously to conserve part of the forest outright and to switch to multistrata agroforestry for the rest of the forest. Farmers were first asked their WTA compensation (from electric utilities engaged in carbon offset projects) for the combined preservation/agroforestry package, and were then asked by how much they would discount the stated WTA to secure access to the environmental services of the conserved part of the land. The difference between the two WTAs measures gives a WTP measure (ie in terms of forgone compensation) for the environmental services. The results were, in average terms:

- US$218 compensation required for forgoing 1 ha of forest that would be converted to outright preservation;
- US$138 compensation for forgoing 1 ha of forest that would be converted to agroforestry;
- US$67 WTP for environmental services for forest preservation;
- US$41 WTP for environmental services for agroforestry.

The difference between the two WTP estimates reveals that farmers are aware of the difference in the value of environmental services from agroforestry compared to preservation. The study is significant in that it elicits (a) the compensation farmers would need to fill the 'gap' between the returns to agroforestry and the returns to slash-and-burn agriculture, and (b) the WTP of farmers for forest services of which they are well aware. The gap between agroforestry and slash-and-burn returns is a perceived one and is highly influenced by farmers' discount rates, that is returns are higher to agroforestry over a long period but lower if the time horizon is limited to a few years. Interestingly, the stated WTA to switch to agroforestry compared very well with the difference in the annual stream of returns when viewed from this short-term perspective, indicating that farmers were well aware of the returns from different systems.

The relevance of the study lies mainly in the information provided about farmers' perceptions about the forest, but insofar as agroforestry can be thought of as a form of SFM, it suggests that the social returns to SFM are higher than slash-and-burn provided the 'rest of the world' is WTP for carbon services from the forest.

An Economic Model of Sustainable Forestry

One way of encapsulating the previous discussion is to place it in the context of an economic model of forested land use. Sustainable forestry can only exist if returns to it exceed those of alternative uses of the land and exceed the costs of management, including the costs of preventing entry by colonists. Hyde (1999) suggested that these conditions will tend not to prevail in the earlier stages of development so that, generally, the poorer the nation the less likely it is for sustainable forestry to emerge as a viable land-use option. But the analysis also suggests that if non-market values are high, there could be substantial returns from managing forests on a sustainable basis. The additional condition, of course, is that the returns must be capable of 'capture' by the forest owner, whether it is a private individual or the state.

Our analysis suggests that most non-market values will not be high enough to change the underlying and somewhat pessimistic conclusion of Hyde's approach, that is that sustainable forestry is potentially viable but risky in areas where development is still at the early stages. The fairly clear exception, however, is carbon, and the few case studies that are relevant seem to confirm that carbon values from carbon trading could produce the situation where a significant sustainable forest sector emerges based on non-market values. Additionally forest certification offers considerable promise, depending on the extent to which stated WTP is confirmed by actual WTP.

Conclusions

In this chapter we have reviewed the available literature in an attempt to cast some light on the issue of the type of forest management regime that is best suited to the overall aim of slowing the rate of loss of the world's forests and biodiversity. The traditional argument that 'sustainable forestry' is the most preferred option has recently come under criticism from those who argue that it is neither profitable nor necessarily environmentally preferable to conventional logging.

Finding general conclusions is complex, not least because the terminology in the literature is confusing and often value-laden, even regarding 'logging' as an undesirable activity per se. While not entirely satisfactory, we adopt the language of 'conventional logging', 'STM', 'SFM' and 'protection'. The essential differences are that sustainable systems pay more regard to longer-term outcomes than do conventional systems, and that sustainable systems are likely to involve far more regulatory supervision than conventional systems.

The model adopted proceeds from a comparison of financial rates of return to differing forest management systems, through to economic rates of return, and from there to wider rate-of-return concepts that include non-market values, for example biodiversity conservation and carbon storage. There are then at least three stakeholder perspectives on these rates of return: those of the logger, those of the nation and those of the world as a whole. In reality, there are many different divisions of interest, from those of illicit forest users, indigenous peoples, enforced migrants and so on. The rough benchmark is that forested land should be used for the highest social value use, that is the use that maximizes the broad concept of rate of return indicated above. This notion requires that any values not embodied in the marketplace be 'captured' through various incentive mechanisms. Those mechanisms – such as debt-for-nature swaps, carbon offsets, green image investments and so on – are not discussed here. The idea of maximizing a rate of return also does not embrace the crucially vital question of the distribution of gains and losses. While important, these concerns lie outside the scope of this paper.

A survey of financial rates of return reveals that sustainable systems appear capable of earning returns in excess of some 'modest' discount rate (5 per cent and in some cases 10 per cent), but cannot compete financially with other systems. Given the nature of the management process for sustainability, this is not unexpected and conforms with the critics' view of sustainable management.

Are there any factors that mitigate this inequality? We looked at the various arguments that have arisen, from improving concessionaire property rights, to the future of timber prices, and the valorization of non-commercial species. None appears to give sustainable timber management any financial edge over conventional systems. All have some role to play, but it is not significant. The evidence on discount rates tends to reinforce the critics' arguments. Recent studies suggest that discount rates in poor countries are very high indeed, so high that few investments of any kind, let alone in forestry, would seem to be economically justified. But if the focus is on sustainable and unsustainable forest systems, then high discount rates simply reinforce the initial preference for conventional systems based on rapid liquidation of the timber and other resources without regard for future harvests or other impacts.

Some of the critics of SFM as a conservation strategy have argued that logging should be permitted so as to get an initial period of damage over and done with, leaving the way open for protection. The argument rests on the assumptions that future logging threats are minimized, and that land is cheaper once the loggers have gone. There are doubts about this argument: logging once may simply lead to subsequent visits from the same or other loggers; loggers open the way for colonists; and damage may be so extensive that protection ceases to have much of a conservation justification. Sustainable systems are also open to threat since they too open up forests to colonists. The extent to which they will avoid being converted by colonists rests heavily on their financial viability, which, as we have seen, may not be very strong.

The focus therefore shifts to non-market values. Are these higher under sustainable management than conventional logging? Our review suggests that they are. It is true that sustainable forestry loses some environmental benefits relative to the pre-intervention period. But there is no necessary link between sustainable forestry and environmental damage. Part of the problem arises from extrapolating from limited experience, for example with mahogany, to tropical forestry in general. This said, research on 'biodiversity impact profiles' is not strong enough yet to reach firm conclusions. For carbon storage, the picture seems fairly clearly in favour of sustainable systems.

The final stage of the analysis asks if these NTVs are sufficiently important that they outweigh the financial deficit of sustainable forestry when compared to conventional logging. While there is only a limited number of studies to guide us in this respect, those that exist seem fairly uniform in finding that the non-market benefits of sustainable systems are significant. All tend to acknowledge that timber yields are less on a comparative basis but that NTVs more than offset the relatively lower yield. The role of carbon is highlighted because a survey of non-market values suggests that carbon values dominate the non-market values overall, a conclusion echoed in the case studies reviewed here. Other indirect evidence is also marshalled, for example there appears to be a marked WTP by consumers for natural regeneration of forests and for sustainable managed systems.

The prospects for sustainable forest management is low in the early stages of development, and increases as the values attached to the forest and its services rise over time. Extended to include carbon and biodiversity values, it is arguable that the potential for sustainable forestry is far greater, even in the early stages of development, than might be thought.

Acknowledgements

This chapter is based on a report prepared for the Natural Resources International, UK and UK Department for International Development (CSERGE Working Paper GEC 99-15). The Centre for Social and Economic Research on the Global Environment (CSERGE) is a designated research centre of the UK Economic and Social Research Centre (ESRC). We are deeply indebted to Marcus Robbins and Trevor Abell of NRI for comments on earlier drafts, and to Bill Hyde of Virginia Polytechnic Institute for valuable discussions.

Note

1   We use the term forested land rather than forestland to make it clear that we are dealing with land that still has forest on it, rather than land which has a potential to be used for forest in one form or another.

References

1   Ashton, M. S. and Peters, C. (1999) 'Even-aged silviculture in mixed tropical forests with special reference to Asia: Lessons learned and myths perpetuated'. *Forestry*, vol 97, no 1, pp14–19.
2   Bach, C. and Gram, S. (1996) 'The tropical timber triangle'. *Ambio*, vol 25, no 3, pp166–170.
3   Barbier, E., Burgess, J., Bishop, J. and Aylward, B. (1994) *The Economics of the Tropical Timber Trade*. Earthscan, London.
4   Bawa, K. and Seidler, R. (1998) 'Natural forest management and conservation of bio-diversity in tropical forests'. *Conservation Biology*, vol 12, pp46–55.
5   Boscolo, M. and Vincent, J. (1998) *Promoting Better Logging Practices in Tropical Forests: A Simulation Analysis of Alternative Regulations*. Development Discussion Paper 652. Harvard Institute for International Development, September.
6   Bowles. I., Rice, R., Mittermeier, R. and da Fonseca, A. (1998) 'Logging and tropical forest conservation'. *Science*, vol 280, pp1899–1900.
7   Brooks, D., Pajuoja, H., Solberg, B., Wardle, P. and Peack, T. (1996) *Long-term Trends and Prospects in World Supply and Demand for Wood*. Paper to the UN Intergovernmental Panel

on Forests. European Forest Institute, Joensuu and Norwegian Forest Research Institute, Australia.

8   Bruenig, E. F. (1996) *Conservation and Management of Tropical Rainforests*. CAB International, Wallingford.

9   Bruner, A. G., Gullison, R. E., Rice, R. E. and da Fonseca, G. A. B. (2001) 'Effectiveness of parks in protecting tropical biodiversity'. *Science*, vol 291, pp125–128.

10 Buschbacher, R. (1990) 'Natural forest management in the humid tropics: ecological, social, and economic considerations'. *Ambio*, vol 19, no 5, pp253–258.

11 Cannon, J., Gullison, R. and Rice, R. (1998) 'Conservation and logging in tropical forests'. *Conservation International*, for the World Bank, Washington, DC.

12 Chomitz, K. and Kumari, K. (1996) *The Domestic benefits of Tropical Forests: a Critical Review*. Global Environmental Change (GEC) Series, pp96–109. Centre for Social and Economic Research on the Global Environment (CSERGE), University College London and University of East Anglia.

13 Crossley, R. and Points, J. (1998) *Investing in Tomorrow's Forests: Profitability and Sustainability in the Forest Products Industry*. WWF, Godalming.

14 Dickinson, M., Dickinson, J. and Putz, F. (1996) 'Natural forest management as a conservation tool in the tropics: divergent views on possibilities and alternatives'. *Commonwealth Forestry Review*, vol 75, pp309–315.

15 Dixon, R. (1997) 'Silvicultural options to conserve and sequester carbon in forest systems: Preliminary economic assessment'. *Critical Review Environmental Science Technology*, vol 27 (special), pp139–149.

16 Godoy, R. and Lubowski, R. (1992) 'Guidelines for the economic valuation of non-timber tropical-forest products'. *Current Anthropology*, vol 33, no 4, pp423–433.

17 Godoy, R., Lubowski, R. and Markandya, A. (1993) 'A method for the economic valuation of non-timber tropical forest products'. *Ecology and Botany*, vol 47, no 3, pp220–233.

18 Grut, M. (1990) *Economics of Managing the African Rainforest*. World Bank, Washington, DC.

19 Gullison, R. (1995) *Conservation of Tropical Forests Through the Sustainable Production of Forest Products*. PhD Thesis. Princeton University.

20 Howard, A., Rice, R. and Gullison, R. (1996) 'Simulated financial returns and selected environmental impacts from four alternative silvicultural prescriptions applied to the neotropics: A case study of the Chimanes Forest Bolivia'. *Forest Ecology and Management*, vol 89, pp43–57.

21 Howard, A. and Valerio, J. (1996) 'Financial returns from sustainable forest management and selected agricultural land use options in Costa Rica'. *Forest Ecology and Management*, vol 81, pp35–49.

22 Hunter, I. (1998) 'Multiple-use forest management systems', in See, L., May, D., Gauld, I. and Bishop, J. (eds) *Conservation Management and Development of Forest Resources*. Forest Research Institute, Kuala Lumpur, pp169–177.

23 Hyde, W. (1999) *Patterns of Forest Development*. Lecture given at IIED, London, April 1999.

24 Intergovernmental Forum on Forests (1999) *Valuation of Forest Goods and Services: Report of the Secretary General*. United Nations Commission on Sustainable Development, New York. Report E/CN.17/IFF/1999.12.

25 Kaimowitz, D. and Angelsen, A. (1998) *Economic Models of Deforestation: A Review*. CIFOR, Bogor, Indonesia.

26 Kumari, K. (1995) *An Environmental and Economic Assessment of Forest management Options: A Case Study of Malaysia*. Environment Department Working Paper, Series 026. World Bank, Washington, DC.

27 Kumari, K. (1996) 'Sustainable forest management: myth or reality. Exploring the prospects for Malaysia'. *Ambio*, vol 25, no 7, pp459–467.

28 Lampietti, J. and Dixon, J. (1995) *To See the Forest for the Trees: a Guide to non-Timber Forest Benefits*. World Bank, Washington, DC.

29 Manokaran, N. (1998) 'Effect 34 years later, of selective logging in the lowland dipterocarp forest of Pasoh. Peninsular Malaysia and implications of present day logging in the hill forests', in See, L., May, D., Gauld. I. and Bishop, J. (eds) *Conservation, Management and Development of Forest Resources*. Forest Research Institute, Kuala Lumpur, pp41–60.

30 Mourato, S. (1999) 'Do slash and burn farmers value forest preservation?' Evidence from the Peruvian Amazon, Chapter 2 of PhD thesis, *Essays in Contingent Valuation*. University College London.

31 Pearce, D. W. (1986) *Cost-benefit Analysis*. Macmillan, Basingstoke.

32 Pearce, D. W. (1993) *Economic Values and the Natural World*. Earthscan, London.

33 Pearce, D. W. (1996) 'Global environmental value and the tropical forests: Demonstration and capture', in Adamowicz, W., Boxall, P., Luckett, M., Phillips, W. and White, W. (eds) *Forestry, Economics and the Environment*. CAB International, Wallingford.

34 Pearce, D. W. (1998) 'Can non-market values save the tropical forest?' in Goldsmith, B. (ed) *Tropical Rain Forest: a Wider Perspective*. Chapman & Hall, London, pp255–268.

35 Pearce, D. W. (1999) *Economics and Environment: Essays on Ecological Economics and Sustainable Development*. Edward Elgar, Cheltenham.

36 Pearce, D. W. and Moran, D. (1994) *The Economic Value of Biodiversity*. Earthscan, London.

37 Pearce, D. W., Day, B., Newcombe, J., Brunello, A. and Bello, T. (1998) *The Clean Development Mechanism: Benefits of the CDM for Developing Countries*. Centre for Social and Economic Research on the Global Environment (CSERGE). University College London for the Department for International Development.

38 Pearce, D. W., Krug, W. and Moran, D. (1999) *The Global Value of Biodiversity*. Report to UNEP, Nairobi.

39 Perman, R., Ma, Y. and McGilvray, J. (1999) *Natural Resource and Environmental Economics*. 2nd Edition. Longman, Harlow.

40 Pinedo-Vazquez, M. and Rabelo, F. (1999) 'Sustainable management of an Amazonian forest for timber production: myth or reality?' *PLEC News and Views*, vol 12, pp20–28.

41 Putz, F. E., Dykstra, D. P. and Heinrich, R. (2000) 'Why poor logging practices persist in the tropics'. *Conservation Biology*, vol 14, pp951–956.

42 Putz, J., Sirot, L. K. and Pinard, M. A. (2001) 'Tropical forest management and wildlife: silvicultural effects on forest structure, fruit production and locomotion of non-volant arboreal animals', in Fimbel, R., Grajal, A. and Robinson, J. (eds) *Conserving Wildlife in Managed Tropical Forests*. Columbia University Press, New York.

43 Putz, F. E., Blate, G. M., Redford, K. H., Fimbel, R. and Robinson, J. (2001) 'Tropical forest management and conservation of biodiversity: An overview'. *Conservation Biology*, vol 15, pp7–20.

44 Reid, J. and Rice, R. (1997) 'Assessing natural forest management as a tool for tropical forest conservation'. *Ambio*, vol 26, no 6, pp382–386.

45 Repetto, R. (1990) 'Deforestation in the tropics'. *Scientific American*, vol 262, pp36–42.

46 Rice, R., Gullison, R. and Reid, J. (1997) 'Can sustainable management save tropical forests?' *Scientific American*, vol 276, pp34–39.

47 Rice, R., Sugal, C. and Bowles, I. (1998a) *Sustainable Forest Management: A Review of the Current Conventional Wisdom*. Conservation International, Washington, DC.

48 Rice, R., Sugal, C., Frumhoff, P. and Losos, E. (1998b) 'Options for conserving bio-diversity in the context of logging in tropical forests', in Prickett, G. and Bowles, I. (eds) *Footprints in the Jungle: Natural Resource Industries*. Infrastructure and Biodiversity Conservation, Oxford University Press, Oxford.

49 Romero, C. (1999) 'Reduced impact logging effects on commercial non-vascular pendant epiphyte biomass in a tropical forest in Costa Rica'. *Forest Ecology and Management*.

50 Salick, J. (1995) 'Non-timber forest products integrated with natural forest management'. *Ecology Applications*, vol 5, pp922–954.
51 Schneider, R. (1995) *Government and the Economy on the Amazon Frontier*. Environment Paper no 11, World Bank, Washington, DC.
52 Smith, J., Mulongoy, K., Persson, R., Sayer, J. (1999a) *Harnessing Carbon Markets for Tropical Forest Conservation: Towards a More Realistic Assessment*. Centre for International Forestry Research (CIFOR), Jakarta (mimeo).
53 Smith, J., Mourato, S., Veneklaas, E., Labarta, R., Reategui, K. and Sanchez, G. (1999b) *Willingness to Pay for Environmental Services Among Slash and Burn Farmers in the Peruvian Amazon: Implications for Deforestation and Global Environmental Markets*. Centre for Social and Economic Research on the Global Environment, University College London and University of East Anglia (mimeo).
54 Sohngen, B., Mendelsohn, R. and Sedjo, R. (1998) *The Effectiveness of Forest Carbon Sequestration Strategies with System-wide Adjustments*. Resources for the Future, Washington, DC (mimeo).
55 Southgate, D. (1996) *What Roles Can Ecotourism, Non-limber Extraction, Genetic Prospecting, and Sustainable Timber Play in an Integrated Strategy for Habitat Conservation and Local Development?* Department of Agricultural Economics, Ohio State University. Report to Inter-American Development Bank (mimeo).
56 Southgate, D. (1998) *Tropical Forest Conservation: An Economic Assessment of the Alternatives in Latin America*. Oxford University Press, Oxford.
57 Southgate, D., Coles-Ritchie, M. and Salazar-Canelos, P. (1996) *Can Tropical Forests Be Saved by Harvesting Non-Timber Products?* Global Environmental Change (GEC) Series 96-02. Centre for Social and Economic Research on the Global Environment (CSERGE), University College London and University of East Anglia.
58 Stephens, M. (1999) *The Marginal Costs of a Safe Minimum Standard for an Endangered Arboreal Species*. Department of Forestry, Australian National University, Canberra (mimeo).
59 Toman, M. and Ashton, M. (1994) *Sustainable Forest Ecosystems and Management: A Review Article*. Discussion Paper 94–42. Resources for the Future, Washington, DC.
60 Vanclay, J. K. (1994a) *Modeling Forest Growth and Yield: Applications to Mixed Tropical Forests*. CAB International, Wallingford.
61 Vanclay, J. K. (1994b) 'Sustainable timber harvesting: Simulation studies in the tropical rainforests of north Queensland'. *Forest Ecology and Management*, vol 69, pp299–320.
62 Vanclay, J. K. (1996a) 'Assessing the sustainability of timber harvests from natural forests: Limitations of indices based on successive harvests'. *Journal of Sustainable Forestry*, vol 3, pp47–58.
63 Vanclay, J. K. (1996b) 'Lessons from the Queensland rainforests: Steps towards sustainability'. *Journal of Sustainable Forestry*, vol 3, pp1–17.
64 Vanclay. J. K. (2001) 'The effectiveness of parks'. *Science*, vol 293, no 5532, p1007.
65 Vincent, J. (1992) 'The tropical timber trade and sustainable development'. *Science*, vol 256, p1651.
66 Whitmore, T. C. (1999) 'Arguments on the forest frontier'. *Biodiversity Conservation*, vol 8, pp865–868.
67 World Commission on Forests and Sustainable Development (1999) *Our Forests Our Future*. Cambridge University Press, Cambridge.

15

Forest Fragmentation and Biodiversity: The Case for Intermediate-sized Conservation Areas

Pieter A. Zuidema, Jeffrey Sayer and Wim Dijkman

Introduction

Isolation of forest patches is caused by human activities such as logging, conversion to agriculture and road construction (McCloskey, 1993; Skole and Tucker, 1993; FAO, 1995; Vogelmann, 1995). The resulting forest fragments are surrounded by agriculture, urban landscapes, plantation forests, secondary forests or wastelands. In general, forest fragmentation can be expected to cause local extinctions of original forest species, and fragmented forests will contain fewer of the original forest species than continuous forests. Efficient programmes to conserve forest biodiversity require an ability to predict the scale of losses of biodiversity which will occur as a result of the fragment's representativeness, size and degree of isolation. Currently, the scientific basis for predictions of species extinction rates resulting from deforestation and fragmentation is weak (Simberloff, 1992). Models based on island-biogeography theory (MacArthur and Wilson, 1967) predict much higher extinction rates than empirical studies have reported (Heywood and Stuart, 1992; Heywood et al, 1994; Turner and Corlett, 1996).

Forest biodiversity conservation is an area of major international concern (ITTO, 1993; Heywood and Watson, 1995). Despite this, only limited financial resources are available for conservation programmes and these need to be used efficiently, especially in developing tropical countries. Many conservation plans advocate the establishment of extremely large protected forest areas, but observation suggests that socioeconomic constraints limit the success of such ambitious schemes. Many 'paper parks' exist only in plans and on maps. A set of strategically placed reserves of a size which is consistent with resources available for their protection may be a more realistic option (Boyle and Sayer, 1995). Since these areas will inevitably be 'fragments', knowledge of the effects of forest fragmentation on rates of species loss is essential. However, research has dealt almost exclusively with fragments of 100ha or less. The protected areas which are established to conserve forest biodiversity are several orders of magnitude larger than this. We have extracted information on

Note: Reprinted from *Environmental Conservation*, vol 23, Zuidema, P. A., Sayer, J. and Dijkman, W., 'Forest fragmentation and biodiversity: The case for intermediate-sized conservation areas', pp290–297, Copyright © (1997), with permission from Cambridge University Press

the size of nature conservation areas in tropical forest countries from Collins et al, *(1991)* for Asia, Sayer et al (1992) for Africa, and Harcourt and Sayer (1996) for South America. These publications give partial lists of areas established by the late 1980s. They indicate that the average size of National Parks in Brazil is 340,000ha, in Indonesia is 345,000ha and in Zaire is 1.2 million ha. There are 118 National Parks and equivalent strict Nature Reserves in the three countries, of which only 18 are less than 10,000ha.

In this chapter we attempt to use the existing literature on the biological impacts of fragmentation to draw conclusions about the viability of different size classes of conservation area. We review 58 original papers reporting on studies in forest fragments. The aggregate number of fragments in these studies was 1488, but several studies presented data from the same fragments so the actual number of fragments covered by this review is somewhat lower than this total. The studies focused on various changes and processes in forest fragments, which may affect rates of species loss.

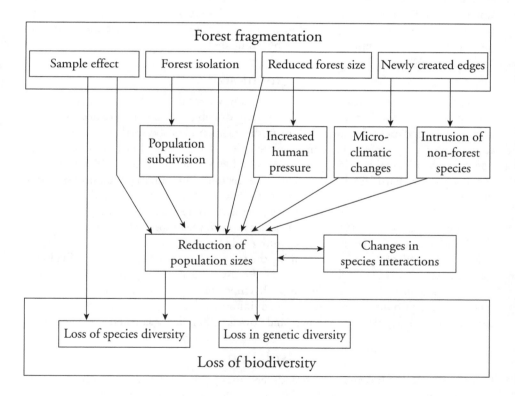

Note: Arrows indicate causal relations

Figure 15.1 *A schematic representation of the theoretical ways in which four forest fragmentation factors (sample effect, forest isolation, reduced forest size and newly created forest edges) may cause reduction of population sizes, which may result ultimately in loss of biodiversity*

Processes Affecting Species Persistence in Forest Fragments

Forest fragmentation may negatively influence the forest's original biodiversity at the levels of genes, species and species associations. The effect of forest fragmentation on biodiversity can either be a direct result of non-random 'sampling' of a certain forest area (representativeness), or indirect, resulting from chains of causes and effects. Different 'fragmentation factors' may affect species persistence in different ways (Figure 15.1).

Sample effects

Forest fragments are 'samples' of a larger area of forest and may exclude patchily distributed species that were present in the original area (Wilcox, 1980). The extent to which the forest fragment represents the original forest depends on the proportion of the landscape converted into non-forest, the spatial arrangement of the remaining fragments, the size of the fragments and the spatial distribution of original forest species. Selectivity of deforestation will affect the representativeness of forest fragments. For example, lowland forests in flat areas are generally cleared before forests on steep slopes in the uplands are cleared, as is the case on Java (Thiollay and Meyburg, 1988). As a result, groups of species, or complete communities, may be eliminated.

Fragment size

The reduction in forest area due to fragmentation will result in a decrease in population sizes of forest species. For species with a patchy distribution, the abundance in a fragment will depend on the location and size of the fragment. For species that naturally occur at high densities, population size may not be reduced to critically low numbers in forest fragments of reasonable size (eg Klein, 1989; De Souza and Brown, 1994). Species occurring at low densities will suffer from considerable reduction in population size, and may become vulnerable to local extinction as a result of stochastic events or reduced genetic fitness. Reduced forest size will also make fragments more accessible for logging, hunting and gathering (Janzen, 1986), which can also contribute to species loss (Turner, 1996).

Transformation of a large forest area into several fragments results in population subdivision. The nature of the habitat separating the fragments, and the capacity of individuals, seeds or pollen to cross gaps, will determine the effective size of resulting population(s) (Hanski, 1989). Altered microclimatic conditions (high temperature, low moisture, strong wind), and increased susceptibility to predation, may inhibit movement, or even completely impede crossings of inter-fragment areas (Powell and Powell, 1987; Bierregaard and Lovejoy, 1989).

Isolation

Clearly, the effect of fragment isolation differs among species, depending on their mobility, dispersal mechanism or pollination agent (Wilcox, 1980). Laurance (1991a) has reported on the effects of an animal's degree of habitat specialization and behavioural avoidance of open habitats on extinction proneness. Some species, or their propagules, can cross large areas of non-forest vegetation (eg raptors; Thiollay and Meyburg, 1988), whereas for other species an inter-fragment distance of 80–100m or less can act as a strong barrier (Mader, 1984; Powell and Powell, 1987; Temple and Cary, 1988; Klein, 1989).

Open vegetation outside forest fragments (grassland, agricultural fields, plantation forest) results in higher air temperatures, wind speeds and light availability, and a drier air and soil moisture regime, in the edge zones of forest fragments (for review see Murcia, 1995). The depth to which microclimatic influences extend depends on the steepness of microclimatic gradients, and on the structure of the edge vegetation; a more dense vegetation buffers the intrusion of microclimatic changes (Lovejoy et al, 1986; Williams-Linera, 1990; Saunders et al, 1991). In temperate, as well as tropical regions, these microclimatic changes were found to penetrate over 50m into the forest (Ranney et al, 1981; Kapos, 1989; MacDougall and Kellman, 1992; Young and Mitchell, 1994). Pioneer and agricultural species may benefit from the altered microclimate in the edge zone, and may out-compete original forest species (Janzen, 1983; Lovejoy et al, 1984; Laurance, 1991b; Brothers and Spingarn, 1992). Laurance (1991b) found non-forest species invading up to 500m into fragments surrounded by pasture. Non-forest species may also introduce pests and diseases into fragments, and, in this way, affect viability of original forest species (Janzen, 1986).

The processes described above may result in a reduction in the population sizes of forest species. Clearly, the time scale of such changes differs between species, depending on their life cycle; long-lived species such as tropical trees will show a decline in adult population only a long time after forest fragmentation (Lovejoy et al, 1983; Turner and Corlett, 1996).

A drastic population reduction may result in a population size below a minimum viable level. The size of the minimum viable population will depend on the species' life cycle, demography and breeding system (Schaffer, 1981; Soule, 1987; Nunney and Campbell, 1993), but is ultimately determined by vulnerability to stochastic fluctuations in population size and reduced genetic fitness (Schaffer, 1981; Gilpin and Soule, 1986; Lande, 1993). Permanent local extinction is avoided when individuals or propagules are exchanged with other fragment populations, that is when a metapopulation is established (Hanski, 1989, 1994). Young et al (1996) reviewed literature on the population genetic consequences of habitat fragmentation for plants. They concluded that, for the species examined so far, genetic variation usually is reduced by fragmentation. They attribute this to genetic 'bottlenecks' at the time of fragmentation and subsequent inbreeding in small populations, but they note that there is no direct evidence of the latter. Significantly, they point to evidence of fragmentation thresholds above which genetic variation is not lost, and to situations where fragmentation may lead to increased gene flow amongst remnant populations. Their overall conclusion is that the genetic effects of fragmentation appear to be more varied than simple population genetics models would predict, and that remnant populations can play a significant role in maintaining the genetic diversity of a species.

Subjects and Findings of Fragmentation Studies to Date

The theoretical literature on fragmentation is quite large (eg Simberloff and Abele, 1982; Wilcox and Murphy, 1985; Boecklen and Simberloff, 1986; Zimmerman and Bierregaard, 1986; Murcia, 1995), but there are relatively few empirical studies of what actu-

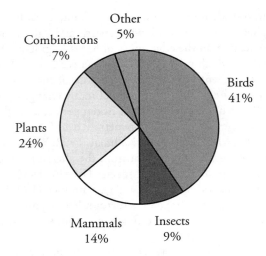

Figure 15.2 *Distribution of the species groups studied in the 58 original forest fragmentation papers reviewed (see Table 15.1)*

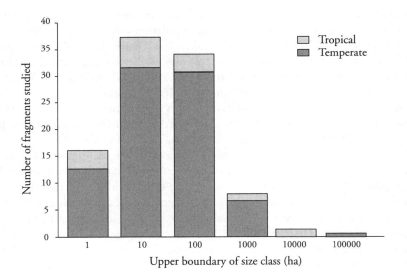

Note: Several papers refer to studies carried out in the same fragment and the actual aggregate number of fragments covered by the studies cited is less than 1488.

Figure 15.3 *Size distribution of the 1488 forest fragments studied in the 58 papers reviewed (see Table 15.1) for temperate and tropical forest areas*

ally has happened in isolated pieces of forests. We have reviewed 58 of the studies which measured the impact of fragmentation on species persistence (Table 15.1).

Table 15.1 *List of 58 reviewed fragmentation studies classified per taxonomic group with information on size distribution of studied fragments, studied factors and their findings*

Studies per taxonomic group	Upper limit of size classes of studied fragments (ha)							Factors	Findings
	10^0	10^1	10^2	10^3	10^4	10^5	10^6		
Birds									
Askins et al, 1987		8	18	16	4			S	PD
Bierregaard & Lovejoy, 1989	5	4						SI	PD
Blake & Karr, 1984		4	6	2				SI	PD
Diamond et al, 1987		1						S	PD
Estrada et al, 1993	3	6	9	12				S	D
Forman et al, 1976	3	5	2					S	I
Gibbs & Faaborg, 1990		1	4	1				S	P
Haila et al, 1993	2	11						S	U
Lynch & Whigham, 1984*		35	216	18	1			SI	P
Mills, 1995	3	10						S	P
Newmark, 1991	5	3	2	1				SI	D
Nores, 1995					2	4	3	SI	D
Opdam et al, 1984	17	19						SI	D
Porneluzi et al, 1993		1	7	3				S	P
Robinson et al, 1995*			6	20	6	1		SI	I
Schieck et al, 1995		4	12	4	1			S	U
Stouffer & Bierregaard, 1995	5	4						S	U
Telleria & Santos, 1992	14	5	6	4				S	I
Telleria & Santos, 1995	11	5	7	4				S	P
Thiollay & Meyburg, 1988				1		4		S	PD
Van Dorp & Opdam, 1987	10	185	40					SI	D
Villard et al, 1995		7	40	3				S	PD
Whitcomb et al, 1981		15	7	8				SI	D
Willson et al, 1994	3	5		3				S	PD
Plants									
Brothers & Spingarn, 1992*	1	6						E	P
Dzwonko & Gawronski, 1994		4						I	PD
Dzwonko & Loster, 1989	59	4	3					SI	PD
Esseen, 1994	5							O	P
Levenson, 1981	7	31	5					S	U
Norton et al, 1995		5	4	3	2			S	P

Table 15.1 *List of 58 reviewed fragmentation studies classified per taxonomic group with information on size distribution of studied fragments, studied factors and their findings*

Studies per taxonomic group	Upper limit of size classes of studied fragments (ha)							Factors	Findings
	10^0	10^1	10^2	10^3	10^4	10^5	10^6		
Rankin-de Merona cited in Bierregaard et al (1992)		1.	1					E	P
Scanlon, 1981		16	6	1				S	D
Simberloff & Gotelli, 1984	31	10	9	1				S	P
Sizer cited in Bierregaard (1992)		1						E	U
Van Dongen et al, 1994	3	4	4	1				I	G
Weaver & Kellman, 1981		10						SI	U
Young & Mitchell, 1994	2	3						O	P
Young et al, 1993		8						S	U
Mammals									
Bennett, 1990	5	19	15					I	P
Laurance, 1990		4	5	1				SI	D
Laurance, 1994		4	5	1				S	PD
Malcolm, 1994	4	1	1					SI	PD
Matthiae & Stearns, 1981	4	16	2					S	D
Pahl et al, 1988		4	7					S	D
Rylands & Keuroghlian, 1988		4	1					S	D
Van Apeldoorn et al, 1994	7	30	12					SI	P
Insects									
Baz & Garcia-Boyero, 1995		3	4	5	1			SI	D
Becker et al, 1991	2	2	1					S	U
De Souza & Brown, 1994	1	1						S	D
Klein, 1989	3	3						S	PD
Powell & Powell, 1987	1	1	1					S	P
Combinations									
Aizen & Feinsinger, 1994	5	4	1					S	I
Howe, 1984*	25	14						S	PD
Rosenberg & Raphael, 1986		6	29	11				S	D
Santos & Telleria, 1994	3	4	2	2				S	I

Table 15.1 *List of 58 reviewed fragmentation studies classified per taxonomic group with information on size distribution of studied fragments, studied factors and their findings*

Studies per taxonomic group	Upper limit of size classes of studied fragments (ha)							Factors	Findings
	10^0	10^1	10^2	10^3	10^4	10^5	10^6		
Others									
Kapos, 1989	2		2					E	U
Laurance, 1991b		4	5	1				E	U
Ranney et al, 1981		8	1					E	U

Note: S = size; I = isolation; SI = size and isolation; E = edge; O = other; P = population size; D = species diversity; PD = population size and species diversity; G = genetic diversity; I = species interaction; U = uncertain or no effect.
Numbers in columns indicate the number of fragments studied in that size class.
Number of fragments in size classes is approximate for this reference.

Species studied

Birds are the most intensively studied species group (Figure 15.2; see review by Andren, 1994). This can be attributed to the abundance of ornithologists and the advanced state of knowledge of bird taxonomy and geographical distribution (eg Whitcomb et al, 1981; Opdam et al, 1984). Plants are the second most studied taxonomic group. Plant population sizes and species diversity have often been studied in relation to edge effects (eg Ranney et al, 1981; Esseen, 1994; Young and Mitchell, 1994). Insects have received relatively less attention, in spite of their important contribution to total species diversity (eg Powell and Powell, 1987; Klein, 1989; Didham et al, 1996). The over-representation of bird studies may result in biased general conclusions, since effects of fragmentation on birds will differ from those on groups of less mobile species (Wilcox, 1980). Extrapolation between organisms with different capacities of movement and dispersal will not normally be possible.

Almost all fragmentation studies describe the response of a single species or several species from one taxonomic group, for example beetles, euglossine bees, understorey birds and small mammals. Information on changes in interactions of plants and pollinators, or prey and predator species, is scarce (Aizen and Feinsinger, 1994; Santos and Telleria, 1994). Further research on inter-species' dependencies and ecological processes in fragmented forests will be of importance in predicting species extinctions resulting from forest fragmentation (Terborgh, 1992; Estrada et al, 1993; Harrington et al, 1997).

Size of fragments studied

A relatively small proportion of fragmentation research has been conducted in the tropics and of all forest fragments that were studied in the papers reviewed, more than 50 per cent were smaller than ten hectares (Figure 15.3). Species that occur at low densities and cannot disperse between fragments will usually fail to maintain a minimum viable population in fragments of this size. Also, species requiring a home range larger than a few hectares, and which are unable to expand their home range to incorporate several fragments, will risk extinction in these small areas.

The large edge-to-core ratio in small forest fragments dramatically reduces the area of suitable habitat for many obligate forest species. Several studies indicated that in one-ha forest fragments edge conditions predominate, and no core area remains in which the original forest microclimate is maintained (Esseen, 1994; Young and Mitchell, 1994). Assuming a 50m penetration depth of microclimatic changes, almost half of the area of a ten-ha circular forest fragment is affected. For edge effects penetrating 200–500m, as found by Laurance (1991b) for forest fragments with a non-circular shape, the fraction of forest area affected by edge conditions is even greater. Furthermore, small fragments rarely are representative of the original forest, since they cannot include much habitat variability. It seems reasonable to conclude that fragments smaller than c.100 ha should not be a main focus of biodiversity conservation, although they may still be valuable as components of a matrix habitat (Turner and Corlett, 1996).

The sizes of the forest fragments that have been the object of these studies are several orders of magnitude smaller than the major protected areas in tropical countries. Species persistence in small fragments is determined largely by edge effects and extreme reduction in population size. These may not be the most important processes determining species persistence in the protected areas which are the mainstay of real-world forest conservation programmes. There is much empirical evidence to suggest that extrapolation between the two scales is not justified.

Fragmentation factors studied

Of the fragmentation factors that may affect biodiversity (see Figure 15.1), fragment size has been the focus of the majority of the studies (Figure 15.4). Effects of isolation are studied less frequently, presumably since isolation is more difficult to quantify, as it depends on both the distance to surrounding fragments and the characteristics of the inter-fragment vegetation. Since different measures of isolation have been used in fragmentation studies (eg Opdam et al, 1984; Laurance, 1990), comparison of isolation effects is difficult.

Effects on species survival and diversity

The fragmentation studies reviewed emphasize decreased population sizes and reduction of species diversity as the two most significant effects of forest fragmentation (Figure 15.5). For many studies in which an effect on species diversity was detected, the findings were based solely on presence or absence data (Figure 15.5), and not on population sizes (eg Baz and Garcia-Boyero, 1995; Nores, 1995). Lack of information on population sizes of species present may lead to underestimates of longer-term fragmentation effects. Haila et al (1993) noted that a reduction in population size, or of species diversity, of mobile bird species in temperate forest fragments may be caused by natural variability in patch occupancy (eg shifting territory locations), rather than by reduced forest size or increased isolation.

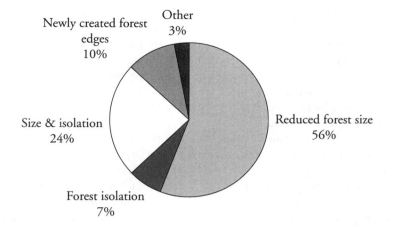

Figure 15.4 *Distribution of the fragmentation factors studied in the 58 original forest fragmentation papers reviewed (see Table 15.1)*

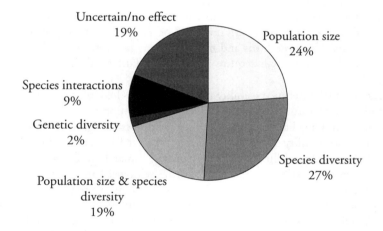

Figure 15.5 *The effects of fragmentation identified in the 58 original forest fragmentation papers reviewed (see Table 15.1)*

Conclusions

From the overview of studies presented here, it is clear that knowledge of effects of forest fragmentation on biodiversity is inadequate for effective planning of conservation programmes. It is apparent that it is extremely difficult to isolate different fragmentation effects in empirical studies, and many authors have actually looked at combined edge, area and isolation effects, without genuinely controlling each factor. The overall picture is based on studies in small fragments, focusing on particular taxonomic groups, testing for effects of fragment size more than isolation, and for effects on short-term survival of species rather than changes in factors such as gene frequency, levels of heterozygosity and interactions among species which will determine their long-term viability.

The general findings of the fragmentation studies presented here probably cannot be extrapolated to larger areas. The implicit assumption that forest biodiversity can be conserved only in very large fragments (eg Meffen and Carroll, 1994) is neither supported nor refuted by the empirical studies reviewed.

Perceptions of conservation managers and decision makers may have been influenced excessively by unjustified extrapolations of the results of research on small fragments. Protected area plans overestimate fragmentation effects and underestimate the social, political and logistic difficulties inherent in the protection of extensive reserves located in remote frontier areas (Sayer, 1995). Strategically targeted systems of Reserves, whose management needs to correspond more closely to the resources available to support conservation, probably represent a better option for conservation. Empirical evidence suggests that Reserves as small as 1000ha can play an important role in biodiversity conservation (Heywood and Stuart, 1992; Turner and Corlett, 1996). The relationship of 50 per cent species loss for 90 per cent habitat reduction postulated by MacArthur and Wilson (1967) is an inadequate basis for conservation planning. There is almost certainly an inflection in the species/area curve above which the rate of species loss declines. The habitat size at which this inflection occurs will differ between organisms, but Turner and Corlett (1996) make a strong case for increased persistence of plant species in fragments as small as 100ha. However, birds and mammals may be more vulnerable in fragments of this size, and certainly will require significantly larger habitat areas (Corlett and Turner, 1997).

In order to further refine the ability to predict species loss in isolated protected areas, it will be necessary to study the impact of fragmentation on the ecological interactions between species and on changes in gene frequency and heterozygosity in fragmented populations. The susceptibility of species in isolated protected areas to random demographic factors and environmental fluctuations may prove ultimately to be the most important factor in determining extinction risk. In countries with weak nature conservation institutions, there is abundant anecdotal evidence that the impacts of hunting and other anthropogenic influences are much greater in fragments. However, to some extent this may be a function of the fact that smaller fragments are not perceived to be important for conservation. It is interesting to note that of the nine National Parks in Indonesia having an area of more than 100,100ha, six have major internationally funded conservation projects. Of the seven National Parks with areas between 10,000 and 100,000ha, only one benefits from significant international support. The conclusion of this chapter is that the greatest area of uncertainty lies in understanding the species conservation potential and risks for areas in the size range of 10,100–100,000ha, but that very little of the current research is relevant to this issue. It may be possible to demonstrate that protected areas in this medium-size range could support viable populations of most species for periods of at least several hundreds of years. Focusing more conservation resources on areas in this size range could allow major cost savings and improvements in efficiency in conservation programmes. Pragmatic decisions will then have to be taken as to when the theoretical, very long-term risks to biodiversity in forest fragments are outweighed by the financial, social and logistic advantages of smaller, more manageable reserves.

Acknowledgements

We thank T. Boyle, R. Condit, F. E. Putz, M. Rijks, A. Souren, R. Didham and W. H. Laurance for valuable comments on drafts of this chapter. Inna Bangun provided invaluable help in preparing the manuscript.

References

1   Aizen, M. A. and Feinsinger, P. (1994) 'Forest fragmentation, pollination, and plant reproduction in a chaco dry forest, Argentina'. *Ecology*, vol 75, pp330–351.

2   Andren, H. (1994) 'Effects of habitat fragmentation on birds and mammals in landscapes with different proportions of suitable habitat: a review'. *Oikos*, vol 1, pp355–366.

3   Askins, R. A., Philbrick, M. J. and Sugeno, D. S. (1997) 'Relationship between the regional abundance of forest and the composition of forest bird communities'. *Biological Conservation*, vol 39, pp129–152.

4   Baz, A. and Garcia-Boyero, A. (1995) 'The effects of forest fragmentation on butterfly communities in central Spain'. *Journal of Biogeography*, vol 22, pp129–140.

5   Becker, P., Moure, J. S. and Peralta, F. J. A. (1991) 'More about euglossine bees in Amazonian forest fragments'. *Biotropica*, vol 23, pp586–591.

6   Bennett, A. F. (1990) 'Habitat corridors and the conservation of small mammals in a fragmented forest environment'. *Landscape Ecology*, vol 4, pp109–122.

7   Bierregaard, R. O. J. and Lovejoy, T. E. (1989) 'Effects of forest fragmentation on Amazonian understorey bird communities'. *Acta Amazonica*, vol 19, pp215–241.

8   Bierregaard, R. O. J., Lovejoy, T. E., Kapos, V., Dos Santos, A. A. and Hutchings, R. W. (1992) 'The biological dynamics of tropical rainforest fragments'. *Bioscience*, vol 42, pp859–866.

9   Blake, J. G. and Karr, J. R. (1984) 'Species composition of bird communities and the conservation benefit of large versus small forests'. *Biological Conservation*, vol 30, pp173–187.

10   Boecklen, W. J. and Simberloff, D. (1986) 'Area-based extinction models in conservation', in Elliot, K. (eds) *Dynamics of Extinction*, John Wiley, New York, pp247–276.

11   Boyle, T. J. B. and Saver, J. A. (1995) 'Measuring, monitoring and conserving biodiversity in managed tropical forests'. *Commonwealth Forestry Review*, vol 74, pp20–25.

12   Brothers, T. S. and Spingarn, A. (1992) 'Forest fragmentation and alien plant invasion of central Indiana old-growth forests'. *Conservation Biology*, vol 6, pp91–100.

13   Collins, N. M., Sayer, J. A. and Whitmore, T. C. (eds) (1991) *The Conservation Alias of Tropical Forests, Asia and the Pacific*. IUCN/Macmillan, London.

14   Corlett, R. T. and Turner, I. M. (1997) 'Long-term survival in tropical forest remnants in Singapore and Hongkong', in Laurance, W. F. and Bierregaard, R. O. (eds) *Tropical Forest Remnants Ecology, Management and Conservation of Fragmented Communities*, University of Chicago Press, Chicago.

15   De Souza, O. F. F. and Brown, V. K. (1994) 'Effects of habitat fragmentation on Amazonian termite communities'. *Journal of Tropical Ecology*, vol 10, pp197–206.

16   Diamond, J. M., Bishop, K. D. and Balen, S. V. (1987) 'Bird survival in an isolated Javan woodland: Island or mirror?' *Conservation Biology*, vol 1, pp132–142.

17   Didham, R. K., Ghazoul, J., Stork, N. E. and Davis, A. J. (1996) 'Insects in fragmented forests: a functional approach'. *Trends in Ecology and Evolution*, vol 11, pp255–259.

18 Dzwonko, Z. and Gawronski, S.(1994) 'The role of woodland fragment, soil types, and dominant species in secondary succession on the western Carpathian foothills'. *Vegetatio*, vol 11, pp49–160.

19 Dzwonko, Z. and Loster, S. (1989) 'Distribution of vascular plant species in small woodlands on the West Carpathian foothills'. *Oikos*, vol 56, pp77–86.

20 Esseen, P. A. (1994) 'Tree mortality patterns after experimental fragmentation of an old-growth conifer forest'. *Biological Conservation*, vol 68, pp77–86.

21 Estrada, A., Coates-Estrada, R., Meritt, D. J., Montiel, S. and Curiel, D. (1993) 'Patterns in frugivore species richness and abundance in forest island and in agricultural habitats in Los Tuxtlas, Mexico'. *Vegetatio*, vol 107/108, pp245–257.

22 FAO (1995) *State of the World's Forests*. Forestry Department, FAO, Rome.

23 Forman, R. T. T., Galli, A. E. and Leek, C. F. (1976) 'Forest size and avian diversify in New Jersey woodlots with some land use implications'. *Oecologia*, vol 26, pp1–7.

24 Gibbs, J. P. and Faaborg, J. (1990) 'Estimating the viability of ovenbird and Kentucky Warbler populations in forest fragments'. *Conservation Biology*, vol 4, pp193–196.

25 Gilpin, M. E. and Soule, M. E. (1986) 'Minimum viable populations: processes of species extinction', in Soule, M. E. (ed) *Conservation Biology: The Science of Scarcity and Diversity*, Sinauer Associates. Sunderland, Massachusetts, pp19–34.

26 Haila, Y., Hanski, I. K. and Raivio, S. (1993) 'Turnover of breeding birds in small forest fragments: The "sampling" colonization hypothesis corroborated'. *Ecology*, vol 74, pp714–725.

27 Hanski, I. (1989) 'Metapopulation dynamics: Does it help to have more of the same?' *Trends in Ecology and Evolution*, vol 4, pp113–114.

28 Hanski, I. (1994) 'Patch-occupancy dynamics in fragmented landscapes'. *Trends in Ecology and Evolution*, vol 9, pp131–135.

29 Harcourt, C. S. and Sayer, J. A. (eds) (1996) *The Conservation Atlas of Tropical Forests, the Americas*. IUCN/Simon & Schuster, New York.

30 Harrington, G. N., Irvine, A. K., Crome, F. H. J. and Moore, L. A. (1997) in Laurance, W. F. and Bierregaard, R. O. (eds) *Tropical Forest Remnants: Ecology, Management Conservation and of Fragmented Communities*, University of Chicago Press, Chicago.

31 Heywood, V. H. and Stuart, S. N. (l992) 'Species extinction in tropical forests', in Whitmore, T. C. and Sayer, J. A. (eds) *Tropical Deforestation and Species Extinction*, Chapman and Hall, London, IUCN, pp91–117.

32 Heywood, V. H., Mace, G. M., May, R. M. and Stuart, S. N. (1994) 'Uncertainties in extinction rates'. *Nature*, vol 368, pp105.

33 Heywood, V. H. and Watson, R. T. (eds) (1995) *Global Biodiversity Assessment*. United Nations Environment Programme and Cambridge University Press, Cambridge.

34 Howe, R. W. (1984) 'Local dynamics of bird assemblages in small forest habitat islands in Australia and North America'. *Ecology*, vol 65, pp1585–1601.

35 ITTO (1993) ITTO *Guidelines on the Conservation of Biological Diversity in Tropical Production Forests*. Supplement to ITTO Guidelines for the sustainable management of Natural Tropical Forests. ITTO, Yokohama.

36 Janzen, D. H. (1983) 'No park is an island: increase in interference from outside as park size decreases'. *Oikos*, vol 41, pp402–410.

37 Janzen, D. H. (1986) 'The eternal external threat', in Soule, M. E. (ed) *Conservation Biology. The Science of Scarcity and Diversity*, Sinauer Associates, Sunderland, Massachusetts, pp182–204.

38 Kapos, V. (1989) 'Effects of isolation on the water status of forest patches in the Brazilian Amazon'. *Journal of Tropical Ecology*, vol 5, pp173–185.

39 Klein, B. C. (1989) 'Effects of forest fragmentation on dung and carrion beetle communities in Central Amazonia'. *Ecology*, vol 70, pp1715–1725.

40   Lande, R. (1993) 'Risks of population extinction from demographic and environmental stochasticity and random catastrophes'. *American Naturalist*, vol 142, pp911–927.

41   Laurance, W. F. (1990) 'Comparative responses of five arboreal marsupials to tropical forest fragmentation'. *Journal of Mammalogy*, vol 71, pp641–653.

42   Laurance, W. F. (1991a) 'Ecological correlates of extinction proneness in Australian tropical rain forest mammals'. *Conservation Biology*, vol 5, pp79–89.

43   Laurance, W. F. (1991b) 'Edge effects in tropical forest fragments: application of a model for the design of nature reserves'. *Biological Conservation*, vol 57, pp205–219.

44   Laurance, W. F. (1994) 'Rainforest fragmentation and the structure of small mammal communities in tropical Queensland'. *Biological Conservation*, vol 69, pp23–32.

45   Levenson, J. B.(1981) 'Woodlots as biogeographic islands in Southeastern Wisconsin', in Burgess, R. L. and Sharpe, P. M. (eds) *Forest Islands in Man-Dominated Landscapes*, Springer-Verlag, New York, pp13–39.

46   Lovejoy, T. E., Bierregaard, R. O., Rankin, J. M. and Schubart, H. O. R. (1983) 'Ecological dynamics of tropical forest fragments', in Sutton, S. L., Whitmore, T. C. and Chadwick, A. C. (eds) *Tropical Rain Forests: Ecology and Management*, Blackwell, Oxford, pp377–384.

47   Lovejoy, T. E., Rankin, J. M., Bierregaard, R. O., Brown, K. S., Emmons, L. H. and Van de Voort, M. E. (1984) 'Ecosystem decay of Amazon forest remnants', in Nitecki, L. H. (ed) *Extinctions*, University of Chicago Press, Chicago, pp295–326.

48   Lovejoy, T. E., Bierregaard, R. O., Rylands, A. B., Malcolm, J. R., Quintela, C. E., Harper, L. H., Brown, K. S., Powell, A. H., Powell, G. V. N., Schubart, H. O. R. and Hays, M. B. (1986) 'Edge and other effects of isolation on Amazon forest fragments', in Soule, M. E. (ed) *Conservation Biology: The Science of Scarcity and Diversity*, Sinauer Associates, Sunderland, Massachusetts, pp257–285.

49   Lynch, J. F. and Whigham, D. F. (1984) 'Effects of forest fragmentation on breeding bird communities in Maryland, USA'. *Biological Conservation*, vol 28, pp287–324.

50   MacArthur, R. H. and Wilson, E. O. (1967) *The Theory of Island Biogeography*. Princeton University Press, Princeton, New Jersey.

51   MacDougall, A. and Kellman, M. (1992) 'The understorey light regime and patterns of tree seedlings in tropical riparian forest patches'. *Journal of Biogeography*, vol 19, pp667–675.

52   Mader, H. D. (1984) 'Animal habitat isolation by roads and agricultural fields'. *Biological Conservation*, vol 29, pp81–176.

53   Malcolm, J. R. (1994) 'Edge effects in central Amazonian forest fragments'. *Ecology*, vol 75, pp2438–2445.

54   Matthiae, P. E. and Stearns, F. (1981) 'Mammals in forest islands in Southeastern Wisconsin', in Burgess, R. L. and Sharpe, D. M. (eds) *Forest Islands in Man-Dominated Landscapes*, Springer-Verlag, New York, pp55–66.

55   McCloskey, M. (1993) 'Note on the fragmentation of primary rainforest'. *Ambio*, vol 22, pp250–251.

56   Meffe, O. K. and Carroll, C. R. (eds) (1994) *Principles of Conservation Biology*. Sinauer Associates, Sunderland, Massachusetts.

57   Mills, L. S. (1995) 'Edge effects and isolation: Red-backed Voles on forest remnants'. *Conservation Biology*, vol 9, pp395–403.

58   Murcia, C. (1995) 'Edge effects on fragmented forests: Implications for conservation'. *Trends in Ecology and Evolution*, vol 10, pp58–62.

59   Newmark, W. D. (1991) 'Tropical forest fragmentation and the local extinction of understorey birds in the Eastern Usambara Mountains, Tanzania'. *Conservation Biology*, vol 5, pp67–78.

60   Nores, M. (1995) 'Insular biogeography of birds on mountain-tops in north western Argentina'. *Journal of Biogeography*, vol 22, pp61–70.

61   Norton, D. A., Hobbs, R. J. and Atkins, L. (1995) 'Fragmentation, disturbance, and plant distribution: Mistletoes in woodland remnants in the Western Australian Wheatbelt'. *Conservation Biology*, vol 9, pp426–438.

62   Nunney, L. and Campbell, K. A. (1993) 'Assessing minimum viable population size: demography meets population genetics'. *Trends in Ecology and Evolution*, vol 8, pp243–249.

63   Opdam, P. F. M., Van Dorp, D. and Ter Braak, C. J. F. (1984) 'The effect of isolation on the number of woodland birds in small woods in the Netherlands'. *Journal of Biogeography*, vol 11, pp473–478.

64   Pahl, L. I., Winter, J. W. and Heinsohn, G. (1988) 'Variation in response of arboreal marsupials to fragmentation of tropical rainforest in North Eastern Australia'. *Biological Conservation*, vol 46, pp71–72.

65   Porneluzi, P., Bednarz, J. C., Goodrich, L. J., Zawada, N. and Hoover, J. (1993) 'Reproductive performance of territorial ovenbirds occupying forest fragments and the contiguous forest in Pennsylvania'. *Conservation Biology*, vol 7, pp618–622.

66   Powell, A. H. and Powell, G. V. N. (1987) 'Population dynamics of male euglossine bees in Amazonian forest fragments'. *Biotropica*, vol 19, pp176–179.

67   Ranney, J. W., Bruner, M. C. and Levenson, J. B. (1981) 'The importance of edge in the structure and dynamics of forest islands', in Burgess, R. L. and Sharpe, D. M. (eds) *Forest Islands in Man-Dominated Landscapes*. Springer-Verlag, New York, pp67–96.

68   Robinson, S. K., Thompson, F. R., Donovan, T. M., Whitehead, D. R. and Faaborg, J. (1995) 'Regional forest fragmentation and the nesting success of migratory birds'. *Science*, vol 267, pp1987–1990.

69   Rosenberg, K. V. and Raphael, M. G. (1986) 'Effects of forest fragmentation on vertebrates in douglas-fir forests', in Vemer, J., Morrison, M. L. and Ralph, C. J. (eds) *Wildlife 2000. Modelling Habitat Relationships of Terrestrial Vertebrates*, University of Wisconsin Press, Wisconsin, pp263–273.

70   Rylands, A. B. and Keuroghlian, A. (1988) 'Primate populations in continuous forest and forest fragments in central Amazonia'. *Ada Amazonica*, vol 18, pp291–307.

71   Santos, T. and Telleria, J. T. (1994) 'Influence of forest fragmentation on seed consumption and dispersal of Spanish Juniper Juniperus thrujera'. *Biological Conservation*, vol 70, pp129–134.

72   Saunders, D. A., Hobbs, R. J. and Margules, C. R. (1991) 'Biological consequences of ecosystem fragmentation: A review'. *Conservation Biology*, vol 5, pp18–32.

73   Sayer, J. A. (1995) *Science and International Nature Conservation*. CIFOR Occasional Paper. Bogor, Indonesia.

74   Sayer, J. A., Harcourt, N. M. and Collins, N. M. (eds) (1992) *The Conservation Atlas of Tropical Forests, Africa*. IUCN/MacMillan, London.

75   Scanlon, M. J. (1981) 'Biogeography of forest plants in the prairie-forest eco-tone in Western Minnesota', in Burgess, R. L. and Sharpe, D. M. (eds) *Forest Islands in Man-Dominated Landscapes*, Springer-Verlag, New York, pp79–124.

76   Schaffer, M. L. (1981) 'Minimum viable population sizes for species conservation'. *Bioscience*, vol 31, pp131–134.

77   Schieck, J., Lertzman, K., Nyberg, B. and Page, R. (1995) 'Effects of patch size on birds in old-growth montane forests'. *Conservation Biology*, vol 9, pp1072–1084.

78   Simberloff, D. (1992) 'Do species-area curves predict extinction in fragmented forests?' in Whitmore, T. C. and Sayer, J. A. (eds) *Tropical Deforestation and Species Extinction*, Chapman & Hall, London, IUCN, pp75–89.

79   Simberloff, D. and Abele, L. G. (1982) 'Refuge design and island biogeographic theory: effects of fragmentation'. *American Naturalist*, vol 120, pp41–50.

80   Simberloff, D. and Gotelli, N. (1984) 'Effects of insularisation on plant species richness in the prairie-forest ecotone'. *Biological Conservation*, vol 29, pp27–46.

81 Skole, D. and Tucker, C. (1993) 'Tropical deforestation and habitat fragmentation in the Amazon, satellite data from 1978 to 1988'. *Science*, vol 260, pp1905–1910.

82 Soule, M. E. (1987) *Viable Populations for Conservation*. Cambridge University Press, Cambridge, UK.

83 Stouffer, P. C. and Bierregaard, R. O. J. (1995) 'Effects of forest fragmentation on understorey hummingbirds in Amazonian Brazil'. *Conservation Biology*, vol 9, pp1085–1094.

84 Telleria, J. L. and Santos, T. (1992) 'Spatiotemporal patterns of egg predation in forest islands: An experimental approach'. *Biological Conservation*, vol 71, pp29–33.

85 Telleria, J. L. and Santos, T. (1995) 'Effects of forest fragmentation on a guild of wintering passerines: The role of habitat selection'. *Biological Conservation*, vol 71, pp61–67.

86 Temple, S. A. and Gary, J. R. (1998) 'Modeling dynamics of habitat-interior bird populations in fragmented landscapes'. *Conservation Biology*, vol 2, pp340–346.

87 Terborgh, J. (1992) 'Maintenance of diversity in tropical forests'. *Biotropica*, vol 24, pp283–392.

88 Thiollay, J. M. and Meyburg, B. U. (1988) 'Forest fragmentation and the conservation of raptors: A survey on the island of Java'. *Biological Conservation*, vol 44, pp229–250.

89 Turner, I. M. (1996) 'Species loss in fragments of tropical rain forests: A review of the evidence'. *Journal of Applied Ecology*, vol 33, pp200–209.

90 Turner, I. M. and Corlett, R. T. (1996) 'The conservation value of small, isolated fragments of lowland tropical rain forest'. *Trends in Ecology and Evolution*, vol 11, pp330–333.

91 Van Apeldoorn, R. C., Celada, C. and Nieuwenhuizen, W. (1994) 'Distribution and dynamics of the red squirrel (*Sciurus vulgaris* L.) in a landscape with fragmented habitat'. *Landscape Ecology*, vol 9, pp227–235.

92 Van Dongen, S., Backeljau, T., Matthysen, E. and Dhondt, A. A. (1994) 'Effects of forest fragmentation on the population structure of the winter with Operophtera brumata L (Lepidoptera, Geometridae)'. *Acta Oecologica*, vol 15, pp193–206.

93 Van Dorp, D. and Opdam, P. P. M. (1987) 'Effect of patch size, isolation and regional abundance on forest bird communities'. *Landscape Ecology*, vol 1, pp59–73.

94 Villard, M. A., Merriam, G. and Maurer, B. A. (1995) 'Dynamics in subdivided populations of neotropical migratory birds in a fragmented temperate forest'. *Ecology*, vol 76, pp27–40.

95 Vogelmann, J. E. (1995) 'Assessment of forest fragmentation in Southern New England using remote sensing and geographic information systems technology'. *Conservation Biology*, vol 9, pp439–449.

96 Weaver, M. and Kellman, M. (1981) 'The effects of forest fragmentation on woodlot tree biotas in Southern Ontario'. *Journal of Biogeography*, vol 8, pp199–210.

97 Whitcomb, R. F., Robbins, C. S., Lynch, J. F., Whitcomb, B. L., Klimkiewicz, M. K. and Bystrak, D. (1981) 'Effects of forest fragmentation on avifauna of the eastern deciduous forest', in Burgess, R. L. and Sharpe, D. M. (eds) *Forest Islands in Man-Dominated Landscapes*. Springer-Verlag, New York, pp125–206.

98 Wilcox, B. A. (1980) 'Insular ecology and conservation', in Soule, M. E. and Wilcox, B. A. (eds) *Conservation Biology*. Sinauer Associates, Sunderland, Massachusetts, pp95–117.

99 Wilcox, B. A. and Murphy, D. D. (1985) 'Conservation strategy: the effects of fragmentation on extinction'. *American Naturalist*, vol 125, pp879–877.

100 Williams-Linera, G. (1990) 'Vegetation structure and environmental conditions of forest edges in Panama'. *Journal of Ecology*, vol 78, pp356–373.

101 Willson, M. F., Santo, T. L. D., Sabag, C. and Armesto, J. J. (1994) 'Avian communities of fragmented south-temperate rainforests in Chile'. *Conservation Biology*, vol 8, pp508–520.

102 Young, A. G., Merriam, H. G. and Warwick, S. I. (1993) 'The effects of forest fragmentation on genetic variation in *Acer saccharum* Marsh, (sugar maple) populations'. *Heredity*, vol 71, pp277–289.

103 Young, A. and Mitchell, N. (1994) 'Microclimate and vegetation edge effects in a fragmented podocarp-broadleaf forest in New Zealand'. *Biological Conservation*, vol 71, pp63–72.
104 Young, A., Boyle, T. and Brown A. (1996) 'The population genetic consequences of habitat fragmentation for plants'. *Trends in Ecology and Evolution*, vol 11, pp413–418.
105 Zimmerman, B. L. and Bierregaard, R. O. (1986) 'Relevance of the equilibrium theory of island biogeography and species-area relations to conservation with a case from Amazonia'. *Journal of Biogeography*, vol 13, pp133–143.

Harnessing Carbon Markets for Tropical Forest Conservation: Towards a More Realistic Assessment

J. Smith, K. Mulongoy, R. Persson and J. Sayer

Introduction

Under the Kyoto Protocol to the Convention on Climate Change (1997) industrialized countries and economies in transition (Annex B countries) undertook to reduce their greenhouse gas emissions by around 5 per cent below 1990 levels (in terms of CO_2 equivalent) by 2008–2012 (Najam and Page, 1998). For the forestry sector, a highly significant development is that the Protocol incorporates a Clean Development Mechanism (CDM) under which Annex B countries can obtain credit for funding projects that enhance carbon sequestration or reduce emissions in the forestry or energy sectors of developing countries. This paves the way for international financial and technological transfers to support forest conservation in developing countries. CDM also differs from other emission trading mechanisms in that its stated objective is not only to lower the cost of emission reduction, but also to promote sustainable development in countries hosting CDM projects.

The concept of using forest sinks for mitigating climate change has been intensely controversial since its inception. It has been enthusiastically supported by those who believe that conservation of tropical forests will be difficult unless forest owners and managers are compensated for the environmental services their forests provide (Pearce, 1996). COM is seen as a mechanism for achieving this, without resorting to subsidies. On the other hand, others have been concerned that large financial transfers oriented towards forest conservation for 'carbon-farming' may ignore social concerns and the full range of products and environmental services of forests. CDM-induced increases in forested land may also be incompatible with the development objectives of host countries (Cullet and Kameri-Mbote, 1997).

It has also been argued that climate change mitigation would be more effective if focused on reduction in fossil fuel usage, rather than reductions in tropical deforestation. Emissions from tropical deforestation were estimated to be about 29 per cent of emissions from fossil fuel and cement production between 1980 and 1989 (Schimel,

Note: Reprinted from *Environmental Conservation*, vol 27, Smith, T., Mulongoy, K., Persson, R. and Sayer, J., 'Harnessing carbon markets for tropical forest conservation: Towards a more realistic assessment', pp300–311, Copyright © (2003), with permission from Cambridge University Press

1995; IPCC, 1996a) and about 34 per cent in the early 1990s (calculated from Malhi et al, 1999 and Fearnside, 1999). Brown (1997) suggests that if existing carbon budgets are adjusted to take account of carbon uptake by tropical forests, the estimated net flux from tropical forests could be as low as 9 per cent of emissions from fossil fuel and cement production. Although these estimates should be interpreted cautiously because of methodological difficulties, they indicate that industrial emissions are primarily responsible for increases in atmospheric carbon. CDM may thus enable developed countries to 'export' their industrial pollution, without changing unsustainable patterns of energy use in developed countries (La Rovere, 1998; Najam and Page, 1998; TERI, 1998).

The objective of this chapter is to provide insights into how and to what extent carbon markets could contribute to tropical forest conservation and sustainable use. Specifically, we assess the potential for trade in carbon from tropical forests, highlight situations where trade could result in perverse outcomes, but also point out opportunities where trade could enhance the effectiveness of conventional approaches to forest conservation. We draw on recent research as well as insights from a policy dialogue organized by the International Academy of the Environment (IAE), Geneva, and the Center for International Forestry Research (CIFOR) that took place in Geneva in 1998 (Mulongoy, 1998; Mulongoy et al, 1998).

Why Expectations Should Be Scaled Down

Initial estimates of the contribution forests could make to climate change mitigation assumed that large areas of land could be allocated for carbon forestry. For example, estimates quoted by the Intergovernmental Panel on Climate Change (IPCC, 1996a) indicate that in Latin America, Africa and Asia 251 to 551 million ha of land could be available for forest plantations. Sedjo and Solomon (1989) estimated that 465 million ha of fast-growing plantations could compensate for the expected increase in carbon emissions during the next 30 to 50 years. These estimates, which focused on technical suitability and largely ignored social implications, implied that large-scale plantation forestry could play a major role in mitigating climate change. Other studies (Nilsson and Schopfhauser, 1995; Trexler and Haugen, 1995; Brown, 1997) estimated that, with aggressive policy changes, 485 million ha of land in the tropics would be available for supplying 45–50 GtC by 2050, with forest regeneration and avoided deforestation accounting for over 70 per cent of the available area. These estimates led to expectations of very large North–South financial transfers. Many foresters, for example, are hopeful that CDM can contribute towards filling the gap of around US$30 billion between current Official Development Assistance (ODA) for forestry and annual funding requirements estimated by the United Nations Conference on Environment and Development (UNCED) (Mulongoy et al, 1998). The magnitude of these potential transfers heightened the optimism of those who saw these funds as a means of saving tropical forests, while simultaneously fuelling the fears of those who saw massive transfers directed single-mindedly at carbon-farming, as derailing the efforts to incorporate social and environmental concerns into sustainable forest management. This led to a marked polarization of views on the advantages and disadvantages of incorporating forests in the Kyoto Protocol. Recent studies that have looked more carefully at the economic and political realities of implementing the mechanism imply, however, that the potential for trade in forest carbon may be considerably less than previously believed.

Cost-effectiveness

In a pure market-based system, the extent to which industrialized countries invest in CDM forestry will depend on the cost of tropical forestry projects relative to the cost of other options, such as emission reduction energy projects in industrialized and developing countries and forestry projects in industrialized countries. Reliable estimates of relative costs are hard to come by, primarily because of methodological differences related primarily to the following issues:

- Studies vary in the comprehensiveness of carbon sources included (above and below-ground biomass, soil carbon, end-use carbon).
- Estimates may be based on carbon stocks or on carbon flows (ie year to year differences in carbon stock) and may be averaged over the period analysed or discounted according to when sequestration occurs. Some studies also weight carbon estimates by the increasing damage from emissions over time.
- Studies may be based on either average or marginal costs, actual or projected costs. The comprehensiveness of costs included varies widely.
- Very few studies consider that the duration of a forestry project affects its contribution to global warming mitigation. (This aspect is discussed in greater detail in a later section.)

Subject to these caveats, the limited estimates available indicate that most estimates of the cost of tropical forestry projects fall within a relatively narrow range of around US$2–25/tC (Table 16.1). Amongst these, Boscolo et al (1997) estimate discount carbon flows. Ridley (1998), in addition to discounting, incorporates the impact of increasing damage from emissions over time. Smith et al (1999) extend Ridley's (1998) methodology by also incorporating the effect of project duration.

The cost of emission reduction varies widely amongst countries because of inter-country variations in industrial structure, differences in current energy sources and economic development patterns. Most countries, however, follow the general pattern of a number of low-cost opportunities, followed by a sharp increase in emission reduction costs (IPCC 1996a).

Overall these patterns indicate that the cost-effectiveness of tropical forestry depends on emission reduction targets. For emission reduction targets above about 20M>, tropical forests appear to be substantially more cost-effective than energy projects in either industrialized or developing countries (Table 16.1). For less ambitious targets, a number of low-cost opportunities appear to exist in the energy sector and the cost-effectiveness of tropical forests appears to be far less clear cut. Forestry in industrialized countries appears to be more expensive than in tropical countries, although some overlap in costs is seen.

The above discussion indicates that in a fully fledged market, a blend of different mitigation options is likely to be traded. Jepma and Munasinghe (1998) estimate the cost-effective combination of mitigation options for achieving a global emission reduction of 2.4 GtC. Almost half of the reduction (IGtC) would come from energy and forestry projects in industrialized countries and countries in transition. Tropical forestry

Table 16.1 *Cost estimates of carbon sequestration forestry projects compared to emission-reduction energy projects*

Projects	Cost(US$/tC)
Forestry projects	
Developing countries	2–25[a]
Industrialized countries and economies in transition	5–81[b]
Energy projects: developing countries	
Level of emission reduction:	
5–10%	negligible[c]
10–25%	~100[c]
Not specified	36–376[d]
Energy projects: industrialized countries	
Level of emission reduction:	
10–15%	negligible[e]
35–40%	100–200[f]

Source: a Swisher and Masters (1992); Source: IPCC (1996a); Boscolo et al (1997); Ridley (1998); Smith et al (1999)
b Source: Ridley (1998); IPCC (1996a)
c Halsnaes (1996)
d Ridley (1998)
e Swisher and Masters (1992); IPCC (1996a)
f Swisher and Masters (1992)

would contribute 700 million tC and energy projects in developing countries another 100 million tC. These results indicate that once cost considerations are taken into account, tropical forests are likely to account for less than a third of targeted emission reductions. This still constitutes a considerable contribution to mitigation, but implies that even in a purely market-based system, a very substantial part of mitigation activities would take place within the countries that currently have emission reduction targets.

Potential market size

The size of the market for carbon sequestration services from tropical forests is likely to be lower than Jepma and Munasinghe's (1998) estimates indicate. Their estimates are based on a global emission reduction target of 2.4 GtC. This is equivalent to a reduction of about 55 per cent relative to 1990 emissions for Annex B countries (Bolin, 1998). A 55 per cent reduction in emissions is above the level the US may have to achieve by 2010 to meet its emission reduction commitment of 7 per cent below 1990 levels. The USA Energy Information Administration estimates that US emissions will be 33 per cent and 47 per cent above 1990 levels in 2010 and 2020, respectively (Carpenter, 1998b).

In addition, countries in the former Soviet Union (FSU) are unlikely to be purchasing carbon sequestration services, because emissions in 2010 are predicted to be 13 per cent below the aggregate commitments of these countries (Ellerman et al, 1998) primarily because of the collapse of Soviet-era heavy industries. About eight industrialized countries had also reduced emissions below 1990 levels by 1995 (Bolin, 1998).

Based on projected emission levels for 2010, Pearce et al (1998) estimated that developed countries may have to reduce emissions by around 730 million tC and that market price could range from US$5–US$23/tC. Ellerman et al (1998) estimated exports and imports of carbon emissions by industrialized and developing countries by using estimates of the marginal cost of reducing emissions by different amounts by different countries. They concluded that in a fully developed market, net carbon imports (ie imports minus exports) of industrialized countries would be around 935 million t, with a market price of around US$24/tC. Of this, net exports from developing countries would be around 723 million t, with the balance coming from the FSU. If credits that could be obtained through CDM projects were limited to around 50 per cent of emission reduction requirements, Ellerman et al (1998) estimated that net imports of industrialized countries would fall to 656 million tC, at a market price of US$13/tC. Net exports from developing countries would fall to 473 million t. These estimates should clearly be interpreted with caution, given their sensitivity to yet-to-be-decided national and international carbon trading policies. Also, neither of the above studies estimate the extent to which developing country exports would come from tropical forests, as opposed to the energy sector. Given these caveats and assuming that tropical forestry supplies 30 per cent of the amount demanded by industrialized countries (Jepma and Munasinghe, 1998), the size of the market for tropical forestry could range from around 200 to 300 million tC with a value of about US$2.5 billion to US$7 billion. Although sizeable, this would clearly be well below initial optimistic estimates.

Additionality and leakage

According to the Kyoto Protocol, projects that qualify for credits have to satisfy the 'additionally' requirement, that is reductions in emissions must be additional to any that would occur in the absence of the project. This implies that forest conservation qualifies only if the conserved forest is under threat of conversion to other uses. Sequestration projects, such as reforestation, qualify only if the project is not financially viable without CDM, or if CDM funding is required to overcome other barriers to implementation. Projects may also turn out to be 'non-additional' because of 'leakage' which occurs when the emission reduction achieved within the project causes increased emissions outside the project boundary, or at a later period of time. Leakage could occur for example if local communities agree to preserve a forested area, with the intention of increasing deforestation in other areas as compensation. Leakage could also occur if, for example, a forest protection project, by forbidding logging, increased the price of timber, which in turn increased logging outside project boundaries.

Additionality and leakage may not reduce the cost-effectiveness of forestry versus energy projects, because these problems also frequently apply to the energy sector (Chomitz, 1998). Considerable progress has also been made in developing methodologies for screening out 'non-additional' projects (Chomitz, 1998). On the other hand, a survey of pioneer investors in carbon projects shows that investors consider the risk of leakage to be considerably higher in forestry projects relative to energy projects (Newcombe, 1998). Overall the implication is that initial estimates of the land available for CDM forestry may need to be scaled down because of additionality and leakage problems.

Duration of forestry projects

A further complication that could reduce the competitiveness of forestry is that in forestry projects (with the exception of those producing bioenergy), carbon is sequestered only while the forest or harvested forest products exist. By comparison, in an energy project, when a new clean technology is adopted, the emissions avoided are prevented from entering the atmosphere in perpetuity, irrespective of the duration of the project. This implies that if a tonne of carbon sequestered in a forestry project is to offset a tonne of industrial emissions, the forest and products will have to exist in perpetuity (Pearce et al, 1998) or at least as long as the tonne to be offset has a radiative forcing effect, in other words an impact on global warming (Moura Costa and Wilson, 1999). Long-term obligations are, however, likely to increase considerably the price of supplying forestry carbon sequestration services. Also, long-term forest conservation may not be consistent with the host country's development objectives, particularly if political, market and social conditions change.

The concept of tonne-years has been proposed as an alternative to long-term obligations (Fearnside, 1997; Chomitz, 1998). In this approach, the temporal implications of energy and forestry projects can be made comparable by computing the number of tonne-years of reduction in atmospheric concentration achieved by the forestry project. The concept can be illustrated simplistically as follows: if we assume that an emitted tonne impacts global warming for 100 years, a tonne of emission reduction from an energy project could be offset by 100 t sequestered in a forestry project lasting one year, or (equivalently) 2 t sequestered for 50 years.

This concept has been used in various ways. In one approach, the declining impact of emitted emissions on global warming over time is taken into account over a time horizon of 100 years. Applying this concept, Moura Costa and Wilson (1999) calculated that a tonne of carbon sequestered for one year in a forestry project is equivalent to 0.02 of a tonne of emissions avoided in an energy project.

Under another approach, conversion factors between forestry and energy projects are calculated as the ratio of the Present Value (PV) of a tonne of carbon displaced for the duration of the forestry project to the PV of a tonne of carbon displaced in perpetuity (Pearce et al, 1998). This approach incorporates two refinements, one which favours earlier emission reductions, while another favours later emission reductions (Ridley, 1998). First, a discount rate is used which values later emission reductions at a lower rate than earlier reductions. The use of a discount rate is justified because earlier emission reductions avoid the damage caused by global warming for a longer period than reductions at a later date (Fearnside, 1997). The second adjustment favours later emission reductions because the damage from emissions increases over time as a result of increasing population and atmospheric concentrations. This is incorporated by weighting the incremental flow of carbon sequestered in each year as a result of the forestry project (Boscolo et al, 1997) by increases in damage from emissions over time (Fankhauser, 1995). Applying this concept, Smith et al (1999) show that in the case of a 15-year project for slash-and-burn farmers, taking account of project duration increases the cost of sequestering carbon through agroforestry almost fourfold and more than doubles the cost of forest conservation, at a 3 per cent discount rate. Thus, taking account of temporal implications could reduce the cost-effectiveness of forestry versus energy projects (with the exception of bioenergy projects).

The tonne-year approach has the advantage of reducing the risk of investing in forestry projects and permitting more flexibility in land use to forest owners and policy-makers in host countries. At the same time, it should be recognized that the tonne-year approach is based on the concept that non-permanent forestry projects slow down the build-up of atmospheric concentrations, unlike energy projects, which actually reduce emissions. Non-permanent forestry projects should therefore be regarded as an intermediate policy option or a way of 'buying time' until more permanent ways of reducing emissions become viable.

It should be noted that some authors (eg Fearnside, 1999) have claimed that emissions avoided by protection of forests under threat are comparable to emission avoidance in energy projects. Others (eg Smith et al, 1999) contend that since an area of forest can only be cut down once, emissions avoided through forest protection would be equivalent to emission reduction in energy projects only if an additional area of forest were protected each year.

Future trends in forestry's cost-effectiveness

Forestry's cost-effectiveness may be likely to decline in future if the cost of emission reduction in the energy sector declines. A consensus appears to be emerging that the cost of emission reduction in the energy sector will depend on how soon and how deeply emissions have to be cut (IPCC, 1996a; Richels and Sturm, 1996). In the next few decades, costs are likely to remain high because of the lack of economical fuel-switching technologies. Although certain low-cost technological options exist, such as switching building heating systems to natural gas, their adoption is currently impeded by pre-existing capital stock based on high emission technologies. The World Energy Council (1998) estimates that fossil fuels will remain dominant in 2050 and that their share will remain close to the 1990 level of 76 per cent. Coal prices are also expected to fall by 30 per cent between 1997 and 2020 (Carpenter, 1998b). Thus, in the next few decades, certain countries with rapidly increasing emissions, such as the US, Japan, Canada, Portugal, Spain and the Scandinavian countries (Bolin, 1998), may be major players in the market for carbon sequestration services.

In the longer term, the situation is more ambiguous. While economic growth is expected to increase energy demand, the emission intensiveness of economic growth (ie the amount of emissions required to produce a particular level of income) has been falling due to improved energy efficiency, the declining share of manufacturing in industrialized economies and decreasing coal use (IPCC, 1996b). In the US, for example, emissions increased by 0.4 per cent in 1998 while economic growth was 3.9 per cent (Flavin, 1999). Models illustrate that the key determinant of future trends in emission reduction costs will be the rate of innovation in low emission technologies (IPCC, 1996b). Jepma and Munasinghe's (1998) analysis of the physical and economic potential of a range of technologies concludes that renewable energy may have a 15–30 per cent share in the next 30–40 years and that natural gas and hydrogen could allow for a massive switch-over in energy sources by 2100. Johansson et al (1996) conclude that renewable energy in electric power generation could meet more than half of global energy needs by 2050. A study commissioned by the IPCC (1996b) concludes that biomass fuels would account for 31 per cent of primary energy in 2025 and would rise to 50 per cent by 2100 under the most optimistic scenarios. Other promising opportunities are improved efficiency of fuel conversion in the transport industry through hybrid and fuel-cell technologies and co-generation of heat and electricity using (mainly) natural gas (Carpenter, 1998a).

Technology development will depend on whether or not energy policies create the right signals for the private sector. Recent developments indicate that major changes in attitude are occurring in some companies in the sectors conventionally opposed to emission reduction commitments. In the auto industry, Toyota released the world's first mass produced hybrid (gasoline-electric) car in December 1997 (Carpenter, 1999). A heavily financed research and development race amongst the world's biggest auto makers to release commercially viable fuel-cell powered vehicles within five years is underway. General Motors is replacing steel with aluminium to promote fuel efficiency. BP-AMOCO has set itself an internal target to cut emissions by 10 per cent by 2010 and has introduced an internal emissions trading programme (Carpenter 1998a, b). While increasing demand growth in the transport sector may erode a part of these gains, it appears likely that technological progress will reduce the cost of emission reduction in the industrial sector. Thus the demand for forest carbon sequestration services may decline after the next few decades, unless emission reduction commitments are progressively increased.

Political realities

The incorporation of forests into the Kyoto Protocol impacts on and could be influenced by a wide range of stakeholders in investor and host countries including developing and industrialized country governments, forest-dwelling communities, the private sector including forestry industries, pilot carbon project developers and intermediaries, environmental and social NGOs, specialists in forestry, biodiversity, social sciences, law, energy and climate change. The IAE-CIFOR policy dialogue entitled 'Are joint implementation and the clean development mechanism opportunities for sustainable forest management through carbon sequestration projects?' brought together 42 representatives of stakeholders from academia, the business sector, governments, international organizations and NGOs. The 'pair interview' technique known as Round Robin was used to ensure active and equal contributions from all stakeholders (Mulongoy, 1998). The results confirmed the potential for tension between the different constituencies. Discussions indicated that while under the right conditions CDM forestry had the potential to contribute to forest conservation and the livelihoods of local communities, there was also a very real risk of unfavourable outcomes for forests and the poor. Participants were divided in their assessments of whether overall the benefits would outweigh the potential risks. The gaps in scientific information about these issues and the uncertainty about future CDM modalities contributed to the polarization of views. Some of these issues will be discussed in greater detail later in this chapter.

The earlier North–South divide, with developing countries opposing flexibility mechanisms such as CDM and developed arguing in its favour, appears to have been replaced by a debate on the role of forestry in the CDM, with significant divergences in views amongst both industrialized and developing countries. Certain industrialized countries emphasize the importance of reducing emissions cost-effectively and therefore argue for the inclusion of forestry. Other industrialized countries emphasize the importance of fundamentally changing patterns of energy use and argue against the inclusion of forestry. In the case of developing countries, some are concerned that ad-hoc carbon sequestration contracts between private parties may result in land-use changes that are not compatible with the host country's land-use priorities (K. L. Johnson, personal communication, 1998). On the other hand, others see CDM as an instrument for achieving their own forestry objectives. Costa Rica, for example, has set up an incentive scheme to increase forest cover and an institutional and financial structure to facilitate CDM projects, financed by taxing its people, thus providing evidence of political will to conserve its forests (Maldonado Ulloa et al, 1998). The Costa Rican model may however be unattractive to countries where forest conservation is not high on the list of national priorities. These may be countries that still have very large forested areas or countries where the pressure to convert forest to other uses is high, due to factors such as high population densities (Lopez, 1996). These countries may be unwilling to participate in carbon markets. CDM forestry projects are likely to have greater support from host countries (and may therefore be more effective) where forest conservation coincides with national priorities and where host countries have provided concrete evidence of political will to conserve forests.

Investor priorities

Currently, very little is known about the priorities of potential CDM investors. A recent survey of pioneer investors in carbon markets (Newcombe, 1998) indicates that forestry projects are regarded as somewhat riskier than energy projects, primarily because of leakage problems. Investors are likely to be selective in their choice of host countries. Amongst the various types of risks associated with carbon projects, investors were most anxious to avoid countries prone to economic and political instability (ibid). These results are consistent with the current distribution of pilot projects, the majority of which are in the energy sector and concentrated in Latin America. On the positive side, the next most important concern was to provide other environmental and social benefits, in addition to carbon sequestration. Amongst non-carbon benefits, biodiversity conservation and poverty alleviation were rated most highly (ibid). These results indicate that while investor priorities could limit the number of countries that are able to attract investors, within these countries well-designed projects that provide social and environmental benefits may, at least to a segment of the market, be more attractive than cost-efficient carbon-farming projects.

Summary of arguments for scaling down expectations

Initial estimates about the share of tropical forestry in mitigation options and the magnitude of financial transfers to developing countries from forestry carbon projects may need to be scaled down. The cost-effectiveness of forestry may have been overestimated. Also, political realities and investor priorities may not have been sufficiently understood. Initial perceptions that CDM could provide long-term funding to support the forestry sector may also need to be modified. The most significant opportunities for CDM forestry may occur during the next few decades. During this period, forestry projects could make a valuable contribution towards slowing down emission concentrations in the atmosphere and lowering the cost of meeting emission targets. After this period, the situation is more ambiguous. Cheaper low emission technologies and autonomous improvements in energy efficiency are likely to reduce the cost-effectiveness of forestry, although increases in energy demand and more stringent emission reduction targets may dilute this effect. This revised assessment of the role of forestry in CDM could have major implications for the successful implementation of CDM forestry projects.

Implications for Forests and Forest Stakeholders

The above discussion indicated that forestry projects (with the exception of bioenergy projects) are likely to be an intermediate climate change mitigation strategy for buying time, until more permanent options become available. The implication is that the most important justification for including forests in CDM may lie in the contribution CDM could potentially make towards stimulating forest conservation and sustainable use in developing countries. CDM projects could influence where forests exist, which types of forests exist and who benefits from forests. Thus, judicious use of this mechanism may be able to enhance the effectiveness of more conventional approaches, while misguided application could derail current forest conservation efforts.

Many uncertainties remain about CDM. While in principle CDM projects should be compatible with 'sustainable development', it is unclear how this is to be implemented and what aspects of sustainable development are to be addressed. Should CDM provide social benefits? How will equity issues be addressed? Should CDM require non-contravention of the Biodiversity Convention (1992)? Should CDM projects be compatible with the recommendations of international fora, such as the International Forum on Forests (IFF, an intergovernmental panel that started in 1997 and was relabelled a forum in 1998)?

The wording of the Kyoto Protocol is unclear on which types of forestry activities (such as plantations, forest protection, forest management) qualify for credits under CDM, or indeed whether forestry activities will qualify at all for credit. Whether or not CDM funds could be used only for forestry projects or could also be used for stimulating changes that support forest conservation (such as policy or institutional changes, technology development, capacity-building or enforcement of environmental regulations) has also not been clarified.

Many of these decisions could be taken by the end of 2000. Thus, considerable opportunity exists for informing global debates on CDM modalities. Preparation of a special report on Land Use Change and Forestry by the IPCC is currently under way to provide policy makers with scientific information relevant to forestry issues in the Kyoto Protocol. Simultaneously there has been a massive proliferation of literature about

CDM. In this section, we contribute to this effort by drawing on forestry experience to present insights about the use of CDM and raise a number of issues that need to be resolved if CDM is to be effectively harnessed for the benefit of forests and forest stakeholders.

Potential CDM forestry activities

Forestry activities contribute to lowering carbon concentrations either by preventing emissions or by sequestering carbon. A wide range of forestry activities, including the following, can prevent emissions:

- Conserving primary or secondary forests that are under the threat of being converted to other land uses with a lower carbon stock, such as slash-and-burn agriculture with short fallows, permanent agriculture, pasture or agricultural plantations.
- Improved forest management such as shifting from conventional logging to Reduced Impact Logging (RIL) in natural forests. RIL retains more carbon than conventional logging, by reducing damage to residual trees and surrounding vegetation. RIL also sequesters carbon, because forests recover faster (Pinard and Putz, 1997).
- Production of bioenergy by, for example, sustainable fuel wood management or biomass energy plantations. These hybrid forestry-energy options are particularly interesting candidates for CDM projects. They are carbon-neutral land uses: the carbon being sequestered during growth of fuelwood or other biomass is released into the atmosphere when the biomass is burned for fuel. Unlike other forestry activities that benefit the carbon cycle only while the forest lasts, bioenergy projects reduce emissions permanently in the same way as energy projects, if biomass fuels replace fossil fuels. Carbon benefits can be augmented by simultaneously increasing the efficiency of energy use by, for example, improved stoves for cooking or processing of agricultural products. Thus bioenergy projects may overcome some of the concerns about other forestry projects: their spread may result in more sustainable patterns of energy use, while simultaneously their cost-effectiveness may not be adversely affected because they could be capable of permanent emission reduction.

Forestry activities contribute to carbon sequestration when the new forestry-based land use sequesters more carbon than the previous land use. Examples are:

- Establishment of forest plantations on previously deforested land (reforestation) or on land not recently forested (afforestation).
- Regeneration of secondary forests on previously deforested land.
- Agroforestry systems on agricultural land.
- Lengthening of primary or secondary forest rotation cycles.

Equity implications

If only a subset of the above forestry activities is eventually eligible for CDM funds, this may introduce economic biases against those that are excluded. Even if CDM is open to all forestry land uses, market forces are likely to favour activities which deliver carbon at the lowest cost, where 'additionally' is most easily established, where the risk of failure is low and where transactions costs, such as organizational and verification costs, are lowest. Although pioneer investors appear to prefer projects that also provide social and environmental benefits (Newcombe, 1998), this may not reflect investor attitudes in a fully-fledged market. It is quite plausible that the factors given above may favour specialized blocks of simplified forestry ecosystems, such as large-scale plantations. Carbon monitoring costs may be lower in these projects because of their scale and uniform nature. Investors may prefer deals with a few large-scale operators, such as improved forest management by logging concession holders, instead of improved community forestry by local communities, in the interests of lowering transaction costs. Large-scale operators are also likely to be better informed about CDM and to see it as a ready source of subsidy for their operations. If CDM favours large-scale operators, it could have serious equity implications, by putting at an even greater economic disadvantage more diversified land uses that may be more appropriate for local communities. Multiple use forestry by smallholders may also be more environmentally sound and may thus be more compatible with the 'sustainable development' clause of CDM. Investor preference for large-scale forestry could also increase claims by large-scale operators to landholdings of smallholders, which often lack formal land titles. Forest protection projects in areas characterized by uncertain, overlapping use rights could result in compensations for only the most visible and vocal stakeholders, ignoring local communities with informal use rights and less ability to articulate demands. Outcomes such as these would increase disparities in political power, incomes and assets.

Thus, a better understanding of investor priorities is required. Evaluation of the different forestry activities on the basis of investors' criteria and the trade-off with social and environmental benefits is required. This could be used to formulate safeguards, such as criteria for the selection of forestry activities eligible for CDM. This could also indicate whether special mechanisms would be required to support activities that provide social and environmental co-benefits.

The complex and highly localized nature of forestry problems is well known. Evaluation of the social and environmental benefits is therefore likely to be contingent on a range of political, social, institutional and biophysical conditions. A considerable amount of literature now exists on the lessons to be learnt from experiences of success and failure in forestry projects (Byron, 1997; Dawkins and Philip, 1998). Pilot projects where payments for sequestration services are being made could also provide useful insights. These lessons could contribute to the development of principles for the implementation of CDM.

Should CDM be used to subsidize unprofitable forestry activities?

The possibility of trade in carbon sequestration services initially raised the hopes of those who believed that forests were being converted to other land uses because of the lack of mechanisms for compensating forest owners for the environmental services provided by their forests. Doubts about the financial viability of forested land compared to alternative land uses are widespread in the literature, whether in relation to sustainable logging (Kishor and Constantino, 1993) or non-timber forest products (Southgate and

Clark, 1993). Payments for carbon sequestration services, it was hoped, could funda-mentally alter the economics of forested land compared to other land uses. We argue below that, in many cases, this may not be the best use of CDM funds.

Research on the underlying causes of unsustainable land uses shows that the poor economics of forested land is only part of the problem and, in many cases, is not neces-sarily the key factor. For example, in the case of unsustainable logging, Sunderlin and Resosudarmo (1996) highlight political economy aspects, such as the political power of the forestry sector and rent-seeking behaviour, which they attribute to the excessively high profitability of logging concessions, caused by factors such as low royalty fees. In an analysis of the underlying causes of deforestation, Kaimowitz and Angelsen (1998) emphasize the role of policies outside the forestry sector, such as agricultural subsidies which stimulate agricultural expansion at the forest margins and the construction of penetration roads which provide access to forests. Others have emphasized institutional obstacles to the sustainable use of forest, such as the nature of property rights (Ostrom, 1990) and the need to devolve decision-making authority over forest resources from gov-ernmental authorities to community-based organizations (Saxena, 1997). These studies indicate that in many cases the impact of negative policy and institutional arrange-ments may overwhelm the positive incentives provided by CDM funds. Under these conditions, even if improved land use and carbon benefits occur within project bounda-ries, high levels of leakage are likely. Where, for example, deforestation is driven by road construction, a carbon project may control the symptom (ie deforestation) within the project area, but carbon payments may be used to open up additional areas of forest outside the project area, unless the underlying cause (ie road construction) is halted. The use of CDM to subsidize unprofitable forestry may also provide perverse incentives against policy reform, such as the removal of agricultural subsidies. Countries with pol-icies that favour high rates of deforestation may find it easier to establish additionality and therefore attract CDM funding than countries where sound policies are lowering deforestation rates (Fearnside, 1997). In these situations policy and institutional changes may be more efficient than carbon projects.

The possibility of the cost-effectiveness of forests declining over time raises additional complexities. Declining cost-effectiveness implies that the demand for forestry seques-tration projects is likely to decline and in many cases sequestration payments are likely to be forthcoming only for a limited period. If the decision to change land use is driven primarily by the subsidy provided by sequestration payments, the implication is that land will revert to its previous use, once the payments cease. While even a temporary improvement in land use would delay the build-up of emission concentrations in the atmosphere and contribute to the lowering of emission reduction costs, it might not achieve the host country's objective of stimulating a lasting improvement in land-use patterns.

The temporary nature of CDM projects indicates that rather than using CDM funds to improve the economics of unprofitable forestry systems, a more effective strategy may be to use them for removing non-economic impediments to the adoption of economi-cally viable forestry activities. An example of a non-economic impediment could be high start-up costs such as access to information, training, technologies and markets. Invest-ing CDM funds in mechanisms for overcoming such barriers is more likely to result in lasting improvements in land use than cash handouts to forest owners.

Should CDM credit be given for policy and institutional reform?

Given the key role of policies and institutions in stimulating land-use change, Lopez (1999) proposes that carbon credits be used to stimulate policies favourable to forest conservation. Clearly, however, the conditions under which policies and institutional changes could be made eligible for credit would have to be specified. Moral hazard problems could arise, for example, if credit is given for policies which, while favourable to forests, also increase the nation's economic efficiency, as is the case with removal of agricultural subsidies. Pearce et al (1998) also point out the difficulty of establishing the quantity of emission reduction that could be attributed to policy changes. One possibility would be to make CDM projects conditional on the removal of distorting policies (Pearce et al, 1998). While this would considerably limit the use of CDM, it would reduce the risk of leakage and project failure and would be more likely to result in lasting improvements in land use. If multilateral institutions made loans and other assistance conditional on policy and institutional reform, this could expand the pool of CDM projects that meet preconditions. Clearly, therefore, there is a need to establish principles for guiding the types of interventions which will receive credit, the preconditions which CDM projects will have to meet, and mechanisms for catalysing policy and institutional changes favourable to forest conservation.

Lessons from the forestry literature

Plantations

Plantations are widely perceived as the archetypal candidate for CDM forestry but R. Persson (personal communication, 1995) found that only 70 per cent of the plantation area established according to official figures could actually be found on the ground and emphasized the high rate of plantation failures in the tropics. On average, plantations have given only 20–30 per cent of expected production (R. Persson, personal communication, 1995). Using evidence from several countries Evans (1999) shows that poor silvicultural practices and the lack of holistic management are largely responsible for plantation failures. R. Persson (personal communication, 1995) on the other hand contends that the main reason for widespread failure is that plantations were established without a clear market in mind. Large-scale plantation projects also failed because, in many cases, they had little support from local communities who had not been involved in the project and had often suffered from tenure conflicts resulting from plantation establishment (Fearnside, 1996). In many cases, plantations were established on land perceived by outsiders as 'degraded' or 'abandoned' but which was subject to a variety of uses by local people (Fearnside, 1996; Tomich et al, 1997). On the other hand, plantations have been successfully established in Laos to supply the Thai market and successful cases have also been reported from Chile and South Africa. An analysis of the causes of success and failure could provide useful guidance for the development of CDM guidelines on plantation projects.

Lessons could also be learnt from cases where incentives have been provided for short periods and then withdrawn. The literature suggests that if CDM funds were used to stimulate plantation activities that would otherwise be unprofitable, this could stimulate a temporary expansion in plantations and therefore in processing facilities. Withdrawal of CDM funds when the project ends could then cause plantations to be abandoned. Processing facilities may then turn to unsustainable logging in natural forest in order to procure raw materials (Fearnside, 1996). A more effective use of CDM funds may be to develop mechanisms for coordinating raw material supplies with processing facilities, in cases where both activities are perceived to be economically viable. For example, in certain cases, smallholder plantations providing biomass fuel for electricity generation in rural areas may be considered economically viable, but are not established because adequate local processing facilities are not available. At the same time, the private sector may be unwilling to invest in processing facilities due to the lack of raw material supplies and technology. CDM funds could be used to remove 'chicken and egg' impediments of this nature and to make technology available for both plantations and electricity generation from biomass fuels. Land-use changes of this nature, which provide local benefits (such as improvements in the quality of life in this case), are less likely to suffer from leakage during the project and to persist after CDM funding ceases.

CDM funds may also be effective if directed at situations that are on the brink of a transition to a carbon-friendly land use. Sawmills which have traditionally relied on supplies from natural forests may be receptive to the establishment of timber plantations, when increasing distance to timber supplies increases log prices and cuts into profits. In this situation, CDM funds may be able to speed up the transition to plantations. This is likely to be less risky than using CDM to finance plantations where forests are accessible, in an attempt to leapfrog the phase when logs are obtained from natural forests.

Improved management of natural forests

Technologies such as reduced impact logging (RIL) have been little adopted so far. Clearly, as mentioned above, many of the reasons include policy failures and institutional weaknesses. While the economics of RIL versus conventional logging is controversial, Putz et al (2000) show that the economic benefits of RIL depend on biophysical conditions, with benefits being highest in areas characterized by level terrain with well-drained soils and pronounced rainfall seasonality. D. Dykstra (personal communications, 1996) claims that training and the correct use of the right equipment is an important determinant of the profitability of RIL. D. Dykstra (ibid) reports that a growing list of studies demonstrates that if logging operations are correctly planned, well supervised and carried out by well-trained crews with the proper equipment, harvesting costs under RIL can be substantially lower than under conventional logging. Clearly, in addition to these factors, the political and institutional context has to be considered, if CDM funds are to catalyse better logging practices and deliver carbon benefits. This implies that there is a need to identify the niches where these conditions apply, although they may be fairly limited in geographical area.

In these niches, if CDM funds are used for removing non-economic barriers to adoption, such as the start-up costs of training and capacity-building for concessionaires, logging contractors and workers, RIL techniques may continue to be implemented even after the CDM project ends. If, on the other hand, CDM funds are used to subsidize operational costs in areas where RIL is unlikely to be profitable without incentives, RIL would probably be abandoned once the project ends and the logged area would be likely to be converted to agricultural use. Thus, if CDM funds are not judiciously used, they may help to reduce emission reduction costs for industrialized countries but may contribute little towards lasting improvements in land use in developing countries.

Conservation of forests under threat

As explained earlier, forest conservation CDM projects are likely to fail or suffer from a high degree of leakage where policy and institutional reforms are required to address the underlying causes of deforestation. Opportunities for CDM forest conservation projects exist where the forces driving forest conversion can be largely addressed through project activities. Where, for example, deforestation is driven by the need for food and income by small-scale farmers or forest-dwelling communities, projects that support diversified livelihood strategies integrating sustainable forest management with improvements in other livelihood sources could reduce deforestation and benefit the poor. The type of improved forest management would have to be carefully tailored to local conditions and requirements, by drawing on the forestry literature. While multiple-use community forestry, for example, may be appropriate where markets for forest products exist, population densities are low and land is managed under Common Property Regimes, agroforestry may be more successful under medium to high population densities and private property regimes (G. Shepherd, personal communication, 1999). Projects such as these which incorporate forests as a livelihood strategy are likely to be more successful than projects which manage forests as protected areas, because local communities benefit less from protected forests and have to make greater changes in their lifestyles (ibid). If therefore, CDM rules on eligible projects exclude improved forest management but include forest protection, there may be fewer opportunities for benefiting local communities.

Non-timber forest products

A number of authors have studied the conditions under which extraction of non-timber forest products (NTFPs) by local populations could contribute to forest conservation. The importance of both an appropriate policy and institutional framework, such as forest dwellers' rights and local empowerment, as well as biophysical and socioeconomic factors, such as sound management practices and transparent markets, has been raised by some (Fearnside, 1996; Ruiz-Perez and Byron, 1999). Others, such as Crook and Clapp (1998) and Coomes (1995), have pointed out the dangers of over-exploitation resulting from boom-and-bust cycles and the negative relationship that usually prevails between NTFP profitability and forest diversity. NTFPs may therefore be able to contribute to forest conservation and the resulting carbon benefits under a relatively narrow range of biophysical and socioeconomic conditions. CDM may be able to support forest conservation if used for creating these conditions and thus expanding the niche for NTFPs.

Summary of Implications for Forests and Forest Stakeholders

The importance of involving forest stakeholders more closely in CDM debates has been highlighted. CDM could determine where forests exist and who benefits from them. At the same time, lessons from forestry successes and failures could make important contributions to the design of effective CDM projects. CDM funds may best be seen as a mechanism that could enhance the effectiveness of ongoing efforts to stimulate a change to more sustainable systems. Crook and Clapp (1998) point out the importance of targeting market-based mechanisms to areas where sufficient understanding exists of complex local decision making, as well as the broader political and institutional context. Within these areas, certain niches may exist where CDM could make a valuable contribution. Where the knowledge base is inadequate, CDM projects are likely to suffer from leakage problems and have unintended perverse outcomes.

Conclusions

The history of the international debate on forests shows that 'magic solutions' have periodically been put forward. Examples are intensification of slash-and-burn agriculture, sustainable forest management for timber and non-timber forest products, and integrated conservation and development projects. Each has met with only limited success. Trade in environmental services also raised the hope that it could fundamentally alter the economics of forested land compared to alternative land uses. This chapter shows that:

- Expectations should be scaled down and payments for carbon sequestration services may in reality be received in a far more limited area than initially expected and for only a limited period.
- The use of CDM funds to subsidize unprofitable forestry activities is unlikely to result in lasting improvements in land use. CDM may be more effective if used to remove non-economic impediments to the adoption of forestry activities that are economically viable and meet local needs. This implies that the concept of additionality in CDM should be broad enough to accommodate such possibilities.
- A comparative evaluation of potential CDM forestry activities will be required to investigate whether investor preferences are likely to exacerbate existing disparities between large- and small-scale operators or reinforce trends towards simplified ecosystems.
- To prevent perverse outcomes and reduce leakage of emission reduction, CDM projects will probably have to be limited to niches which meet certain preconditions such as favourable policies and institutions and where sufficient understanding exists of local decision making and the broader political and institutional framework. Lessons from the forestry literature and experiences with pilot carbon sequestration projects thus need to play a more prominent role in the formulation of CDM guidelines. Mechanisms for catalysing policy and institutional reforms could broaden the niche for CDM forestry.

Carbon markets should be seen as one more tool in the battle to achieve sustainable use of forested land. This chapter illustrates the complexity of using CDM and provides some examples of the dangers of misusing it, but it also gives examples of how it could be harnessed to increase the effectiveness of more conventional ways of promoting forest conservation and sustainable use.

Acknowledgements

The authors are grateful to Dennis Dykstra, Jack Putz and David Pearce for helpful comments.

References

1   Biodiversity Convention (1992) *United Nations Environment Programme (UNEP). Convention on Biodiversity, June 1992*. UNEP, Environmental Law and Institutions Programme Activity Centre, Nairobi.
2   Bolin, B. (1998) 'The Kyoto negotiations on climate change: A science perspective'. *Science*, vol 276, pp330–331.
3   Boscolo, M., Buongiorno, J. and Panayotou, T. (1997) 'Simulating options for carbon sequestration through improved management of a lowland tropical rainforest'. *Environment and Development Economics*, vol 2, pp241–263.
4   Brown, S. (1997) 'Forests and climate change: role of forest lands as carbon sinks', in *Forestry for Sustainable Development: Towards the 21st Century. Forest and Tree Resources*. Proceedings of the XI World Forestry Congress, 13–22 October 1997, Antalya, vol 1, FAO, Rome, pp117–130.
5   Byron, N. (1991) 'International development assistance in forestry and land management: the process and the players'. *Commonwealth Forestry Review*, vol 76, pp61–67.
6   Carpenter, C. (1998a) *Climate News: 23 November*. International Institute for Sustainable Development (USD). www.iisd.ca/linkages
7   Carpenter, C. (1998b) *Climate News: 1 December*. International Institute for Sustainable Development (USD). www.iisd.ca/linkages
8   Carpenter, C. (1999) *Climate News: 30 March*. International Institute for Sustainable Development (USD). www.iisd.ca/linkages
9   Chomitz, K. (1998) *Baselines for Greenhouse Gas Reductions: Problems, Precedents, Solutions*. Unpublished report, Carbon Offset Unit, World Bank, Washington, DC, USA.
10 Coomes, O. T. (1995) 'A century of rain forest use in Western Amazonia: Lesson for extraction-based conservation of tropical forest resources'. *Forest and Conservation History*, vol 39, pp108–120.
11 Crook, C. and Clapp, R. A. (1998) 'Is market-oriented forest conservation a contradiction in terms?' *Environmental Conservation*, vol 25, no 2, pp131–145.
12 Cullet, P. and Kameri-Mbote, A. P. (1997) 'Activities implemented jointly in the forestry sector: conceptual and operational fallacies'. *The Georgetown International Environmental Law Review*, vol 10, pp97–118.
13 Dawkins, C. and Philip, M. S. (1998) *Tropical Moist Forest Silviculture and Management. A History of Success and Failure*. CAB International, Oxford.
14 Ellerman, A. D., Jacoby, H. D. and Decaux, A. (1998) *The Effects on Developing Countries of the Kyoto Protocol and Carbon Dioxide Emissions Trading Policy Research Working Paper 2019*. The World Bank, Washington, DC.

15 Evans, J. (1999) *Sustainability of Forest Plantations. The Evidence.* The Department for International Development (DFID), London, p64.

16 Fankhauser, S. (1995) *Valuing Climate Change: The Economics of the Greenhouse.* Earthscan, London.

17 Fearnside, P. M. (1996) 'Socioeconomic factors in the management of tropical forests for carbon', in Apps, M. and Price, D. (eds) *Forest Ecosystems, Forest Management and the Global Carbon Cycle*, NATO ASI Series I, vol 40. Springer-Verlag, Berlin, pp. 349–61.

18 Fearnside, P. M. (1991) 'Monitoring needs to transform Amazonian forests maintenance into a global-warming mitigation option'. *Mitigation and Adaptation Strategies for Global Change*, vol 2, pp285–302.

19 Fearnside, P. M. (1999) 'Forests and global warming mitigation in Brazil: Opportunities in the Brazilian forest sector for responses to global warming under the "clean development mechanism"'. *Biomass and Bioenergy*, vol 16, pp171–189.

20 Flavin, C. (1999) *World Carbon Emissions Fall.* Worldwatch News Brief 27 July. www.worldwatch.org/alerts/990727.html. Worldwatch Institute, Washington, DC.

21 Halsnaes, K. (1996) 'The economics of climate change mitigation in developing countries'. *Energy Policy*, vol 224, pp917–926.

22 IPCC (1996a) *Climate Change. Economic and Social Dimensions of Climate Change.* Contribution of Working Group III to the Second Assessment Report of the Intergovernmental Panel on Climate Change. Cambridge University Press, Cambridge.

23 IPCC (1996b) Climate Change 1995. *The Science of Climate Change. Contribution of Working Group I to the Second Assessment Report of the Intergovernmental Panel on Climate Change.* Cambridge University Press, Cambridge.

24 Jepma, C. J. and Munasinghe, M. (1998) *Climate Change Policy. Facts, Issues, and Analyses.* Cambridge University Press, Cambridge.

25 Johansson, T. B., Williams, R. H., Ishitani, H. and Edmonds, J. A. (1996) 'Options for reducing CO_2 emissions from the energy sector'. *Energy Policy*, vol 24, pp985–1003.

26 Kaimowitz, D. and Angelsen, A. (1998) *Economic Models of Tropical Deforestation: A Review.* CIFOR (Center for International Forestry Research), Bogor, Indonesia.

27 Kishor, N. and Constantino, L. (1993) *Forest Management and Competing Land Uses: An Economic Analysis for Costa Rica.* Unpublished report, Latin America and the Caribbean Technical Department, Environment Division, Dissemination Note 7, World Bank, Washington, DC.

28 Kyoto Protocol to the Convention on Climate Change (1997) The Climate Change Secretariat, Bonn, Germany.

29 La Rovere, E. L. (1998) *The Challenge of Limiting Greenhouse Gas Emissions Through Activities Implemented Jointly in Developing Countries: A Brazilian Perspective.* Unpublished report, LBNL-41998. Ernest Orlando Lawrence Berkeley National Laboratory, Environmental Energy Technologies Division, University of California.

30 Lopez, R. (1996) *Policy Instruments and Financing Mechanisms for the Sustainable Use of Forests in Latin America.* Unpublished report, Resources for the Future (RFF), Washington, DC.

31 Lopez, R. (1999) *Incorporating Developing Countries Into Global Efforts for Greenhouse Gas Reduction.* Resources for the Future (RFF) Climate Issue Brief #16. RFF, Washington, DC.

32 Maldonado Ulloa, T., de Camino, T., Campos, J. J., Ortiz, R., Finegan, B. and Miranda, M. (1998) *Capacidad y Riesgos de Actividades Forestales en el Almacenamiento de Carbono y Conservacion de Biodiversidad en Fincas Privadas.* Unpublished report, Area de Conservacion Cordillera Volcanica Central, Costa Rica (Investigation colaborativa CIFOR-CATIE).

33 Malhi, Y., Baldocchi, D. D. and Jarvis, P. O. (1999) 'The carbon balance of tropical, temperate and boreal forests'. *Plant, Cell and Environment*, vol 22, pp15–40.

34 Moura Costa, P. and Wilson, C. (1999) *An Equivalence Factor Between CO_2 Avoided Emissions and Sequestration – Description and Applications in Forestry*. Unpublished report, Ecosecurities Ltd, Oxford, UK.

35 Mulongoy, K. (1999) 'Introducing forestry sector concerns in the development of joint implementation and the clean development mechanism', in Linguri (ed) *Climate Change in the Global Economy: Policy Dialogues of the International Academy of the Environment*, International Academy of the Environment (IAE), Geneva, pp21–30.

36 Mulongoy, K., Smith, J., Alirol, P. and Muehlmann, A. W. (1998) *Are Joint Implementation and the Clean Development Mechanism Opportunities for Forest Sustainable Management Through Carbon Sequestration Projects?* Unpublished report, International Academy of the Environment (IAE), Geneva.

37 Najam, A. and Page, T. P. (1998) 'The Climate convention: deciphering the Kyoto commitments'. *Environmental Conservation*, vol 25, pp187–194.

38 Newcombe, J. (1998) *The Risks and Environmental Benefits of Investing in Climate Change Projects under the Kyoto Protocol: An Investor Perspective*. Unpublished report, CIFOR, Indonesia.

39 Nilsson, S. and Schopfhauser, W. (1995) 'The carbon-sequestration potential of a global afforestation programme'. *Climate Change*, vol 14, pp243–262.

40 Ostrom, E. (1990) *Governing the Commons*. Cambridge University Press, Cambridge.

41 Pearce, D. (1996) 'Global environmental value and the tropical forests: demonstration and capture', in Adamowicz, W., Boxall, P., Luckert, M., Phillips, W. and White, W. (eds) *Forestry, Economics and the Environment*. CAB International, Wallingford, UK, pp11–48.

42 Pearce, D., Day, B., Newcombe, J., Brunello, T. and Bello, T. (1998) *The Clean Development Mechanism: Benefits of the Calm for Developing Countries*. Unpublished report, CSERGE (Centre for Social and Economic Research on the Global Environment), University College, London.

43 Pinard, M. A. and Putz, F. E. (1997) 'Monitoring carbon sequestration benefits associated with a reduced-impact logging project in Malaysia'. *Mitigation and Adaptation Strategies for Global Change*, vol 2, pp203–215.

44 Putz, F. E., Dykstra, D. P. and Heinrich, R. (2000) 'Why poor logging practices persist in the tropics'. *Conservation Biology*, vol 14, pp951–956.

45 Richels, R. and Sturm, P. (1996) 'The costs of CO_2 emissions reductions. Some insights from global analyses'. *Energy Policy*, vol 24, pp875–888.

46 Ridley, M. A. (1998) *Lowering the Cost of Emission Reduction: Joint Implementation in the Framework Convention on Climate Change*. Kluwer Academic Publishers, Dordrecht, The Netherlands.

47 Ruiz-Perez, M. and Byron, N. (1999) 'A methodology to analyze divergent case studies of non-timber forest products and their development potential'. *Forest Science*, vol 45, no 1, pp1–14.

48 Saxena, N. C. (1997) *The Saga of Participatory Forest Management in India*. CIFOR Special Publication, CIFOR (Center for International Forestry Research), Bogor, Indonesia.

49 Schimel, D. S. (1995) 'Terrestrial ecosystems and the global carbon cycle'. *Global Change Biology*, vol 1, pp77–91.

50 Sedjo, R. and Solomon, A. M. (1989) 'Climate and forests', in Rosenberg, N. J., Easterling III, W. E., Crosson, P. R. and Darmstadter, J. (eds) *Greenhouse Warming: Abatement and Adaptation*. Resources for the Future, Washington, DC, pp105–119.

51 Smith, J., Mourato, S., Veneklaas, E., Labarta, R., Reategui, K. and Sanchez, G. (1999) *Can Carbon Trading Induce Land Use Change in Slash-And-Burn Systems? Evidence from the Peruvian Amazon*. Unpublished report, CIFOR, Bogor, Indonesia.

52 Southgate, D. and Clark, H. L. (1993) 'Can conservation projects save biodiversity in South America?' *Ambio*, vol 22, pp163–166.

53 Sunderlin, W. D. and Resosudarmo, I. A. P. (1996) *Rate and Causes of Deforestation in Indonesia: Towards a Resolution of the Ambiguities.* CIFOR Occasional Paper 9, CIFOR (Center for International Forestry Research), Bogor, Indonesia.

54 Swisher, J. and Masters, G. (1992) 'A mechanism to reconcile equity and efficiency in global climate protection: International Carbon Emission Offsets'. *Ambio*, vol 21, no 2, pp154–159.

55 TERI (1998) *Clean Development Mechanism: issues and modalities.* Unpublished report, Centre for Global Environment Research, Tata Energy Research Institute, New Delhi.

56 Tomich, T. P., Kuusipalo, J., Menz, K. and Byron, N. (1997) 'Imperata economics and policy'. *Agroforestry Systems*, vol 36, pp233–261.

57 Trexler, M. C. and Haugen, C. (1995) *Keeping it Green: Evaluating Tropical Forestry Strategies to Mitigate Global Warming.* World Resources Institute, Washington, DC.

58 World Energy Council (1998) *Energy for Tomorrow's World-Acting Now, Annex 2:1 Global Energy Perspectives*, Nakicenovic, N., Grubler, A. and McDonald, A. (eds), Cambridge, Cambridge University Press.

Reinventing a Square Wheel: Critique of a Resurgent 'Protection Paradigm'[1] in International Biodiversity Conservation

Peter R. Wilshusen, Steven R. Brechin,
Crystal L. Fortwangler and Patrick C. West

According to the conventional wisdom that emerges in recent literature on international biodiversity conservation, current people-oriented approaches to protecting the world's biologically richest areas are failing miserably. This conclusion is set against the backdrop of a global biodiversity crisis. By even the most conservative estimates, progressive degeneration of ecosystem structure and function with its associated loss of species continues to occur at an alarming pace as a result of human activities ranging from slash-and-burn agriculture to large-scale timber harvesting to oil extraction. Given this situation, many conservation biologists view national parks and other protected areas as the last safe havens for large tracts of tropical ecosystems. Unfortunately, according to several expert observers, protected areas in the 'developing' countries that house these highly valued, species-rich zones are ineffectively managed, if at all, and thus provide little or no protection for biodiversity. Based on this analysis, some members of the conservation community advocate a renewed emphasis on strict protection through authoritarian enforcement practices. In essence, they argue that dire circumstances require extreme measures.

In this policy review chapter we explore this line of thinking by examining four recent works, each of which offers sharp critiques of current conservation approaches. Two books, in particular, stand out for their attempts to distil experiences from around the tropical world as a means to call for a new wave of strictly enforced nature protection. They are *Requiem for Nature* by John Terborgh (1999) and *Myth and Reality in the Rain Forest* by John F. Oates (1999). Two other books, *The Last Stand: Protected Areas and the Defense of Tropical Biodiversity,* edited by Randall Kramer, Carel van Schaik and Julie Johnson (1997), and *Parks in Peril: People, Politics and Protected Areas,* edited by Katrina Brandon, Kent Redford and Steven Sanderson (1998b), offer arguments similar to those of the first two about nature protection but present more nuanced discussions regarding action strategies.

The purpose of this chapter is to analyse the key elements of a resurgent protectionist argument, which form a common thread tying together these four books. We find

Note: Reprinted from *Society and Natural Resources*, vol 27, Wilshusen, P. R., Brechin, S. R., Fortwangler, C. L. and West, P. C., 'Reinventing a square wheel: Critique of a resurgent "protection paradigm" in international biodiversity conservation', pp300–311, Copyright © (2000), with permission from Taylor and Francis

that this argument's underlying reasoning appears rational on the surface but presents problems upon closer examination. In this regard, we identify five main elements of the protectionist argument that leave significant gaps when considered from a social science perspective. We argue that, in the absence of greater attention to these gaps, the authors' core conclusions calling for strictly enforced protection are operationally unrealistic and morally questionable as policy proposals. In effect, the authors provide important observations about the lack of protection in current conservation practices but largely fail to account for the logical implications of their conclusions. Thus, their reasoning is incomplete, not necessarily because of factual oversights but as a result of significant blind spots that overlook the deeply politicized nature of nature protection.

Our intent is to encourage debate on how biodiversity conservation strategies are conceived and carried out in order to increase their chances for achieving the goal of protection. This chapter is not a statement against nature protection. Nor is it an uncritical defense of sustainable use and integrated conservation and development projects (ICDPs). Rather, the paper argues for clarification on *how* nature protection can and should occur in ways that are not just ecologically sound but also pragmatically feasible and socially just. In sum, we believe that the success of international biodiversity conservation rests on our collective ability to negotiate legitimate, enforceable agreements. This would require significantly strengthening existing institutional and organizational arrangements. In cases where immediate emergency intervention is required to protect biodiversity, national and international conservation organizations should be held accountable for the political and social impacts of their actions.

In general, we find that the linear structure and content of the arguments presented conceal much of the social and political complexity of working in the 'developing' world and thus lead to inadequate recommendations. By not responding comprehensively to the human organizational complexity of conservation challenges, we conclude that exclusive reliance on authoritarian protectionism most likely will not achieve the desired end of nature protection. Indeed it may have the opposite effect. If this most recent version of the 'conservation through strict protection' strategy translates simply as government-led, authoritarian enforcement, as the books we reviewed seem to indicate, it would suggest that we have learned little from past failures. For this reason we argue that if the conservation community were to adopt the logic of this approach as policy, it would be tantamount to reinventing a square wheel. If, on the other hand, we can incorporate the important findings of conservation scientists into a broadly inclusive, critical debate on improving nature protection efforts locally, regionally, nationally and internationally, we stand a much better chance of advancing in ways that respond to ecological, political, and social justice parameters for success.

We wish to state at the outset that by looking closely at the more outspoken opinions that appear in the literature, we do not mean to condemn conservation as an activity or the conservation community at large. Rather, our intent is to uncover key points for debate, and these emerge most clearly in the works cited. We share the goal of protecting biodiversity worldwide and are keenly aware of the strong evidence of species loss. By encouraging critical debate we hope to confront head-on certain contentious issues that often impede concerted collective action in favour of nature protection.

Five Core Elements of the Protectionist Argument

Perhaps the best overall summary of the protectionist argument appears in Alan Rabinowitz's (1999, pp70–72) review of *Requiem for Nature*. Rabinowitz, a well-known conservation scientist with the Wildlife Conservation Society, states:

> I marvelled at his [Terborgh's] insights, empathized with his frustrations and applauded his courage in speaking out about the mistakes of the conservation movement . . . He also dismisses key aspects of the popular concept of sustainable development. . . Terborgh warns, sustainable development is unattainable, and the further degradation of parkland is inevitable . . . Biodiversity conservation is doomed to failure when it is based on bottom-up processes that depend on voluntary compliance. Like him, I would also advocate a top-down approach to nature conservation – contrary to much contemporary political and conservation rhetoric – because in most countries it is the government, not the people around the protected areas, that ultimately decides the fate of forests and wildlife.

The underlying urgency and scepticism in Rabinowitz's statement pervades all four books. Overall, these authors conclude that current people-oriented approaches to biodiversity conservation have taken on too broad an agenda that features conflicting objectives – species protection and sustainable development. They find that conservation programmes have become diluted by strategies that promote community development and greater local participation in decision making. While Rabinowitz and others recognize that community development and participation may be noble goals, they argue that 'politically correct' approaches to conservation channel away a significant portion of available funding yet produce minimal results in terms of biodiversity protection. Consequently, proponents believe that conservation programmes should stop trying to be all things to all people and simply focus on the central goal of nature protection. The heightened emphasis on protection evident in these writings suggests strict enforcement and defence of protected areas.

In general, the assertions that emerge in the literature can be organized into five interrelated themes, highlighting (1) the central importance of protected areas, (2) the moral imperative of nature protection, (3) the ineffectiveness of conservation linked to development, (4) the mythical status of harmonious, ecologically friendly local people, and (5) the immediate need for strictly enforced protection measures. In the sections that follow, we critically analyse each of these themes, drawing direct quotes from the books we reviewed. In many cases we agree with the general findings while pointing to key issues that are left out of the protectionist argument. In other words the incomplete nature of the protectionist argument is apparent at times in what the authors say but also in what they do not say.

Protected Areas Require Strict Protection

In the book *Last Stand,* van Schaik et al (1997, p64) open a chapter called 'The Silent Crisis' with the following grim conclusion:

Table 17.1 *Five incomplete aspects of the protectionist argument*

Argument	Counterpoints
1 Protected areas require strict protection • 'The final bulwark erected to shield tropical nature from extinction is collapsing' (van Schaik et al, 1997, p64) • Nature continually loses out in the face of human population increase and economic growth	Greater protection is required but we need to clarify how it can and should occur. Exclusive focus on the ecological maintenance role of protected areas: • masks their political role including territorial control, domination by rival social/ethnic groups, and advancement of elite interests; • hides historical trajectories often associated with colonial domination and/or state-led coercion; • overlooks how they alter social and political 'landscapes'
2 Biodiversity protection is a moral imperative • Utilitarian rationale for preservation emphasizes real and potential use values • However, in most cases tropical forests are 'worth more dead than alive' (Terborgh, 1999, p18).Thus protection mandate rests primarily on moral rationale	Fine as far as it goes but moral argument: • ignores how different cultural groups' perceptions of the natural world might affect dialogue on conservation; • assumes that local and non-local interests are of the same order and carry the same weight; • hides the widely held perception that the 'common good' refers to elite special interests; • presents a series of choices structured around excluding or trading human rights in favour or at the expense of intrinsic rights of nature
3 Conservation linked to development does not protect biodiversity • Sustainable use depletes biodiversity. Not all places should be open to use • ICDPs have not effectively safeguarded protected area core zones	Points are well taken but argument: • ignores social and political realities (ie, preexisting use rights) to which interventions must adapt; • misses possibility that ICDPs' lack of protection 'success' could stem from implementation shortfalls rather than fundamental incompatibility of conservation with development; • overlooks impact of intervening variables like conflict, organization and governance; • ignores numerous cases where state intervention disrupts traditional institutions that govern self-enforcement of resource use
4 Harmonious, ecologically friendly local communities are myths • Because of rapid social change, resource management by indigenous and other traditional peoples cannot guarantee species protection • The 'ecologically noble savage' (Redford, 1990) does not exist	Attempts to counteract stereotypes of local people overgeneralize in the opposite direction. In general, they: • imply that no 'traditional' peoples are able to conserve their resources; • oversimplify rural communities' motivations and cultural practices; • overlook how decision making, organization and governance institutions shape peoples' motivations and abilities to participate; • assume that local institutions cannot adapt to social change

Table 17.1 *Five incomplete aspects of the protectionist argument*

Argument	Counterpoints
5 Emergency situations require extreme measures • Governments have the duty to limit individual freedoms to protect the common good • Governments and aid organizations should encourage industrial development to provide economic opportunities for the rural poor • There may be a role for the national military or international nature protection forces in biodiversity conservation efforts	Recommendations are inadequate from both a pragmatic and moral perspective. They: • assume that governments serve the common good of their citizens; • rely on long-term social 'engineering' that sacrifices some areas to conventional development (urbanization and industrialization) to depopulate and protect other ecologically rich areas; • ignore the possibility that the military might use conservation as an excuse for territorial control, ethnic cleansing

> In the tropical forest realm . . . protected nature preserves are in a state of crisis. A number of tropical parks have already been degraded almost beyond redemption; others face severe threats of many kinds with little capacity to resist. The final bulwark erected to shield tropical nature from extinction is collapsing.

The problem, according to scientists like John Terborgh (1999, pp17–20), is that nature continually loses out in the face of human population increase and economic growth. In *Requiem for Nature* he asserts that:

> Ultimately, the issue boils down to habitat – how much for humans and how much for nature. In the late 1990s, the global balance stands at roughly 5 per cent for nature (counting only parks and other strict nature preserves) and 95 per cent for humans ... Economic forces, driven by population growth and the fervent desire of people everywhere to advance their material well-being, are inexorably eliminating the world's remaining wildlands. Short of radical changes in governmental policy in country after country, all unprotected tropical forests appear doomed to destruction within thirty to fifty years. When that time arrives, the only remaining examples of tropical nature and, consequently, most of what remains of tropical biodiversity will reside in parks. Parks therefore stand as the final bulwark of nature in the Tropics and elsewhere.

The central tenet of the protectionist argument is straightforward. It holds that all existing protected areas restricting human occupation and use (International Union for the Conservation of Nature [IUCN] categories I–IV, which account for 64.8 per cent of all protected areas, (Cox, 2001)) must be enforced.[2] We support this conclusion. Protected areas have been and will continue to be essential elements of global biodiversity conservation strategies. Proponents of strict protection present convincing scientific arguments for dedicating significant time and energy to protected areas management (Terborgh and van Schaik, 1997). In particular, many species – especially large mammals – need extensive, undisturbed tracts of habitat to ensure their survival. Most observers agree that protected areas are fundamental to maintaining ecological structure and function.

From our perspective the issue is not *whether* protected areas should be enforced but *how*. In this regard, advocates quite understandably tend to emphasize the ecological maintenance role of protected areas. This, however, has the effect of downplaying or ignoring their political side (exceptions include Sanderson and Bird (1998) and Sanderson and Redford (1997)). Looking at the politicized nature of protected areas helps explain why conflict and resistance so often develop in response to parks and their management.[3] In other words, the political trajectories of protected areas to a large extent shape how they are perceived by local people and other players, including, most importantly, the degree of legitimacy that management restrictions carry. Protected area managers deal with these political realities on a daily basis and yet the broader policy implications have been explored only superficially.

Conventional deliberations surrounding protected areas focus on zones (core, buffer, multiple-use), categories (I–VI), corridors and boundaries. Discussion of protected areas in these terms does two things. First, it legitimates land use and enforcement on ecological grounds. On one level this is a necessary process from both a scientific and management perspective. The flip side of this, however, is that scientific and managerial explanations mask other political understandings of the roles that protected areas play in the context of wider social and cultural spheres. For outsiders looking in, such as resource-dependent agrarian communities, protected areas are not necessarily understood as a means of providing ecological and economic services but rather as territorial control strategies. In many cases, members of rural communities have not been told that they live near a protected area, and, even if they do know, they may not be familiar with strategies employed by the protected area. In other instances, local reactions are more violent. In Mexico, for example, rural residents in southeastern Campeche only found out about the creation of Calakmul Biosphere Reserve a year after it was gazetted when scientists showed up to start ecological studies. In the context of hotly contested land claims within the new reserve, some farmers threatened to kill anyone in the region claiming to be an ecologist (Haenn, 1997).[4] In still other cases, parks are viewed as playgrounds for the wealthy (an example would be Thailand's Khao Yai National Park with its two golf courses). Recognizing the politicized nature of protected areas is important because it suggests that awareness-raising and attempts at consultation or 'participation' most likely will not change many rural peoples' suspicions and resistance unless dialogue attends to broader social and political factors.

Beyond political considerations, local and regional resistance to protected areas takes on clearer meaning when viewed in historical perspective. Whether fully accurate or not, many outside observers, including communities, activists, government officials and others, view conservation as yet another manifestation of external control that, in some contexts, mirrors earlier eras of imperial domination. For example, the notion that protected areas serve elite interests in places like East Africa and India can be traced to the creation of royal game reserves for trophy hunters (MacKenzie, 1988; Grove, 1990; Neumann, 1997; Neumann, 1998). In another sense, sources claiming that conservation practices are often excessively coercive point to the use of protected areas over time as a means of elite control of territory or rival ethnic groups (Peluso, 1993; Hitchcock, 1995). An historical perspective of protected areas and conservation practices is important for two reasons. First, it helps explain why some local communities view management restrictions as illegitimate, since compliance develops over years of political interaction rather than as a result of outright acceptance of abstract management regulations. Second, a historical understanding of people and park dynamics offers insight on how current scientific management approaches inevitably inherit past legacies (see Zerner, 2000).

As many of the authors associated with the protectionist argument suggest, protected areas managers should not be expected to take on all of these highly complex and often volatile social problems alone (Redford et al, 1998, pp462–463). At the same time, however, the conservation community automatically becomes a key player among a host of others since it contributes heavily to shifts in power dynamics in rural areas that are already highly politicized. This is a result of its relative wealth and influence compared to most local actors. In short, conservation practices are not benign. They alter the local playing field, sometimes drastically.

By touching on these points we do not mean to imply that fair enforcement is impossible in protected areas management. Rather, we wish to point out that the enforcement process requires a great deal more of concerted negotiation with affected parties than is often assumed in order to be viewed as legitimate.

Biodiversity Protection is a Moral Imperative

The most commonly cited rationales for protecting biological diversity constitute a combination of pragmatic and moral arguments. Kramer and van Schaik (1997, p8) summarize the pragmatic argument in *Last Stand* by explaining the utilitarian importance of preserving biodiversity in terms of economic use and non-use values. This type of thinking emphasizes real and potential use values of plant and animal species as sources of new pharmaceuticals, genetic banks for key agricultural crops, and environmental services such as flood control, as well as non-use values, which imply maintaining natural areas for recreation or other reasons.

While economic and life-supporting rationales for preserving biodiversity are compelling, many analysts recognize that, when considered in terms of net present value (the deciding factor for most resource users), tropical forests are, in John Terborgh's (1999, p18) words, 'worth more dead than alive'. If we accept Terborgh's conclusion, then the rationale of protectionist thinking rests primarily on a moral foundation that is succinctly summarized in his *Requiem for Nature* (p19):

> Ultimately, nature and biodiversity must be conserved for their own sakes, not because they have present utilitarian value. Essentially all the utilitarian arguments for conserving biodiversity are built on fragile assumptions that crumble under close scrutiny. Instead, the fundamental arguments for conserving nature must be spiritual and aesthetic, motivated by feelings that well up from our deepest beings. What is absolute, enduring, and irreplaceable is the primordial nourishment of our psyches afforded by a

quiet walk in an ancient forest or the spectacle of a thousand geese against a blue sky on a crisp winter day. There are no substitutes for these things, and if they cease to exist, all the money in the world will not bring them back.[5]

Terborgh's moral argument is based on two basic rights: (1) the intrinsic right of nature to exist (ie humans do not have the right to eradicate other species), and (2) the right of global, regional and local communities to enjoy the aesthetic qualities of nature now and in the future. These rights underlie the belief held by many conservationists that the international community can and should act on behalf of nature in different parts of the planet as 'global citizens'. This typically translates into a justification for foreign involvement in the management of a country's biodiversity (van Schaik and Kramer, 1997, p216).[6] A closely related point focuses on the common tension between local and non-local interests and values. Sanderson and Redford (1997, p129) argue, 'local institutions should be respected wherever possible; but . . . local initiatives cannot automatically supplant national or regional system values; and . . . global biodiversity concerns should be built around landscape-scale system management.' Similarly, van Schaik and Kramer (1997, p217) note:

> devolution of protected area management is likely to be effective, but . . . the interests of other stakeholders, be it the nation or the international community, should always be represented. Thus the strong point of devolution (reduced bureaucracy) should be married to the strong point of state involvement (promotion of the common good).[7]

In many respects the pragmatic and moral arguments underlying the protectionist perspective are fine as far as they go. At the same time, however, they fail to account for two crucial points. First, in general, the arguments do not acknowledge that there are different ways of understanding and appreciating nature that directly affect dialogue on and proposals for biodiversity protection. It makes sense that the conservation community, which is made up largely of scientists and professionals raised and trained in the West, would grant priority to scientific explanations and strategies for protecting biodiversity (see Taylor, 2000). In practical terms, we fully recognize that the discourse of science will continue to be the predominant lens for viewing conservation problems and solutions. However, by failing to recognize intellectual and philosophical traditions besides hypothetico-deductive science, conservationists tend to limit dialogue with other groups, including a wide variety of local communities. At the very least, it behoves us to take the time to explore the complex social and cultural histories that shape different peoples' understandings of and relationships to the natural world. On one level, this approach could allow for more mutually understandable dialogue (even if fundamental conceptual or philosophical contradictions emerge). On a second level, attempts at finding a common language increase the likelihood of uncovering creative solutions that carry greater legitimacy for all parties.[8]

A second oversight stems from the protectionist claim that local interests should not automatically supercede regional, national and (presumably) global interests. The authors correctly point out that nature protection serves the common good, which both justifies local restrictions and impacts as well the stakeholder status of the state and the international community. At the same time, however, the argument assumes that both types of interests (local and non-local) are of the same order and carry the same weight. This is important because it affects the legitimacy of claims. How should we respond to the common 'local' argument that imposed conservation restrictions produce not only economic hardship but also irreversible cultural impacts? For example, representatives of black and indigenous communities in Colombia's Pacific Coastal region posed this question: 'If the forest continues to exist because of our way of life and our way of life is directly tied to the land, why would we sacrifice our livelihoods in the name of the "common good?"' (Wilshusen, 2003). Many people living in and near protected areas perceive their interests as tangible and immediate and the 'common' interest as unclear and intangible. Underlying this view is the issue of distributive justice, including the widely held belief that the 'common good' refers to elite special interests imposed on the rural poor. This argument suggests that dialogue on conservation would need to clarify this balance of interests in specific contexts and attend to the perception that local people carry a disproportionate burden in terms of negative social, cultural and economic impacts.[9]

The pragmatic and moral arguments forwarded by proponents of strict protectionism – even those that contemplate local rights and participation – present a series of choices structured around excluding or trading human rights in favour or at the expense of the intrinsic rights of nature. In other words, this conventional approach tends to create a false moral dilemma in which the rights of humans are pitted against the rights of nature. This type of binary thinking creates a zero-sum decision-making scenario where nature 'loses out' by default when humans 'win'. Win-lose scenarios imply that there is no room for dialogue or negotiation. For example, in reading Terborgh (1999) and Oates (1999), one comes away with the impression that trading in rights is necessary, given the urgency of the situation. The arguments seem to suggest a willingness to trade the rights of what is perceived to be a small percentage of human populations in and around protected areas in favour of the intrinsic rights of nature. Since the issue of nature protection is framed as a moral, win–lose proposition, debate along these lines produces opposing camps that might be described as 'pro-nature' versus 'pro-people'. This confrontational dichotomy forces a separation between humans and nature as a means of protecting 'wildlands' that precludes dialogue on how both human rights and the intrinsic rights of nature can be promoted simultaneously. Since people do not necessarily separate a 'human' or 'culture' realm (Reichel-Dolmatoff, 1976; MacCormack and Strathern, 1980; Croll and Parkin, 1992; Seeland, 1997; Johnson, 2000), dialogue structured by these rigid oppositions will most likely produce, at best, very superficial commitment or, more likely, strong resistance to programmes structured with such a separation in mind.

Regardless of whether one agrees or disagrees with the moral argument concerning nature's intrinsic rights (and humans having no right to destroy it), biodiversity conservation programmes do not necessarily have to hinder the attainment of human rights for all people living in or near protected areas. The moral argument in favour of nature protection is perfectly defensible as far as it goes. But if nature protection occurs at the expense of humans without accountability, based on a separation between humans and nature, then it becomes less defensible. Nor should the debate over international biodiversity conservation end as a simple moral standoff: pro-nature versus pro-people.

Conservation Linked to Development Does Not Protect Biodiversity

The main critique of much of the work we reviewed centres on the perceived failure of conservation with development in protecting species in parks and reserves. In this light, Kramer and van Schaik (1997, p7) expose what they see as a fundamental conflict between conservation and development at the local level. This incompatibility takes the form of two main conclusions. One is that sustainable use depletes biodiversity (Robinson, 1993; Redford and Richter, 1999). The second is that integrated conservation and development projects have not effectively safeguarded protected area core zones. Regarding sustainable use, the editors of *Parks in Peril* (Brandon et al, 1998a, p2) state:

> There are limits on sustainable use as a primary tool for biodiversity conservation. Serious questions as to whether sustainable use is axiomatically compatible with biodiversity conservation have been raised. The trend to promote sustainable use of resources as a means to protect these resources, while politically expedient and intellectually appealing, is not well grounded in biological and ecological knowledge. Not all things can be preserved through use. Not all places should be open to use. Without an understanding of broader ecosystem dynamics at specific sites, strategies promoting sustainable use will lead to substantial losses of biodiversity.

This statement is perfectly reasonable as far as it goes. It is important to note that the authors associated with this statement critique sustainable use as a means of achieving species protection in and around parks and reserves. In particular, the books *Last Stand* and *Parks in Peril* are careful to point out that use strategies may be worthwhile community development pursuits *away from* protected areas (Kramer and van Schaik, 1997, pp7–8). However, while it seems reasonable to propose that not all places should be open to use, it is important to clarify whether all protected areas (IUCN categories I–IV) should be closed to use and to justify these findings to policy makers. A recent debate in the journal *Conservation Biology* regarding sustainable use in Amazonia suggests that the evidence against controlled resource use is not as conclusive as proponents of the protectionist argument suggest (Redford and Sanderson, 2000; Schwartzman, et al, 2000a; Schwartzman et al, 2000b; Terborgh, 2000).

Even if we accept the view that sustainable use will not maintain species diversity, most critics ignore the important role that it plays as one component of a broader landscape conservation strategy. Many of the arguments on the futility of sustainable use take their cue from the article 'The Limits to Caring', in which John Robinson (1993) of the Wildlife Conservation Society asserts that exclusive reliance on this approach – as proposed in the policy statement *Caring for the Earth* (IUCN et al, 1991) – will almost always lower biological diversity. Unlike Terborgh (1999) and Oates (1999), however, he concluded that it may be an important conservation tool in certain circumstances. By grounding the sustainable use debate entirely in ecological terms, critics hide the fact that controlled use may be the only viable political and economic alternative for large tracts of tropical forest and other ecosystems. In a recent critique of the sustainable use argument already described, Schwartzman et al (2000a, p1352) drew on their experiences in

Amazonia to suggest that 'the real choice in large parts of the tropics . . . is between forests inhabited and defended by people and cattle pastures or industrial agriculture'. As in Amazonia, sustainable use strategies have played a central role in reversing large-scale land-use conversion of tropical dry forests in Quintana Roo, Mexico. Without the long-term intervention of a Government of Mexico/German aid agency (GTZ)-led programme in community forest management, it is highly likely that most of the newly created agrarian reform settlements in the southern part of the state would have chosen indus-trial agriculture and cattle ranching for economic production. Currently, the commu-nity-managed forests in central and southern Quintana Roo form a densely forested corridor linking the Sian Ka'an and Calakmul biosphere reserves.

In this sense, the majority of arguments against sustainable use in and around pro-tected areas tend to leave out the fact that, in most cases, parks overlap with or adjoin areas with pre-existing land-use rights. In other cases, landless migrants arrive from other regions in search of resources or political refuge. One strategy for dealing with these political realities has been to work proactively with communities in buffer zones to pur-sue local development in ways that direct resource use away from core protected zones. If this strategy has not been effective, as critics claim, it would appear to be more a prob-lem of implementation rather than concept. While those groups with pre-existing land-use rights may have greater legal and moral standing for maintaining resource access and use in buffer zones (or in some cases within core zones), questions of strict protection versus sustainable use cannot be answered in the abstract. For example, in situations where large numbers of landless migrants seek resources and refuge in and around a protected area, strict boundary maintenance combined with basic relief services may be the best option. In other circumstances, protected area managers may be able to work proac-tively with well-established communities to negotiate legitimate and binding agreements for controlled use. In any case, we would argue that exclusive reliance on authoritarian protection – just as with universal application of sustainable use as critiqued by Robinson (1993) – will not protect species. More likely, it would prove politically suicidal for most government officials and produce violent resistance by resource-dependent populations, which has been the case historically (West and Brechin, 1991; Fortwangler, 2003).

The second major assertion underlying the protectionist argument that conserva-tion linked to development has failed focuses on integrated conservation and develop-ment projects (ICDPs). This finding appears in several different forms and each offers some well-grounded observations. One such theme argues that conservation programmes have become diluted with social goals like poverty reduction and social justice. In *Last Stand*, Brandon (1997, 104–105) wrote:

> At both the field and policy levels, the links between poverty and environment remain ill defined. To push at the forefront of the poverty and environment nexus would require defining the policies that can protect biodiversity most effectively while also helping the poor. What is known is that alleviating poverty will not necessarily lead to improve-ments in biodiversity conservation.

This statement makes some important points, including the widely held perception that ICDPs and other participatory strategies have emphasized development much more than nature protection. Katrina Brandon's work in particular has uncovered a number of assumptions or 'myths' associated with joining conservation and development, of which the supposition that poverty alleviation will protect biodiversity is but one example (Brandon, 1996, 1997, 1998b). The core idea of the ICDP approach, that local people will stop exploiting resources within parks if they achieve increased incomes or are otherwise economically compensated for 'opportunity costs', relies on a number of tenuous assumptions about human behaviour.

In a similar vein, John Terborgh (1999, pp165, 169) asserted that ICDPs actually increase rather than reduce human-use pressures on protected areas (see also Oates, 1995; Brandon, 1997).

> Despite the gilded rhetoric presenting them as conservation endeavours, ICDPs represent little more than wishful thinking. Project objectives typically have little direct relevance to the protection of biodiversity. To the contrary, project managers who successfully innovate and invigorate the local economy risk aggravating the very problem they are trying to solve. By stimulating the local economy, an ICDP attracts newcomers to a park's perimeter, thereby increasing the external pressure on the park's resources.

Terborgh argued further that by focusing entirely on local level processes, ICDPs do not effectively attend to larger political economic processes that affect how local people make resource use decisions; these forces, Terborgh contended, are the root causes of some of the main pressures on protected areas.

> Misconception embodied in the ICDP concept is the conviction that social change can be brought about through bottom-up processes. The ICDP approach assumes that the destiny of parks lies in the hands of local people, an assumption that is only partly correct. What ICDPs do not take into account is that local people are only minor players in a much larger theater. The lives of village people are strongly influenced by decisions of the central government and conditions determined by it: construction of roads; availability of rural credit, subsidies, and tax incentives; inflation versus stability of the national currency; raising or lowering trade barriers; laws governing labour practices; receptivity to foreign capital; and so forth. Against powerful forces such as these, ICDPs pale into utter insignificance. Now we have come full circle. Unable to grasp the stick of enforcement, conservation organizations turned to the carrot of economic assistance, but they must now come to grips with the failure of that approach as well. Bottom-up processes initiated at the village level will not improve the security of parks because they rely 100 per cent on voluntary compliance.

Published reviews of ICDPs touch on these and other shortcomings, concluding that projects from around the world present mixed results vis-à-vis their objectives (Wells and Brandon, 1992; Larson et al, 1997). Indeed, ICDPs appear not to offer biodiversity protection at a sufficient scale or degree to guarantee species survival. A recent evaluation of ICDPs in Indonesia presents this finding as one of its main conclusions (Wells et al, 1999). Another comparative study of community-based natural resource management projects with cases from Nepal and Kenya presents similar findings (Kellert et al, 2000). It is vitally important to continue to assess ways of instituting better protection measures for protected areas. At the same time it would be a mistake to assume out of hand that if ICDPs do not sufficiently address biodiversity protection then we have nothing to learn from the approach and thus it should be tossed out as a policy tool.[10] Both Brandon (1997, 1998b) and Terborgh (1999) stop short of such a statement, but it is a point worth clarifying. If we briefly revisit the linear argument in both commentaries we will find that the reasoning is fine as far as it goes. Yet, there are also misconceptions and hidden aspects that have a role to play in debates over how conservation should be carried out.

In general, the authors from all four books find that the pursuit of human development goals at or near protected areas is axiomatically opposed to nature protection. This type of logic is appealing when framed as a direct, one-to-one relationship but ignores a whole series of 'intervening variables' related to social process including conflict, organization and governance. This is critical when it comes time to evaluate programmes. We may be able to establish that ICDPs in most cases have not led to species protection. If, however, we ask why this is so, the task becomes more complicated. Brandon's observations on the overly 'diluted' condition of ICDPs are well taken, and yet what can we learn as a result? Certainly local protected area managers are overwhelmed by the array of tasks that come with multiple objectives. At the same time, an integrated approach may respond more comprehensively to the highly complex array of social, economic, political and ecological factors that shape the challenges of nature protection. Both *Parks in Peril* and *Last Stand* note this possibility but argue that other parties besides conservation organizations should be responsible for social development. While conservation programmes cannot be expected to 'do it all' (poverty reduction, social justice, sustainable development), they also cannot simply disengage from the social and political context by leaving the 'social work' to others as these arguments imply. Does the presumed ineffectiveness of ICDPs to protect biodiversity mean that the approach is a complete failure, or could it indicate that we have yet to develop adequate political and organizational arrangements to pursue all goals equally well?

Terborgh's comments on ICDPs illustrate how viewing conservation problems with lenses that filter out politics can lead to important observations that breed inadequate solutions for saving species. He is quite correct to point out that wider political economic forces strongly influence local actions. And although he does not mention it directly, most observers further recognize that timber, oil, mineral and other large resource exploitation enterprises (the so-called 'resource pirates') with strong political connections often represent much greater threats to protected areas than rural communities (van Schaik and Kramer, 1997, pp224–226). Regarding local processes, however, his comments present two important misconceptions. First, he suggests that local communities have no influence over national political and economic decisions. While local people are often hampered by national development policy, as Terborgh points out, it is incorrect to assume that they are powerless in all cases. In Mexico, for example, community forestry associations played a strong role in shaping national forestry policy in both 1986 and 1997.

Second, Terborgh appears to think that voluntary compliance (associated in this case with bottom-up processes) precludes self-enforcement. Indeed, self-enforcement may not occur in many cases, and it would be naive to assume that outside enforcement is unnecessary. At the same time, many traditional societies, groups of rural communities and so on operate under complex sets of rules and responsibilities that govern resource use and provide for differing degrees of self-enforcement. In fact, modern legal institutions often directly conflict with customary legal institutions, and a complex challenge is understanding how and why this occurs. By ignoring questions of social and political process associated with 'doing conservation', Terborgh and other authors fail to recognize the important lessons that practitioners have learned about working with a wide range of groups including communities, cooperatives, local government and state agencies. By not asking how groups are organized and why they may or may not respect protected area management restrictions, Terborgh inevitably falls back on the simplistic conclusion, which suggests that the stick must once again supplant the carrot.

In the absence of greater clarification regarding the specific role that ICDPs might play within a broad conservation strategy, we may find some policy makers promoting the idea that conservation with development is useless and deciding to throw out the good with the bad. If we were to uncritically revert entirely to protectionism based on this reasoning, it could effectively derail existing attempts to build alliances, strengthen organizations and negotiate programmes with people at all geographic scales.

Harmonious, Ecologically Friendly Local Communities are Myths

The core of the critique just presented on sustainable use and development is rooted in two related observations about so-called 'traditional' people. As with the discussion on use, the four books we reviewed react to much of the grey literature and promotional materials produced by the large, international conservation organizations such as the World Conservation Union (IUCN) publication *Caring for the Earth*. The first observation concludes that community-based natural resource management by indigenous and other traditional peoples cannot guarantee species protection. This is due to rapid social change, which is causing these groups to lose the very 'traditional' qualities that historically allowed them to live in relative harmony with nature compared to modern societies (eg Terborgh, 1999, p51). Kramer and van Schaik (1997, pp6–7) wrote:

> It is often claimed that forest resources would be well managed if only the traditional users were allowed to maintain control. It is, indeed, widely believed that traditional communities use their resources in a sustainable manner. This belief is based on the fact that traditional communities lived at low densities, had limited technology, and practiced subsistence rather than commercial utilization. Unfortunately, given growing population pressure, increased access to modern technology, increasing market orientation, and steady erosion of traditional cultures, there no longer are guarantees that biodiversity objectives will be any more likely to be achieved if resource control is placed in the hands of indigenous groups.

Closely related to this argument is the second observation, which debunks the myth of the idyllic native living in perfect harmony with other community members and with nature (Robinson 1993; Redford and Mansour, 1996; Brandon, 1997; Redford et al, 1998; Redford and Richter, 1999). In *Myth and Reality in the Rain Forest,* John Oates (1999, p55) asserted:

There is little robust evidence that traditional African societies (or indeed 'traditional' societies anywhere in the world) have been natural conservationists. On the contrary, wherever people have had the tools, techniques, and opportunities to exploit natural systems they have done so. This exploitation has typically been for maximum short-term yield without regard for sustainability; unless the numbers of people have been very low, or their harvesting techniques inefficient, such exploitation has usually led to marked resource depletion or species extinction. There are instances where strict hunting controls have existed, but these have typically been in hierarchical societies where leaders have wished to control the access of others to resources, especially the rarest and most prized resources.

It is fair to say that much of the conservation literature does tend to glorify indigenous peoples specifically and 'traditional' communities more generally. Like Borrini-Feyeraband (1996), Agrawal and Gibson (1999) cautioned community-based conservation advocates against ignoring the complex interests and processes within communities, and between communities and other social actors. They point out that 'community-based conservation reveals a widespread preoccupation with what might be called "the mythic community". . .[this] vision fails to attend to differences within communities' (see also Brosius et al, 1998; Belsky, 1999; Belsky, 2003). As Redford et al (1998, p458) perceptively pointed out in *Parks in Peril,* 'It is clearly not that communities are "bad" but rather that they must not be stereotyped. Some will actively work to conserve some components of biodiversity; others will not, and have not'.

However, by attempting to counteract this trend of stereotyping local people, several critics overgeneralize in the opposite direction. Their arguments seem to imply that since all 'traditional' peoples (whomever they may be) are not the 'natural conservationists' they are made out to be, then conservationists should abandon feel-good, bottom-up approaches and get back to the business of nature protection. Just as with the related critiques of sustainable use, the scepticism of traditional peoples' status as conservationists betrays numerous simplistic assumptions about communities' motivations and cultural practices. For example, Oates (1999) creates the impression that resource-dependent peoples act only to maximize short-term gains independently of wider political and economic factors that might encourage them to do so. In general, even though *Last Stand* and *Parks in Peril* provide qualifying remarks in several chapters, the authors' criticism of 'myths' associated with traditional peoples, sustainable use and community-based management function to discredit participatory strategies as management tools for parks and reserves. Without further guidance, some policy makers might conclude that sustainable use, community-based conservation and co-management are flawed and thus have no role in protected areas management. This type of reasoning largely ignores decision making, organizational and governance processes – both customary and modern – that structure resource use within and among rural communities. Scholars of common property resource management document how these complex institutions develop and change over time (Ostrom, 1990). By toppling the myth of the 'ecologically noble savage', authors such as Redford (1990) point to the significant changes that traditional societies face in the modern era. At the same time, they tend to ignore the possibility that, even in the face of rapid change, these groups might be able to adapt patterns of use to a more sustainable pathway, especially with outside support.

What should one conclude from these arguments? True, there should be no room for 'noble savage' imagery in today's conservation. And true again, some communities may be too weak and divisive to participate in much of anything let alone complicated conservation and development interventions. How shall we work with local people in light of these social and political challenges? While some of the authors cited favour continued local involvement under certain circumstances (MacKinnon 1997; Brandon 1998b), more outspoken commentators such as Terborgh (1999), Oates (1999) and Rabinowitz (1999) advocate authoritarian enforcement that amounts to circling the wagons to keep the 'natives' out.

Emergency Situations Require Extreme Measures

Given the often strong attacks on biodiversity conservation and protected areas management as it is currently practised, what alternatives are left? All four books adopt a tone of urgency regarding the rapid extinction of species and frustration stemming from a lack of focus in contemporary conservation efforts. The most detailed proposals for protecting biodiversity emerge in *Last Stand*. In the concluding chapter, van Schaik and Kramer (1997) discussed action strategies that consider two broad sets of causes for protected area degradation – those brought about by 'small players' and those precipitated by 'big players'. The majority of attention centres on 'small' or local players. In this regard, they focus on (1) the state's role in limiting personal freedom for the public good, (2) economic development and incentives and (3) possible military intervention. Referring to the state's role in protecting the public interest, van Schaik and Kramer (1997, pp218–220) observed:

> Governments of civilized nations have the duty to ask their citizens to accept restraints on their freedom of action when it serves the common good. And governments have established enforcement mechanisms in implicit or explicit recognition of the underlying conflict of interest. . . In the case of tropical forest parks, governments can claim forest lands as national property because they serve national and international interests.

On the face of it, the notion that the governments of 'civilized nations' have a duty to impose restrictions on individual freedoms in the name of the common good seems perfectly reasonable. Indeed, one of the strongest arguments in favour of nature protection, in general, invokes government's role as steward of resources in public trust for the benefit of all citizens. While this is certainly an important doctrine in many contexts, one must be careful not to assume that in reality governments actually serve the public interest. In many cases, elite groups working within government agencies use their power to favour special-interest groups such as logging and mineral exploitation industries. This is important because it helps to explain why, when rural communities in developing countries are asked to accept restrictions on their individual freedoms, they often cite past government abuses as reasons for viewing government regulations as illegitimate. Beyond the inevitable conflicts of interest that must be governed in these situations, we cannot assume that government enforcement mechanisms are necessarily legitimate.

Although outside enforcement is necessary in many situations, is it appropriate to impose restrictions on those at-risk groups that bear a disproportionate burden of impacts associated with conservation when governments clearly are not acting to serve the public welfare?

In line with the state's strong protection mandate is the second alternative, which might be called the 'social engineering' approach since it favours increased industrial development to encourage even greater rural to urban migration than already exists, mirroring demographic trends of 'developed' nations. The editors of *Last Stand* wrote:

> The most effective long-term solution is to provide aid aimed at improving urban infra-structure elsewhere that encourages industrial development. This development would act as a source of employment for the supernumerary rural poor, much in the same way that in the northern hemisphere the rural population surplus was absorbed into the developing cities. Industrialization and urbanization in the tropical world are proceeding apace with this historical trend, and should lead to significant reductions of pressure, if the options available to people are considered more attractive than subsistence farming or extraction (p222).

In the absence of further clarification, one could conclude that this proposal promotes 'First World' conventional economic growth – and its associated environmental impacts – in one place to protect nature in another. If this is the case, the concept fails to account for the resource consumption needs of the 'supernumerary rural poor' as they move to urban areas. Would not these resources most likely come from the countryside? Beyond the high potential for ecological degradation inherent in this second proposal, a number of questions arise. Is it reasonable to think that depopulating the countryside is a viable alternative for guaranteeing conservation success? Is it appropriate to encourage processes that more than likely would lead to the further enlargement of urban poverty belts surrounding megacities? Would not the process of rural to urban migration simply make it easier for large enterprises with little interest in conservation to buy up property for large-scale resource exploitation?

The third alternative presented in *Last Stand* raises the possibility of including the military in nature protection.[11] Van Schaik and Kramer (1997, p224) wrote:

> Some . . . have suggested that there is a role for the national military in this regard. This is not as far fetched as it sounds, since the role of the military is to protect the nation's interest, usually against outsiders but in case of emergency also against rebellious insiders. Moreover, the military is often the only power with authority and is the best-organized and equipped institution in the country. Use of the military, however, may cause resentment among local residents and reduce local conservation support, so it should be considered only as a means of last resort.

In *Requiem for Nature,* John Terborgh (1999, pp199, 201) also mentions increasing the role of armed forces in protecting biodiversity and suggests the possibility of creating 'internationally financed elite forces within countries'. He writes:

If peacekeeping has been widely accepted as an international function, why not nature keeping? If local park guards are too weak or too subject to corruption and political influence to carry out their duties effectively, internationally sponsored guards could be called in to help.

Given histories of military abuse of power, it is questionable to promote use of the military or 'nature-keeping forces' without clarifications. Although authors such as van Schaik and Kramer (1997) are careful to suggest that the military be used only as a last resort, they overlook the possibility that the military or authoritarian governments may use conservation to further their own political ends in tenuously controlled rural areas (Peluso, 1993). Moreover, it is possible that the military would use nature protection as an excuse to wage war against particular ethnic groups.[12] While conservation interests might be advanced in the short term by employing the military in special 'emergency' situations, the long-term deployment of armed forces to enforce protected areas boundaries and management regulations could easily turn into or support a repressive regime that encourages violent resistance on the part of rural communities. The statement by van Schaik and Kramer ignores the fact that in most 'developing' countries the military serves elite interests, not the public. It leaves unanswered questions with much left to the imagination. For example, who qualifies as a 'rebellious insider' and who decides? Moreover, the authors are not clear about which 'last resort' situations would necessitate military intervention.

While military intervention might cause significant negative social impacts, there may be situations where the military might appropriately enforce protected areas. For example, might the military help maintain resource use at sustainable levels for local people in the face of heavily armed poachers from outside the community? Or in more extreme situations such as civil war, perhaps the use of armed personnel such as UN Peacekeepers would be appropriate to protect critically endangered species or resources (ie mountain gorillas in Rwanda) until a peaceful outcome can be achieved.[13]

A Call for Greater Clarification

By critiquing the protectionist argument found in recent writings on international biodiversity conservation, we point to key issues in an ongoing and at times contentious debate regarding the future of the world's most biologically rich regions. The fact that most of these zones occur in areas high in poverty and political instability makes the challenge of doing conservation even more complex. Our intent in writing this article is to encourage the conservation community, broadly construed, to constructively debate *how* nature protection can and should occur in specific places. We support the conclusion that greater protection measures need to be adopted in order to safeguard rapidly disappearing tropical landscapes. We also find it reasonable to conclude that ICDPs feature important shortcomings and that conservation with development as a singular strategy most likely will not provide sufficient protection of biodiversity. However, to take these conclusions in isolation from other political, social, and economic factors and then recommend that conservation policy revert to strict protectionism based on government-led, authoritarian practices makes little sense from both a moral and prag-matic perspective.

How can we build upon the findings of conservation scientists without reinventing a square wheel in the process? We suggest that greater recognition of the deep political and social complexity inherent in conservation work in developing countries necessitates significant clarification on the issues we highlight in this article, among others. Scientific reasoning and solutions alone will not be enough to safeguard biodiversity. We must also apply political analyses and responses. Although most of the authors of the works we cite recognize this fact, their recommendations for authoritarian nature protection efforts fail to comprehensively account for the dense web of social and political processes asso-ciated with doing conservation.

The concluding chapter of *Parks in Peril* argues that 'burdening parks with an over-whelming set of social goals' sets 'the achievement bar at an impossible height' and thus represents 'a recipe for ecological and social failure' (Redford et al, 1998, p546). This point is well taken and needs to be considered at the next World Parks Congress. At the same time, it raises several questions about the future of international biodiversity con-servation. In focusing the majority of our attention on protected areas (as important as they are), are we largely ignoring wider landscapes inhabited and worked by diverse peoples that also contain significant biological diversity? In decrying the unattainability of sustainability in general terms (as Terborgh, Rabinowitz, and Soule and Lease have stated), should we give up working toward more sustainable rural economies? By point-ing to the ineffectiveness of ICDPs in protecting biodiversity, should we conclude that they have failed on all counts and are thus useless in all contexts? By arguing that 'tradi-tional peoples' or local communities do not necessarily manage their resources in ways that protect biodiversity, should we discontinue or reduce efforts at capacity-building and sustainable development? In general, should we treat rural people as potential allies or as potential enemies? In the end, we have two broad choices. We can promote a policy shift toward authoritarian protectionism that would most likely alienate key allies at local, regional and national levels and thus precipitate resistance and conflict. Alter-natively, we can build on past experience and constructively negotiate ecologically sound, politically feasible and socially just programmes in specific contexts that can be legiti-mately enforced based on strong agreements with all affected parties.

Acknowledgements

We thank Lisa Curran, John Vandermeer, Ivette Perfecto, Tom Dietsch, Charles Benjamin, Grant Murray, Juliet Erazo, Christopher Thorns, Elizabeth McCance, Ann-Marie Finan, Heather Plumridge, and Curran's Tropical Ecosystem Dynamics seminar at the Univer-sity of Michigan for their comments on this chapter. We are very grateful to Katrina Brandon, Tom Rudel, Chuck Geisler, Max Pfeffer, David Bray, Jill Belsky, John Hough, Gonzalo Oviedo, Sally Jeanrenaud, Michel Pimbert, Anthony Anderson, Peter Schachenmann, Gordon Claridge, Mike Lara, Elery Hamilton-Smith, Sally Timpson, Andy Willard, Toddi Steelman, Kate Christen, Andrew Mittelman and Cristina Egh-enter for sending their critiques and observations. Thanks also to the approximately 150 practitioners, programme managers, academics and other people around the world who expressed interest in reading this chapter.

Notes

1   The term 'protection paradigm' appears most prominently in *Last Stand*, Chapter 10, in which van Schaik and Kramer (1997) called for a 'new protection paradigm'. Chapter 10 of John Terborgh's (1999) book *Requiem for Nature* is entitled, 'Why Conservation in the Tropics Is Failing: The Need for a New Paradigm'. In Chapter 9 of *Myth and Reality in the Rain Forest*, entitled 'Conservation at the Close of the Twentieth Century', John Oates called for a renewed emphasis on strict protection of national parks and reserves and less focus on community development. In *Parks in Peril*, the final chapter, called 'Holding Ground', by Redford, Brandon and Sanderson, made a similar argument for greater protection.

2   The editors of *Last Stand* offered four basic principles for designing solutions to biodiversity conservation problems. The first of these principles holds that 'protected areas will always be in need of active defence, no matter how great their benefits are to local communities or to society at large' (van Schaik and Kramer, 1997, p228).

3   Van Schaik and Kramer (1997) recognized that local communities show hostility toward protected area restrictions but assumed that it results from lost 'opportunity costs' that can simply be economically compensated

4   The book *Parks in Peril* picks up on some of these points in its discussions regarding park creation and management (Brandon, 1998a).

5   Van Schaik and Kramer (1997, p213) offered a different perspective: '[P]rotected areas are needed, not to satisfy some Western romantic ideal about paradisal nature unspoiled by humankind's uncouth hands but because a considerable number of species are vulnerable to extinction due to overexploitation or disturbance'.

6   The claim justifying international involvement is part of the third principle (out of four) for designing biodiversity policy solutions found in *Last Stand*. '[Effective] solutions require the involvement of all stakeholders, including representatives of both the local and the international community' (van Schaik and Kramer, 1997, p228).

7   This is the fourth principle for long-term success in protected areas management and biodiversity conservation: '[W]hile delegation of management to local communities is to be encouraged, there is always a role for the national government as the representative of the nation or the international community' (van Schaik and Kramer, 1997, p22).

8   Our critique of arguments that assume a universal knowledge of and experience of nature relies on a critical or 'postmodern' perspective. By offering this type of critique, we do not mean to imply that scientific understandings of the natural world are invalid, nor do we want to suggest that groups cannot reach agreement about mutually understood 'truths' (the so-called 'morass of relativism'). Critical perspectives simply posit that our understandings and explanations of natural and human phenomena are contingent historically and may vary culturally. In response, some conservationists claim that critical perspectives on the human–nature relationship

represent a politically motivated attack on the project of nature protection (Soule and Lease, 1995; see specifically pp137–138 on the 'social siege of nature').

9   Van Schaik and Kramer (1997) argued that economic compensation for opportunity costs takes care of this problem. While this solution may cover economic impacts, most likely it does not adequately attend to negative social and cultural impacts.

10   Issues related to the political and social challenges of conservation with development have produced a number of insights on participatory approaches (Hough, 1988; Lehmkuhl et al, 1988; Hough and Sherpa, 1989; Gibson and Marks, 1995; Naughton-Treves and Sanderson, 1995; Albers and Grinspoon, 1997; Freudenberger et al, 1997; Mehta and Kellert, 1998; Ulfelder et al, 1998; Maikhuri et al, 2000).

11   For related discussion on the relationships between the military and nature protection, see D'Souza (1995), Harbottle (1995) and Westing (1992). Bruce Albert (1992) has explored how the military 'manipulated environmental legislation and ecological rhetoric in order to perpetuate military hegemony over the development of Amazonia to the benefit of mining interests'.

12   Military intervention linked to conservation could serve other political ends, such as the removal of minority ethnic groups that the military decides should not continue to dwell in a particular area. The role of the military in helping to establish the Myinmoletkat Nature Reserve in Burma is especially troubling. See Associated Press (1997), Faulder (1997) and Levy and Scott-Clark (1997).

13   Thanks to Tom Dietsch for raising the possibilities listed here.

References

1   Agrawal, A. and Gibson, C. C. (1999) 'Enchantment and disenchantment The role of community in natural resource conservation'. *World Development*, vol 27, pp629–649.

2   Albers, H. J. and Grinspoon, E. (1997) 'A comparison of the enforcement of access restrictions between Xishuangbanna Nature Reserve (China) and Khao Yai National Park (Thailand)'. *Environmental Conservation*, vol 24, pp351–362.

3   Albert, B. (1992) 'Indian lands, environmental policy and military geopolitics in the development of the Brazilian Amazon: The case of the Yanomami'. *Development and Change*, vol 23, pp35–70.

4   Associated Press (1997) *Burma Using Forced Labor to Build Tourist Park, Exiles Say*. AP Wire, October 20.

5   Belsky, J. M. (1999) 'Misrepresenting communities: The politics of community-based rural ecotourism in Gales Point Manatee, Belize'. *Rural Sociol.*, vol 64, pp641–666.

6   Belsky, J. M. (2003) 'Unmasking the "local": Gender, community and the politics of rural ecotourism in Belize', in Brechin, S. R., Wilshusen, P. R., Fortwangler, C. L. and West, P. C. (eds) *Contested Nature: Power, Protected Areas and the Dispossessed Promoting International Biodiversity Conservation with Social Justice in the 21st Century*, SUNY Press, Albany, NY.

7   Borrini-Feyerabend, G. (1996) *Collaborative Management of Protected Areas: Tailoring the Approach to the Context*. IUCN, Gland, Switzerland.

8   Brandon, K. (1996) 'Traditional people, nontraditional times: Social change and the implications for biodiversity conservation', in Redford, K. H. (ed) *Traditional Peoples and Biodiversity Conservation in Large Tropical Landscapes*. The Nature Conservancy, Rosslyn, VA, pp219–265.

9   Brandon, K. (1997) 'Policy and practical considerations in land-use strategies for biodiversity conservation', in Kramer, R. A., van Schaik, C. P. and Johnson, J. (eds) *Last Stand: Protected Areas and the Defense of Tropical Biodiversity*, Oxford University Press, New York, pp90–114.

10   Brandon, K. (1998a) 'Comparing cases: A review of findings', in Brandon, K., Redford, K. H. and Sanderson, S. E. (eds) *Parks in Peril: People, Politics and Protected Areas*. Island Books, Washington, DC, pp375–414.

11   Brandon, K. (1998b) 'Perils to parks: The social context of threats', in Brandon, K., Redford, K. H. and Sanderson, S. E. (eds) *Parks in Peril: People, Politics and Protected Areas*. Island Press, Washington, DC, pp415– 440.

12   Brandon, K., Redford, K. H. and Sanderson, S. E. (1998a) 'Introduction', in Brandon, K., Redford, K. H. and Sanderson, S. E. (eds) *Parks in Peril: People, Politics and Protected Areas*. Island Press, Washington, DC, pp1–26.

13   Brandon, K., Redford, K. H. and Sanderson, S. E. (eds) (1998b) *Parks in Peril: People, Politics and Protected Areas*. Island Press, Washington, DC.

14   Brosius, J. P., Tsing, A. L. and Zerner, C. (1998) 'Representing communities: Histories and politics of community-based natural resource management'. *Society and Natural Resources*, vol 11, pp157–168.

15   Cox, M. C. (2001) *Protected Areas and Population Growth: Implications for Protected Area Policy in the 21st Century*. Master's thesis, University of Michigan, Ann Arbor.

16   Croll, E. and Parkin, D. (eds) (1992) *Bush Base: Forest Farm Culture, Environment and Development*. Routledge, London.

17   D'Souza, E. (1995) 'Redeeming national security: The military and protected areas', in McNeely, J. A. (ed) *Expanding Partnerships in Conservation*. Island Press, Washington, DC, pp156–165.

18   Faulder, D. (1997) *In the Name of Money: SLORC, the Thais and Two Multinational Oil Giants are Building a Gas Pipeline. The Karen are in the Way and That's Just Too Bad*. Asiaweek, 9 May, p42.

19   Fortwangler, C. L. (2003) 'Environmental justice and conservation', in Brechin, S. R., Wilshusen, P. R., Fortwangler, C. F. and West, P. C. (eds) *Contested Nature: Power, Protected Areas and the Dispossessed Promoting International Biodiversity Conservation with Social Justice in the 21st Century*. SUNY Press, Albany, NY.

20   Freudenberger, M. S., Carney, J. A. and Lebbie, A. R. (1997) 'Resiliency and change in common property regimes in West Africa: The case of the Tongo in the Gambia, Guinea and Sierra Leone'. *Society Nat. Resources*, vol 10, pp383–402.

21   Gibson, C. C. and Marks, S. A. (1995) 'Transforming rural hunters into conservationists: An assessment of community-based wildlife management programs in Africa'. *World Development*, vol 23, pp941–957.

22   Grove, R. H. (1990) 'Colonial conservation, ecological hegemony and popular resistance: Towards a global synthesis', in MacKenzie, J. M. (ed) *Imperialism and the Natural World*, Manchester University Press, Manchester, pp15–50.

23   Haenn, N. M. (1997) *The Government Gave Us This Land: Political Ecology and Regional Culture in Campeche, Mexico*. PhD thesis, University of Indiana, Bloomington.

24   Harbottle, M. (1995) *New Roles for the Military: Humanitarian and Environmental Security*. Research Institute for the Study of Conflict and Terrorism, London.

25   Hitchcock, R. K. (1995) 'Centralizations, resource depletion and coercive conservation among the Tyua of the Northeastern Kalahari'. *Human Ecology*, vol 23, no 2, pp169–198.

26   Hough, J. L. (1988) 'Obstacles to effective management of conflicts between national parks and surrounding human communities in developing countries'. *Environmental Conservation*, vol 15, pp129–136.

27 Hough, J. L. and Sherpa, M. N. (1989) 'Bottom-up vs. basic needs: Integrating conservation and development in the Annapurna and Michura conservation areas of Nepal and Malawi'. *Ambio*, vol 18, pp435–441.
28 IUCN, UNEP and WWF (1991) *Caring for the Earth: A Strategy for Sustainable Living.* IUCN, Gland, Switzerland.
29 Johnson, L. M. (2000) 'A place that's good: Gitksan landscape perception and ethnoecology'. *Human Ecology*, vol 28, no 2, pp301–325.
30 Kellert, S. R., Mehta, J. N., Ebbin, S. A. and Lichtenfeld, L. L. (2000) 'Community natural resource management: Promise, rhetoric and reality'. *Society Natural Resources*, vol 13, pp705–715.
31 Kramer, R. A. and van Schaik, C. P. (1997) 'Preservation paradigms and tropical rain forests', in Kramer, R. A., van Schaik, C. P. and Johnson, J. (eds) *Last Stand: Protected Areas and the Defense of Tropical Biodiversity.* Oxford University Press, New York, pp3–14.
32 Kramer, R. A., van Schaik, C. P. and Johnson, J. (eds) (1997) *Last Stand: Protected Areas and the Defense of Tropical Biodiversity.* Oxford University Press, New York.
33 Larson, P., Freudenberger, M. and Wyckoff-Baird, B. (1997) *Lessons from the Field: A Review of World Wildlife Fund's Experience with Integrated Conservation and Development Projects 1985–1996.* World Wildlife Fund, Washington, DC.
34 Lehmkuhl, J. F., Upreti, R. K. and Sharma, U. R. (1988) 'National parks and local development: Grasses and people in Royal Chitwan National Park, Nepal'. *Environmental Conservation*, vol 15, pp143–148.
35 Levy, A. and Scott-Clark, C. (1997) 'Burma's junta goes green: Save the rhino, kill the people, The Observer', in MacCormack, C. P. and Strathern, M. (eds) (1980) *Nature, Culture and Gender.* Cambridge University Press, Cambridge, p9.
36 MacKenzie, J. M. (1988) *The Empire of Nature: Hunting, Conservation and British Imperialism.* Manchester University Press, Manchester.
37 MacKinnon, K. (1997) 'The ecological foundations of biodiversity protection', in Kramer, R. A., van Schaik, C. P. and Johnson, J. (eds) *Last Stand: Protected Areas and the Defense of Tropical Biodiversity*, Oxford University Press, New York, pp36–63.
38 Maikhuri, R. K., Nautiyal, S., Rao, K. S., Chandrasekhar, K., Gavali, R. and Saxena, K. G. (2000) 'Analysis and resolution of protected area–people conflicts in Nanda Devi Biosphere Reserve, India'. *Environmental Conservaton*, vol 27, pp43–53.
39 Mehta, J. N. and Kellert, S. R. (1998) 'Local attitudes toward community-based conservation policy and programmes in Nepal: A case study of Makalu-Barun conservation area'. *Environmental Conservation*, vol 25, pp320–333.
40 Naughton-Treves, L. and Sanderson, S. E. (1995) 'Property, politics and wildlife conservation'. *World Development*, vol 23, pp1265–1275.
41 Neumann, R. P. (1997) 'Primitive ideas: Protected area buffer zones and the politics of land in Africa'. *Development and Change*, vol 28, pp559–582.
42 Neumann, R. P. (1998) *Imposing Wilderness: Struggles over Livelihood and Nature Preservation in Africa.* University of California Press, Berkeley.
43 Oates, J. F. (1995) 'The dangers of conservation by rural development. A case study from the forests of Nigeria'. *Oryx*, vol 29, pp115–122.
44 Oates, J. F. (1999) *Myth and Reality in the Rainforest: How Conservation Strategies Are Failing in West Africa.* University of California Press, Berkeley.
45 Ostrom, E. (1990) *Governing the Commons: The Evolution of Institutions for Collective Action.* Cambridge University Press, Cambridge.
46 Peluso, N. L. (1993) 'Coercing conservation: The politics of state resource control'. *Global Environmental Change*, vol 3, no 2, pp199–218.
47 Rabinowitz, A. (1999) 'Nature's last bastions: Sustainable use of our tropical forests may be little more than wishful thinking'. *Natural History*, vol 108, pp70–72.

48   Redford, K. H. (1990) 'The ecologically noble savage'. *Cultual Survival Quarterly*, vol 15, pp46–48.
49   Redford, K. H. and Mansour, J. (eds) (1996) *Traditional Peoples and Biodiversity Conservation in Large Tropical Landscapes*. The Nature Conservancy, Rosslyn, VA.
50   Redford, K. H. and Richter, B. (1999) 'Conservation of biodiversity in a world of use'. *Conservation Biology*, vol 13, pp1246–1256.
51   Redford, K. H. and Sanderson, S. E. (2000) 'Extracting humans from nature'. *Conservation Biology*, vol 14, pp1362–1364.
52   Redford, K. H., Brandon, K. and Sanderson, S. E. (1998) 'Holding ground', in Brandon, K., Redford, K. H. and Sanderson, S. E. (eds) *Parks in Peril: People, Politics and Protected Areas*. Island Press, Washington, DC, pp455–464.
53   Reichel-Dolmatoff, G. (1976) 'Cosmology as ecological analysis: A view from the rain forest'. *Man*, vol 11, no 3, pp307–318.
54   Robinson, J. C. (1993) 'The limits to caring: Sustainable living and the loss of biodiversity'. *Conservation Biology*, vol 7, pp20–28.
55   Sanderson, S. and Bird, S. (1998) 'The new politics of protected areas', in Brandon, K., Redford, K. H. and Sanderson, S. E. (eds) *Parks in Peril: People, Politics and Protected Areas*, Island Press, Washington, DC, pp441–454.
56   Sanderson, S. E. and Redford, K. H. (1997) 'Biodiversity politics and the contest for ownership of the world's biota', in Kramer, R. A., van Schaik, C. P. and Johnson, J. (eds) *Last Stand: Protected Areas and the Defense of Tropical Biodiversity*, Oxford University Press, New York, pp115–132.
57   Schwartzman, S., Moreira, A. and Nepstad, D. (2000a) 'Rethinking tropical forest conservation: Perils in parks'. *Conservation Biology*, vol 14, pp1351–1357.
58   Schwartzman, S., Nepstad, D. and Moreira, A. (2000b) 'Arguing tropical forest conservation: People versus parks'. *Conservation Biology*, vol 14, pp1370–1374.
59   Seeland, K. (ed) (1997) *Nature Is Culture: Indigenous Knowledge and Socio-Cultural Aspects of Trees and Forests in Non-European Cultures*. Intermediate Technology, London.
60   Soule, M. E. and Lease, G. (eds) (1995) *Reinventing Nature?: Responses to Postmodern Deconstruction*. Island Press, Washington, DC.
61   Taylor, D. A. (2000) 'The rise of the environmental justice paradigm: Injustice framing and the social construction of environmental discourses'. *American Journal of Behavioral Science*, vol 43, pp508–580.
62   Terborgh, J. (1999) *Requiem for Nature*. Island Press–Shearwater Books, Washington, DC.
63   Terborgh, J. (2000) 'The fate of tropical forests: A matter of stewardship'. *Conservation Biology*, vol 14, pp1358–1361.
64   Terborgh, J. and van Schaik, C. P. (1997) 'Minimizing species loss: The imperative of protection', in Kramer, R. A., van Schaik, C. P. and Johnson, J. (eds) *Last Stand: Protected Areas and the Defense of Tropical Biodiversity*, Oxford University Press, New York, pp15–35.
65   Ulfelder, W. H., Poats, S. V., Recharte, B. J. and Dugelby, B. (1998) Participatory conservation: Lessons of the PALO MAP study in Ecuador's Cayambe-Coca Ecological Reserve. Rep. no 1b. The Nature Conservancy, Latin America and Caribbean Division, Arlington, VA.
66   van Schaik, C. P. and Kramer, R. A. (1997) 'Toward a new protection paradigm', in Kramer, R. A., van Schaik, C. P. and Johnson, J. (eds) *Last Stand: Protected Areas and the Defense of Tropical Biodiversity*, Oxford University Press, New York, pp212–230.
67   van Schaik, C. P., Terborgh, J. and Dugelby, B. (1997) 'The silent crisis: The state of rain forest nature preserves', in Kramer, R. A., van Schaik, C. P. and Johnson, J. (eds) *Last Stand: Protected Areas and the Defense of Tropical Biodiversity*, Oxford University Press, New York, pp64–89.
68   Wells, M. and Brandon, K. (1992) *People and Parks: Linking Protected Area Management with Local Communities*. World Bank, Washington, DC.

69 Wells, M., Guggenheim, S., Khan, A., Wardojo, W. and Jepson, P. (1999) *Investing in Biodiversity: A Review of Indonesia's Integrated Conservation and Development Projects*. World Bank, Washington, DC.

70 West, P. C. and Brechin, S. R. (eds) (1991) *Resident Peoples and National Parks: Social Dilemmas and Strategies in International Conservation*. University of Arizona Press, Tucson.

71 Westing, A. (1992) 'Protected natural areas and the military'. *Environmental Conservation*, vol 19, pp343–348.

72 Wilshusen, P. R. (2003) 'Territory, nature and culture: Negotiating the boundaries of biodiversity conservation in Colombia's Pacific coastal region', in Brechin, S. R., Wilshusen, P. R., Fortwangler, C. L. and West, P. C. (eds) *Contested Nature: Promoting International Biodiversity Conservation with Social Justice in the 21st Century*. SUNY Press, Albany, NY.

73 Zerner, C. (ed) (2000) *People, Plants and Justice: The Politics of Nature Conservation*. Columbia University Press, New York.

Part 5

The Way Forward: Forestry for the Future

We start this last section with a reminder, in the paper by Sheil and van Heist (Chapter 18), that forest management still requires a strong body of ecological knowledge. This is important as so much of the public debate focuses uniquely on the policy and institutional aspects of sustainable forests. Sheil and van Heist show how we still face major uncertainties regarding the ecology of the forests, and that this traditional skill of the forester and conservationist must not be neglected.

The second simple message of this last section is that we need new institutions for forests and that these must have the capacity to influence forest-related decisions at a much broader scale than we have attempted in the past. Placing forests in the hands of private individuals and local communities offers some part of the answer but this will not diminish the need for government oversight of large-scale landscape issues. We may devolve many responsibilities for forests to local people and we may allow these people to exercise considerable discretion over how they use their forestland, but checks and balances will need to be retained to ensure that the whole forest estate adds up to more than the sum of its parts. The environmental benefit that the whole of society gains from forests will have to be maintained in the face of the shorter-term market-driven demands of local owners and stakeholders.

In Chapter 19 Alden Wily shows how the first generation of devolved forest management programmes often provided people with responsibilities but denied them the benefits of the forest. Decentralized forest management is important, but it may be difficult to achieve in practice. The transition from highly centralized management to decentralized systems has sometimes been a period of rapid degradation of forest resources.

Edmunds and Wollenberg (Chapter 20) show that getting people to participate effectively and equitably in the management of forests is not as simple as it may seem. Certainly the rather rigid participatory frameworks that many international agencies have advocated in recent years may not work in the best interests of the poorest of the poor – people can participate in different ways and on varying terms. Ensuring equitable outcomes requires a lot of sensitivity and is not favoured by dependence on a single model of participation.

Ultimately different forests will always have to be managed for different purposes. In many situations the best outcomes will be achieved through mosaic landscapes where a balance is achieved between conservation and use at larger spatial scales. Restoration will often be needed in those areas where forests have already been seriously degraded. Plantations will inevitably play a bigger role in the future, in providing both fibre and environmental benefits. Sayer et al (Chapter 21) underline the need to find the right balance between protection, management and restoration – in effect saying that the integration of conservation and development has to be moved to a larger spatial scale than in the past.

In the concluding chapter Sayer spells out what this all means for the forest institutions and forest professionals of the future. We are embarking on a period of great challenges for forests and the people who are entrusted with their stewardship. In the past, forestry was a conservative discipline sharply focused on the management unit and enjoying a high level of predictability regarding the expectations of society for forest products and services. In the future it is likely to be exactly the opposite. Foresters will be dealing with uncertainty and continuing change. They will need to be eclectic – responsive to a diversity of opinions and interests. They will have to be good at building relationships and facilitating debates. But the fundamental knowledge of the ecology, economics and social needs of the forest system will still be important. The wise management of forest environments will require a new generation of professionals and will provide them with exciting and stimulating jobs of ever-greater significance to society.

18

Ecology for Tropical Forest Management

Douglas Sheil and Miriam van Heist

Introduction

Ecology provides the foundation for forest management. However, the relationship between management and ecology can sometimes appear obscure. Lowe (1995) for example recently claimed that ecology 'has contributed little of practical use to tropical silviculture'. We disagree (eg see Sheil and Hawthorne, 1995), but the relationship between forestry and ecology can certainly be improved. The prominence of 'ecological issues' and ecologists in the mounting opposition to forest harvesting in many parts of the world has lead to a view that ecology is somehow 'anti-production', and has been symptomatic of deterioration in the relationship between biological scientists and forest managers. At the same time, there is an ever-growing pressure on foresters to review their practice and adapt to changing demands. We strongly believe that ecology is fundamental to progress, and that a cooperation between ecologists and foresters is needed. Here we are particularly addressing those concerned with the practical aspects and implications of tropical forest management.

Tropical ecologists, seeking research funding, frequently emphasize how little is known about tropical forests. This can be misinterpreted – 'lack of knowledge' does not excuse the prevalence of poor practices, and enough is known to make improvements. There is a considerable body of potentially useful information. A problem remains that many of those involved in the management of tropical forests have limited access to such ecological knowledge and insights, or indeed lack the means to translate such information into real actions. Recognizing this concern is a first step in developing remedial activities, and it is the purpose of this chapter to show that this goal is important and realistic.

It would be ideal if we could simply present ecological facts and the recommendations they inspire. However, applications will always depend on many factors and while we do not examine policy, economics, regulations, enforcement or management-capacity here, we recognize the demanding context in which most managers operate. What we provide are insights that demonstrate how ecology helps us understand forests and their behaviour. Stakeholders must decide what changes are acceptable or preferred. We emphasize the broad range of factors requiring forest managers to take a holistic, long-term landscape-level view. An in-depth account of everything is impossible in this short note and some crucial aspects (eg

Note: Reprinted from *International Forestry Review*, vol 2, Sheil, D. and van Heist, M., 'Ecology forest management', pp261–270, Copyright © (2000), with permission from the Commonwealth Forestry Association

soil ecology and genetics) cannot be considered. Nor are we able to discuss important related themes such as further research, indigenous knowledge and the role of local communities. Our intention then is to emphasize the very practical need for ecology by outlining a range of topics, and discussing some principles and recommendations.

The Ecology of Silviculture

Harvesting as disturbance

Disturbance as an ecological term is generally used for any rapid release or reallocation of community resources such as light, water or soil nutrients (eg Clark, 1990; Glenn-Lewin and van der Maarel, 1992; Roberts and Gilliam, 1995). Much ecological work has examined the effects of *disturbance* in plant communities. Intensity, scale and frequency of disturbances are thought to be key aspects and can be readily interpreted in harvesting terms.

A useful distinction can be made between intrinsic and extrinsic disturbances. Intrinsic disturbances are inherent to all forest formations, that is tree falls, whereas logging is an extrinsic disturbance. It is sometimes suggested that if processes of harvesting would create disturbance patterns similar in size and intensity to such intrinsic processes, this would cause least change (eg Skorupa and Kasenene, 1984). However, 'replicating natural processes' would require that only dead or vulnerable trees were impacted, whereas harvesting is always an addition to intrinsic processes. Change after harvesting is therefore inevitable and 'naturalness', and any of the values uniquely associated with it (eg wilderness value), cannot be sustained. The responsibility of managers is rather to minimize deleterious changes while maintaining future stand productivity and management options.

In forestry, the intensity of harvesting generally determines the reduction in basal area (Johns, 1988; Skorupa, 1988; Favrichon, 1998), which is closely correlated to changes in canopy cover and thus understorey light. As foresters know, and ecologists have examined, increased illumination affects tree growth and regeneration, but responses can be greatly influenced by species and context (Silva et al, 1996; Whitmore and Brown, 1996). Guilds provide convenient classes of species responses to such changes in illumination. Hawthorne's (1996) system is particularly useful as it is explicitly related to silvicultural characteristics (Table 18.1). It is at this level, in combination with knowledge of regeneration requirements (see below), that foresters have modified forest composition and productivity through the control of canopy opening.

Foresters typically distinguish between monocyclic and polycyclic harvesting systems where ecologists might emphasize the types of regeneration involved (eg Swaine, 1996). In a monocyclic system, all the standing timber in a compartment is cut at one time. Future harvests depend on regeneration from seeds in the soil prior to cutting (the seed-bank, dominated by pioneer species) – or from seeds arriving from outside (the seed-rain), often depending on animals for dispersal, or remnant mother trees. Such seed rain is also important in the recovery of open and damaged areas such as log-collection areas (Hladik and Miquel, 1990).

On the other hand, in a highly selective or polycyclic system only a limited proportion of stems, usually the largest, are cut. The same compartment will be revisited after

Table 18.1. *Some guild definitions, based on Hawthorne (1995)*

		Seedlings found	
		Almost exclusively in non deep-shade areas	In shaded (gap-free forest)
Adults found	Almost exclusively in non deep-shade areas	*Pioneers* Seedlings develop most profusely in open areas and canopy gaps. Seedlings may be found in shade but seldom survive long in such locations. Includes many low light-timber species	*(Cryptic Pioneers)*
	In shaded (gap-free forest)	*Non Pioneer Light Demanders (NPLDS)* Seedlings may be found under closed canopy, but illumination is needed for further development. Includes many high value (medium density) timber species	*Shade bearers* All sizes > 5cm dbh can be found under closed canopy, and will persist in these conditions. Generally very high density timbers.

the younger trees (the 'advanced regeneration' or 'seedling bank') that survived the harvesting process have grown sufficiently (Hartshorn, 1978; Clark, 1994; Whitmore and Brown, 1996; Brown and Jennings, 1998). This advanced regeneration is especially important for many timber species with seeds that are only briefly viable and germinate rapidly under natural circumstances (Ng, 1983; Vazquez-Yanes and Orozco-Segovia, 1993). An additional form of regeneration is presented by species that coppice. Such regeneration is probably much more common than generally recognized, particularly amongst small stems, and is particularly important in the ecology of areas of the world that suffer intermittent storm damage (eg Bellingham et al, 1994). Although it has not attracted much attention in tropical silviculture (aside from teak, *Tectona grandis*), some timber species do coppice well (eg *Milicia*).

Canopy gaps have dominated ecological ideas on rainforest dynamics for much of the last decade and more (eg Brokaw, 1987; Denslow, 1987; Brandani et al, 1988). It now appears that in most circumstances the composition and richness of natural forests is little influenced by the nature of natural gaps which are generally small and quickly filled by advanced regeneration (eg Brown and Jennings, 1998; Hubbell et al, 1999). However, in harvested forest, gaps are generally larger and in higher densities than in unlogged forest, and advanced regeneration is often destroyed. In such cases pioneer vegetation, germinating from seed, can dominate initial regrowth (Swaine and Hall, 1983; Denslow et al, 1990; Silva et al, 1996; Pelissier et al, 1998). According to an extensive review by Hawthorne et al (in press) a few broad generalizations are available: NPLDs (see Table 18.1) generally benefit from some canopy opening, while shade-bearers generally decline in all canopy-opening processes and regeneration far from undisturbed forest suffers. Loading bays and logging roads favour pioneers over NPLDs, but skid trails and felled tree gaps support a higher concentration of NPLDs. Areas close to intact forest usually recover more quickly from any clearance. The maintenance of scattered 'reserves' within a managed forest is justified as a practical precaution.

Excessive canopy opening can lead to regeneration problems especially in exposed conditions where soils dry out rapidly and nutrient loss through run-off is common. Herbaceous vegetation associated with severe opening can interfere with regeneration and impedes recovery (Epp, 1987; Hawthorne, 1993, 1994), for example *Pennisetum* (Kasenene and Murphy, 1991) and *Imperata* grass in Uganda, and *Chromolaena* in Ghana (Hawthorne, 1993, 1994, 1996). Areas of low regrowth attract ground herbivores which may also damage regrowth and maintain open areas (Laws et al, 1975; Struhsaker et al, 1996).

Tree reproduction and survival

Thus far, we have considered familiar silvicultural issues; we now extend our considerations to tree reproduction, a comparatively neglected topic. Trees do not generally have the ability to flower and set seed until they have reached some minimum size, and greatest fecundity is normally found in the largest stems (Appanah and Mohd.-Rasol, 1990; Thomas, 1996; Chapman and Chapman, 1997). Plumptre (1995) found that the seedling densities of four canopy species, including *Khaya anthotheca*, were strongly dependent on the abundance of potential parent trees (diameter > 50cm). Thus, removal of larger stems can impair subsequent regeneration due to loss of fruit and seed sources. The use of a 'minimum felling size' as sole silvicultural control is inadequate to protect seed production.

Even when potential parent trees are present, reproduction processes may require further protection. The role of many animal species in pollination and seed dispersal is underlined by research. Though a complex body of ecological knowledge exists on these roles, it is often species specific and thus not readily generalized (Gautier-Hion et al, 1985). Knowledge of timber tree reproduction remains surprisingly limited, for example the primary pollinators of the 'African mahogany' species (eg *Khaya*, *Entandrophragma*, *Lovoa* spp.) remain uncertain, while bats are required for the effective seed dispersal of the probably wind-pollinated and dioecious *Milicia* (Osmaston, 1965). Recent research has also indicated that seed dispersal distances are characteristically much lower than previously thought even for animal dispersed species (eg Hubbell et al, 1999), a realization that suggests the need for denser or at least more even distribution of retained mother trees. Guiding principles should be available soon when such assessment is coupled with advances in seed and fruit classification (eg Gautier-Hion, 1985; Howe and Westley, 1988).

Ultimately dispersal and pollination syndromes and potential fecundity must help determine choice of mother tree retention and associated site management (Baur and Hadley, 1990; Schupp, 1990). As an example, dioecious trees should probably be at twice the density of equivalent hermaphrodite species (Lawton, 1955; Kigomo et al, 1994 note that important timber species such as *Milicia*, and possibly *Entandrophragma*, are dioecious), but there are few data to define adequate pollination densities for any tropical tree species (eg Stacy et al, 1996; Ghazoul et al, 1998).

Protecting mother trees and unharvested 'reserves' within the forest landscape offers some insurance, but the maintenance of key animal populations is also necessary to ensure long-term viability and this may require additional attention (eg Howe and Westley, 1988; Parren, 1991; Bakuneeta et al, 1995; Hawthorne and Parren, 2000). Gordon et al (1990) have shown that crucial Central-American forest pollinators can themselves be dependent on non-forest areas outside of gazetted reserves, implying that such areas and functions need to be included in long-term management.

Remnant trees are influential as more than seed sources. Most trees survive and grow better near sources of mycorrhizae, such sources often being other trees (Alexander et al, 1992; Hogberg and Alexander, 1995; Moyersoen et al, 1998). Some fruit trees provide a strong catalytic function by drawing in fruit dispersing vertebrates and enriching the local seed-rain (Guevara, 1986). Protecting or planting these species can accelerate forest recovery in degraded sites.

Forest Food, Habitat and Wildlife

Animals play a major role in maintaining forest vegetation through pollination and seed dispersal. Wildlife *can* be compatible with managed forest (Johns, 1997) and is often a major value in itself, providing vital food for local communities, and/or ensuring a high conservation status for the forest.

Manipulation of forest composition has implications for hungry wildlife. While the chemistry of forest vegetation is complex (eg Gartlan et al, 1980; Waterman, 1983; Waterman et al, 1988), it is generally true that heavy timbered species are generally better defended against browsers and support fewer animals than faster growing, lighter species (eg Janzen, 1979; Gartlan et al, 1980; Loehle, 1988). Fruit availability shows a roughly similar pattern: pioneer species generally have wind-dispersed or small-fruited (bird-dispersed) seeds (Loiselle et al, 1996), NPLDs (Table 18.1) include many species with larger fleshy fruits (eg Sapotaceae, Moraceae), while many shade-tolerant species have gravity-dispersed seeds. There is thus often a low abundance of edible fruit and vegetation in dense old-growth forest. These patterns, and the occurrence of herbaceous growth in clearings, explain why young or disturbed areas frequently support higher densities of wildlife than old-growth forests. However, it must be emphasized that 'specialist' species associated with old-growth forests may have a higher conservation status and are more vulnerable to eradication. Some trees, most notably figs *(Ficus* spp.), are especially important for wildlife as they provide fruit throughout the year, and fulfil vital nutritional needs, such as the calcium needed by vertebrates living on otherwise mineral-poor diets (see O'Brien et al, 1998).

Large, old and hollow trees have considerable significance for many forest taxa that utilize or are dependent on them, for example hornbills (Datta, 1998; Whitney and Smith, 1998; Whitney et al, 1998), woodpeckers (McNally and Schneider, 1996), hyraxes and other hollow tree nesters (Zahner, 1993). This includes important pollinators like bees (Kerr et al, 1994; McNally and Schneider, 1996) and seed dispersers (Whitney and Smith, 1998; Whitney et al, 1998). The loss of such large stems can thus have long-term influences (Gordon et al, 1990) and, although not well documented, is potentially involved in otherwise inexplicable decline or failure in forest regeneration in various parts of the world (H. C. Dawkins and J. Palmer, personal communication).

Lianas have been shown to exacerbate the effects of logging, and cutting prior to felling is often advised (eg Liew, 1973; Appanah and Putz, 1984; Putz, 1984). However, these can also provide habitat and food for a number of specialized fauna. Protected liana and climber species already occur in some silvicultural prescriptions, for example in recognition of a key fruit source (F. E. Putz, personal communication).

Hunting frequently poses a greater threat to larger forest fauna than does timber harvesting. However, hunting pressure itself is often related to forestry roads and the provisioning of logging camps (Robinson et al, 1999). Many large mammals have been exterminated from areas well within their historical ranges even when suitable habitats remain. The widespread loss of larger forest wildlife throughout the tropics has been called the 'empty forest' syndrome, and some already view such forests as 'ecologically dead' (Redford, 1992). The longer-term impacts of animal loss remain hard to specify since we lack basic information, but they are likely to be substantial involving changes in forest composition and structure (De la Cruz and Dirzo, 1987; Campbell, 1991; Sheil, 1998). Forest managers need to consider their responsibility to control excessive hunting and ensure the protection of important habitat features, such as sites that provide vital mineral nutrients directly through soil or salt springs (eg Klaus et al, 1998).

Risks and Threats

Loss of animals is not the only looming problem in tropical forests; fire, exotic species and forest fragmentation are all increasingly familiar concerns. Until recently fire had not been a concern in many tropical forest countries, but this is changing. For deciduous and semi-deciduous forest fire is now seen as the greatest risk associated with selective harvesting in some regions (Hawthorne, 1994). Risk factors are dry debris and improved access (Buechner and Dawkins, 1961; Phillips, 1987). Even under moist conditions fire risk is substantially higher in logged forest than in primary forest, especially after periods of drought, and in areas of fragmented cover (Malingreau et al, 1985; Bertault, 1990). Managers should look at ways to maintain evergreen undergrowth where possible, especially at forest edges.

Another global problem is invasion by non-native plants and animals (exotics). While the ability of exotic plants to colonize intact continental rainforest is debated, the problem on islands is often considerable (Whitmore, 1991; Cronk and Fuller, 1994). The presence of such plants can severely reduce options for maintaining productive natural vegetation after harvesting (Sheil, 1994; Rejmanek et al, 1996). Invasion can progress very slowly; an interesting example of which is the spread of the exotic tree *Broussonetia papyrifera* (L.) Vent. (Moraceae), a growing problem in both Uganda (Sheil, 1994) and Ghana (Hawthorne, 1995). This species is unusual in having the tendency to sucker vegetatively from roots or fallen stems. These young sucker shoots dominate the understorey around adult trees and pre-empt other regeneration – while physical control is difficult (the species also coppices well), the existence of a linked root network suggests that poisoning may be an efficient control option. If not recognized in time, such 'slow problems' may nonetheless become ultimately catastrophic and control options unaffordable.

Recent ecological research has begun to clarify how invasive species might be identified from general principles. Rejmanek (1999) provides a scheme for screening out high-risk woody exotics and notes that early maturity associated with high seed outputs and long-distance dispersal is particularly dangerous. The threat is not restricted to plants, but involves a wide range of organisms. Introduced fungi are already seen as a potential threat to native forests in some locations (eg *Phytophera* in Australia, Brown, 1976) and measures are needed to ensure minimal transfer of soil and other potentially infected material between sites. General recommendations are to eliminate local exotics before they spread over large areas, to be aware of which species have caused problems elsewhere, not to transfer known pest species between sites, and to monitor recovering areas.

The fragmentation of forest cover has profound ecological significance. This is a controversial topic with much recent research (eg Turner, 1996; Laurance and Bierregaard, 1997), but there are some generally accepted relationships. Small isolated areas of forest cannot maintain as many species in the longer term as the same area would if part of some larger tract. Small populations, such as arise in fragmented or heavily harvested landscapes, run much greater risks of reduced reproduction, genetic deterioration and extinction (Nason and Hamrick, 1997). The effects will be least if the distances between forest patches are low, and recent research has highlighted the importance of maintaining 'forest-like' or patchy 'forest stepping stone' habitats in the intervening landscape (Laurance and Bierregaard, 1997; Gascon et al, 2000). Some wildlife is especially vulnerable to habitat fragmentation, and will not cross open areas, even avoiding forest margins. For example, Newmark (1991) indicates that forest corridors need to be at least 200m wide to allow the free movement of sensitive East African forest birds. Some more general patterns are also emerging based on species characteristics. It appears that nocturnal flying animals (including pollinators and seed dispersers) tend to be less affected by fragmentation than are species which are active in the day, that is moths are less sensitive than butterflies, bats than birds (Daily and Ehrlich, 1996). Forest fragments are also especially vulnerable to fire (Buechner and Dawkins, 1961), invasion by weedy species, and other processes of habitat erosion (Gascon et al, 2000). There is a real need to maintain forest connectivity, minimize road width, and avoid unnecessary edge creation. Regulations requiring maintenance of forest cover along stream and river margins (theoretically this should provide a widely linked corridor network) are clearly useful.

Balancing Choices

Timber production versus conservation

There is a conflict between common silvicultural objectives (leading to 'high disturbance') and generally stated conservation goals ('low disturbance'). This choice is well illustrated in Uganda where the majority of preferred timber species grow best in open or disturbed environments (eg *Maesopsis eminii* Engl.), and many do not regenerate adequately without significant canopy opening (Meliaceae). A long-standing aim of forestry in Uganda has been to deplete and eradicate the originally widespread but low-yielding old-growth formations in favour of the earlier successional timber forests (Dawkins, 1958; Dawkins and Philip, 1998). The elimination of native species is no longer considered compatible with modern environmental standards, and as already noted, many non-timber species previously controlled by poisoning (eg *Ficus*) are now known to serve important ecological roles.

Similarly, the maintenance of animal populations is not without cost. For example elimination of browsing animals can lead to a substantial improvement in regeneration densities of some timber species – this was the primary reason that the Uganda Forest Department invested heavily in the control of elephants (Laws et al, 1975). The acceptability of such measures is a separate matter.

The benefits of encouraging a broader range of harvestable species are equivocal: yields and management options are theoretically increased, but there is also potential for greater forest degradation when regulations cannot be enforced. R. A. Plumptre (1996) provides a useful review showing how higher timber volumes can be extracted from tropical forests if more species are accepted (this being an issue of marketing and treatment). He shows how this would greatly improve the economics of harvesting, but emphasizes that this should not go beyond the ecological limits.

Despite many debates, there is little doubt that many conservation values can theoretically be maintained in managed forest (eg Johns, 1997) even though such forest is not going to be the 'pristine' entity that some conservationists seek (eg Struhsaker, 1997). Many areas that have been harvested for timber maintain high conservation significance (eg the Bwindi National Park, Uganda, still contains a high proportion of the world's mountain gorillas). While quantitative information is scarce, all harvesting appears to reduce large-scale tropical forest diversity (Struhsaker, 1997; Bawa and Seidler, 1998). There are vulnerable species that are likely to require protected areas or special management. Hawthorne and Abu-Juam (1995) have made intensive reviews and studies, but are still willing to contend that timber harvesting has not caused any plant extinction in Ghana. However, care is needed, because the presence of species does not mean that long-term ecological viability is assured. For example, trees may live for many centuries despite not being able to regenerate – hence the 'living dead' may be recorded simply 'present' in surveys (see Turner et al, 1996; Hawthorne and Parren, 2000). The 'empty forest' syndrome already discussed poses an unknown threat, and the effects of forest fragmentation, fires, exotic species, arid climate change (see eg Markham, 1998) pose many additional concerns that urge against complacency.

A common misunderstanding caused by ecologists and conservationists arises in equating concern about 'biodiversity' with the ecologists' classical measures of species richness, and hence into some form of management criteria. However, species are not equal and species counting does not reflect value. Conservation needs to ensure the long-term protection of vulnerable and threatened taxa – not to maximize the number of species recorded in some discrete area. Disturbance in old-growth forest will often promote increased species richness – the added species are generally robust common species, while old-growth forest is rare. Thus, while an increase in the number of tree species after harvesting or disturbance is not unusual (Plumptre, 1996; Sheil, 1998; Cannon et al, 1998), the conservation relevance is unclear. A more detailed study in Malaysia has indicated that unharvested forest had a much higher 'conservation value' defined by weighting tree species according to their rarity and vulnerability – globally rarer species appear to become disproportionately rarer in areas that have been harvested (Chua et al, 1998). The generality of these patterns remains unclear but species counting is definitely a poor way to assess management objectives (Sheil et al, 1999).

Ecologists frequently compare managed forests with 'undisturbed', 'pristine' or 'natural' vegetation but caution is required when any notions of a 'stable unchanging natural state' are invoked. Ecologists now recognize that all forests are changing and will continue to change, with or without interventions, fire and exotic species (eg Sheil, 1996; Phillips and Sheil, 1997; Whitmore and Burslem, 1998). One of the clearest explanations for this instability lies in the volatile nature of the world's climate which has always been changing and is now doing so even more rapidly as a result of human activities. Even areas set aside for conservation may ultimately require active management if values are to be maintained in the longer term.

Drawing It All Together: A Basis for Better Management

Presenting available ecological information like the examples presented above into a more accessible and practical form remains a challenge. Ecologists need to derive and develop guidelines and recommendations that are useful and appropriate to local circumstances. Consider three example questions that might arise naturally from our previous account: How do managers know (1) how to define an 'adequate' retention of mother trees, (2) which exotic species to control, and (3) which management practices are needed to sustain vital pollinators? If we are willing to aim at management improvement rather than perfection, incremental gains are readily available. Regarding the first question: a minimum density for mother trees would be derived as a series of recommendations based on current knowledge of the species (or similar species). Information to be summarized would include minimum size of reproductive stems, normal density and spacing (as well as pollination biology, type of fruit and seed, and requirements of known dispersal agents). This would then be translated back into practical guidelines for forest harvesting operations (eg 'over 40 per cent of class A species in the over 80cm dbh class will be maintained, and over 25 per cent of class B in the over 50cm dbh class, total canopy opening should not surpass 30 per cent in any given 10ha block etc').

For exotic species there could be a list of species known to cause trouble or with attributes that make problems likely (possibly based on an international agreement and updated as needed). Associated guidelines would be required for managers to carefully assess any introductions, identify problems rapidly and ensure effective control whenever needed.

To maintain 'key pollinators', management plans should include specific reference to what is known about the life history of the important trees and their dependencies (ie against a predefined checklist). Planning should then take this knowledge into account by ensuring key aspects are protected (eg 'pollinators require neighbouring grasslands within 500m of the forest area to be maintained; these will be gazetted as . . . and so on') – or indicating collaboration with national forest bodies in targeted research to clarify further

Table 18.2 *Ecology for tropical forest management: some general propositions*

- Ecology is vital to good forest management, not opposed to it. Any definition of 'good practice' or 'sustainability' that neglects a comprehensive treatment of ecological principles will be incomplete

- There is already a vast amount of ecological information that can be used to guide management, but managers require various types of support to be able to make better use of this

- Wider availability of information (Internet etc) may help, but more importantly, ecologists and foresters must seek a constructive alliance in which the best ecology is made available as a basis for management

- Ecologists must not merely criticize management practices but should seek ways by which they can be improved. Nor should ecologists expect to tell forest managers what they can and cannot do – there is a need for dialogue to find the most acceptable compromises

- An alliance between ecologists, foresters and others is needed to combat the many threats facing forests

- We support the many initiatives looking at codes of conduct, criteria and indicators, certification, forestry education etc but suggest that there is a need to provide adequate ecological support (information, training, materials, incentives, access to expertise) to those responsible for forest management.

what is needed. Lists, and derived guidelines, could similarly be used to specify the resources and species to which managers need pay special attention in order to fulfil local and regional conservation and cooperative management objectives.

The development of programmes such as 'codes of practice', 'criteria and indicators' and certification schemes may provide a potential context in which the more demanding ecological issues can be brought to the mainstream. In this way, the likely costs of improved management may not disadvantage the conscientious managers. It is not realistic that either foresters or ecologists working alone could do all this – it requires partnership and informed consultation.

The problem of information access remains. Most information is found in academic books and journals, and is not easily accessible to the general forestry community. This may be improved by Internet libraries and abstracting services (eg the CABI tree CD service) and increasing education and training activities by international environmental organizations and cooperative programmes. However, in many tropical countries, ecologists are too often seen as opponents rather than providers of positive advice, and working on a better mutual understanding is an important step.

We have discussed how improvements are attainable, but we do not wish to give the impression that the best management can be attained through prescriptions, regulations and bureaucratic procedures alone. The best management practices will only be achieved if trained, experienced and motivated forest managers are available on site to address ecological concerns on a day-to-day basis. Our suggestions require the acceptance of a greater emphasis on ecological information and ecologists in the planning, implementation, and control of forest management and a genuine willingness to improve current practices. Certainly more research will uncover new insights and factors that may be relevant, but this should not stop the development of a range of guidelines, and local management codes can be useful now.

Conclusion

Ecology provides a vital component of the understanding needed to reconcile the long-term viability of tropical forests with human needs. Though tropical forests remain incompletely understood, this cannot excuse the current prevalence of poor management practices – enough is known to manage tropical forests much better than is generally the case. Forest managers and ecologists must work together to face the considerable challenges to tropical forests in the new millennium.

While we have presented a number of ecological principles and ideas through the text, these serve as illustrations only, and are neither definitive nor comprehensive. We conclude with some more general propositions (Table 18.2).

Acknowledgements

We thank Andrew Plumptre, Derek Pomeroy, Henry Osmaston, Michael Spilsbury, Dennis Dykstra, Francis (Jack) Putz, Carol Colfer, Jeff Sayer, Arnold Grayson, Alan Pottinger and referees for their comments, encouragement and criticisms, and Melinda Wan, Rosita Go, and the CIFOR library staff for their help.

References

1   Alexander, I., Ahmad, N. and See, L. S. (1992) 'The role of mycorrhiza in the regeneration of some Malaysian forest trees. Philosophical Transactions of the Royal Society of London'. Series B, *Biological Sciences*, vol 335, pp379–388.

2   Appanah, S. and Mohd.-Rasol, A. M. (1990) 'Smaller trees can fruit in logged dipterocarp forests'. *Journal of Tropical Forest Science*, vol 3, pp80–87.

3   Appanah, S. and Putz, F. E. (1984) 'Climber abundance in virgin dipterocarp forest and the effect of pre-felling climber cutting on logging damage'. *The Malaysian Forester*, vol 47, pp335–342.

4   Bakuneeta, C, Johnson, K., Plumptre, R. and Reynolds, V. (1995) 'Human uses of tree species whose seeds are dispersed by chimpanzees in the Budongo forest'. *African Journal of Ecology*, vol 33, pp276–278.

5   Baur, K. S. and Hadley, M. (eds) (1990) *Reproductive Ecology of Tropical Forest Plants*. Man and the Biosphere series, UNESCO, Paris.

6   Bawa, K. S. and Seidler, R. (1998) 'Natural forest management and conservation of biodiversity in tropical forests'. *Conservation Biology*, vol 12, pp46–55.

7   Bellingham, P. J., Tanner, E. V. J. and Healey, J. R. (1994) 'Sprouting of trees in Jamaican montane forests, after a hurricane'. *Journal of Ecology*, vol 82, pp747–758.
8   Berthault, J. G. (1990) 'Comparaison d'écosystèmes forestiers naturels et modifiés après incendies en Côte d'Ivoire', in Puig, H. (ed) *Atelier Sur L'aménagement et la Conservation de L'écosystème Forestier Topicale Humide*. Cayenne, 1990. UNESCO, Paris.
9   Brandani, A., Hartshorn, G. S. and Orians, G. H. (1988) 'Internal heterogeneity of gaps and species richness in Costa Rican tropical wet forest'. *Journal of Tropical Ecology*, vol 4, pp99–119.
10 Brokaw, N. (1987) 'Gap phase regeneration of three pioneer tree species in a tropical forest'. *Journal of Ecology*, vol 75, pp9–20.
11 Brown, B. N. (1976) 'Phytophthora cinnamomi associated with patch death in tropical rain forests in Queensland'. *APPS-Newsletter*, vol 5, pp1–4.
12 Brown, N. D. and Jennings, S. (1998) 'Gap-size niche differentiation by tropical rainforest trees: a testable hypothesis or a broken-down bandwagon?' in Newbery, D. M., Prins, H. H. T. and Brown, N. D. (eds) (1996) *Dynamics of Tropical Communities. The 37th Symposium of the British Ecological Society*, Cambridge University. Blackwell Science, Oxford, pp79–94.
13 Buechner, H. K. and Dawkins, H. C. (1961) 'Vegetation change induced by elephants and fire in Murchison Falls National Park, Uganda'. *Ecology*, vol 42, pp752–766.
14 Campbell, D. G. (1991) 'Gap formation in tropical forest canopy by elephants, Oveng, Gabon, Central Africa'. *Biotropica*, vol 23, pp195–196.
15 Cannon, C. H., Peart, D. R. and Leighton M. (1998) 'Tree species diversity in commercially logged Bornean rainforest'. *Science*, vol 281, pp1366–1368.
16 Chapman, C. A. and Chapman, L. J. (1997) 'Forest regeneration in logged and unlogged forests of Kibale National Park, Uganda'. *Biotropica*, vol 29, pp396–412.
17 Chua, L. S, Hawthorne, W. D., Guan, S. L. and Seng, Q. E. (1998) 'Biodiversity database and assessment of logging impacts', in See, L. S. et al (eds) *Conservation, Management and Development of Forest Resources*. Proceedings of the Malaysia-United Kingdom Programme Workshop, 21–24 October 1996, Kuala Lumpur, Malaysia, pp30–40.
18 Clark. D. A. (1994) 'Plant demography', in McDade, L. A., Bawa, K. S., Hespenheide, B. H. and Hartshorn, G. S. (eds) *La Selva: Ecology and Natural History of a Neotropical Rainforest*. University of Chicago Press, Chicago, pp90–105.
19 Clark, D. B. (1990) 'The role of disturbance in the regeneration of neotropical moist forests', in Bawa, K. S. and Hadley, M. (eds) *Reproductive Ecology of Tropical Forest Plants*. MAB series 7, UNESCO, Paris, pp291–315.
20 Cronk, Q. C. B. and Fuller, J. L. (1994) *Invasive Plants: The Threat to Natural Ecosystems Worldwide*. A WWF Handbook. Chapman and Hall, London.
21 Daily, G. C. and Ehrlich, P. R. (1996) 'Nocturnality and species survival'. *P.N.A.S*, vol 93, no 21, pp11709–11712.
22 Datta, A. (1998) 'Hornbill abundance in unlogged forest, selectively logged forest and a forest plantation in Arunachal Pradesh, India'. *Oryx*, vol 32, pp285–294.
23 Dawkins, H. C. (1958) *The Management of Natural Tropical High-Forest with Special Reference to Uganda*. Imperial Forestry Institute, University of Oxford, Oxford.
24 Dawkins, H. C. and Philip, M. S. (1998) *Tropical Moist Forest Silviculture and Management. A History of Success and Failure*. CAB International, Wallingford.
25 De la Cruz, M. and Dirzo, R. (1987) 'A survey of the standing levels of herbivory in seedlings from a Mexican rain forest'. *Biotropica*, vol 19, pp98–106.
26 Denslow, J. S. (1987) 'Tropical rainforest gaps and tree species diversity'. *Annual Review of Ecological Systematics*, vol 18, pp431–451.
27 Denslow, J. S., Gomez Dias, A. E. and Spies, T. A. (1990) 'Seed rain to tree fall gaps in a neotropical rainforest'. *Canadian Journal of Forest Research*, vol 20, pp642–648.

28 Epp, G. A. (1987) 'The seed bank of Eupatorium odoratum along a successional gradient in a tropical rain forest in Ghana'. *Journal of Tropical Ecology*, vol 3, pp139–149.

29 Favrichon, V. (1998) 'Modelling the dynamics and species composition of a tropical mixed-species uneven-aged natural forest – effects of alternative cutting regimes'. *Forest Science*, vol 44, pp113–124.

30 Gartlan, J. S., McKey, D. B., Waterman, P. G., Mbi, C. N. and Struhsaker, T. T. (1980) 'A comparative study of the phytochemistry of two African rain forests'. *Biochemical Systematics and Ecology*, vol 8, pp401–422.

31 Gascon, C., Williamson, G. B. and da Fonseca, G. A. B. (2000) 'Receding forest edges and vanishing reserves'. *Science*, vol 288, pp1356–1358.

32 Gautier-Hion, A., Duplantier, J.-M., Quris, R., Feer, F., Sourd, C., Decoux, J. P., Emmons, L., Dubost, G., Erard, C., Hechestweiler, P., Moungazi, A., Roussilhon, C. and Thiollay, J.-M. (1985) 'Fruit characters as a basis of fruit choice and seed dispersal in a tropical forest verte-brate community'. *Oecologia*, vol 65, pp324–337.

33 Ghazoul, J., Liston, K. A. and Boyle, T. J. B. (1998) 'Disturbance-induced, density depend-ent seed selection in Shorea siamensis (Dipterocarpaceae), a tropical forest tree'. *Journal of Ecology*, vol 86, pp462–473.

34 Glenn-Lewin, D. C. and van der Maarel, E. (1992) 'Patterns and processes of vegetation dynamics', in Glenn-Lewin, D. C, Peet, R. K. and Veblen, T. T. (eds) *Plant Succession: Theory and Prediction*, Chapman and Hall, London, pp11–59.

35 Gordon, W. F., Vinson, S. B., Newstrom, L. E., Barthell, J. F., Haber, W. A. and Frankie, J. K. (1990) 'Plant phenology, pollination ecology, pollinator behaviour and conservation of pol-linators in Neotropical dry forest', in Bawa, K. S. and Hadley, M. (eds) *Reproductive Ecology of Tropical Forest Plants*. MAB series 7, UNESCO, Paris, pp37–48.

36 Guevara, S. S. (1986) *Plant Species Availability and Regeneration in Mexican Tropical Rain For-est*. Acta Universitatis Upsaliensis, Comprehensive Summaries of Uppsala Dissertations for the Faculty of Science, 48.

37 Hartshorn, G. S. (1978) 'Tree falls and tropical forest dynamics', in Tomlinson, P. B. and Zimmerman, M. H. (eds) *Tropical Trees as Living Systems*. Cambridge University Press, New York, pp617–638.

38 Hawthorne, W. D. (1993) *Forest Regeneration after Logging: Findings of a Study in the Bia South Game Production Reserve, Ghana*. ODA Forestry Series. Natural Resources Institute, Chatham.

39 Hawthorne, W. D. (1994) *Fire Damage and Forest Regeneration in Ghana*. ODA Forestry Series. Natural Resources Institute, Chatham.

40 Hawthorne, W. D. (1995) *Ecological Profiles of Ghanaian Forest Trees*. Tropical Forestry Papers 29. Oxford Forestry Institute, UK.

41 Hawthorne, W. D. (1996) *Holes and the Sums of Parts in Ghanaian Forest: Regeneration, Scale and Sustainable Use*. Proceedings of the Royal Society of Edinburgh, 104B, pp75–176.

42 Hawthorne, W. D. and Abu-Juam, M. (1995) *Forest Protection in Ghana*. IUCN, Gland, Switzerland and Cambridge, UK.

43 Hawthorne, W. D., Agyeman, V. K., Abu Juam, M. and Foli, E. G. (in press) *Taking Stock: An Annotated Bibliography of Logging Damage and Recovery in Tropical Forests, and the Results of New Research in Ghana*. Tropical Forestry Papers. Oxford Forestry Institute, UK.

44 Hawthorne, W. D. and Parren, M. P. E. (2000) 'How important are forest elephants to the survival of woody plant species in Upper Guinean forests?' *Journal of Tropical Ecology*, vol 16, pp133–150.

45 Hladik, A. and Miquel, S. (1990) 'Seedling types and plant establishment in an African rain forest', in Bawa, K. S. and Hadley, M. (eds) *Reproductive Ecology of Tropical Forest Plants*. MAB series 7, UNESCO, Paris, pp37–47.

46   Hogbero, P. and Wester, J. (1998) 'Root biomass and symbioses in Acacia mangium replacing tropical forest after logging'. *Forest Ecology and Management*, vol 102, pp333–338.

47   Howe, H. F. and Westley, L. C. (1988) *Ecological Relationships of Plants and Animals*. Oxford University Press, Oxford.

48   Hubbell, S. P., Foster, R. B., O'Brien, S. T., Harms, K. E., Condit, R., Wechsler, B., Wright, S. J. and Loo de Lao, S. (1999) 'Light-gap disturbances, recruitment limitation, and tree diversity in a neotropical forest'. *Science*, vol 283, pp554–557.

49   Janzen, D. H. (1979) 'New horizons in the biology of plant defenses', in Rosenthal, G. A. and Janzen, D. H. (eds) *Herbivores: Their Interaction with Secondary Plant Metabolites*. Academic Press, London, pp331–350.

50   Johns, A. D. (1988) 'Effects of "selective" timber extraction on rain forest structure and composition and some consequences for frugivores and foliovores'. *Biotropica*, vol 20, pp31–37.

51   Johns, A. D. (1997) *Timber Production and Biodiversity Conservation in Tropical Rain Forests*. Cambridge University Press, Cambridge.

52   Kasenene, J. M. (1987) 'The influence of mechanized selective logging, felling intensity and gap-size on the regeneration of a tropical moist forest in the Kibale Forest, Uganda'. *Tropical Ecology*, vol 25, pp179–195.

53   Kasenene, J. M. and Murphy, P. G. (1991) 'Post-logging tree mortality and major branch loss in Kibale forest reserve, Uganda'. *Forest Ecology and Management*, vol 46, pp295–307.

54   Kerr, W. E., Nascimento, V. A. and Carvalho, G. A. (1994) 'Ha salvaçao para os meliponineos? (Is there salvation for meliponine bees?)', in *Anais do 1 deg encontro sobro abelhas*, de Ribeirao Preto, 2–4 June 1994, USP-Campus de Ribeirao Preto, Brazil, pp60–65.

55   Kigomo, B. N., Woodell, S. R. and Savill, P. S. (1994) 'Phenological patterns and some aspects of the reproductive biology of Brachylaea huillensis O. Hoffrh'. *African Journal of Ecology*, vol 94, pp296–307.

56   Klaus, G., Klaus-Hugi, C. and Schmid, B. (1998) 'Geophagy by large mammals at natural licks in the rain forest of the Dzanga National Park, Central African Republic'. *Journal of Tropical Ecology*, vol 14, pp829–739.

57   Laurance, W. F. and Bierregaard, R. O. (eds.) (1997) *Tropical Forest Remnants – Ecology Management and Conservation of Fragmented Communities*. University of Chicago Press, Chicago and London.

58   Laws, R. M., Parker, I. S. C. and Johnstone, R. C. B. (1975) *The Ecology of Elephants in North Bunyoro, Uganda*. Clarendon Press, Oxford.

59   Lawton, R. M. (1955) 'The relationship between crown form and sex in (Chlorophora excelsa)'. *Empire Forestry Journal*, vol 34, pp192–193.

60   Liew, T. C. (1973) 'The practicability of climber cutting and tree marking prior to logging as a silvicultural tool in the management of dipterocarp forest in Sabah'. *The Malaysian Forester*, vol 36, pp80–122.

61   Loehle, C. (1988) 'Tree life history strategies: the role of defences'. *Canadian Journal of Forest Research*, vol 18, pp209–222.

62   Loiselle, B. A., Ribbens, E. and Vargas, O. (1996) 'Spatial and temporal variation of seed rain in a tropical lowland wet forest'. *Biotropica*, vol 28, pp82–95.

63   Lowe, R. G. (1995) 'How can ecology contribute to the silviculture of natural tropical high forest and its regeneration?' *Commonwealth Forestry Review*, vol 74, pp162–163.

64   Malingreau, J. P., Stephens, G. and Fellows, L. (1985) 'Remote sensing of forest fires: Kalimantan and North Borneo in 1982–83'. *Ambio*, vol 14, pp314–321.

65   Markham, A. (ed.) (1998) *Potential Impacts of Climate Change on Tropical Forest Ecosystems*. Kluwer Academic, Netherlands (reprinted from Climatic Change Volume 39).

66 McNally, L. C. and Schneider, S. S. (1996) 'Spatial distribution and nesting biology of colonies of the African honey bee Apis mellifera scutellata (Hymenoptera: Apidae) in Botswana, Africa'. *Environmental Entomology*, vol 25, pp643–652.

67 Moyersoen, B., Fitter, A. H. and Alexander, I. J. (1998) 'Spatial distribution of ectomycorrhizas and arbuscular mycorrhizas in Korup national park rain forest, Cameroon, in relation to edaphic parameters'. *New Phytologist*, vol 139, pp311–320.

68 Nason, J. D. and Hamrick, J. L. (1997) 'Reproductive and genetic consequences of forest fragmentation: two case studies of neotropical canopy trees'. *Journal of Heredity*, vol 88, pp264–276.

69 Newmark, W. D. (1991) 'Tropical forest fragmentation and the local extinction of understorey birds in the Eastern Usambara Mountains, Tanzania'. *Conservation Biology*, vol 5, pp67–78.

70 Ng, F. S. P. (1983) 'Ecological principles of tropical lowland rain forest conservation', in Sutton, S. L., Whitmore, T. C. and Chadwick, A. C. (eds) *Tropical Rain Forest Ecology and Management*. Special Publication Number 20 of the British Ecological Society, Blackwell Scientific Publications, Oxford, pp359–375.

71 O'Brien, T. G., Kinnaird, M. F., Dierenfeld, E. S., Conklin-Brittain, N. L., Wrangham, R. W. and Silver, S. C. (1998) 'What's so special about figs?' *Nature*, vol 392, p668.

72 Osmaston, H. A. (1965) 'Pollen and seed dispersal in Chlorophora excelsa and other Moraceae, and in Parkia filicoidea (Mimosaceae) with special reference to the role of the fruit-bat, Eidolon helvum'. *Commonwealth Forestry Review*, vol 44, pp96–103.

73 Parren, M. E. (1991) *Forest Elephant (Loxodonta Africana Cyclotis Matschie) Messenger-Boy or Bulldozer? The Possible Impact on the Vegetation, with Special Reference to 41 Tree Species of Ghana*. A.V. 90/51. Department of Forestry, Wageningen Agricultural University, The Netherlands.

74 Pelissier, R., Pascal, J. P., Houllier, F. and Laborde, H. (1998) 'Impact of selective logging on the dynamics of a low elevation dense moist evergreen forest in the Western Ghats (South India)'. *Forest Ecology and Management*, vol 105, pp107–119.

75 Phillips, C. (1987) 'Preliminary observations on the effects of logging on a hill forest in Sabah', in Yusuf Hadi, Kamis Awang, Nik Muhamad Majid and Shukri Mohamed (eds) *Impact of Man's Activities on Tropical Upland Forest Ecosystems*. Universiti Pertanian Malaysia, Serdang, Malaysia, pp187–215.

76 Phillips, O. L. and Sheil, D. (1997) 'Forest turnover, diversity and CO_2'. *Trends in Ecology and Evolution*, vol 12, p404.

77 Plumptre, A. J. (1995) 'The importance of "seed trees" for the natural regeneration of selectively logged tropical forest'. *Commonwealth Forestry Review*, vol 74, pp253–258.

78 Plumptre, A. J. (1996) 'Changes following 60 years of selective timber harvesting in the Budongo Forest Reserve, Uganda'. *Forest Ecology and Management*, vol 89, pp101–113.

79 Plumptre, R. A. (1996) 'Links between utilisation, product marketing and forest management in tropical moist forest'. *Commonwealth Forestry Review*, vol 75, no 4, pp316–325.

80 Putz, F. E. (1984) 'The natural history of lianas on Barro Colorado Island, Panama'. *Ecology*, vol 65, pp1713–1724.

81 Redford, K. H. (1992) 'The empty forest'. *BioScience*, vol 42, pp412–422.

82 Rejmanek, M. (1999) 'Invasive plant species and invasible ecosystems', in Sandlund, O. T. (ed) *Invasive Species and Biodiversity Management*. Kluwer Academic, The Netherlands, pp79–102.

83 Rejmanek, M., Reimankova, E. and Katende, T. (1996) *Invasive Plants in Ugandan Protected Areas*. Res Explor.

84 Roberts, M. R. and Gilliam, F. S. (1995) 'Patterns and mechanisms of plant diversity in forested ecosystems: implications for forest management'. *Ecological Modelling*, vol 86, pp37–50.

85   Robinson, J. G., Redford, K. H., and Bennett, E. L. (1999) 'Wildlife harvest in logged tropical forests'. *Science*, vol 284, pp595–596.

86   Schupp, E. W. (1990) 'Annual variation in seedfall, post-dispersal predation and recruitment of a neotropical tree'. *Ecology*, vol 71, pp504–515.

87   Sheil, D. (1994) 'Naturalised and invasive plant species in the evergreen forests of the East Usambara Mountains Tanzania'. *African Journal of Ecology*, vol 32, pp66–71.

88   Sheil, D. (1996) 'Species richness, forest dynamics and sampling: questioning cause and effect'. *Oikos*, vol 76, pp587–590.

89   Sheil, D. (1998) 'A half century of permanent plot observation in Budongo Forest, Uganda: Histories, highlights and hypotheses', in Dallmeier, F. and Comiskey, J. A. (eds) *Forest Biodiversity-Research, Monitoring and Modelling: Conceptual Background and Old World Case Studies*. Proceedings from the 1995 Smithsonian MAB Washington Symposium. MAB, UNESCO, Paris, pp399–428.

90   Sheil, D. and Hawthorne, W. (1995) 'How can ecology contribute to the silviculture of natural tropical high forest and its regeneration? A reply to Lowe'. *Commonwealth Forestry Review*, vol 74, pp181–182.

91   Sheil, D., Sayer, J. A. and O'Brien, T. (1999) *Tree Diversity and Conservation in Logged Rainforest*.

92   Silva, J. N. M., Carvalho, J. O. P. de, Lopes, J. do C. A., Oliveira, R. P. de K. and Oliveira, L. C. de (1996) 'Growth and yield studies in the Tapajos region, Central Brazilian Amazon'. *Commonwealth Forestry Review*, vol 75, pp325–329, 350–352.

93   Skorupa, J. P. (1988) *The Effects of Selective Timber Harvesting on Rain Forest Primates in Kibale Forest, Uganda*. Ph.D. thesis, University of California, Davis.

94   Skorupa, J. P. and Kasenene, J. M. (1984) 'Tropical forest management: Can rates of natural treefalls help guide us?' *Oryx*, vol 18, pp96–101.

95   Sousa, W. P. (1984) 'The role of disturbance in natural communities'. *Annual Review of Ecology and Systematics*, vol 15, pp353–359.

96   Stacy, E. A., Hamrick, J. L., Nason. J. D., Hubbell. S. P., Foster, R. B. and Condit, R. (1996) 'Pollen dispersal in low-density populations of three neotropical tree species'. *The American Naturalist*, vol 148, pp275–298.

97   Struhsaker, T. T. (1997) *Ecology of an African Rain Forest: Logging in Kibale and the Conflict Between Conservation and Exploitation*. University Press of Florida, Gainesville.

98   Struhsaker, T. T., Lwanga, J. S. and Kasenene, J. M. (1996) 'Elephants, selective logging and forest regeneration in the Kibale forest, Uganda'. *Journal of Tropical Ecology*, vol 12, pp45–64.

99   Swaine, M. D. (ed) (1996) *The Ecology of Tropical Forest Tree Seedlings*. Man and Biosphere Series, Volume 17. Unesco and Parthenon, Paris.

100 Swaine, M. D. and Hall, J. B. (1983) 'Early succession on cleared forest land in Ghana'. *Journal of Ecology*, vol 71, pp601–627.

101 Thomas, S. C. (1996) 'Relative size at onset of maturity in rain forest trees: a comparative analysis of 37 Malaysian species'. *Oikos*, vol 76, pp145–154.

102 Turner, I. M. (1996) 'Species loss in fragments of tropical rainforests: a review of evidence'. *Journal of Applied Ecology*, vol 33, pp200–209.

103 Turner, I. M., Chua, K. S., Ong, J. S. Y., Soong, B. C. and Tan, H. T. W. (1996) 'A century of plant species loss from an isolated fragment of lowland tropical rain forest'. *Conservation Biology*, vol 10, pp1229–1244.

104 Vazquez-Yanes, C. and Orozco-Segovia, A. (1993) 'Patterns of seed longevity and germination in the tropical rainforest'. *Annual Review of Ecology and Systematics*, vol 24, pp69–77.

105 Waterman, P. G. (1983) 'Distribution of secondary metabolites in rain forest plants; towards an understanding of cause and effect', in Sutton, S. L., Whitmore, T. C. and Chadwick,

A. C. (eds) *Tropical Rain Forest Ecology and Management*. Special Publication Number 20 of the British Ecological Society. Blackwell Scientific Publications, Oxford, pp167–179.

106 Waterman, P. G., Ross, J. A. M., Bennett, E. L. and Davies, E. G. (1988) 'A comparison of the floristics and leaf chemistry of the tree flora in two Malaysian rain forests and the influence of leaf chemistry on populations of colobine monkeys in the Old World'. *Biological Journal of the Linnean Society*, vol 34, pp4–32.

107 Whitmore, T. C. (1991) 'Invasive woody plants in perhumid tropical climates', in Ramakrishnan, P. S. (ed) *Ecology of Biological Invasions in the Tropics*. International Scientific Publications, New Delhi.

108 Whitmore, T. C. and Brown, N. D. (1996) 'Dipterocarp seedling growth in rain forest canopy gaps during six and a half years'. *Philosophical Transactions of the Royal Society of London. Series B, Biological Sciences*, vol 351, pp1195–1203.

109 Whitmore, T. C. and Burslem, D. F. R. P. (1998) 'Major disturbances in tropical rainforests', in Newbery, D. M., Prins, H. H. T. and Brown, N. D. (eds) *Dynamics of Tropical Communities*. The 37th symposium of the British Ecological Society, Cambridge University 1996. Blackwell Science, Oxford, pp549–565.

110 Whitney, K. D., Fogiel, M. K., Lamperti, A. M., Holbrook, K. M., Stauffer, D. J., Hardesty, B. D., Parker, V. T. and Smith, T. B. (1998) 'Seed dispersal by Ceratogymna hornbills in the Dja Reserve, Cameroon'. *Journal of Tropical Ecology*, vol 14, pp351–371.

111 Whitney, K. D. and Smith, T. B. (1998) 'Habitat use and resource tracking by African Ceratogymna hornbills: Implications for seed dispersal and forest conservation'. *Animal Conservation*, vol 1, pp107–117.

112 Zahner, V. (1993) 'Hohlenbaume und Forstwirtschaft. (Hollow trees and forestry)'. *Allgemeine Forst Zeitschrift*, vol 48, pp538–540.

19

From Meeting Needs to Honouring Rights: The Evolution of Community Forestry

Liz Alden Wily

Introduction: From Needs to Rights

Acceptance of the concept of community forestry is widespread around the world today. In more than 100 states, ordinary citizens are being acknowledged and encouraged as forest conservators.[1] Many of the roles and powers that governments have previously kept for themselves are being handed over to citizens, including, unusually, the remote rural poor. The trend is uneven. Intentions are mixed, ranging from what is initially little more than burden and cost sharing to real power sharing. Doubt, contestation, backtracking and backlash are common in an evident evolution of one to the other.

That overall change in forest management relations is afoot, and that this is broadly devolutionary in character, cannot be denied. Nor can it be denied that this development is gathering a momentum of its own – and one that goes well beyond the kind intentions of the Rio Earth Summit in 1992 (the United Nations Conference on Environment and Development – UNCED) to pay attention to the needs of people. Through community forestry initiatives and the policies and legal paradigms being built to support them, what is changing is more than how forests are managed and by whom. As it matures, community forestry is reaching deep into questions of rights, local rights to regulate local forests, and rights of ownership over forestland, in particular. Social relations as a whole are being affected. Whilst the pivotal relationship between State and people is most acted upon, relations internal to the community are also altered, ultimately usually towards more inclusive and fairly managed norms.

An important driver in this is the fact that forests generally exist in relation to people ultimately as a common property interest. In defining community-level actors and powers, clarification of community membership normally results, although not without contestation. Previously excluded poorer or institutionally weaker members tend to gain, if only by virtue of residency. Outsiders are also more specifically defined, and their interests made subordinate. Where outsiders are neighbouring communities, this may catalyse their own definition of 'our forests' to broadly positive effect. Where commer-

Note: Reprinted from *Proceedings of the 12th World Forestry Congress, Quebec, Canada*, Alden Wily, L., 'From meeting needs to honouring rights: The evolution of community forestry', Copyright © (2003), with permission from the World Forestry Congress

cial interests are restrained, the resulting tensions and effort required to balance interests are generally greater and more complex.

Such shifts are integral to wider shifts in power relations in contemporary global agrarian society. What is startling about the forestry case is just how far it is helping to drive, as well as be driven by changing norms, and – given the tangible nature of the resource at stake – in concrete, not just declamatory ways. Whilst it may be premature to claim that improved governance results, the signs are that this will be the case; that community forestry is proving a potent route to the empowerment of ordinary rural people, enabling them to gain more control over resources that support their livelihood and environment and the way in which they organize themselves to act upon their circumstances. Forests and society gain.

Discussion: The African Case

The case of Africa is used to illustrate this social transformation.[2] Happily, this may also serve as a partial antidote to the Africa of global publicity, the 'basket case'continent, riven by tribal dispute, malign dictatorship, rampant corruption, falling standards of living, extreme indebtedness and now failing health. For there is another face of Africa that bodes more positively and from which changes in the forestry sector draw strength; rejection of corrupt administrations, the emergence of multiparty democracy and poverty-centred planning with new, majority-centred targets. The strengthening of devolutionary governance, allowing for higher levels of mass participation in decision making, is equally powerful. And widespread land reform, pledging, inter alia to improve the tenure security of agrarian majorities, is equally promising (Toulmin and Quan (eds), 2000; Alden Wily and Mbaya, 2001).

What drives these changes? Common drivers appear to be: frustration with the failure of central regimes to raise standards of living or keep the peace, the ending of the natural life of European-introduced norms that did little to secure progress for the majority or to conserve the precious natural heritage of the continent, and international and often specific donor support towards finding new ways forward.

Rescue through reconstruction

Crucially, these developments look increasingly to 'community' as the platform of change at the periphery. Despite a century of formal suppression or benign neglect, community identity and norms in Africa remain relatively robust. Through necessity or common sense, sector after sector is beginning to build upon the basis of community. In the process, community norms are tending to be reconstructed along more inclusive and less hierarchical lines than has customarily been the case. Thus, we begin to see local governments, land administration, dispute resolution and natural resource management bodies at more and more local levels, including the village level.

New law as indicator

Where is the evidence of this transformation? Partly it is in practical delivery in the forestry sector, as described below. A more general indicator is substantial policy and legal reform across the continent, which is evidence of at least seriously *intended* change. Since 1990 alone, some 20 new National Constitutions, 20 new local government laws, 30 new land laws, 15 environmental management laws, 15 wildlife laws and at least 30 new forestry laws have been enacted across Africa (FAO, 2002; Alden Wily, 2003). When the substance of new legislation is examined and the common concern of *governance* noted – along with stronger positioning of the interests of communities, women, pastoralists, the dispossessed, the untenured and the very poor – it may be concluded that a more people-centred, enabling environment is being put in place. It is within this context that community-based forest management is evolving.

Community forestry

In an FAO-supported workshop on participatory forestry management in Africa in February 2002, 22 states were able to report that they had formal community forest management projects underway, a great advance upon the handful of initiatives described at a similar conference three years previously (FAO, 2003). Since then, some ten other states have begun implementation. Today it may be estimated that some 5000 communities are involved, bringing around 3Mha of either nationally or locally owned forests under community or state-community protection and management. Almost without exception, these include forests that would otherwise have remained or become degraded (Alden Wily, op cit).

While these figures represent but tiny proportions of either Africa's forests or communities, a rapid and active start has clearly been made. This gains substantial support from the plethora of guiding new policies and laws. Of 56 mainland and island states, 26 countries have enacted new forest laws since 1990 and another 15 have legislation in draft (FAO, 2002; Alden Wily, op cit). Almost all make commitments to increased public participation in forest management, particularly by those who live within or adjacent to forests.

Differences in approach

The means through which increased public participation in forest management is pursued are various. Salient differences exist, for example, in *where* community participation may be advanced. As is the case elsewhere, one or two states still restrict local participation to forests that are either less biologically diverse, are less important for tourism, have limited commercial importance and/or are degraded or settled. Often this coincides with a distinction between forests that are reserved and those that are not. In Cameroon for example, community forests may be established only in unclassified forest and are also restricted to 5000ha and ten-year agreement periods. In contrast, Uganda, South Africa, Ethiopia and Guinea Conakry have begun community management in forest reserves, including some of highest conservation priority. Most countries have made it a matter of principle that community management may be applied for all forest classes and values. This is the case, for example, in Tanzania and The Gambia. Whilst natural forest is the focus, an intention to involve communities in the privatization of commercial plantations is also gaining ground (eg Tanzania, Malawi, South Africa). Related social responsibility spending is under way in Ghana and Cameroon, where timber-harvesting companies are now required to divert specified small percentages of profit to local-area developments, via local councils or otherwise.

Products or power

The most fundamental differences between strategies stem from different perceptions of the basic rationale behind involving communities in forest management in the first instance. A broad distinction may be drawn between approaches designed to gain local cooperation with existing government-controlled regimes (benefit sharing) and those designed to devolve control to community level (power sharing). Both claim improved local livelihoods as a result. The former focuses on providing alternative sources to forest income (buffer zone developments), employment opportunities, improved legal access to the resource, and/or beneficial shares from revenue earned from the forest, often paid in the form of local social services. The latter assists the community in bringing the source of livelihood (the forest) under its own control and even tenure, on the grounds that only this level of empowerment will enable the community to conserve the forest for lasting livelihood and local environmental benefit. Costs to the State may also plummet, as forestry staff move from positions of controlling manager or co-manager to more remote technical adviser and watchdog.

With each passing year, the balance of practice (and legal provision) is shifting more positively towards the latter power-sharing norms. This may be manifested in the endowment of community actors with stronger rights to regulate forest use themselves (eg Nigeria, Ethiopia, Namibia), the relaxation of Government veto powers over the content of community-designed management plans (eg Cameroon), or the extension of contractual arrangements to longer periods (eg Madagascar, Senegal).

Community forests

As might be expected, power sharing is much more advanced where the forest is off-reserve, and often (but not always) on land tacitly accepted as belonging to the community. Broadly, the designation 'Community Forests' is being applied to such estates. Significantly, it is this development, rather than co-management initiatives associated with National Forests, that is seeing most growth. In the lead cases of The Gambia, Tanzania and Cameroon, for example, only a small proportion of State Forests are coming under any level of community management compared to the several hundreds of Community Forests being declared in these states.

Today, at least 20 countries provide for the legal establishment of Community Forests. In practice, most are established only through formal agreement with central forestry administrations on the basis of an agreed management plan and often with formal survey and mapping of the forest area required, significantly delaying finalization (The Gambia) or raising costs for the community (Cameroon). Tanzania is an exception, where administrative and legal acceptance of a Community Forest is through community declaration, witnessed agreement of boundaries with neighbouring communities and registration of these details along with a simple management plan and the by-laws that will be used to enforce adherence, at district government level.

At this point, less than half of the Community Forests provided for in new policies and laws enable the community to be formally recognized as owner-manager and to manage the forest in largely autonomous ways. The Gambia and Tanzania are strong examples. The remainder either limit recognition of local tenure (eg Cameroon, Senegal, Ethiopia) or as commonly, acknowledge local tenure but circumscribe local level jurisdiction (eg Nigeria, Mali, Burkina Faso). A typical arrangement is for Government to retain control over licensing and its enforcement. The ability to determine how the forest may be accessed and used in the first place, including the right to define and exclude outsiders, remains elusive for many community managers. Nor have more than a handful of states yet made it definitively possible for communities to enact forest use and management regulations that courts are obliged to uphold, should they be challenged. In this respect, few communities have the advantages of those in Tanzania, where village governments are elected and endowed with real decision-making and legislative power. Nonetheless, the act of creating a Community Forest in most of these countries does appear to establish a platform from which communities may slowly but surely garner stronger roles and powers that are eventually reflected in law. This is a transition that is visible in countries as far apart as Cameroon, Senegal, Nigeria, Malawi, Namibia and Madagascar.

Prompting institutional change

Institutional change is important because community forest management of necessity involves the creation or revitalization of a local-level institutional base. Even where the committee or association is formed initially only for the purpose of distributing benefits from state- or private-sector-run developments (eg Mozambique, Ghana) or as a framework through which local subsistence forest use is organized (eg Zimbabwe, Namibia, Kenya), such actions tend to catalyse organized action and thence demands.

In the process, the character of the local institution itself changes. Questions of representation, inclusion and accountability begin to arise and be dealt with. One trend is movement away from arrangements that are pinned solely upon traditional authorities, towards the widening of this base to included representatives of a broader source of interest groups (eg Nigeria, Malawi, Mozambique). A more common shift is away from associations formed solely by and on the basis of user groups towards more community-wide organizations constituted as management authorities (eg Ethiopia, Uganda, Zambia, Mali). This may be regarded as a critical maturation of community forestry, allowing for fuller devolution of jurisdiction to community levels and their repositioning as custodians, not just users. It also allows for a useful separation at the local level of the vested interests of community members, who are forest users or dependants, from more general environmental support interests of the community as a whole.

The crucial role of forest land tenure and reform

Issues of forest tenure are often central to these changes. This is typically the case when one sector of the community disputes the gathering control or benefit of another over what is perceived as a shared asset of the community. Clearer definition of the constitution of the community, and the nature of its rights over the resource, tend to follow.

In such considerations, community forestry is gaining substantially from land reform developments, which are again being most concretely expressed in new land laws. An important emerging theme is new state law support for majority customary land interests. The new land laws of Uganda, Tanzania, Mozambique, Niger, Mali, Namibia and South Africa are exceptional examples, with similar prescriptions promised in proposed legislation in Lesotho, Swaziland and Malawi.

The relevance to community forestry is that these developments not only provide for formal recognition of individually held customary rights to be upheld as private rights, but also for properties held in common to gain this new legal support. In such circumstances, communities may, for the first time ever, secure local forests as group-held private property and even register their ownership as such. Helpfully, definition of what is 'customary' is being defined less by tradition than by prevailing community-supported norms. In countries like Eritrea, Ethiopia, Rwanda and Burkina Faso, where customary rights are eschewed in principle, ample provision is gradually being made to enable existing, if not customary communal holding to similarly gain registrable form (Alden Wily and Mbaya, op cit).

A corollary development is seeing greater restraint being placed upon the use of the routine right by Governments to appropriate land for public purposes – including the creation of Government Forest Reserves. Procedures are being made more publicly accountable and almost everywhere require fuller consultation with those affected. The now much higher rates of compensation that must be paid to those who lose rights acknowledged as private rights in the new laws are a special disincentive to wanton appropriation of local commons like forested areas.

The upshot of these developments is that many new forestry laws of necessity now lay out more cautious procedures for declaring or classifying forests (as Government Reserves) and are encouraged to provide alternative routes to securing still unreserved or un-demarcated forests as formally dedicated to the purposes of forest conservation and sustained use. Community Forests provide this route. Even where Governments remain determined to bring a forest under their own jurisdiction, the agreement of local communities is practically and sometimes legally more essential than in the past. In such cases in Tanzania for example, the State is legally bound to consider whether declaration of a community forest would not be 'a more efficient, effective and equitable route to balance the maintenance of existing rights with the protection and sustainable use of forest resources'. Such provisions clearly illustrate the changing tenor of forest management.

The difficulties of process

It would be incorrect to give the impression that evolutionary community forestry around the world is problem-free. On the contrary, as we have seen in recent years in countries as far apart as Nepal, Zambia and Brazil, two steps forward in policy support may be followed by one step backwards. Or at the local level, replication may founder upon rather too expensively developed models. Or strictures in new tenure, wildlife and mineral legislation may contradict the directions community forestry promises. As is typically the case with social transformation, problems may inhibit progress and stimulate effort at one and the same time. What does seem to be clear is that democratization in forest management relations is gradually occurring. There is, in fact, a chance that by the end of the 21 century, conservation management around the world will be predominantly in the hands of people, with the conservators of the 20th century – national governments – playing only advisory and monitoring roles. If this is achieved, then it should be registered as a success. The sector will have shown itself able to grasp the nettle and recognize that the resistant problems it faces are ultimately problems of governance, and are resolvable only through transformation in governance relations. This includes sounder and fairer determination as to how control over the forest is vested and exercised and by whom, and who owns the forest. That the future of forests, and the bridge between human and environmental needs, lies in democratic transformation seems an essential conclusion to be drawn.

Notes

1   The 46 issues of the FAO-funded *Forests, Trees and People Newsletter* provide an invaluable record of the evolution of community forestry up until 2002.
2   Stringent word limitations mean that documentation is deliberately avoided in this chapter. Readers are asked to refer to these publications for details and sources: Alden Wily & Mbaya, 2001; Alden Wily, 2003; FAO, 2003.

References

1   Alden Wily, L. (2003) *Community Forest Management in Africa Progress and Challenges in the 21st Century*, in FAO (2003).
2   Alden Wily, L. and Mbaya, S. (2001) *Land, People and Forests in Eastern & Southern Africa at the Beginning of the 21st Century. The Impact of Land Relations of the Role of Communities in Forest Future*. IUCN, Nairobi.
3   FAO (2002) *Law and Sustainable Development Since Rio: Legal Trends in Agriculture and Natural Resource Management*. FAO Legislative Study 73, Rome.
4   FAO (2003) *Proceedings of the Second International Workshop on Participatory Forestry in Africa. Defining the Way Forward: Sustainable Livelihoods and Sustainable Forest Management Through Participatory Forestry*, February 2002, Arusha, United Republic of Tanzania, FAO, Rome, pp18–23.
5   Toulmin, C. and Quan, J. (eds) (2000) *Evolving Land Rights, Policy and Tenure in Africa*. London, DFID/IIED/NRI.

20

A Strategic Approach to Multistakeholder Negotiations

David Edmunds and Eva Wollenberg

Introduction

Professionals working in environment and development have focused recently on formal stakeholder identification and negotiation processes to address competition among different groups for natural resources (Borrini-Feyerabend, 1997; Roling and Wagemakers, 1998; FAO, 1999; O'Faircheallaigh, 1999; Ramirez, 1999; Steins and Edwards, 1999). Stakeholder identification and negotiation have promised to bring visibility, compromise and democratic decision making to stakeholder relations.

In the absence of negotiations, competing claims to forests have often been settled by force. Disadvantaged groups have used force successfully through protest and resistance (Peluso, 1992; Melucci, 1996; Severino, 1998). More often, however, powerful stakeholders take the upper hand using the legal and extra-legal means available to them. The results for disadvantaged groups have often been devastating, including displacement, cultural disintegration, dramatic material deprivation and violence (Anderson and Grove, 1987; Hecht and Cockburn, 1989; Parajuli, 1998). Negotiations would seem to benefit disadvantaged groups in particular by publicly acknowledging their claims, creating a forum to reach compromise between them and other stakeholders, and legitimating compromises with formal agreements.

Yet the benefits of multistakeholder negotiations to disadvantaged groups depend on *how* negotiations are undertaken. We believe many approaches to multistakeholder negotiations mask abuses of power and more structural, enduring inequity. In doing so, they are prone to exaggerate the level of consensus reached through negotiations and expose disadvantaged groups to greater manipulation and control by more powerful stakeholders. These approaches share one or more of the following assumptions:

- A neutral or objective space for negotiation can and should be created.
- Consensus is desirable.
- All stakeholders need to be involved for the process to be effective.
- Negotiations can be considered in isolation from other strategies employed by stakeholders.
- Generally, the principal barrier to effective collective action is poor communication.

Note: Reprinted from *Development and Change*, vol 32, Edmunds, D. and Wollenberg, E. 'A strategic approach to multistakeholder negotiations', pp231–253, Copyright © (2001), with permission from Blackwell Publishing

These assumptions are either explicit or implicit in the documentation of negotiation experiences and in guidelines for negotiation and multistakeholder collaboration (Fisher, 1995; Porter and Salvesen, 1995; Borrini-Feyerabend, 1996; Allen et al, 1998; Kearney et al, 1999; Roling and Maarle-veld, 1999; Roling and Wagemakers, 1998). We suggest that rather than try to eliminate or temporarily neutralize political differences within negotiations to achieve broad agreement, practitioners should use negotiations to build alliances, gather information and test ideas strategically, with the explicit goal of increasing the decision-making power of disadvantaged groups.

Of course, this argument assumes that disadvantaged groups wish to expand their decision-making authority. More broadly, it assumes that those of us who are not members of disadvantaged groups can find a way to understand adequately what the diverse, sometimes shifting and contradictory desires of such groups are, and develop strategies for effectively helping groups as they try to realize those desires. These assumptions are certainly problematic. They are, however, no more problematic than assuming that practitioners can act as neutral observers or facilitators. Moreover, as we discuss below, feminist post-structuralism and radical pluralism encourage continuous reflection on these relationships, and modesty in making claims about them. This we feel is sometimes missing among practitioners grounded in other theoretical work.

To make our case for a strategic approach to negotiations, we examine both the theoretical literature and experiences in the co-management of protected areas, forests and other common-pool resources by states, communities and other stakeholders. We critique the assumptions listed above, and suggest how a strategic approach to negotiations can yield benefits for disadvantaged groups without exposing them to a range of risks we describe in the remainder of the paper.

In this article, we define disadvantaged groups of people[1] as those with limited power to influence decisions in multistakeholder settings. Their power is limited by their social status, their representation in public fora or their negotiating capacities. We are especially interested in the case of disadvantaged groups of people living in forests (such as collectors of forest products, indigenous peoples living in or near forests, swidden farmers)[2] who are consistently the weakest players in negotiations with powerful corporations, international environmental NGOs, government interests and even local elites.

The article is aimed at environment and development practitioners who are interested in multistakeholder negotiations – including, among others, those who seek to reduce conflict among stakeholders in a resource, those who wish to promote sustainable development through collaborative agreements among stakeholders, and those who seek greater social justice for disadvantaged groups through negotiations – with a view to promoting a greater alertness and responsiveness when disadvantaged groups raise concerns pertaining to multistakeholder negotiations.

Theoretical Underpinnings of Multistakeholder Negotiation Processes

It is important to recognize that the assumptions we discuss are not always explicit in the work on multistakeholder negotiations. Nor are they all necessarily found together in the various approaches to multistakeholder negotiations.[3] One or more, however, are implied in: the critique of taking positions and other strategic behaviour (Ramirez, 1999; Roling and Maarleveld, 1999); the occasional failure to differentiate among stakeholders and to define their historical relationships (Roling and Maarleveld, 1999); and the focus on developing communication and information exchange institutions, such as platforms (Roling and Wagemakers, 1998; Maarleveld and Dangbegnon, 1999), techniques for facilitation and rules for convening negotiations (Borrini-Feyerabend, 1997), criteria and indicators as a common language for stakeholders (CIFOR, 1999), and rules of order for interactions among stakeholders (see Sinclair and Smith, 1999).

We briefly describe two bodies of thought that influence this approach to negotiations: communicative rationality and liberal pluralism. We then discuss feminist poststructuralist and radical pluralist theories as foundations for an alternative, strategic approach to negotiations. Of course, each of these bodies of work is substantially more complex and subtle than we can acknowledge here. Our purpose, however, is to draw out the elements that seem to inform the current practice of multistakeholder negotiations and suggest where alternatives might better support disadvantaged groups struggling to defend their interests in forest management.

Communicative rationality

The assumption that negotiations are more valid and effective to the degree that they achieve a politically neutral space has roots in the work of Jürgen Habermas on communicative rationality (Habermas, 1984, 1987, 1995). Habermas sought to create a public sphere in which 'critical, reflexive, activist modes of thought' promote social justice (Flyvbjerg, 1998). To do so, he developed an 'intersubjective' theory of communicative rationality (Habermas, 1984, pp12–15). Communicative rationality would restore to the life-world of shared understandings and unquestioned presuppositions (which serve as the basis for common definitions of the situations people confront) its rightful role in cultural reproduction, social integration and socialization, rather than leaving these tasks to the reified and pathological 'system imperatives' of a 'monetary-bureaucratic complex' (Habermas, 1987, pp374–375). According to Habermas, communicative rationality is achieved when:

> under the pragmatic presuppositions of an inclusive and noncoercive rational discourse among free and equal participants, everyone is required to take the perspective of everyone else, and thus project herself into the understandings of self and world of all others; from this interlocking of perspectives there emerges an ideally extended we-perspective from which all can test in common whether they wish to make a controversial norm the basis of their shared practice; and this should include mutual criticism of the appropriateness of the languages in terms of which situations and needs are interpreted. In the course of successively undertaken abstractions, the core of generalizable interests can emerge step by step (Habermas, 1995, pp117–118).

For Habermas, and implicitly for many involved in stakeholder negotiations, the emphasis is on an ideal, formal process (Habermas, 1984, p25; Rawls, 1995; Parkin, 1996); a process that can produce valid and effective agreement in virtually any context (Habermas, 1984, p31). The process would be formalized in procedures, rules, laws, policies and constitutions. The content of any communication would then be a matter of lesser consequence for the cause of social justice – the process would assure the proper outcome (Habermas, 1992, 1995).

Habermas clearly recognizes the substantial barriers to achieving such a process in practice: inequities in status and power among participants and an absence of a culture of freedom are among the most important (Habermas, 1984, 1992). In response to critiques of feminists, among others, his work more explicitly addresses unequal power relations within public spheres, and acknowledges the existence of multiple publics that are not easily grouped into a general public sphere because communication among them is exceedingly difficult (see Benhabib, 1992; Fraser, 1992). He has called for empirical verification of the conditions specified for rational communication and identification of specific barriers to achieving these conditions (Habermas, 1992, p448).

Liberal pluralism

Theories of pluralism are frequently referred to in discussions of multistakeholder negotiations (Doherty and de Geus, 1996; Hirst, 1997; FAO, 1999). Pluralism, however, exists in multiple forms that affect the empowerment of disadvantaged groups differently (Rescher, 1993; Bickford, 1999). Liberal pluralism treats groups as distinct from one another and as having particular interests. The public interest is created from bargaining and synthesis of particular views, and is ethically superior to and 'contains' those particular views (Jakobsen, 1998, pp126–127). Decision outcomes are syncretic, 'all the alternatives should be accepted: all those seemingly discordant positions are in fact justified; they must, somehow, be conjoined and juxtaposed' (Rescher, 1993, p80). Crucially, individual interests that have no reference point within the ethical framework of the public interest struggle for legitimacy. It is also more difficult to make the claim that particular groups, such as forest dwellers, should have more say than others in forest management decisions based on longstanding historical claims or entitlements (Leach et al, 1997).

How does the bargaining and synthesis necessary for the creation of the public interest occur? Much of pluralist political theory has, until recently, focused on interest group pluralism (Bickford, 1999). In this view, groups are a central focus of a person's political life and are formed around relatively well-defined and stable interests. Groups focus on efforts to influence the state, the embodiment of the public interest, often through well-known and regulated mechanisms such as lobbying or contributing to campaigns. Of particular relevance to our argument, liberal pluralism does not focus on inequities among different groups in access to or influence over decision-making processes (Bickford, 1999).

We suggest that an unhealthy combination of communicative rationality and liberal pluralism underpins many attempts to organize formal multistakeholder negotiations in forest management. The influence of liberal pluralist ideas on the practice of multistakeholder negotiations has meant that the qualifications of Habermas's theory are not taken as seriously as they should be. At the same time, Habermas's own critiques of strategic action serve to delegitimate such action on the part of disadvantaged groups participating in negotiations. In the context of extreme inequalities in power and diversity of cultures, stakeholders in tropical and subtropical forests are brought together to hammer out binding agreements in short-term, facilitated sessions organized in supposedly apolitical fora. The results have often been disastrous. In the following section we analyse why we should explore an alternative that better serves the interests of disadvantaged groups.

Feminist Post-Structuralist and Radical Pluralist Theory and Multistakeholder Negotiations

Issues of identity and representation are central to feminist and radical pluralist work (Hooks, 1990; Haraway, 1991; Benhabib, 1992; Fraser, 1992; Hawkesworth, 1997; Hekman, 1997; Hill-Collins, 1997) and suggest that any 'neutral' representation of a group within a negotiation process would be extremely difficult to achieve, if not impossible. Feminist discussions of affinities and alliances point to the possibility of collective action, but emphasize the limited, contingent and strategic quality of such connections. Feminists have also developed critiques of Western rationalism as a basis for creating mutual understanding across cultures, genders, races and classes (Harding, 1986, 1997; Jakobsen, 1998). Feminist post-structuralists generally make firm commitments to linking theory and practice (Morris, 1997), and thus express scepticism of 'ideal' formulations that appear difficult to operationalize in people's everyday experience (see Parkin, 1996). Finally, many feminists are less willing than Habermas to give up substantive claims entirely in favour of procedural justice (SIGNS, 1997).[4] The emphasis among feminist post-structuralists, then, is on building strategic, self-avowedly contingent and reflexive alliances among disadvantaged groups and those sympathetic to their claims in order to achieve greater justice for those groups (Haraway, 1991; Morris, 1997; Jakobsen, 1998).

In contrast to liberal pluralist assumptions, feminists and radical pluralists argue that groups are not easily bounded, identified or cohered (Bickford, 1999) and group formation involves complex and unstable processes of self-identification and representation. Political action, including multistakeholder negotiations, works on many levels to address a plurality of concerns, and is not necessarily targeted at influencing the 'public interest' as defined by the state (Fraser, 1992; Jakobsen, 1998; Bickford, 1999). Radical pluralists place more emphasis on the differences among groups in their access to and influence over decision-making processes, and assert that negotiations are profoundly influenced by this fact (Bickford, 1999). Consensus is likely to mask continuing differences in perspective and discount the input of disadvantaged groups, and so should be viewed with some suspicion.

This literature suggests that in most of the contexts where forest management is debated in the tropics and subtropics, the interests of disadvantaged groups are better served by negotiations when participants 'take a position' about the substance as well as the process of negotiations and acknowledge the often insurmountable barriers to rational communication. Negotiations focus on strategic alliances rather than full consensus, and where agreement *is* reached, participants 'situate' the outcome politically, acknowledging the degree to which agreement is contingent and partial. Finally, negotiations are set in the context of the full range of strategies pursued by each interest group. Below, we examine some of the practical problems associated with the dominant approach to negotiations among environment and development practitioners, and suggest why we believe a more strategic approach is necessary.

Problems with pursuing neutrality

Institutions may be able to assure a high degree of communicative rationality in settings where the power to influence forest management is relatively well-balanced among stakeholders, and where cultural and social heterogeneity is low. Such settings are rare, however, and are not the subject of this paper. We are interested in the potential for negotiations to help disadvantaged groups which, by definition, do not operate in settings of balanced power. In situations of intense historical conflict among many diverse stakeholders, situations that characterize many forest settings today, the risks to disadvantaged groups of falling short of the ideal of rational, non-strategic communication are especially great. We now examine some of the practical limitations on such communication.

A language for rational communication

Postmodern and post-colonial scholars have made a compelling argument that language and information carry power and politics with them (Said, 1978; Foucault, 1980; Ashcroft et al, 1995). Language itself constrains the imagination and limits the types of decisions that can be made, often in intensely political ways. There are many examples of language and information used in political ways in discussions about forests and the people that use them (Field, 1994; Parajuli, 1998; Sioh, 1998).

Science, of course, has been identified as a neutral language that could help frame a rational discussion of stakeholder interests. Yet, it is well accepted that the nature of the problem, the methods and interpretation of results can all be biased by political prejudices (Foucault, 1980; Harding, 1986). The choice to use science is a political decision in itself. Disadvantaged groups of people often feel that scientific methods are not transparent to them and do not make use of their experiential knowledge. Recourse to science does not then eliminate the political quality of language.

One could still search for a rational consensus on the kinds of language and information to be used in communication through a process of argumentation, and Habermas recommends just that (Habermas, 1984). Legare (1995, pp348–351) points out, however, that even widely accepted language contains within it contradictory meanings that can be used by powerful groups to their advantage. She provides an example by examining the history of multiculturalism in Canada. After a long process of consultation the Canadian government adopted a formal policy of multiculturalism in 1971, one that reflected an ideal gaining broad currency in Canada in the 1970s and 1980s. Aboriginal groups in Canada have generally embraced the language of multiculturalism. This discourse brought a measure of recognition to Aboriginals as a group with a distinct history and culture. At the same time, however, it was interpreted by white Canadians as an argument for the denial of any special claims and rights for Aboriginals as prior inhabitants of the land. This was done in two ways: by creating images of authentic Aboriginals and then finding most living individuals to be 'compromised' and so not entitled to special claims; and by suggesting that in a liberal democracy, multiculturalism means all cultures are equal, and no single culture can enjoy rights beyond any other.

For their part, Aboriginals found it difficult to counter these arguments within a discourse of multiculturalism, at great personal and political costs to themselves (Legare, 1995). These contradictory aspects of language, subject to political deployment by different groups of people, suggest that the process of rationally debating the meaning of specific terms could never reach a conclusion, and that a neutral language is unlikely to be available to us.

Full disclosure of information

An ideal negotiation process implies, according to Habermas (1984), a full and transparent disclosure of information on the part of all stakeholders. This may, however, force groups to give up important aspects of their privacy. This was an important issue in Kakadu National Park in Australia. Traditional land owners in the Mirrar area declined to participate in collaborative management of the park, especially in the identification of sacred sites (Government of Australia, 1999). Even when the park offered to allow the group to retain control over sacred sites in exchange for information about their location, the Aboriginal elders remained silent.

> This is tantamount to suggesting that aboriginal heritage must be fully 'declared' in conformity with contemporary western heritage conservation mechanisms (both formal inventory and related descriptive listing systems) to be recognized as heritage. . . There are many global examples of cultural groups who desire to maintain information on cultural properties as private and whose desire to do so does not diminish the cultural values of the property (Bureau of the World Heritage Committee, 1999, p33).

The assumption that information should be shared freely also overlooks the strategic value of control over information. If disadvantaged groups are expected to divulge information freely in multistakeholder negotiations, they also can be put in the difficult position of having to choose whether to be supportive of the process, versus giving potentially valuable information to those who could use that information against them. Scott (1998, pp11–52) discusses a number of examples where states have acquired information about forests and people's use of them, and then have used that information to the disadvantage of weak groups. Driven by high-modernist ideologies and practising authoritarian politics – as in many places where multistakeholder negotiations over forest resources are proposed today – states have suppressed, distorted and exploited the 'metis' knowledge that is important in the daily lives of many forest dwellers (ibid, pp335–336).

It might be argued that the payoff for such disclosures is a rational negotiation process to the benefit of all stakeholders. Powerful stakeholders, such as forest departments and local elites, have not always stated their intentions clearly (Malla, 1999), particularly those that are extra-legal, and there might be real benefits to having them do so. Yet the likelihood of full disclosure is not great, and the risks to disadvantaged groups of failure are high, in all likelihood higher than for more powerful interests.

Representation of interests

The relationship to their constituents of the people participating in any multistakeholder process is also problematic. Haraway suggests accepting that the contradictions – and the politics – of representation are only partially and provisionally resolved when representatives act on behalf of a group in negotiations, and that only such 'situated' acts of representation 'build in accountability' (Haraway, 1991, p111; see also Fraser, 1992, pp134–135). If we agree that the representation is always 'situated' and problematic rather than given, in what sense are the positions of the representatives within a nego-tiation process ever 'objective' and free from strategic action?

The relationship of a representative to his/her constituency is perhaps most politi-cally charged when representatives of a group are designated by outsiders or are account-able to them (Ribot, 1998), as is often the case in multistakeholder negotiations. From the start, outside convenors and facilitators influence representation by the selection of stakeholder groups, the people to represent each group and how the expression of inter-ests is facilitated in the meeting. These decisions rarely meet everyone's objectives. The compromises made, in turn, are political choices that reflect to whom convenors and facilitators are accountable. Unfortunately, facilitators and convenors are rarely account-able to disadvantaged groups, even where local government representatives are selected in relatively democratic states (McCreary and Adams, 1995). In assuming relatively neutral roles for convenors and facilitators, we make it more difficult to trace lines of accountability between representatives and those they represent.

Democratic, internally-directed processes of group representation, however, also encounter problems of representation (Hooks, 1990; Cerulo, 1997). As an example, attempts by a local tenants' organization to represent members in a land dispute in Uganda were complicated by a complex array of tenant identities and interests (Edmunds, 1996). Land-poor families for whom there was little chance of finding new land in the area and for whom wage labour was not attractive feared eviction. Those who had other lands outside the disputed area or who, based on kin ties, could find some were less fearful. Widows with little strength for farming and youth anxious to move to town saw oppor-tunity in the compensation and potential wage labour associated with the eviction. Membership in any of these interest groups was fluid, as new opportunities arose for individuals based on personal networks and shifting public alliances (as when some local opponents of the land transfer changed sides when offered managerial positions in a new commercial agriculture plantation). Even a carefully planned effort to represent a constituency – with frequent and broad consultations and mechanisms of accountabil-ity – may founder on the enormous number of interests associated with such contradic-tory and fast-changing identities.

The relationship between disadvantaged groups and facilitators, convenors and even democratic representatives poses significant risks to disadvantaged groups. In the process of representation, for example, there is a risk that disadvantaged people become 'hyperreal'. Ramos (1994) suggests that advocates for Indians in Brazil have grown uncomfortable with the 'flesh-and-blood' Indians who follow a path of practical politics rather than adhering to standards of ideologically pure and self-sacrificing action. This latter, hyperreal Indian provided an emotional reward for advocates and a more compel-ling image for potential financial supporters of groups working with Indians.[5] Examples abound of NGOs that employ a romanticized or incomplete notion of disadvantaged people to serve their own professional and personal ends (Conklin and Graham, 1995; Froehling, 1997; Veber, 1998; Li, 1999). As a consequence the voices and interests of disadvantaged groups of people may be lost in debates about the needs of these hyper-real versions of themselves.

At the same time, personal perks can tempt representatives of groups to maintain their personal position in negotiations, even at the cost of failing to achieve benefits for their constituents. Members of disadvantaged groups may be especially vulnerable to such opportunities to increase their status. Where negotiations involve knowledge of a language not widely known within the disadvantaged group, or rules of behaviour not usually practised, representatives may become part of a specialized and privileged class of people within the group, and more distanced from their constituency (Melucci, 1996). Singh reports that the problem afflicts federations of forest-user organizations in Orissa, India (Neera Singh, Vasundhara, personal communication). Donors have found such federations a convenient, one-stop source of legitimizing 'participatory' public consultation, and have funded federation activities as a result. In the process, villagers complain that their representatives have been co-opted and are no longer critical of donor initiatives or willing to raise uncomfortable complaints from villagers at federation meetings. Such divisiveness can be especially damaging to disadvantaged groups' interests as collective action is often one of their strongest assets for achieving their objectives.

Establishing appropriate conditions for rational communication

When Habermas's critique of strategic action is wedded to a de-politicizing liberal pluralism, it becomes too easy to organize negotiations before (or in spite of the unlikelihood that) conditions necessary for negotiations to succeed can be realized (Flyvbjerg, 1998, p218). In negotiations about forests, it is common for stakeholders to have different levels and kinds of education, speak different languages, differ in access to formal politics, and hold different beliefs about how nature and society function.

If rationality is assumed before the necessary preconditions are in place (assuming they can ever be in place) it will be more difficult to see, analyse and critique relations of power. Groups may be labelled biased or ideological for not accepting the rationality of a negotiation process, particularly when facilitators of negotiations have invested heavily in the neutrality of the process and their credibility is at stake. Dissident groups are then forced to remain outside that process and miss the possible benefits of engagement with other stakeholders, or dampen their criticism of the persistent politics within the process. If they choose the latter course, they may give up a powerful tool for mobilizing their supporters. Debates over co-optation into nominally de-politicized negotiations with states and other powerful interests are common within social movements (Melucci, 1996). The conflict among environmental movements in North America over whether and how to negotiate over the North American Free Trade Association (NAFTA) is a case in point (Lynch, 1998).

In addition, disadvantaged groups may have limited capacity to challenge dominant images of themselves, and 'rational' negotiations can reinforce such images. Post-colonial scholars have described how experience can be appropriated and abused by others and then, crucially, become normalized in ways that make it difficult for disadvantaged groups to generate alternative images of themselves (Said, 1978; Hooks, 1990; Parajuli, 1998). Swidden farmers in Indonesia, for example, have struggled mightily for many years against their image as destroyers of forest, and now must simultaneously fight characterizations as noble ecological natives that are equally misrepresentative (Li, 2000). In negotiation processes in which the gaps in power and experience can be enormous, the potential for such appropriations and normalizations is large.

We believe that when our professional careers are associated with creating a rational negotiation process or developing particular methods or techniques for achieving such a process, there is a temptation to ascribe greater efficacy to negotiations than is warranted, and to expose disadvantaged groups to the risks described above. We prefer an approach that is explicitly political. This approach would, drawing on the writing of feminist scholars, call for an open discussion of the politics at work in negotiations and situate actors and actions in their political contexts (Fraser, 1992; Jakobsen, 1993; Bickford, 1999). This allows a deeper analysis and critique of relations of power without demanding full disclosure on the part of disadvantaged groups of their particular strategies. Problems associated with visibility and vulnerability, representation and normalization are not eliminated in this approach. They are, however, placed at the centre of the analysis, and their treatment becomes a standard of professional achievement. This, it seems to us, is the best way to assure that disadvantaged groups achieve benefits from negotiations without exposing themselves to significant risks.

The search for agreement

Consensus has been explicitly rejected by many practitioners as a goal for multistakeholder negotiations (Daniels and Walker, 1999). Yet the expectations of governments, donors, NGOs and others that negotiations will generate resource management agreements that bind stakeholders to a coherent course of action is still widespread (Borrini-Feyerabend, 1996; Lewis, 1996; Ayling and Kelly, 1997; O'Faircheallaigh, 1999). As such, there are significant rewards and penalties for practitioners associated with achieving agreement among stakeholders. We will thus take a brief look at problems associated with a search for agreement, whether explicitly stated or implicit in the response to institutional pressures, within multistakeholder negotiations.

The agreement reached through negotiation is never a single thing to all stakeholders, but a more or less workable conglomeration of meanings. Schroeder (1993) describes a project plan for tree planting in the Gambia where expatriate advisers understood the plan as a means of promoting conservation, while local men and women interpreted the project as an opportunity to advance their different visions of a productive landscape. Social forestry programmes in India were sometimes undermined by different interpretations of what it meant to plant a tree in a garden – a source of income for wealthy male residents or a source of subsistence products for poorer villagers (Saxena, 1997). When the Kayan Mentarang Nature Reserve (which did not permit human use) in Indonesia became a national park (which does permit human use), there was intense debate among local villagers about whether they had just gained or given up their rights to the forest. Similar stories of negotiations with multiple meanings and purposes abound in the literature on environment and development (Fortmann and Bruce, 1988; Field, 1994; Rocheleau and Ross, 1995).

Nor are agreements the same for a single stakeholder in all contexts. In Krui, Sumatra, local farmers declined to participate in a model agreement on local rights to agroforestry gardens that they themselves helped to create (Fay and Sirait, forthcoming). They now want ownership rights, and explain that the agreement to use rights was the best option available to them in earlier negotiations. The effective life of an agreement can be very short, and is often subject to external events beyond the control of the stakeholders. Much of the work in adaptive management seems inspired by the inability to construct agreements that work on a time scale necessary for confronting relevant threats (Lee, 1993; Borrini-Feyerabend, 1996).

There is thus a tendency to overestimate both the degree and durability of agreements reached through negotiation processes. Moreover, a focus on agreement can encourage participants in negotiations to gloss over dissenting views, whether of the weak or the powerful (Rescher, 1993; Doherty and de Geus, 1996). Much of the work in Integrated Conservation and Development Programmes, for example, appears to have underestimated the difficulty in 'conjoining' divergent views on environment and development (Wells and Brandon, 1992; Western and Wright, 1994). Agreements reached under the Joint Forest Management Programme in India and the Community Forestry Programme in Nepal have disappointed community members, as Forest Departments have continued to emphasize protection over local use (Saxena, 1997; Sarin, 1998; Malla, 1999) – this often after considerable investments of time and resources by disadvantaged groups.

If groups are pressured to reach 'unsituated' agreement, they also may be (wrongly) accused of bad faith when tensions among interpretations arise, and subjected to sanctions despite their own conviction of having respected the agreement's intent. Malla (1999) reports of community forestry projects started in Nepal's lowlands following the announcement of several government policies on community forestry, themselves the product of consultations with NGOs and donors. Lowland communities interpreted the policies (as did many other stakeholders) as applying to all regions, while the government's unstated assumption was that the policies applied only to hill regions. The government closed down these projects and accused community members of breaking the law. Confusion over what trees may be planted, cut and sold on 'private hills' in Yunnan Province, China, has also led to wasted effort, lost revenue and even fines for farmers, who are accused of behaving in a backward or selfish manner (Edmunds, field notes, 1999). Signing up to 'consensual' agreements makes disadvantaged groups vulnerable to accusations of immoral or anti-social motivations when their interpretation of agreements is opposed by more powerful stakeholders.

Positive outcomes for disadvantaged groups require careful consideration of the politics at work in particular circumstances. Negotiations can be treated as a moment of interface where empathy and an expanded sense of solidarity (Rorty, 1989) can be achieved among a set of stakeholders, but the solidarity is always partial, provisional and unstable, especially when it involves groups with very different levels of power. As such, it must be approached strategically, rather than with the goal of bracketing out stakeholder politics in order to reach a complete, mutually intelligible and durable agreement.

Bringing everyone to the table

In co-management of protected areas, forests and common pool resources, there is often considerable emphasis placed on the importance of including all affected stakeholders as well (Porter and Salvesen, 1995; Borrini-Feyerabend, 1997; Allen et al, 1998). Of course, if negotiations fail to achieve a level playing field, as we suggest they might, then powerful groups are likely to exert more influence over the course of negotiations and the implementation of agreements. The story of timber companies wielding their control over less powerful, local communities, for example, is common from Surinam (Sizer and Rice, 1995) to Sarawak (Bevis, 1995). There are less obvious problems for disadvantaged groups, however, that also warrant consideration.

When scholars and practitioners discuss the need to include all stakeholders in negotiations, they generally stress that this means including disadvantaged groups. Yet securing an invitation to the negotiation table does not mean such groups will have their issues heard by other stakeholders. Stephen (1997) describes how negotiations between various elements of the government and organizations representing the Zapatistas have been channelled away from discussions of rights to land and other resources (issues that are at the heart of the Zapatista struggle) and towards formal rights of association. During fieldwork in a forest reserve in Negros Oriental, the Philippines, Wollenberg observed that shifting cultivators' requests for a road to their village were never discussed earnestly, as local administrators did not want the road to open up the area for settlement. In such cases, disadvantaged groups waste valuable time and resources on negotiations that do not treat issues of concern to them. They may be better off pursuing other means for pressing their claims.

This is why Habermas suggests that all *topics* must be available for consideration and criticism in rational communication. There are, however, issues that disadvantaged groups refuse to open to negotiations because they are so fundamental to their livelihoods, senses of self or security. Li (1999) notes that community forestry agreements in the Philippines appear to make the rights of uplanders to a home and a livelihood contingent upon meeting government-defined environmental standards for land use. She argues that basic human rights should not be the *subject* of negotiations over the management of forests. The issues of tribal sovereignty and rights to religious and cultural freedom are also examples of 'non-negotiables' (Melucci, 1996; Karlsson, 1999). There are many such issues of interest to disadvantaged groups that do not lend themselves easily to agreements, contracts or policy changes. Requiring that all subjects be open to rational debate, even implicitly by asking stakeholders to move away from 'positions' towards more abstract and generalized 'interests' (Ramirez, 1999), may work against disadvantaged groups.

Disadvantaged groups could simply stay away. Unfortunately, by linking legitimacy to the inclusion of *all* stakeholders, people or groups that refuse to participate can be left with no legitimate place from which to criticize the outcomes. Wilson (1999) describes a case from the US where stakeholder meetings to determine a management strategy for public lands excluded crucial stakeholders, such as local Native American tribes and environmental groups. When these outsiders later challenged the basis for the management strategy, their claims were denounced as illegitimate, as they had not participated in an 'open' stakeholder process. The focus on involving all stakeholders in a process, especially a process with a goal of achieving agreement, encourages us to regard all those who refuse to participate as having relinquished their moral authority to speak on the issues. How can we claim to have met the standards of a rational agreement among all stakeholders if 'legitimate' stakeholders were absent?

Disadvantaged groups must be free to *not* participate, to not be made visible to powerful and potentially hostile others. In fact, withdrawal is a powerful form of protest against 'the tendency of the system to exact participation, communication, the acceptance of one's assigned place in society as an effective processor of information' (Melucci, 1996, p183). Is the call to involve all stakeholders – particularly in negotiation processes fraught with historical animosities, wide differences in culture, and uneven power relations – also a call to disadvantaged groups to take their 'assigned place' within an inequitable political system? This is a question too seldom asked in the work on multistakeholder negotiations.

We favour a strategic approach to the questions of who should participate in negotiations and how each group should be treated. For example, in a meeting of villagers, NGOs and government officials to discuss claims to traditional forest areas in the Kayan Mentarang National Park, the World Wide Fund for Nature chose not to invite timber companies to participate for fear they would dominate decisions. They did, however, invite a number of villagers (about 40) to negotiate with an equal number of government officials. Initial proposals were worked out in separate sessions and then negotiated with a facilitator. Though the agreements should continue to be checked for their impact on different disadvantaged groups, particularly those who chose not to participate in these initial discussions, the process is promising because relations among stakeholders were considered in advance. This strategic approach to negotiations, consistent with the lessons of radical pluralism and feminist post-structural theory, promises substantial benefits for disadvantaged groups while decreasing the risk of wasting precious resources or making themselves vulnerable to manipulation and control by more powerful groups.

Negotiations in the context of other stakeholder strategies

Much of the literature on multistakeholder negotiations refers to the principal of BATNA (the best alternative to no agreement) (Ramirez, 1999). While a useful concept for clarifying trade-offs that stakeholders make between negotiations and other strategies for advancing their interests, BATNA encourages us to think of negotiations as separate from these other strategies, or at best sequentially (ibid). In fact, negotiations are often best understood in relationship to other strategies, including strategies of confrontation or competition, which help both practitioners and stakeholders interpret the meaning of negotiations for different participants.

It is possible, for example, that people participate in negotiations precisely because they are also engaged in other activities for pressing their interests. Negotiations may provide political cover for more aggressive tactics, such as media campaigns, sabotage, legal challenges, peaceful protest or even violence. Zapatista negotiations with the Mexican government have taken place while the government has waged a 'quiet war' against the indigenous peoples of Chiapas, including the destruction of local natural resources (Froehling, 1997; Stephen, 1997). Disadvantaged groups, on the other hand, may seek visibility through participation in negotiations as much as they seek a formal (and perhaps unenforceable) agreement. Residents of a village in Uganda seized every opportunity to meet with staff of the Kibale Forest Park. While genuinely concerned with environmental issues, they were also keen to be recognized by the government as legitimate actors in local politics, even if the specific projects they undertook had little local relevance (Edmunds, 1996). Understanding the strategies other stakeholders use helps disadvantaged groups put negotiations in perspective and assess the different meanings stakeholders attribute to the process.

Setting negotiations in context can also lessen the pressure on disadvantaged groups to restrict their activities to negotiations. Disadvantaged groups might well need to pursue other strategies such as lobbying, capacity-building, networking and protesting to expand their decision-making authority (Sherraden, 1991; Froehling, 1997; Turner, 1998). In forested areas of India, for example, people's organizations have become frustrated with their inability to negotiate effectively with the Indian Forest Department on a range of issues such as gender equity, benefit sharing, and regulation of marketing of non-timber forest products (Singh, 1996; Sarin, 1998). As a result, many of these organizations have formed apex bodies and federations to put more political pressure on the Forest Department. These lobbying efforts are not meant to replace negotiations, but to balance power *within* negotiations. Similar links between political pressure and negotiation are found in Nepal:

> [As] forest-user networks become more formalized, they are proving more effective at engaging stakeholders in negotiations about resource-related policies and activities. With a more formalized institutional structure networks become respected representative entities with an identifiable constituency of potential voters. This places them in a better bargaining position for promoting dialogue – between individual actors, and by extension, to forest-user groups, various line agencies, and policy-makers (Shrestha and Britt, 1997, p3).

In particular, negotiations should not be separated from capacity-building among disadvantaged groups. Certainly, politically less-powerful people have not separated the two. Many examples can be found in Latin America of indigenous peoples and Andean farmers simultaneously engaging with government in formal participatory processes *while* building their capacity to deal with the legal system, party politics, public relations and technical/managerial issues (Froehling, 1997; Stephen, 1997; Bebbington, 1998; Veber, 1998). Analyses that treat negotiations as a strategy in isolation from others – as the best available option – will misinterpret the meaning of the negotiations to participants.

Conclusion

The point we are making is *not* that multistakeholder negotiations should always be avoided by disadvantaged groups and their supporters. Experience has shown that they can be useful under many circumstances. We believe, however, that communications are not the sole barrier to collective action, especially among many stakeholders of vastly different levels of political power. Negotiations will therefore achieve more just and equitable outcomes for disadvantaged groups if we approach them strategically, rather than trying to create neutral platforms where all stakeholders can discuss all issues – irrespective of their historical relations and the full range of their present political activities – with the goal of reaching rational, mutually intelligible and universally recognized agreements. Such a strategic approach to multistakeholder negotiations would involve:

- Seeking out possibilities for alliances within negotiations among select stakeholders, rather than trying to achieve an apolitical agreement among all possible stakeholders. Working 'in, with, through' alliances, disadvantaged groups can achieve significant gains for themselves (Jakobsen, 1998, p2) while maintaining greater control over the types and amounts of information made available to historical antagonists.

- 'Situating' the legitimacy of all decisions and agreements. This means analysing the reasons for participation or non-participation by each group in negotiations, how groups are represented, the roles of convenors and facilitators, and the historical context for agreements. It also means treating legitimacy as partial and contingent, rather than assuming that an unproblematic legitimacy is assured through open negotiations among all stakeholders.
- Approaching negotiations as one strategy among several that may be pursued simultaneously by disadvantaged groups, not as *a singular* strategy when all else fails. These other strategies can make negotiations more effective. They can also secure greater decision-making authority for disadvantaged groups in their own right, in complementarity with negotiations.
- Improving the preconditions for successful negotiations, particularly the capacity of disadvantaged groups to participate effectively. It is especially important now to focus on capacity to act at a global scale. There is a growing body of literature on global civil society and social movements that cross international borders (Melucci, 1996; Cerulo, 1997; Turner, 1998). Such efforts are clearly in response to the increasing number of decisions about daily life that are either made or heavily constrained by exchanges of information at a global scale. Yet many organizations that claim to represent disadvantaged groups are just now developing the skills to use these media effectively (Turner, 1992; Froehling, 1997). Making advances in this area, where the outcome of symbolic struggles affects the mobilization of political and material support for organizations, will greatly impact on the success of disadvantaged groups in pushing their claims, both within the context of negotiations and separately from them.

A strategic approach to negotiations, with intellectual roots in radical pluralism and feminist post-structuralism, better reflects the experiences of disadvantaged groups as they have interacted with other 'stakeholders' over issues of forest management. It exposes disadvantaged groups to less risk of regulation or oppression by other stakeholders. It is also a flexible approach, encouraging the consideration of alternative actions that might be better suited to the circumstances of a particular forest and the groups making claims on it. Yet a strategic approach still shows the way to make tangible gains from negotiations. Negotiations can help disadvantaged groups understand the goals and interests of other stakeholders, improving their strategic planning. Negotiations can be a place to build empathy for the positions of disadvantaged groups. They can be a place for these groups to exert influence over other stakeholders. And they can be a place where disadvantaged groups work through 'situated' alliances and agreements among some, if not all stakeholders.

Notes

1   There is considerable debate about the usefulness of the term 'group' when referring to people sharing particular aspects of identity or a set of historical experiences. Some authors continue to use the term group to highlight the political potential of

its members (Hill-Collins, 1997; Bickford, 1999). Others prefer to use 'categories' of people to remind us to be cautious about who is defining the group and how, an issue we take up in a brief discussion of representation. As we are emphasizing the importance of negotiations as a form of political action, we will use the term group. We recognize, however, that disadvantaged people making claims on forests may not always see themselves that way.

2 Many groups are disadvantaged in debates about forest management. We will use the term disadvantaged groups to refer to these three large categories: small-scale forest-product collectors, indigenous peoples, swidden farmers. Li has pointed out that migrants displaced from other agricultural areas to the forest edge are often severely disadvantaged (1997). Others may argue that urban dwellers can be disadvantaged in the debates over forest management, depending on investment and price distortions. Who is disadvantaged in a debate about particular forests at a given point in time is best defined by looking at the historical record of management for those forests.

3 Daniels and Walker, for example, argue against the search for consensus (1999), yet outline a number of methods designed to create a neutral space where all stakeholders can negotiate openly. Steins and Edwards (1999) note that gender inequities, among others, limit the effectiveness of negotiations. They seem to suggest, however, that inequities can be dealt with through further development of platforms for negotiations.

4 There are debates among feminists, however, as to the necessity of taking an ontological position in favour of particular disadvantaged groups. See Morris (1997) and SIGNS (1997) for debates among those arguing for taking a stance or a standpoint, and those who argue that this is unnecessary, and even limiting to the feminist movement. Feminist treatments of Habermas also express the tension. Benhabib (1992, p95), addressing Habermas on communicative rationality, says: 'the radical proceduralism of this model is a powerful criterion for demystifying discourses of power and their implicit agendas. However, in a society where reproduction is going public, practical discourse will have to be "feminized". Such feminization of practical discourse will mean first and foremost challenging, from the standpoint of their gender context and subtext, unexamined normative dualisms as those of justice and the good life, norms and values, interests and needs.'

5 Though there is a risk of certain members of disadvantaged groups creating such a gulf between themselves and useful constructions of who they are (Ramos, 1994; Conklin and Graham, 1995), the risk is probably most pronounced among outsider advocates for disadvantaged groups.

References

1 Allen, W J., Brown, K., Gloag, T., Morris, J., Simpson, K., Thomas, J. and Young, R. (1998) *Building Partnerships for Conservation in the Waitaki/Mackenzie Basins*. Landcare Research Contract Report LC9899/033. Landcare, Lincoln, New Zealand

2   Anderson, D. and Grove, R. (eds) (1987) *Conservation in Africa: People, Policies and Practices.* Cambridge University Press, New York.

3   Ashcroft, B., Griffiths, G. and Tiffen, H. (eds) (1995) *The Postcolonial Studies Reader.* Routledge, New York.

4   Ayling, R. D. and Kelly, K. (1997) 'Dealing with conflict: Natural resources and dispute resolution'. *Commonwealth Forestry Review*, vol 76, no 3, pp182–185.

5   Bebbington, A. (1998) 'Sustaining the Andes? Social capital and policies for rural regeneration in Bolivia'. *Mountain Research and Development*, vol 18, no 2, pp173–181.

6   Benhabib, S. (1992) 'Models of public space: Hannah Arendt, the liberal tradition, and Jurgen Habermas', in Calhoun, C. (ed) *Habermas and the Public Sphere.* Massachusetts Institute of Technology, Cambridge, MA, pp73–98.

7   Bevis, W. W. (1995) *Borneo Log: The Struggle for Sarawak's Forests.* University of Washington, Seattle, WA.

8   Bickford, S. (1999) 'Reconfiguring pluralism: Identity and institutions in the inegalitarian polity'. *American Journal of Political Science*, vol 43, no 1, pp86–108.

9   Borrini-Feyerabend, G. (1996) *Collaborative Management of Protected Areas: Tailoring the Approach to the Context in Issues in Social Policy.* IUCN, Gland, Switzerland.

10   Borrini-Feyerabend, G. (1997) *Beyond Fences: Seeking Social Sustainability in Conservation.* Volumes 1 and 2. IUCN, Gland, Switzerland.

11   Bureau of the World Heritage Committee (1999) *UNESCO Convention Concerning the Protection of the World Cultural and Natural Heritage.* 23rd Session, Paris, 5–10 July. Information Document WHC-99/CONF. 204/INF.9.D. UNESCO, Paris.

12   Center for International Forestry Research (CIFOR) (1999) *Strategy Document for Local People, Devolution and Adaptive Co-Management Programme.* CIFOR, Bogor.

13   Cerulo, K. (1997) 'Identity construction: New issues, new directions'. *Annual Review of Sociology*, vol 23, pp385–409.

14   Conklin, B. and Graham, L. (1995) 'The shifting middle ground: Amazonian Indians and eco-politics'. *American Anthropologist*, vol 97, no 4, pp695–710.

15   Daniels, S. and Walker, G. (1999) 'Rethinking public participation in natural resource management: Concepts from pluralism and five emerging approaches', in *Pluralism and Sustainable Forestry and Rural Development.* Proceedings of an International Workshop, Rome, 9–12 December 1997, FAO, Rome, pp29–48.

16   Doherty, B. and de Geus, M. (eds) (1996) *Democracy and Green Political Thought: Sustainability, Rights, and Citizenship (European Political Science Series).* Routledge, New York.

17   Edmunds, D. (1996) *Green Imperialism? Conservation Refugees and Kibale Forest National Park.* Paper presented at the Association of American Geographers Annual meeting, Charlotte, NC, 5–9 March.

18   FAO (1999) *Pluralism and Sustainable Forestry and Rural Development.* Proceedings of an International Workshop, Rome, 9–12 December 1997. FAO, Rome.

19   Fay, C. and Sirait, M. (forthcoming) 'Reforming the forests: Challenges to government forestry reform in Post-Soeharto Indonesia', in Colfer, C. J. P. and. Resosudarmo, I. A. P. (eds) *Which Way Forward? Forests, Policy and People in Indonesia.* Resources for the Future (RFF), Washington, DC.

20   Field, L. W. (1994) 'Harvesting the bitter juice: Contradictions of Paez resistance in the changing Colombian nation state'. *Identities*, vol 1, no 1, pp89–108.

21   Fisher, R. J. (1995) *Collaborative Management of Forests for Conservation and Development.* IUCN and WWF, Gland, Switzerland.

22   Flyvbjerg, B. (1998) 'Habermas and Foucault: Thinkers for civil society?' *British Journal of Sociology*, vol 49, no 2, pp 210–233.

23   Fortmann, L. and Bruce, J. (eds) (1988) *Whose Trees? Proprietary Dimensions of Forestry.* Westview Press, Boulder, CO.

24 Foucault, M. (1980) *Power Knowledge: Selected Interviews and Other Writings, 1972–1977*, edited by Colin Gordon. Pantheon, New York.

25 Fraser, N. (1992) 'Rethinking the public sphere: A contribution to the critique of actually existing democracy', in Calhoun, C. (ed) *Habermas and the Public Sphere*. Massachusetts Institute of Technology, Cambridge, MA, pp109–142.

26 Froehling, O. (1997) 'The cyberspace "War of Ink and Internet" in Chiapas, Mexico'. *The Geographical Review*, vol 87, no 2, pp291–307.

27 Government of Australia (1999) *Australia's Kakadu. Protecting World Heritage. Response of Government of Australia to UNESCO World Heritage Committee re Kakadu National Park.* Government of Australia, Canberra.

28 Habermas, J. (1984) *The Theory of Communicative Action.* Volume 1. Beacon Press, Boston, MA.

29 Habermas, J. (1987) *The Theory of Communicative Action.* Volume 2. Beacon Press, Boston, MA.

30 Habermas, J. (1992) 'Further reflections on the public sphere', in Calhoun, C. (ed) *Habermas and the Public Sphere*. Massachusetts Institute of Technology, Cambridge, MA, pp421–461.

31 Habermas, J. (1995) 'Reconciliation through the public use of reason: Remarks on John Rawls's political liberalism'. *The Journal of Philosophy*, vol 17, no 3, pp109–131.

32 Haraway, D. (1991) *Simians, Cyborgs, and Women: The Reinvention of Nature.* Routledge, New York.

33 Harding, S. (1986) *The Science Question in Feminism.* Cornell University Press, Ithaca, NY.

34 Harding, S. (1997) 'Comment on Hekman's "Truth and Method: Feminist Standpoint Theory Revisited": Whose standpoint needs the regimes of truth and reality?' *SIGNS*, vol 22, no 2, pp382–391.

35 Hawkesworth, M. (1997) 'Confounding gender'. *SIGNS*, vol 22, no 3, pp649–676.

36 Hecht, S. and Cockburn, A. (1989) *Fate of the Forest: Developers, Destroyers and Defenders of the Amazon.* Verso, London.

37 Hekman, S. (1997) 'Truth and method: Feminist standpoint theory revisited'. *SIGNS*, vol 22, no 2, pp341–365.

38 Hill-Collins, P. (1997) 'Comment on Hekman's "Truth and Method: Feminist Standpoint Theory Revisited": Where's the power?' *SIGNS*, vol 22, no 2, pp375–381.

39 Hirst, P. (1997) *From Statism to Pluralism.* UCL Press Ltd, London.

40 Hooks, B. (1990) *Yearning: Race, Gender, and Cultural Politics.* South End Press, Boston, MA.

41 Jakobsen, J. R. (1998) *Working Alliances and the Politics of Difference: Diversity and Feminist Ethics.* Indiana University Press, Bloomington, IN.

42 Karlsson, B. G. (1999) 'Ecodevelopment in practice: The Buxa tiger reserve, the World Bank and indigenous forest people in Northeast India'. *Forest, Trees and People Newsletter*, vol 38, pp39–45.

43 Kearney, A. R., Bradley, G., Kaplan, R. and Kaplan, S. (1999) 'Stakeholder perspectives on appropriate forest management in the Pacific Northwest'. *Forest Science*, vol 45, no 1, pp62–73.

44 Leach, M., Mearns, R. and Scoones, I. (1997) 'Challenges to community based sustainable development: Dynamics, entitlements, institutions'. *IDS Bulletin*, vol 28, no 4, pp4–14.

45 Lee, K. (1993) *Compass and Gyroscope. Integrating Science and Politics for the Environment.* Island Press, Washington, DC.

46 Legare, E. I. (1995) 'Canadian multiculturalism and aboriginal people: Negotiating a place in the nation'. *Identities*, vol 1, no 4, pp347–366.

47 Lewis, C. (1996) *Managing Conflicts in Protected Areas.* IUCN, Gland, Switzerland.

48 Li, T. M. (ed) (1999) *Transforming the Indonesian Uplands: Marginality, Power and Production.* Harwood Academic Publishers, Amsterdam.

49   Li, T. M. (2000) *Migrants, Locals, and Government Powers: Sulawesi's Cocoa-Forest Frontier*. Paper presented at the Center for International Forestry Research, Bogor, Indonesia, 14 July.

50   Lynch, C. (1998) 'Social movements and the problem of globalization'. *Alternatives*, vol 23, pp149–173.

51   Maarleveld, M. and Dangbegnon, C. (1999) 'Managing natural resources: A social learning perspective'. *Agriculture and Human Values*, vol 16, no 3, pp267–280.

52   McCreary, S. T. and Adams, M. B. (1995) 'Managing wetlands through advanced planning and permitting: The Columbia river estuary study taskforce', in Porter, D. R. and Salvesen, D. A. (eds) *Collaborative Planning for Wetlands and Wildlife: Issues and Examples*. Island Press, Washington, DC, pp103–137.

53   Malla, Y. (1999) 'Stakeholders' responses to changes in forest policies', in *Pluralism and Sustainable Forestry and Rural Development*. Proceedings of an International Workshop, Rome, 9–12 December 1997, Rome, FAO, pp253–273.

54   Melucci, A. (1996) *Challenging Codes: Collective Action in the Information Age*. Cambridge University Press, Cambridge.

55   Morris, D. (1997) 'The feminist-postmodernist debate over a revitalized public philosophy'. *Social Theory and Practice*, vol 23, no 3, pp479–507.

56   O'Faircheallaigh, C. (1999) 'Making social impact assessment count: A negotiation-based approach for indigenous peoples'. *Society and Natural Resources*, vol 12, pp63–80.

57   Parajuli, P. (1998) 'Beyond capitalized nature: Ecological ethnicity as an arena of conflict in the regime of globalization'. *Ecumene*, vol 5, no 2, pp186–217.

58   Parkin, A. (1996) 'On the practical relevance of Habermas' theory of communicative action'. *Social Theory and Practice*, vol 22, no 3, pp417–441.

59   Peluso, N. L. (1992) *Rich Forests, Poor People, Resource Control and Resistance in Java*. University of California Press, Berkeley, CA.

60   Porter, D. R. and Salvesen, D. A. (1995) 'Conclusion', in Porter, D. R. and Salvesen, D. A. (eds) *Collaborative Planning for Wetlands and Wildlife: Issues and Examples*. Island Press, Washington, DC, pp275–283.

61   Ramirez, R. (1999) 'Stakeholder analysis and conflict management', in Buckles, D. (ed) *Conflict and Collaboration in Natural Resource Management*. IDRC Community-based Natural Resource Management Program. IDRC, Ottawa, pp101–126.

62   Ramos, A. R. (1994) 'The hyperreal Indian'. *Critique of Anthropology*, vol 14, no 2, pp153–171.

63   Rawls, J. (1995) 'Reply to Habermas'. *The Journal of Philosophy*, vol 17, no 3, pp132–180.

64   Rescher, N. (1993) *Pluralism: Against the Demand for Consensus*. Oxford University Press, Oxford.

65   Ribot, J. (1998) *Decentralization and Participation in Sahelian Forestry: Legal Instruments of Central Political-administrative Control*. Harvard Center for Population and Development Studies Working Paper Series no 98.06. Harvard University, Boston, MA.

66   Rocheleau, D. and Ross, L. (1995) 'Trees as tools, trees as text: Struggles over resources in Zambrana-Chaucey, Dominican Republic'. *Antipode*, vol 27, no 4, pp407–428.

67   Roling, N. G. and Maarleveld, M. (1999) 'Facing strategic narratives: An argument for interactive effectiveness'. *Agriculture and Human Values*, vol 16, pp295–308.

68   Roling, N. G. and Wagemakers, M. A. E. (eds) (1998) *Facilitating Sustainable Agriculture: Participatory Learning and Adaptive Management in Times of Environmental Uncertainty*. Cambridge University Press, Cambridge.

69   Rorty, R. (1989) *Contingency, Irony, and Solidarity*. Cambridge University Press, Cambridge.

70   Rossi, J. (1997) 'Participation run amok: The costs of mass participation for deliberative agency decisionmaking'. *Northwestern University Law Review*, vol 92, no 1, pp173–247.

71   Said, E. (1978) *Orientalism*. Vintage Press, New York.

72 Sarin, M. (1998) *Who is Gaining? Who is Losing? Gender and Equity Concerns in Joint Forest Management*. Working Paper by the Gender & Equity Sub-Group, National Support Group for JFM. Society for Promotion of Wastelands Development, New Delhi.

73 Saxena, N. C. (1997) *The Saga of Participatory Forest Management in India*. Center for International Forestry Research, Bogor, Indonesia.

74 Schroeder, R. (1993) 'Shady practice: Gender and the political ecology of resource stabilization in Gambian Garden/Orchard'. *Economic Geography*, vol 69, no 3, pp349–365.

75 Scott, J. (1998) *Seeing Like a State*. Yale University Press, New Haven, CT.

76 Severino, H. G. (1998) *Opposition and Resistance to Forest Protection Initiatives in the Philippines: The Role of Local Stakeholders*. Discussion Paper 92. United Nations Research Institute for Social Development, Geneva.

77 Sherraden, M. S. (1991) 'Policy impacts of community participation: Health services in rural Mexico'. *Human Organization*, vol 50, no 3, pp256–263.

78 Shrestha, N. K. and Britt, C. (1997) *Crafting Community Forestry: Networking and Federation-Building Experiences*. Paper presented at the Seminar 'Community Forestry at a Crossroads', 17–19 July, Bangkok, Thailand.

79 SIGNS (1997) *Special Issue*, vol 22, no 2.

80 Sinclair, A. J. and Smith, D. (1999) 'The model forest program in Canada: Building consensus on sustainable forest management?'. *Society and Natural Resources*, vol 12, no 12, pp121–138.

81 Singh, N. (ed) (1996) *Communities and Forest Management in Orissa: Rethinking on Community Rights Over Forests and Forest Products*. Proceedings of a state-level workshop, Puri, Orissa 30–31 July. Vasundhara, Orissa, India.

82 Sioh, M. (1998) 'Authorizing the Malaysian rainforest: Configuring space, contesting claims and conquering imaginaries'. *Ecumene*, vol 5, no 2, pp144–166.

83 Sizer, N. and Rice, R. (1995) *Backs to the Wall in Surinam*. World Resources Institute, Washington, DC.

84 Steins, N. and Edwards, V. (1999) 'Platforms for collective action in multiple-use common-pool resources'. *Agriculture and Human Values*, vol 16, no 3, pp241–255.

85 Stephen, L. (1997) 'Redefined nationalism in building a movement for indigenous autonomy in Southern Mexico'. *Journal of Latin American Anthropology*, vol 3, no 1, pp72–101.

86 Turner, S. (1998) 'Global civil society, anarchy and governance: Assessing an emerging paradigm'. *Journal of Peace Research*, vol 35, no 1, pp25–42.

87 Turner, T. (1992) 'Defiant images: The Kayapo appropriation of video'. *Anthropology Today*, vol 8, no 6, pp5–16.

88 Veber, H. (1998) 'The salt of the Montana: Interpreting indigenous activism in the rain forest'. *Cultural Anthropology*, vol 13, no 3, pp382–413.

89 Wells, M. and Brandon, K. with Hannah, L. (1992) *People and Parks: Linking Protected Area Management with Local Communities*. World Bank, Washington, DC.

90 Western, D. and Wright, R. M. (eds) (1994) *Natural Connections: Perspectives on Community-based Conservation*. Island Press, Washington, DC.

91 Wilson, R. (1999) *Community-based Collaborative Management on the San Juan National Forest: An Analysis of Participation*. Paper presented at the Association of American Geographers Annual Meeting, Honolulu, Hawaii, 23–27 March.

21

Protect, Manage and Restore: Conserving Forests in Multifunctional Landscapes

Jeffrey Sayer, Christopher Elliot and Stewart Maginnis

Introduction

The past decade has seen an unprecedented effort to solve forest problems at the global level. Forests have been prominent on the agendas of the Convention on Biological Diversity and the Framework Convention on Climate Change. We have seen a succession of intergovernmental forest meetings under the aegis of the Intergovernmental Panel on Forests, the Intergovernmental Forum and most recently the United Nations Forum on Forests. But how much impact has all this international activity had on the ground? To the extent that we make any attempt to measure success we use global statistics. The media and environmental activists track the findings of FAO's Forest Resource Assessment or statistics on different categories of protected areas, areas under certified management and so on. This leads to futile debates about the validity of global statistics (Lomborg, 2001). Meanwhile we avoid the issue of finding effective ways of monitoring the flows of goods and services that forests provide. We know little about forestry's impact on the extinction of the world's forest species. We have only very imprecise estimates of how much carbon our forests store. We can only speculate about the trends in forest resources available to the 1.3 billion poor people that the World Bank tells us depend on forests. The reality is that we know very little about how we are doing in solving global forest problems.

On the other hand we are beginning to learn a lot about how civil society and local communities are solving their forest problems. Kuchli (1997) has documented a number of situations from California to West Africa and from the Himalayas to the Amazon where people have organized themselves to conserve and manage forests. Fairhead and Leach (1995) have shown how communities in West and East Africa have organized themselves to ensure a sustained supply of locally important forest goods and services. A network of small private nature reserves in Costa Rica provides local employment and makes valuable contributions to conserving biodiversity. Joint Forest Management has improved the livelihoods of millions of people in parts of India.

Note: Reprinted from *Proceedings of the 12th World Forestry Congress*, Sayer, J., Elliot, C. and Maginnis, S., 'Protect, manage and restore: Conserving forests in multifunctional landscapes', Copyright © (2003), with permission from the World Forestry Congress

There are large numbers of local successes in improving the conservation and management of forests. But most of these successes will not show up in the global forest statistics. When success occurs it almost always takes the form of subtle mosaics of different forest uses that are adapted to the local landscape and meet a diversity of needs of local people (Kuchli, 1997). There are countless examples in both the developed and developing worlds where local people have rebelled against large-scale solutions to forest problems. Vast industrial monocultures have been rejected with as much vigour as vast national parks. Forest values are dependent on local needs and contexts and every forest system should ideally be managed to optimize its contribution to local livelihoods. If local stakeholders are penalized in order to protect global forest values or to provide for the needs of forest industries, then they should be compensated.

The global values of forests are at present at the centre of attention. International conservation groups, heads of forest services, forest specialists in development assistance agencies and the media have all responded to the 1970s alarm at the disappearance of 'the world's' rainforests. We contend that excessive emphasis on global values and a failure to provide for local values is the reason for the failure of many investments in forest conservation. Attempts to apply globally conceived 'cookie-cutter' solutions to local forest problems may have been counterproductive. We are concerned that too much effort is going into the negotiation of global norms and not enough into the processes that will allow complex solutions to local forest situations to emerge.

The 'Norms' of Forestry

Foresters have always allocated forests to categories. We speak of 'production forests', 'protection forests', 'forest reserves' and 'national parks', and various other categories of protected areas. National and global forest assessments tell us how much land is allocated to these categories. We know that 10 per cent of the world's forests are in protected areas, that 30,852,896ha have been certified under the Forest Stewardship Council rules. The media and the public attach a lot of significance to these statistics. The progress that is being made in addressing the world's forest problems is measured according to the trends that are revealed in statistics produced by the FAO (Food and Agriculture Organization), the World Conservation Monitoring Centre and so on. Many interest groups set targets according to these national and global measures of forest status. Our own organizations seek to promote 50 million hectares of new protected areas by 2005, to ensure that 150,000 hectares of sustainably managed forests are certified by this date (WWF, 2003). Such statistical measures of our success in ensuring the conservation and sustainable use of forests have been very useful in structuring debates and giving us a general idea of our successes and failures. But this normative approach to forests needs to be balanced by a critical examination of the question of what societies require from their forests, what are the real outcomes desired for forestlands. The fundamental questions of 'how much forest do we need', 'where should that forest be located' and 'what sort of forest should it be' are often not addressed. This paper will argue that answers to these questions are essential if we are to invest efficiently in forest conservation and management. The answers will have to be negotiated amongst the people most immediately concerned and this will usually have to be done at a scale larger than the conventional management unit but smaller than nations and regions – we are calling this intermediate scale the 'landscape'. We will argue further that the tendency to try and maximize the areas allocated to certain types of forest – for instance protected areas – may be hindering our real objective of optimizing the area, distribution and type of forest in landscapes. We will not seek to rigidly define a landscape and we do not consider that it needs necessarily to be delimited on maps – we see the term more as a metaphor to convey the idea of connectivity and balance.

Protect, Manage and Restore: Institutional Constraints

It appears self-evident that forest institutions should seek to provide society with forest products and services in the most efficient way. But most forest institutions are mandated to deal with specific components of the forest landscape. Thus we have separate institutions dealing with protected areas, production forests, watershed forests and forests on farms or in cities. And the reality is that we are not good at managing these diverse components of the forest estate complementarily. Worse, because these institutions inevitably tend to try and maximize the output of their own component of the system they often compete with other institutions dealing with forests or are insensitive to the opportunity costs that they impose upon other users of the land. The most striking example of this are the vigorous campaigns mounted by some conservation organizations to expand protected areas in poor countries with large populations of land-hungry people. One international conservation group recently called for 40 per cent of Madagascar to be allocated to totally protected areas. One wonders how the poor people who are struggling to survive at the forest margin receive such proposals.

Segregation or Integration: Competing Paradigms for Forest Management

There have always been two competing paradigms for forests. The first is a paradigm of segregation between production and protection functions. This is the approach of economic orthodoxy where areas of pristine forests are set aside for nature conservation and forest products are produced intensively in areas allocated for this purpose.

The second is a paradigm of integration, an approach where near-natural forests are managed carefully to yield both timber and other products and, at the same time to maximize biological diversity and other environmental values. This is the approach that has been the traditional basis of forestry in continental Europe.

The divergence of these two paradigms is reflected in constant arguments over the definitions of protected areas. The term 'national park' is widely used in Europe to describe multifunctional or 'cultural' landscapes. This use is at variance with the definitions used in the United Nations List of National Parks and Protected Areas, which defines parks as areas allocated exclusively for conservation.

Conservationists promote extensive pristine national parks. Those who are focused on poverty in developing countries prefer integrated approaches. We argue that neither of these doctrines is entirely satisfactory. The problem with the integrated model is that it runs counter to economic orthodoxy. In attempting to manage for everything we fail to maximize returns on any single product. The problem with the segregated model is that it is excessively dependent on market forces. The public goods benefits for which markets are imperfect – amenity, carbon sequestration and so on – do not attract the investments that they require.

The Third Way – Forests and Trees in Multifunctional Landscapes

An alternative approach is to cease to focus on integration or segregation at the management unit level and shift our attention to optimizing outcomes through landscape mosaics. Under this third paradigm we recognize that some areas of special significance for biological diversity must be given total protection as inviolate nature reserves. A significant proportion of forest products will be produced most efficiently in intensively managed plantations. But to achieve an optimization of the full range of goods and services that we require from forests we will need two more ingredients. We will need extensive areas of semi-natural forest to provide the matrix or connectivity between protected areas, to sequester carbon and regulate hydrological functions – none of which requires that forests be maintained in a totally natural 'inviolate' state. And we will need a process to ensure that the different elements of the mosaic are complementary. Examples are habitat corridors that must be continuous between protected areas and must actually provide the ecological conditions needed for species movements (Simberloff et al, 1992). Watershed protection forests must be located so as to intercept linear flows of water and soil – thus parallel to the contours. Intensively managed plantations must be located so as to minimize opportunity costs for agriculture. All of these components must combine to provide an environment in which people can enjoy the benefits of employment, recreation and amenity.

Why Multifunctionality is Important

The concept of multifunctionality is more than just a fine-tuning of existing approaches to land-use planning. If forests are distributed optimally in the landscape and if the different elements of the landscape mosaic complement one another then the total area of forest needed to provide a given yield of forest benefits is less. This has profound implications for conservation planning. For instance most of the plans for conserving forest biodiversity advocate maximizing the extent of protected areas. In a multifunctional landscape one would seek to optimize or even minimize the extent of totally protected areas. Such areas would only be needed to provide the additional habitats of those species that would not be adequately protected in the managed, multiple-use forests. A frequent argument is that some species will always require very large tracts of pristine forest, but we would counter this by pointing out that most if not all of these species are large, wide-ranging species (tigers, elephants, eagles etc) that are well adapted to modified habitats. The majority of the species that are obligate inhabitants of pristine or 'old growth' forests are small and relatively sedentary – many can thrive in islands of natural habitat in a modified forest matrix.

The major attraction of a conservation strategy that does not require maximizing protected areas is that it is more likely to find support in poor developing countries where there is a need for more land for agriculture and access to forests for economic gain. Multifunctional landscapes can achieve conservation objectives whilst reducing the opportunity costs that are imposed upon local people. Any area taken out of production in order to protect biological diversity will deprive local people of land and forest products. The history of integrated conservation and development projects (ICDPs) highlights the dangers of failing to recognize the costs that conservation programmes can impose upon local people (McShane and Wells, in press). ICDPs have also taught us that technical assistance alone will rarely cause people to abandon pathways that they perceive to be economically rational in their local context. Attempts to conserve forests in multifunctional landscapes will therefore fail if the full costs and benefits to all of the stakeholders are not subject to negotiations and trade-offs in an open and transparent manner.

Approaches to Achieve Multifunctional Landscapes – Eco-Regional Conservation Planning

Several international conservation organizations are establishing their conservation priorities at the scale of large ecological regions (Dinnerstein et al, 2000). This involves an analysis of conservation options in the context of competing land uses. A range of stakeholders participates in the development of a vision for what an ideal biodiversity programme for the eco-region might be. Thus, instead of focusing exclusively on the individual cells of a matrix – the nature reserves – eco-regional planning aims to design a set of complementary protected areas covering all the major biological features of the region. It seeks to set these in a matrix of other land uses that favour biodiversity – for instance through the maintenance of habitat corridors. But conservation priorities are set in the context of other land-use priorities. The overall objective is to obtain a compromise between what would be ideal for species conservation and what is realistic in terms of local people's needs.

Knowledge Management

The knowledge that is required to achieve multifunctional landscapes is not the exclusive preserve of technical forestry specialists. Much of the knowledge about any landscape will be in the form of the tacit knowledge of local people and this has to be given

as much value as the explicit knowledge from modern sciences, remote sensing etc. Negotiators who arrive with their maps already drawn on the basis of satellite imagery will already
have foreclosed many of the options that local people might have sought to promote (Scott,
1996). Transparency and a level playing field for all interested parties are essential (Wollenberg et al, 2001). Geographic information systems (GIS) will provide valuable tools
for making all information spatially explicit and amenable to scientific analysis but they
can also be abused if they are used to promote an external scientific vision of a situation.

Spatial models can allow the identification of those areas that would provide the
largest incremental benefits for biological diversity from the smallest increase in forest
extent. This approach can help address one of the fundamental issues of designing conservation area networks – at what point does the marginal increase in the extent of
conservation areas, and thus their cost, exceed the marginal increase in the value of biodiversity conserved?

Negotiation Frameworks

Far too often what passes for negotiation is in reality a process in which powerful gov-
ernmental institutions (or development assistance agencies) inform local stakeholders of
what has been decided. Neutral facilitation is a valuable tool for breaking down the power
differentials that so often characterize such negotiations in rural settings. Solutions that
are imposed in the face of explicit or even tacit resistance by local stakeholders will rarely
be sustainable. The rural development and social forestry literature is rich in analysis of
the difficulties of achieving a genuinely egalitarian process of negotiation between all the
different stakeholders who have legitimate interests in forests and land use. Valid processes
require much more time, patience and sensitivity to local cultures than most outside
experts are prepared to allocate. Neutral facilitation and explicit recognition of the trade-
offs between the interests of different stakeholders are important ingredients of success.

Conclusions – Protect, Manage and Restore

There are countless examples of protected areas being too isolated, of poor farmers being
forced onto land that would be better suited to watershed protection, of jobs in forestry
being lost when forests are protected in situations where conservation objectives did not
require total protection. Global planning is not addressing local values (Sheil and
Wunder, 2002). Many of our forest landscapes are dysfunctional. Competition between
management agencies and the need to allocate all forests to a simple set of international
categories are combining to drive us towards a rigid adherence to one or other of the
segregated or integrated paradigms. This is happening at a time when globalization is
also pushing hard in the direction of economic rationality and the segregated model
(Sayer and Campbell, in press).

Forest conservation and management seems to be succeeding best in places where
solutions are being sought at the scale of the landscapes with which the most important

stakeholders are most familiar. In most cases mosaics of complementary land and forest uses best achieve landscape functionality. This may mean that relatively small areas of forest are allocated to total protection – the old-growth forests – or to intensive planta-tions. Areas of multiple-use forest that contribute additional products will often com-plement these. They will provide amenity values and habitat continuity for some forest species. The landscape will be viable to the extent that it yields the greatest benefits, or imposes the fewest costs, on the stakeholders most directly involved in its use on a day-to-day basis. We conclude that functional forest landscapes are difficult to achieve but that the following are important elements of success:

- land-use decisions are subject to fair negotiation amongst all stakeholders;
- effective and fair mechanisms exist to resolve conflicts;
- trade-offs are made explicit and reconciled by legal sanction or payments;
- payments are made to people who forgo local benefits in favour of off-site environ-mental objectives;
- land and resource tenure and access rights are clear and can be defended legally;
- research and especially spatial analysis tools are used to increase the range of options available and perhaps to produce counter-intuitive solutions to conflicts;
- management is continually monitored and desired outcomes renegotiated in a large-scale adaptive management framework.

References

1   Dinnerstein, E., Powell, G., Olson, D., Wickramanayake, E., Abell, R., Loucks, C., Underwood, E., Allnutt, T., Wettengel, W., Ricketts, T., Strand, H., O'Connor, S. and Burgess, N. (2000) *A Workbook for Conducting Biological Assessments and Developing Biodiversity Visions for Eco-region-based Conservation, Part 1: Terrestrial Eco-regions*. World Wildlife Fund, Washington, DC.
2   Fairhead, J. and Leach, M. (1995) 'False forest history, complicit social analysis: Rethinking some West African environmental narratives'. *World Development*, vol 23, pp1023–1035.
3   Kuchli, C. (1997) *Forests of Hope*. Earthscan, London.
4   Lomborg, B. (2001) *The Skeptical Environmentalist*. Cambridge University Press, Cambridge, UK and New York.
5   McShane, T. and Wells, M. (in press) *Integrating Conservation and Development*. Columbia University Press, New York.
6   Sayer, J. and Campbell, B. (in press) *The Science of Sustainable Development: Local Livelihoods and the Global Environment*. Cambridge University Press.
7   Scott, J. C. (1998) *Seeing Like a State*. The Yale ISPS series, Yale University Press, New Haven, Connecticut.
8   Sheil, D. and Wunder. S. (2002) 'The value of tropical forests to local communities: Com-plications, caveats and cautions'. *Conservation Ecology*, www.consecol.org/vol16/iss2/art9
9   Simberloff, D., Farr, J. A., Cox, J. and Mehlman, D. W. (1992) 'Movement corridors: Con-servation bargains or poor investments'. *Conservation Biology*, vol 6, no 4, pp493–504.
10 Wollenberg, E., Edmunds, D., Anderson, J. (eds) 'Accommodating multiple interests in local forest management'. *International Journal of Agricultural Resources, Governance and Ecology. Special issue*, vol 1, no 3/4, pp193–356.
11 WWF (2003) www.panda.org/forests4life

22

Where Next? Adapting Forestry Institutions to Meet New Challenges

Jeffrey Sayer

The papers included in this Reader challenge much of the conventional wisdom of forestry. They suggest that we are embarking upon new approaches to forests, and I hope that they will convince readers that forestry is going to be an exciting and challenging profession. I use the term profession with some circumspection, as it is clear that forestry will no longer be a single profession. The people working on the management and conservation of forests will be part of teams with diverse and complementary skills. They will have to deal with a range of problems and situations that go far beyond those that were encountered by their predecessors. I would like therefore to conclude the Reader with some thoughts on how the world of forests and forestry is evolving and to suggest what it will be like to be professionally active in this arena in the 21st century. One issue that I would like to address directly is that of whether all these new ideas for forestry are just talk or whether real changes are happening on the ground and in the corridors of the institutions charged with the husbandry of our forests.

Attending endless meetings at which solemn commitments are made to strategies and action plans for the world's forests can lead one to scepticism or even cynicism. One always has the impression that these events with their diplomatic protocols and cautious statements are irrelevant to the realities of day-to-day decision making about forests. However, compiling this collection of papers has forced me to look back over the last two decades and compare the sorts of publications that were available to my friend and colleague Simon Rietbergen for his Reader published ten years ago and those available to me today. The profile of the publications today is much broader and richer – the discourse has certainly changed, and I will argue that the realities on the ground have also changed – although perhaps not as much as one would have wished.

The hot topic for discussion at the United Nations Forum on Forests (UNFF) and the conference of the parties to the Convention on Biodiversity (CBD) in 2004 was ecosystem approaches to forest management. The issue was how the ecosystem approaches principles adopted by the state parties to the CBD related to, or duplicated, the latest definitions of sustainable forest management (SFM) being discussed at the UNFF. The answer in my opinion was that both were just another articulation of ideas for broader-based, more holistic and participatory management that have gradually been emerging for over a decade. Almost all the international discussions about forests since the Earth Summit in Rio de Janeiro in 1992 have agreed that forests should be managed for a wide range of ecosystem services and for the multiple products that different sectors of society require. This was not the case before Rio. The Tropical Forestry Action Plan of the mid-1980s was very much a traditional timber and jobs view of forests. The Rio principles on the conservation and sustainable use of all kinds of forests were a major step forward. An important source of this difference was that the Rio principles emerged from a process where for the first time the developing countries of the South were able to articulate their views effectively.

Most of the elements that constitute ecosystem approaches to management were anticipated in the Rio principles. The same ideas recur in the definitions of sustainable forest management that were adopted by the International Tropical Timber Agreement and included in its target for the year 2000. The initiatives to develop criteria and indicators for SFM that proliferated in several regions in the late 1990s all reflected this vision of including broad social and environmental objectives in forest management. The principles that were finally adopted by the CBD for both ecosystem approaches and for sustainable use of biodiversity were not therefore innovative. Their adoption served to codify them and give them legitimacy at an international level.

The reason that I believe that this all goes beyond mere rhetoric is that there are demonstrable changes in many forest institutions all over the world and most of these go in the direction of these same integrative and broad-based approaches to forestry. The bitter debates and difficult negotiations that led to broad changes in the way forestry is organized in the Pacific Northwest of the US and in adjacent British Colombia occurred at the same time as the international forestry world was moving to its new paradigm. Some would argue that ecosystem management in the Pacific Northwest anticipated the Rio principles and the wave of more integrative C&I. The classic study of the science of ecosystem management in the US published by Kohm and Franklin in 1997 resulted from a workshop held in 1993. This study contains valuable analysis of the practical issues relating to the implementation of ecosystem management and in many ways goes beyond the general principles of the CBD or the debates on SFM at the UNFF. The Kohm and Franklin study draws on several years of practical application of ecosystem management.

The other indicator of real change on the ground lies in the reforms of institutional structures and laws that have occurred over the past decade in many different countries. There are numerous examples of forestry being moved from the jurisdiction of ministries of agriculture to that of ministries of the environment. Forestry schools and departments in universities have in many cases been rolled up into departments of environmental sciences or natural resources management. The training and education of foresters has evolved dramatically over the past decade or so. The staffs that join the ranks of the organizations responsible for our forests now come from a range of disciplines and in general have a far broader educational background than their predecessors.

Many countries have changed their forest laws. All five major forest countries of the Congo Basin have adopted new forest laws since 1994. They all require more environmental approaches to forests and most refer to forest ecosystems as the objects of management. Similar legal changes have been introduced in many countries.

Skills and Institutions for Modern Forestry

This Reader is intended for people who are either embarking on a career related to forest conservation and management or those who are already engaged in forestry in some way and would like to access the recent literature. I also expressed the hope that journalists, lobbyists and activists who try to influence forest outcomes might benefit from reading some of this material. So what is the take-home message for such people?

The first is that we have to forget the notion that forests can be managed against formal rules or codes that are imposed from above and apply uniformly to all forest situations. This was the way that it was in the past and some traditionalists would like to see C&I used as a sort of blueprint for all forests. But most people now accept that a pluralistic approach to forests is essential. All forests are different and all require their particularity to be taken into account in deciding on their management. This is why more thoughtful approaches to C&I propose menus of C&I from which those appropriate to any particular situation can be selected. Kohm and Franklin (1997) stress repeatedly the need for detailed local knowledge as a prerequisite for ecosystem management – a need echoed by Sheil and van Heist in their chapter in this Reader.

A number of recent publications explore the science of integrated and more holistic approaches to management. *Barriers and Bridges* by Gunderson, Holling and Light (1995) and accounts of 'Navigating social–ecological systems' by Berkes et al (2003) provide ideas that can be applied to forest management. This new generation of books on natural resources management shows similarities with the modern management literature emerging from the commercial and corporate sectors.

The new vision of forest institutions that emerges from all these publications is of institutions that will no longer be developing expert-driven management prescriptions and then regulating their enforcement. Forest departments will be steering rather than rowing – working with society to establish broadly accepted visions of forest futures and then providing the technical support services to help attain these. So the key roles of forest departments will be:

- Facilitating a dialogue amongst all forest stakeholders to establish a vision for their forests and to determine the limits within which forest owners and managers may operate.
- Establishing and maintaining multiple resource databases on forests to enable trends and patterns to be detected and adjustments to management to be made.
- Providing a problem-solving research capacity to deal with emerging problems of pest and diseases, to determine management requirements for specific targets – the maintenance of an endangered species, for example.
- Developing an early-warning capacity so that emerging threats and issues can be detected and adaptive management measures brought to bear to deal with them – the early signs of climate change impacts are a good example.
- Providing the overview, analysis and verification needed to make environmental service payments effective in supporting the production of the public goods values of forests.
- And, notwithstanding all the above, it will still be necessary to enforce regulations to ensure that the broader interests of society are not compromised by individual self-interest.

The nature of the interactions between forest departments and civil society are destined to change. We are seeing the emergence of a new form of collective action to manage forests at large scales. This has parallels in community forestry but it is forestry where the community includes distant and urban populations and where values include 'existence values' and so on.

Criteria and indicators are emerging as a sort of civil society soft law mechanism by which regulation of forestry activities is subject to broader scrutiny by civil society. Forest departments will no longer have the last say – that role is being taken over by civil society.

What Else Does One Need to Read?

At first I found it difficult to locate enough good quality papers to cover all the topics that I felt needed to be included in a Reader of this sort. Now, having invested a lot of time in the search, I have got to the point where it was difficult to decide what to retain and what to leave out. A number of topics that are of great importance in modern approaches to forests are not covered in the Reader. This was either because I could not find a suitable paper or simply because we ran out of space. I could have included more material on technological innovation. There has been enormous progress in recent years in improving industrial processes for transforming wood fibre into a huge range of end products. These technologies have major implications for how we manage forests in the future. We will have less need of tree species with special technological properties – these properties will be incorporated during processing. Plantations are rapidly becoming the major sources of wood fibre in many parts of the world and this trend will certainly continue. I could have included a more strategic overview of plantation issues – both the opportunities that they present and the problems that they may either create or that their managers will face. This might be the subject of an entire volume as plantation techniques begin to intrude into all aspects of forestry. The whole issue of certification is highly topical and is fundamental to the process of re-examining how we manage forests, but I could not locate a single paper that covered the issues adequately. However some of the issues that underlie certification are dealt with indirectly in several chapters in this volume. Forest restoration is a rapidly emerging issue, but here again I did not find a paper that covered the full breadth of the issues involved.

Emerging Issues

There are a number of issues that will be dominant on the forest agenda in the coming decades but about which we still know far too little. The most notable amongst these is climate change. This will have enormous implications for all aspects of forest management. Biodiversity reserves will no longer support the species for which they were established. Natural forests and plantations will suffer new disease and pest attacks. Forests may, or may not, be impacted by measures to mitigate climate change. Climate change is an excellent example of a threat that will require an adaptive management response. We will have to make our best judgements on how to deal with an emerging situation and then learn from practical experience. Resilience to climate change will be a major management objective.

The future of agriculture will determine to a large extent how much forest we will maintain and where and what type of forest it will be. If genetic engineering and globalization run their course and all of the world's foods are produced in highly specialized high production systems then we may have quite a lot of land for different types of forest. If we move towards a more dispersed, locally adapted system of agriculture then the landscapes will be more heterogeneous and attractive but there may be less land for large expanses of natural forest. If different forms of nuclear energy do not deliver on their promises then vast areas of land may have to be allocated to the growth of bioenergy – raw materials to replace fossil fuels – and this may place serious pressures on forests.

Perhaps the only really safe prediction is that the stewards of our forests in the future will encounter challenges that we cannot anticipate today. Surprises are inevitable. This means that we need a society where the issues of forests are constantly debated, where knowledge is generated and circulated and where forest stewards work closely with civil society to create an environment where societies needs for forest goods and services can be met. Perhaps the single thing that intrigues me most as I conclude the task of assembling this Reader is to speculate on what sorts of topics might be dealt with in a successor volume in a decade from now. In the meantime the chapters that make up this Reader are as good a place as any to start. They represent the state of our present knowledge on many of the key and controversial issues that anyone interested in forests needs to understand. They will also provide the inquisitive reader with the references to the literature that will make it easy to access material on most things that are important in the world of forests.

References

1 Berkes, F., Colding, J. and Folke, C. (eds) (2003) *Navigating Social–Ecological Systems: Building Resilience for Complexity and Change*. Cambridge University Press, Cambridge, pp1–393.
2 Gunderson, L. H., Holling, C. S. and Light, S. (eds) (1995) *Barriers and Bridges to the Renewal of Ecosystems and Institutions*. Columbia University Press, New York.
3 Sheil, D. and van Heist, M. (2000) 'Ecology for tropical forest management'. *International Forestry Review*, vol 2, pp261–270.

Index